Epithelial Transport Physiology

George A. Gerencser
Editor

Epithelial Transport Physiology

Humana Press

Editor
George A. Gerencser
Department of Physiology & Functional
Genomics
University of Florida
College of Medicine
1600 SW. Archer Road
Gainesville FL 32610-0274
USA
gag@phys.med.ufl.edu

ISBN 978-1-60327-228-5 e-ISBN 978-1-60327-229-2
DOI 10.1007/978-1-60327-229-2

Library of Congress Control Number: 2009937721

Printed on acid-free paper

springer.com

Preface

Biological cell membranes regulate the transfer of matter and information between the intracellular and extracellular compartments as basic survival and maintenance functions for an organism. This volume contains a series of reviews that are concerned with how epithelial plasma membranes regulate the transport of solutes between the intracellular and extracellular compartments of a cell. This book is also an attempt to analyze the molecular basis for the movement of various solutes across an epithelial cell membrane.

This volume is devoted to a diversity of epithelial transport mechanisms in representative cell membranes of a variety of living things. The first section of the book (Chapters 1–6) focuses on mechanisms of solute transport in epithelia of invertebrates. The last section which comprises ten chapters (Chapters 7–16) deals with solute transporters in epithelial cell membranes of vertebrates. It is hoped that with this particular ordering the reader can glean a telescopic view of the evolutionary history of the various epithelial solute transporters.

Although this book is designed to bring together a large body of literature dealing with different types of epithelial transport processes in cellular membranes and is aimed at the researcher, I hope this volume will be a valuable contribution to senior-level studies in membrane biology as well as clinical research. As editor, I wish to thank my friends and colleagues who have kindly contributed chapters to this volume. I also wish to thank Humana Press, who assumed the task of producing this volume, and my secretary, Ms. Robyn Edwards, and our department's computer specialist, Kevin Fortin, for their skillful help in various phases of manuscript preparation, and also my wife, Alison, and my sons, Rob and Jeff, whose witty humor and kindness helped so much in lightening the editorial burdens.

Possibly of most importance, I would like to thank my late mentors, Professor William McD. Armstrong of Indiana University, Professor Suk Ki Hong of the State University of New York at Buffalo, and my dearly beloved father, George S. Gerencser. Any virtues that this volume may possess must largely arise from the stimulus, encouragement, and inspiration that I received from these most distinguished, accomplished men. I would also like to personally thank my father for the love, trust, pride, and confidence that he had for me; otherwise my accomplishments, including this book, would be significantly less.

Gainesville, Florida George A. Gerencser
January 2008

Contents

Contributors

Gregory A. Ahearn, Ph.D. Department of Biology, University of North Florida, Jacksonville, FL, USA, gahearn@unf.edu

Dimitri Y. Boudko, Ph.D. Department of Physiology and Biophysics, Chicago Medical School, Rosalind Franklin University of Medicine and Science, North Chicago, IL, USA, dmitri.boudko@rosalindfranklin.edu

Brian D. Cain, Ph.D. Department of Biochemistry and Molecular Biology, University of Florida, Gainesville, FL, USA, bcain@ufl.edu

Chris I. Cheeseman, Ph.D. Membrane Protein Research Group, Department of Physiology, University of Alberta, Edmonton, AB, Canada, chris.cheeseman@ualberta.ca

Yongping Gao, Ph.D. Department of Biological Sciences, Wright State University, Dayton, OH, USA

George A. Gerencser, Ph.D. Department of Physiology and Functional Genomics, College of Medicine, University of Florida, Gainesville, FL 32610-0274, USA, ggerencs@ufl.edu

Christopher M. Gillen, Ph.D. Department of Biology, Kenyon College, Gambier, OH, USA

Michelle L. Gumz, Ph.D. Division of Nephrology, Hypertension and Transplantation, Malcolm Randall Veterans Affairs Medical Center, and University of Florida College of Medicine, Gainesville, FL, USA, michelle.gumz@medicine.ufl.edu

J. W. Hanrahan, Ph.D. Department of Physiology, McGill University, Montréal, QC, Canada, hanrahan@med.mcgill.ca

William A. Harvey, Ph.D. Whitney Laboratory for Marine Biosciences, University of Florida, St. Augustine, FL, USA; Department of Epidemiology and Biostatistics, Emerging Pathogens Institute, University of Florida, Gainesville, FL, USA, wharvey@whitney.ufl.edu

I. Jeannette Lynch North Florida/South Georgia Veterans Health System, University of Florida College of Medicine, Gainesville, FL, USA

Prabir K. Mandal, Ph.D. Department of Biology, University of North Florida, Jacksonville, FL, USA

Daniel Markovich, Ph.D. School of Biomedical Sciences, The University of Queensland, St. Lucia, QLD, Australia, d.markovich@uq.edu.au

David F. Moffett, Ph.D. School of Biological Sciences, Washington State University, Pullman, WA, USA, dmoffett@wsu.edu

Bernard A. Okech, Ph.D. Whitney Laboratory for Marine Biosciences, University of Florida, St. Augustine, FL, USA; Department of Epidemiology and Biostatistics, Emerging Pathogens Institute, University of Florida, Gainesville, FL, USA

Horst Onken, Ph.D. Department of Biological Sciences, Wagner College, Staten Island, NY, USA, horst.onken@wagner.edu

Barbara Roggenbeck Department of Biology, University of North Florida, Jacksonville, FL, USA

Kenneth M. Sterling Department of Biology, University of North Florida, Jacksonville, FL, USA

Bruce R. Stevens, Ph.D. Department of Physiology and Functional Genomics, College of Medicine, University of Florida, Gainesville, FL, USA, stevensb@ufl.edu

Lisa R. Stow Department of Physiology and Functional Genomics, Malcolm Randall Veterans Affairs Medical Center, University of Florida College of Medicine, Gainesville, FL, USA

I. David Weiner, M.D. Division of Nephrology, Hypertension and Transplantation, University of Florida College of Medicine, and Nephrology and Hypertension Section, North Florida/South Georgia Veterans Health System, Gainesville, FL, USA, david.weiner@medicine.ufl.edu

Amanda K. Welch Department of Biochemistry and Molecular Biology, University of Florida College of Medicine, Gainesville, FL, USA

Michele G. Wheatly, Ph.D. Department of Biological Sciences, College of Science and Mathematics, Wright State University, Dayton, OH, USA, michele.wheatly@wright.edu

John F. White, Ph.D. Department of Physiology, School of Medicine, Emory University, Atlanta, GA, USA, jfwhite@physio.emory.edu

Charles S. Wingo, M.D. North Florida/South Georgia Veterans Health System and University of Florida College of Medicine, Gainesville, FL, USA, charles.wingo@va.gov

Shen-Ling Xia, Ph.D. Division of Nephrology, Hypertension and Transplantation, and Department of Physiology and Functional Genomics, Malcolm Randall Veterans Affairs Medical Center, University of Florida College of Medicine, Gainesville, FL, USA

Deborah L. Zies, Ph.D. Department of Biological Sciences, University of Mary Washington, Fredericksburg, VA, USA

Chloride ATPase Pumps in Epithelia

George A. Gerencser

Abstract Five widely documented mechanisms for chloride transport across biological membranes are known: anion-coupled antiport, Na^+ and H(+)-coupled symport, Cl^- channels and an electrochemical coupling process. These transport processes for chloride are either secondarily active or are driven by the electrochemical gradient for chloride. Until recently, the evidence in favour of a primary active transport mechanism for chloride has been inconclusive despite numerous reports of cellular Cl(−)-stimulated ATPases coexisting, in the same tissue, with uphill ATP-dependent chloride transport. Cl(−)-stimulated ATPase activity is a ubiquitous property of practically all cells with the major location being of mitochondrial origin. It also appears that plasma membranes are sites of Cl(−)-stimulated ATPase pump activity. Recent studies of Cl(−)-stimulated ATPase activity and ATP-dependent chloride transport in the same plasma membrane system, including liposomes, strongly suggest a mediation by the ATPase in the net movement of chloride up its electrochemical gradient across the plasma membrane structure. Contemporary evidence points to the existence of Cl(−)-ATPase pumps; however, these primary active transporters exist as either P-, F-, or V-type ATPase pumps depending upon the tissue under study.

Keywords Cl^-stimulated ATPase · Primary active transporter · $C1^-$Pump-type · V-type ATPase · F-type ATPase · Ion-translocating ATPase

1 Introduction

The electrical activity of biological tissue has been a source of intense interest and much scientific study since the early reports of DuBois-Reymond [1] and Galeotti

G.A. Gerencser (✉)
Department of Physiology and Functional Genomics, College of Medicine, University of Florida, Gainesville, FL 32610-0274, USA
e-mail: ggerencs@ufl.edu

G.A. Gerencser (ed.), *Epithelial Transport Physiology*,
DOI 10.1007/978-1-60327-229-2_1,

[2]. However, it was not until the brilliant studies of Ussing and Zehahn [3] and colleagues on isolated frog skin that the nature of the bioelectric potential was defined. Ussing defined the interrelationship between bioelectric potential difference and active Na^+ transport and these scientific statements ushered in the modern era of ion transport study in epithelia and other tissues. Skou [4] molecularly defined the nature of Na^+ transport with his discovery of the (Na^+ + K^+)-stimulated ATPase enzyme. For years thereafter, active Na^+ transport across epithelia occupied the collective focus of transport physiologists with Cl^- assuming the secondary role of passive counterion. However, within the past 30 years there has been an intensive interest in transmembrane Cl^- transport primarily because Cl^- has been found to move actively in a very wide range of species [5, 6].

However, a considerable amount of Cl^- transport data has accumulated in the transport literature that does not conform to any of the five well-established models [7, 8]. For instance, Hanrahan and Phillips [8] have provided evidence for an electrogenic Cl^- accumulative mechanism located in the mucosal membrane of locust rectal epithelium. This mechanism is activated and stimulated directly by K^+ and is also independent of Na^+ and $HCO_3{}^-$. Observations of plant cell membranes [9, 10] as well as bacterial membranes [11] have yielded Cl^--pump activity and associated Cl^- accumulation, which are inconsistent with the five models for Cl^- transport described previously [7]. Perhaps, the strongest and most compelling evidence for a primary active transport mechanism of Cl^- (i.e. an ATPase) resides with the observations of Gerencser [12] and Shiroya and colleagues [13] who have characterised Cl^--ATPase activity and ATP-dependent Cl^- transport in the same plasma membrane system as well as reconstituting these activities in a liposome system [12, 14–16]. Indeed, the speculations by Frizzell, Field and Schultz [5], Schultz [17] and DePont and Bonting [18] that Cl^--stimulated ATPases do not exist as discreet molecular entities and, therefore, are not involved in biological Cl^- transport may have been too premature considering the recent possible evidence to the contrary.

2 General Properties of Anion (Cl^-)-ATPases

2.1 Definition of a Primary Active Transport or Pump Mechanism

A primary active transport mechanism for a particular solute is a process that moves that solute up its thermodynamic gradient by its direct linkage with a source of free energy. The source of metabolic free energy may arise from the capture of photons, the movement of electrons or the hydrolysis of high-energy phosphorylated intermediates such as adenosine triphosphate (ATP). If ATP is the source of metabolic free energy then the "Pump" or Primary Active Transport Mechanism is termed an "ion-motive ATPase" or an "ion-transport ATPase." Therefore, an ion-transport ATPase is an enzyme responsible for the primary coupling of ATP hydrolysis to ion movement across a biomembrane. Ion movement is vectorial; likewise ATP hydrolysis is vectorial at the molecular level; but in the absence of a membrane-transport

ATPase assembly the vectoriality is lost and the hydrolysis is scalar in bulk solution [19]. The role of a transport ATPase, then, is not only to catalyse ATP hydrolysis but to translate the ion movement inherent in the vectorial hydrolysis reaction into a net flux of an ionic species across a membrane. The definition of primary coupling between ATP hydrolysis and ion movement is made explicit by the cross coefficient, R_{kr}, in Eq. (1) a general flux equation from non-equilibrium thermodynamics [19].

$$J_k = \frac{\Delta\overline{\mu_k}}{R_{kk}} - \Sigma\frac{R_{kj}}{R_{kk}}J_j - \frac{R_{kw}}{R_{kk}}J_w - \frac{R_{kr}}{R_{kk}}J_r\,. \quad (1)$$

The term $\Delta\mu_k/R_{kk}$ asserts that the flux of an ionic species, J_k, is driven by and is proportional to, a difference in the electrochemical potential of that species across a membrane, $\Delta\mu_k$, it being understood that this difference has two components, a chemical activity difference given by $RT\Delta\ln a_k$ and an electrical potential difference, given by $zF\Delta\Psi$. The straight coefficient, R_{kk}, is a resistance term; that is, a coefficient which converts the proportionality to an equality. A flux coupled solely to $\Delta\mu_k$ is universally accepted as being a "passive" flux. We will consider a flux coupled to the flow of any other solute, J_j, ($R_{kj} \neq 0$) or a flux coupled to the flow of water, J_w, ($R_{kw} \neq 0$) to be secondary to the flow of the other solute or of water. By primary active ion transport I mean the coupling of an ion movement directly to the flow of a chemical reaction, J_r; for such a flux the cross coefficient in the last term, $R_{kr} \neq 0$. Such a primary active ion flux is coupled to a reaction flux, in this case ATP hydrolysis, by mechanisms whose structural orientation within a biomembrane is crucial. In summary, the cross coefficient, R_{kr}, can be viewed as a formal expression of an ion-transport ATPase.

2.2 ATPases

A major functional class of membrane-bound enzymes includes those categorised as primary active transporters and called ATPases because they catalyse the transport of ions against an electrochemical potential by reactions directly linked to the hydrolysis of ATP. The ATPases that actively transport cations have been extensively studied and have been categorised by Pedersen and Carafoli [20] into three classes: F-type ATPases, V-type ATPases, and P-type ATPases. F-ATPases are located in bacterial plasma membranes, inner mitochondrial membranes and thylakoid membranes of chloroplasts, and they operate as ATP synthases, synthesising ATP from ADP and inorganic phosphate using energy derived from electrochemical gradients of protons [21]. Using reaction mechanisms analogous to F-ATPases, the V-ATPases utilise ATP to create proton electrochemical gradients; they are ubiquitously distributed in eukaryotic vacuoles and lysosomes and in archaebacteria and are also present in plasma membranes of various animal tissues [22].

The P-ATPases are broadly distributed, active cation translocators having the distinctive feature of forming a covalent acylphosphate-enzyme intermediate (hence the P-designation) during the cycle of ATP hydrolysis and cation translocation [20, 23]. Among this class of ATPases are the Ca^{2+}-ATPases of the plasma membrane, sarcoplasmic reticulum and endoplasmic reticulum, the H^+-ATPases of yeast and plants, the K^+-ATPase of bacteria, the Na^+/K^+-ATPase of animal cell plasma membranes, and the H^+/K^+-ATPase of gastric parietal cells [23]. Because of the formation of phosphoenzyme intermediates, the enzymatic cycle of P-ATPases can be divided into steps that include a kinase activity, by which an aspartate residue on the enzyme is phosphorylated, and a phosphatase activity, by which the phosphoenzyme is dephosphorylated [23]. Another common feature of these ATPases is their inhibition by submicromolar concentrations of vanadate, acting as a tightly binding phosphate analog [24, 25]. Furthermore, during the enzymatic cycle, P-ATPases characteristically exhibit two phenomenologically and structurally distinct conformations, E_1 and E_2, which have distinct kinetic variables [26]. For this reason, the P-ATPases are also called E_1–E_2 ATPases.

In addition to their functional similarities, P-ATPases have a number of structural homologies, belonging to a common large gene family. All members have a principal peptide of approximately 100 kDa, designated as the catalytic (sub) unit because it contains the site for ATP binding and phosphorylation [20]. The high degree of sequence homology for amino acids within the ATP binding site and the phosphorylation site attests to the highly conserved nature of protein domains that interact with ATP. Some disagreement remains as to the exact location, and even the number of transmembrane segments [27] and the specific locations for the binding of cations and the paths for their transport are far from resolved.

Recently, there has been some clarification of several of these issues. For example, the crystal structure of skeletal muscle sarcoplasmic-reticulum/endoplasmic-reticulum Ca^{2+}-ATPase has been determined at 2.6 Å resolution [28]. This high-resolution crystal structure of the Ca^{2+}-ATPase is the first reported structure of a primary active transport protein. This enzyme proved to be a complex protein with distinct domains that are thought to undergo major movements in the course of a single turnover [29]. Within the membrane, a pocket is formed by kinks in transmembrane α-helices to make an ion-binding site that simultaneously accommodates two Ca^{2+} ions. There is considerable evidence from lower-resolution images that two related ATPases, the Na^+/K^+-ATPase and the H^+-ATPase of *Neurospora*, have important structural features in common with the above-mentioned Ca^{2+}-ATPase [30–32]. Extensive work on the structures of certain subtypes of P-ATPases has shown a common theme: there is a catalytic or alpha subunit of approximately 100 kDa with usually ten predicted transmembrane spans and four intracellular loops that differ greatly in length. A large fraction of the mass of the protein is found in the L4-5 intracellular loop, which has long been known to be the location of the ATP and phosphorylation sites [33]. Critical residues of the ATP-binding site map some distance away from the covalent phosphorylation site, leading to the inference that these are ATP-binding and phosphorylation domains that come together in the course of the enzyme reaction cycle. These conformation changes are thought to influence the structure of the transmembrane domain, where ion-binding sites must reside, by altering the tilt or depth of critical membrane segments in much the same way that the handles of scissors control the blades [34]. There has been considerable controversy, however, about the extent of similarity of the membrane topology of the Ca^{2+}-ATPase of sarcoplasmic reticulum and even closely related ATPases like the Na^+/K^+-ATPase in the C-terminal third of the protein [35].

3 Location of Anion-ATPases

One of the most controversial issues regarding Cl^--stimulated ATPase activity is its site or anatomical localisation within the microarchitecture of cells. It appears that Cl^--stimulated ATPase activity resides in both mitochondrial and microsomal fractions [36] of cell homogenates. However, DePont and Bonting [18] and Schuurmans, Stekhoven and Bonting [36] have categorically stated that microsomal or plasma membrane localisation of this enzyme is entirely due to mitochondrial contamination. Hence the dispute. If Cl^--stimulated ATPase activity is exclusively of mitochondrial origin it is extremely difficult to conceive how it could drive net Cl^- movement across plasma membranes. Therefore, the Cl^--stimulated ATPase should play no direct role in transcellular Cl^- transport, but could function, in some capacity, in intracellular Cl^- transport. On the other hand, if the Cl^--stimulated ATPase is located in the plasma membrane then primary Cl^- transport by this enzyme would be analogous to the (Na^+ + K^+)-stimulated ATPase which mediates net transport of Na^+ and K^+ across plasma membranes [4].

Perhaps the strongest evidence for the existence of Cl^--ATPase activity in a plasma membrane system free from any possible mitochondrial contaminant ATPase was that by Gerencser and Lee [37, 38]. They presented evidence which indicated that purified basolateral membranes (BLM) of *Aplysia californica* foregut absorptive cells contain Cl^--ATPase activity. Their finding that the BLM subcellular membrane fraction had a high specific activity in (Na^++K^+)-ATPase, but had no perceptible cytochrome c oxidase activity and a significantly reduced succinic dehydrogenase activity, supported this conclusion. The observation that there was very little NADPH-cytochrome c reductase activity in the membrane fraction suggested that the BLM in this fraction were also relatively free from endoplasmic reticulum and Golgi body membrane contamination [39]. The failure of oligomycin to inhibit Cl^--ATPase activity in the BLM fraction was also consistent with the non-mitochondrial origin of the Cl^--ATPase. Supporting this contention was the corollary finding that oligomycin inhibited mitochondrial Cl^--stimulated ATPase activity. The finding that efrapeptin, a direct inhibitor of mitochondrial F_1-ATPase activity significantly inhibited Mg^{2+}-ATPase activity in the mitochondrial and not in the BLM fraction [40] unequivocally supported the notion that the plasma membrane fraction was pure and was free from mitochondrial contamination. Additionally, Gerencser and Lee [38] showed that vanadate (an inhibitor of only "P-type ATPases") inhibited Cl^--ATPase activity in the purified BLM fraction and not in the mitochondrial fraction. Taken together, all of these observations strongly supported the hypothesis that Cl^--stimulated ATPase activity exists in, at least, one subcellular locus other than mitochondria. It appears that in numerous biological cells, which transport Cl^-, Cl^--stimulated ATPase activity forms an integral part of the plasma membrane [37, 41–43].

4 Function of Anion-ATPases

To assign a direct role of Cl^- or HCO_3^- transcellular transport to an ATPase, the enzyme should be shown to be an integral component of the plasma membrane. The energy for active transport of Cl^- or HCO_3^- can, in principle, thus be obtained from the hydrolysis of ATP. Therefore, the following question can be asked. Is the anion-stimulated ATPase identical with a primary active transport mechanism ("pump") for anions? The following discussion deals with this controversial question [6, 7, 18].

It was not until the following observations that HCO_3^--stimulated ATPase activity was linked with Cl^- pumping because no Cl^- activation of this enzyme had been observed. DeRenzis and Bornancin [44] were the first to demonstrate the membrane presence of a (Cl^-/HCO_3^-)-stimulated ATPase in goldfish gill epithelium and suggested that the enzyme could participate in the branchial Cl^-/HCO_3^- exchange mechanism. Bornancin, De Renzis and Maetz [45], confirmed these results in freshwater eel gill epithelium as did Bornancin and his colleagues [46] in freshwater trout gill epithelium. Kinetic studies in these three gill epithelial systems strongly suggested that a (Cl^-/HCO_3^-)-stimulated ATPase is involved in the Cl^-/HCO_3^- exchange mechanism and therefore in the acid–base regulation of freshwater fish. These authors reported a parallelism between the affinities of the ATPase for Cl^- and both the Cl^- affinity for the gill transport mechanism and the Cl^- influx rate. The affinity constants for the Cl^--stimulated ATPase were 1.0, 5.9, and 23.0 meq/l for the goldfish [44], freshwater trout [46], and freshwater eel gill epithelium [45], respectively. The affinity of Cl^- for the transport systems in vivo was 0.07, 0.25, and 1.3 meq/l for the goldfish [44], freshwater trout [46], and freshwater eel gill epithelium [45], respectively, while the corresponding maximal Cl^- influxes were 55.0, 19.6, and 0.36 $\mu eq/h^{-1}(100g)^{-1}$. In addition, the finding that the Cl^- activation of anion-stimulated ATPase activity was inhibited by thiocyanate [44] was consistent with transport studies which showed that Cl^- influxes were inhibited by thiocyanate [45]. These studies on gill epithelium strongly supported the hypothesis that the Cl^--stimulated ATPase is involved in gill anion exchanges that are related to mineral and acid–base homeostasis in freshwater fish.

The hindgut of the desert locust *Schistocerca gregaria* possesses an unusual chloride transport system [8, 47]. The isolated locust rectum absorbs chloride from the mucosal (lumen) to the serosal (haemolymph) side in the absence of an electrochemical potential gradient. Net chloride transport persists in nominally Na^+-free or $HCO^-{}_3(CO_2)$-free saline, is insensitive to normal inhibitors of NaCl co-transport and anion exchange, and is independent of the net electrochemical gradient for Na^+ across the apical membrane. However, active chloride transport is strongly dependent on mucosal K^+ level. Chloride entry across the apical membrane is active, whereas the net electrochemical gradient across the basal membrane favours passive Cl^- exit from the cell. Although mucosal K^+ directly stimulates "uphill" chloride entry, there is no evidence for coupled KCl co-transport, nor would co-entry with K^+ be advantageous energetically. To determine if active Cl^- transport across rectal epithelia might be due to an anion-stimulated ATPase, a microsomal fraction was obtained by differential centrifugation [48]. Microsomal ATPase activity was stimulated in the following sequence: sulphite > bicarbonate > chloride. Maximal ATPase activity was obtained at 25 mmol l^{-1} HCO_3^- or 25 mmol l^{-1} Cl^-. Thiocyanate inhibited 90% of the anion-stimulated ATPase activity. The microsomal fraction was enriched in the plasma membrane markers, leucine aminopeptidase, alkaline phosphatase, 5′-nucleotidase, and gamma-glutamyltranspeptidase, and had little contamination from the mitochondrial enzymes, succinate cytochrome c reductase and cytochrome oxidase. Microscopic examination confirmed that the BLMs were associated with mitochondria following differential centrifugation, while the microsomal fraction contained little mitochondrial contamination. These results indicate the presence of an anion-stimulated ATPase activity that could be responsible for active Cl^- transport across the locust rectum. Recently, however, Phillips and his colleagues have discovered a bafilomycin-sensitive H^+ pump in the apical membrane of locust rectum [49]. Since proton pumping could account for almost one-half of the Cl^- transported, the authors could not rule out a "proton recycling process" as responsible for at least partial net Cl^- absorption [50]. Phillips further state that Cl^--ATPase could be responsible for the remainder of the net Cl^- absorption [49].

The following studies on rat brain motoneurons provided the strongest evidence in vertebrates for the existence and function of a Cl^- pump. Shiroya and his colleagues [13] demonstrated that EDTA-treated microsomes prepared from rat brain consisted mainly of sealed membrane vesicles 200–500 nm in diameter and were rich in both Cl^--ATPase and Na^+/K^+-ATPase activities. Such Cl^--ATPase-rich membrane vesicles accumulated Cl^- in an ATP-dependent and osmotically reactive manner in the presence of ouabain. The Cl^- uptake was maximally stimulated by ATP with a K_m (substrate concentration which gives one-half maximal velocity of a chemical reaction) of 1.5 mmol l^{-1}; GTP, ITP, and UTP partially stimulated Cl^- uptake, but CTP, beta, gamma-methylene ATP, ADP, and AMP did not. The ATP-dependent Cl^- uptake was accelerated by an increase in the medium Cl^- concentration with a K_m of 7.4 mmol l^{-1}. Such stimulation of Cl^- uptake by ATP was dependent on the pH of the medium, with an optimal pH of 7.4, and also on the temperature of the medium, with an optimal range of 37–42°C. Ethacrynic acid dose-dependently inhibited the ATP-dependent Cl^- uptake with a concentration for half-maximal inhibition at 57 mmol l^{-1}. N-ethylmaleimide completely inhibited and vanadate partially inhibited the ATP-dependent Cl^- uptake. The membrane vesicles did not accumulate H^+ in the Cl^- uptake assay medium. The ATP-dependent Cl^- uptake profile agreed with that of Cl^--ATPase activity reported previously [51, 52]. More recently, the Cl^--ATPase from rat brain has been reconstituted into liposomes and has been shown to support an ATP-dependent Cl^- uptake[16]. These data collectively strongly supported the idea that Cl^--ATPase in the brain actively transports Cl^- and does so for the partial maintenance of the nerve cell membrane potential.

Gradmann [9] has provided electrophysiological data and data showing ATP synthesis by the Cl^- pump through reversal of Cl^- electrochemical gradients in *Acetabularia acetabulum* which provided strong evidence for the existence of a Cl^- pump in algae. Buttressing these conclusions were those of Ikeda and Oesterhelt [15] who showed a Mg^{2+}-ATPase, isolated from *A. acetabulum,* reconstituted into liposomes and tested for a Cl^--translocating activity. A significant increase in Cl^- efflux from the negative and neutral liposomes was observed upon addition of ATP in the presence of valinomycin after incorporation of the enzyme by short-term dialysis. The ATP-driven Cl^- efflux was inhibited by addition of azide, an inhibitor of the ATPase. When chloride was replaced by sulphate, no ATP-dependent sulphate efflux was detectable from the proteoliposomes. Proton-translocating activity of the enzyme was also tested and was found to be negative. Moritani and her colleagues [53] provided evidence from *Acetabularia cliftoni* that the subunit composition was similar to that of F-type ATPases and incorporation of this ATPase into liposomes also provided an ATP-dependent Cl^- transport activity. Collectively, these observations strongly suggested the existence of a Cl^- pump in *A. acetabulum.* Regarding the question of the physiological significance of the electrogenic Cl^- pump in *A. acetabulum,* Gradmann [9] favours a "Mitchellian" answer. This primary, electrogenic ion pump would create an electrochemical driving force to fuel secondary, electrophoretic (or electroneutral) transport processes, such as uptake of sugars or amino acids. Graves and Gutknecht [54] have provided evidence for an electrogenic Cl^- pump with similar properties in the membrane of *Halicystis parvula,* another marine alga that is related to *Acetabularia.*

One of the most rigorous and definitive proofs for a Cl^- pump's existence and its mode of operation rests with the following experiments of Gerencser and his colleagues [14, 42, 55, 56] in the next sections.

5 Electrophysiological Characteristics of Aplysia Gut

5.1 Tissue

Generally speaking, intestinal preparations of vertebrates bathed in a substrate-free Na^+-containing Ringer's solution generate spontaneously transepithelial potential differences of the magnitude 1–5 mV, the serosal surface being positive relative to the mucosal surface [57–59]. The total active ion absorption in these vertebrate intestines was accounted for by means of active Na^+ absorption in the absence of, or in excess of, active Cl^- absorption [57, 59].

Aplysia californica foregut (crop) bathed in a Na^+-containing seawater medium elicits a spontaneous transepithelial potential difference such that the serosal surface is negative relative to the mucosal surface [60–62]. The total active ion absorption across *A. californica* gut was accounted for by active absorptive mechanisms for both Na^+ and Cl^-, the net absorptive Cl^- transport exceeding the net absorptive Na^+ transport. These results were qualitatively the same as those observed with *Aplysia juliana* gut [63], and account for the negative serosal transepithelial potential difference observed in this gut preparation.

However, past observations suggested that Cl^- absorption was independent of Na^+ absorption [64]. Therefore, Cl^- absorption would be independent of Na^+-K^+-dependent ATPase activity. The transepithelial potential difference measured in a Na^+-free seawater medium (TrisCl) was stable for 3–5 h and the electrical orientation of the transepithelial potential difference was serosa negative relative to the mucosal solution. In the absence of Na^+ in the bathing medium, the total active ionic flux was identical to the net mucosal to serosal Cl^- flux (Table 1). Additionally, mucosally applied thiocyanate or acetazolamide inhibited the transepithelial potential difference and the active absorptive flux of Cl^-. In the absence of an electrochemical potential gradient for Cl^- across the tissue, these observations suggested that there was an active transport mechanism for Cl^-. However, these observations did not delineate location nor type of mechanism for the Cl^- active transport.

Table 1 Chloride fluxes in Tris-Cl seawater media

	J_{ms}	n	J_{sm}	n	J_{ms}^{NET}	SCC	n
Before thiocyanate addition	216.7 ± 14.2	6	180.1 ± 13.2	6	36.6 ± 6.8	30.8 ± 7.9	6
After thiocyanate addition	175.75 ± 8.1	6	173.9 ± 8.6	6	1.8 ± 1.3	0.8 ± 0.7	6

Na^+-independent, active Cl^- absorption across the Aplysia gut.

5.2 Cellular

Reports of intracellular Cl^- activity (a^i_{Cl}) in vertebrate enterocytes demonstrated that Cl^- was accumulated across the mucosal membrane such that the a^i_{Cl} was two to three times that predicted for electrochemical equilibrium across that membrane [65, 66]. These studies concluded that uphill Cl^- movement across the mucosal membrane was coupled to the simultaneous downhill movement of Na^+ and it was this extracellular to intracellular Na^+ electrochemical gradient across the mucosal membrane that was the driving force responsible for intracellular Cl^- accumulation.

The mean a^i_{Cl} determined in *A. californica* foregut epithelial cells bathed in a NaCl seawater medium devoid of substrate was significantly less than that predicted by the electrochemical equilibrium for Cl^- across the mucosal membrane [62, 67]. In the absence of Na^+ in the extracellular bathing solution, the mean a^i_{Cl} was also less than that predicted for electrochemical equilibrium for Cl^- across the mucosal membrane (Table 2). So one need not postulate an active transport mechanism for Cl^- in the apical or mucosal membrane of the *A. californica* foregut absorptive

Table 2 Intracellular Cl^- activities and mucosal membrane potentials in NaCl and Tris-Cl seawater media

	a^i_{Cl} (mM)	a^{eq}_{Cl} (mM)	ψ_m (mV)	n
NaCl	13.9 ± 0.5 (30)	28.6	-64.6 ± 1.3 (48)	7
Tris-Cl	9.1 ± 0.3 (32)	21.4	-7.2 ± 1.4 (51)	7
	$P < 0.01$		$P < 0.01$	

Electrochemical well for $C1^-$ inside Aplysia gut epithelial cells.

cell because Cl^- transport across this membrane could be driven by the downhill, mucosal to cytosol, electrochemical potential gradient for Cl^-. However, once the Cl^- was in the cytosol, it faced a very steep electrochemical potential gradient in its transit across the BLM into the serosal solution [62, 68, 69]. Therefore, thermodynamically, the active transport mechanism for Cl^- exhibited in the tissue studies [60, 61, 70] had to exist in the BLM of the *A. californica* foregut absorptive cell.

6 Biochemistry of the Cl^- Pump

6.1 ATPase Activity

Gerencser and Lee [37, 38] presented evidence which indicated that the BLM, and only the BLM, of *A. californica* foregut absorptive cells contains true Cl^--ATPase activity. Biochemical properties of the *A. californica* foregut absorptive cells BLM-localised Cl^--stimulated ATPase include the following: (1) pH optimum = 7.8; (2) ATP is the most effective nucleotide hydrolysed; (3) also stimulated by $HCO_3{}^-$, $SO_3{}^{2-}$, and $S_2O_3{}^{2-}$, but inhibited by $NO_2{}^-$, and no effect elicited by $NO_3{}^-$ and $SO_4{}^{2-}$; (4) apparent K_m for Cl^- is 10.3 mmol 1^{-1} while the apparent K_m for ATP is 2.6 mmol 1^{-1}; and (5) an absolute requirement for Mg^{2+} which has an optimal concentration of 3 mmol 1^{-1}. Coincidentally, Cl^- has an intracellular activity [62] in the *A. californica* foregut epithelial cell approximating its apparent K_m for the Mg^{2+}-dependent Cl^--ATPase, which supports the interrelationship of its physiological and biochemical activities.

Additionally, the ATPase activity stimulated by Cl^- was strongly inhibited by thiocyanate, vanadate, and acetazolamide but not inhibited by ouabain. These results with inhibitors strongly suggested a possible participation by the Cl^--stimulated ATPase in net chloride absorption by the *A. californica* gut [38]. The finding that anion-stimulated ATPase is inhibited by thiocyanate, but not by ouabain has also been demonstrated in many tissues known to perform active anion transport and to contain anion-stimulated ATPase activity [71]. This inverse parallelism between ouabain insensitivity and thiocyanate sensitivity to Cl^--stimulated ATPase activity and net active Cl^- absorption in the *A. californica* gut warranted conjecture that the active Cl^- absorptive mechanism could be driven by a Cl^--stimulated ATPase found in the BLM of the foregut absorptive cell. Additional support for this contention rested with the finding that Cl^--stimulated ATPase activity of the BLM was inhibited by vanadate. Vanadate has been shown to inhibit ATPases, which form high-energy phosphorylated intermediates while having no effect on the mitochondrial anion-sensitive ATPase [72]. These results strongly suggested that the Cl^--stimulated ATPase is an ion-transporting ATPase of the "P" variety rather than the "F" or "V" types.

Acetazolamide inhibited Cl^--stimulated ATPase activity in the *A. californica* gut [38, 73]. This finding has also been demonstrated for blue crab gill HCO_3^--ATPase [74]. Although acetazolamide has been shown to be a specific inhibitor of carbonic anhydrase [75] it has also been demonstrated to be a Cl^- transport inhibitor [76]. Additionally, it had been shown by Gerencser [73, 77] that carbonic anhydrase does not exist in the BLM of the *A. californica* gut absorptive cell. Thus, the data further strengthen the notion that the BLM-localised Cl^--stimulated ATPase, which is inhibited by acetazolamide by direct competition with Cl^-, may be involved in net Cl^- transport across the molluscan gut.

6.2 Transport Activity

Gerencser and Lee [78] demonstrated an ATP-dependent Cl^- uptake in *A. californica* inside-out gut absorptive cell BLM vesicles that was inhibitable by thiocyanate, vanadate, and also by acetazolamide. The ATP-driven Cl^- uptake was obtained in the absence of Na^+, K^+, HCO_3^-, or a pH gradient between the intra- and extravesicular space, which is strong suggestive evidence that the Na^+/K^+-ATPase enzyme, Na^+/Cl^- symport, H^+/Cl symport, K^+/Cl^- symport, $Na^+/K^+/Cl^-$ symport, Cl^-/HCO_3^-, or Cl^-/OH^- antiport and K^+/H^+ antiport were not mechanisms that are involved in the accumulation of Cl^- within the vesicles. The finding that Cl^- moved up its thermodynamic energy gradient in the presence of ATP, across the BLM of *A. californica* foregut absorptive cells eliminates either Cl^- channels or an electrochemical coupling process as mechanisms for transporting Cl^- across the plasma membrane. Negation of these possibilities rests with the fact that these transport processes are downhill thermodynamic mechanisms. Further evidence eliminating the secondary active transport mechanisms and the passive transport mechanisms as potential candidates for the ATP-dependent Cl^- transporter in *A. californica* gut are the studies utilising specific inhibitors of these processes in both the purified BLM preparation [79] and the proteoliposomal preparation containing the Cl^--ATPase [80].

To elucidate further the electrogenic nature of the ATP-dependent Cl^- transport process, several experimental investigations were performed by Gerencser [68] as follows. First, an inwardly directed valinomycin-induced K^+ diffusion potential, making the BLM inside-out vesicle interior electrically positive, enhanced ATP-driven Cl^- uptake compared with vesicles lacking the ionophore (Fig. 1). Second, an inwardly directed FCCP (p-trifluoromethoxyphenylhydrazone)-induced

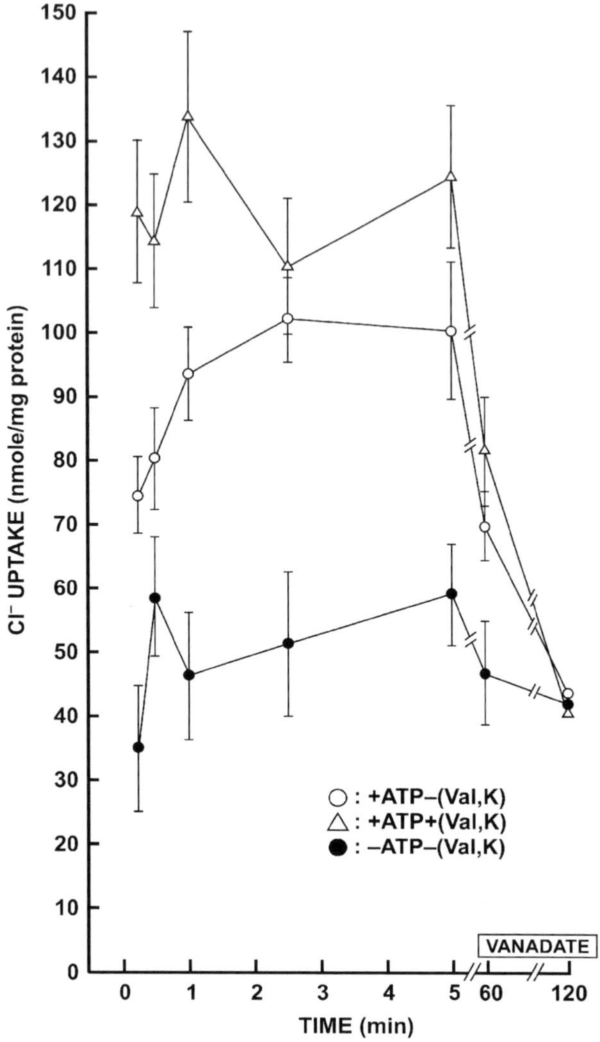

Fig. 1 ATP-driven active chloride transport in a BLM vesicular system

H^+ electrodiffusion potential, making the BLM inside-out vesicle interior less negative, increased ATP-dependent Cl^- uptake compared to controls (Fig. 2). Third, ATP increased intravesicular negativity measured by lipophilic cation distribution across the vesicular membrane. Both ATP and Cl^- appeared to be necessary for generating the negative intravesicular membrane potential, because substituting a non-hydrolysable ATP analog for ATP, in the presence of Cl^- in the extravesicular medium, did not generate a potential above that of control. Likewise, substituting $NO_3{}^-$ for Cl^- in the extra- and intravesicular media, in the presence of extravesicular ATP, caused no change in potential difference above that of control (Table 3).

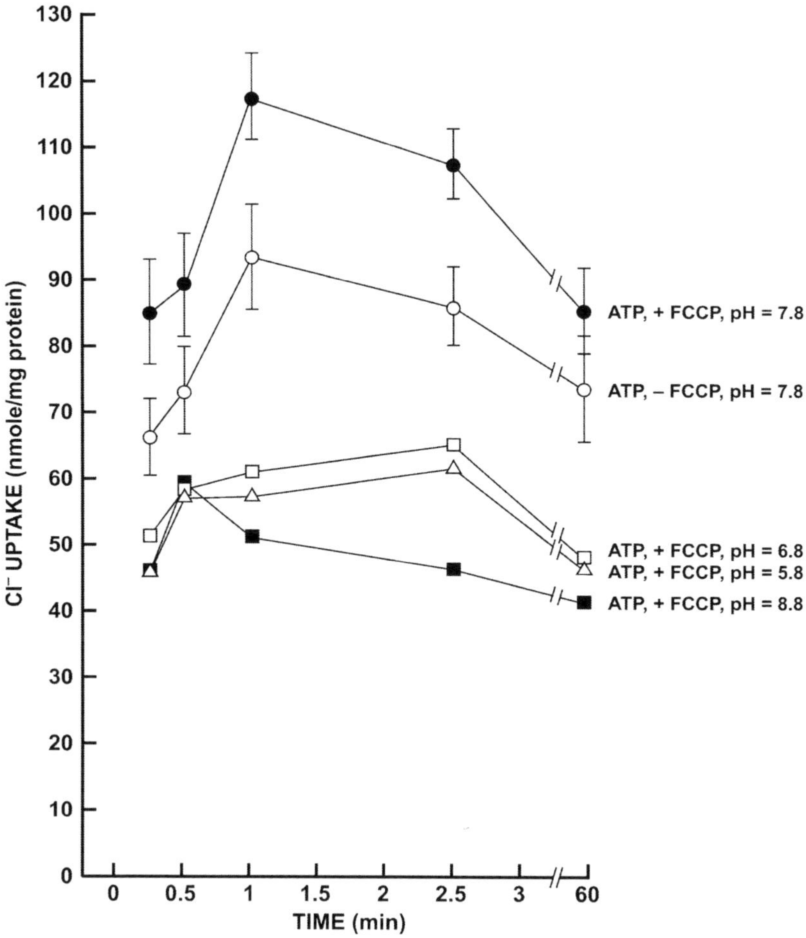

Fig. 2 pH-dependent ATP-driven active chloride transport in a BLM vesicular system

Table 3 Effect of ATP on transport parameters in BLM vesicles

Experimental condition	Cl^- uptake (nmol/mg protein)	n	Vesicular membrane potential difference (mV)	n
+ATP	102.7 ± 7.9	3	−34.9 ± 2.5	12
−ATP	49.7 ± 5.9	3	0.0 ± 5.2	12
+Non-hydrolysable ATP analogue (5′-adenylylimidodiphosphate	59.6 ± 8.3	3	−1.3 ± 0.9	12
NO_3^- for Cl^- (mole for mole)			+3.0 ± 4.6	3

ATP + $C1^-$ generation of BLM potential difference.

These results also suggested that hydrolysis of ATP is necessary for the accumulation of Cl^- in the vesicles. Furthermore, vanadate, acetazolamide, and thiocyanate inhibited the ($ATP+Cl^-$)-dependent intravesicular negativity [7]; and in addition, it had been demonstrated that the pH optimum of the Cl^--stimulated ATPase [38] coincided exactly with the pH optimum of 7.8 of the ATP-dependent Cl^- transport in the *A. californica* foregut absorptive cell BLM vesicles [68]. Therefore, both aspects of the BLM-localised Cl^- pump (ATPase and ATP-dependent Cl^- transport) have the same pH optimum, which suggests that these properties are part of the same molecular mechanism. Bafilomycin had no effect on either ATP-dependent potential change or ATP-dependent Cl^- transport [55], supporting the notion that this transporter was a P-ATPase and not a V-ATPase, since bafilomycin is an inhibitor of V-ATPase activity [81]. Further buttressing this observation was the observation that DCCD (N, N'-dicyclohexylcarbodiimide), an inhibitor of P, V, or F-ATPase proton pumps [20], had no effect on the ATP-dependent transport parameters: Cl^- transport or vesicular membrane potential change. These results negated a proton-recycling mechanism [50] as the means for net Cl^- uptake in BLM vesicles.

7 Reconstitution of the Cl^- Pump

Reconstitution of a membrane protein into a liposome provides one of the few methods that can demonstrate rigorously the existence of a separate and distinct biochemical and physiological molecular entity. This method also provides evidence that all components of the solubilised protein have been extracted intact. With this premise in mind, Gerencser [12] reconstituted both aspects of the Cl^- pump; that is, the catalytic (ATPase) and transport components from the BLM of *A. californica* gut absorptive cells. Cl^--stimulated ATPase activity existed significantly above Mg^{2+}-stimulated ATPase activity found in the proteoliposome population extracted and generated with digitonin. Vanadate inhibited this Cl^--stimulated ATPase activity by 99%. From this digitonin-generated proteoliposome population, there is a significant ATP-dependent Cl^- uptake into these proteoliposomes above that of control, and this ATP-dependent Cl^- uptake is also inhibited by vanadate (Table 4). Not detected in the proteoliposomes solubilised and formed by digitonin were Na^+/K^+-ATPase, alkaline phosphatase, ($Ca^{2+}+Mg^{2+}$)-ATPase, or cytochrome c oxidase activities and, coupled with a previous observation that FCCP (a protonophore) had a stimulatory and not an inhibitory effect on ATP-driven Cl^- accumulation in the BLM vesicles, suggested that none of these enzymes nor eukaryotic vacuolar H^+-ATPases could express Cl^- pump activity. These data also suggested that these two major observations are manifestations of one molecular mechanism: the Cl^- pump. Support of this contention rested with the findings that vanadate (an inhibitor of P-type ATPases) inhibited both Cl^--stimulated ATPase activity and ATP-dependent Cl^- transport in the digitonin-based proteoliposomes. Even though Krogh [82] first coined the term "Cl^- pump" in 1937, it was not until the reconstitution of all of its components into an artificial liposomal system through the study mentioned above [12] that

Table 4 Reconstitution of Cl^- catalytic and transport activities

Proteoliposome ATPase activity Extractive and reconstitutive detergents	Mg^{2+}-ATPase	$(Mg^{2+} + Cl^-)$-ATPase	$(Mg^{2+} + Cl^-)$-ATPase + vanadate
Cholate	n.d.	n.d.	–
Octyl glucoside	n.d.	n.d.	–
Lubrol PX	n.d.	n.d.	–
Digitonin	2.8 ± 0.4	11.2 ± 2.0	2.9 ± 0.4
Cl^- uptake into proteoliposomes			
Extractive and reconstitutive detergents	−ATP	+ATP	+ATP + vanadate
Cholate	85.7 ± 5.6	82.7 ± 6.9	80.6 ± 8.3
Octyl glucoside	82.7 ± 8.0	73.6 ± 9.2	83.9 ± 8.9
Lubrol PX	28.3 ± 11.1	39.9 ± 13.9	39.3 ± 14.0
Digitonin	91.2 ± 6.0	192.5 ± 9.3	93.1 ± 7.9

ATP-dependent Cl^- transport and Cl^--stimulated ATPase activity in a proteoliposomal preparation are inhibited by orthovanadate.

the existence of this mechanism (primary active transport mechanism) was proven. Similar reconstitutions of Cl^- pump activity have since been reported in bacteria [11], alga [15] and rat brain [16, 52, 83]. However, the alga studies [15] are somewhat ambiguous since Cl^- inhibited the Mg^{2+}-ATPase activity despite there being an ATP-dependent Cl^- uptake into the proteoliposome.

7.1 Molecular Mass

Utilising polyacrylamide gel electrophoresis (PAGE) techniques on digitonin-generated proteoliposomes containing the Cl^- pump protein from *A. californica* gut absorptive cells as shown previously [12], the approximate molecular mass of the Cl^- pump was ascertained [14]. Since both aspects of the Cl^- pump were inhibited by vanadate, it was surmised that the approximate molecular mass of the Cl^- pump of *A. californica* should be around 100 kDa since vanadate only inhibited P-type ATPases and not F-, or V-type ATPases [20]. The alpha-subunit of all P-type ATPases approximates 100 kDa in molecular mass. Several major protein bands were eluted through PAGE and, at least one had a molecular mass approximating 110 kDa [14]. Recently, similar molecular masses have been obtained for Cl^- pump catalytic units in alga [15, 53, 84] and rat brain [52, 83] confirming the possible E_1–E_2 nature of the ATPase although the authors of these studies postulate these structural subunits to be part of a V-ATPase assembly despite the ATPases being inhibitable by the P-ATPase inhibitor vanadate [52, 53]. The molecular masses of both the rat brain [83, 85] and alga [53, 84] Cl^--ATPase complexes exceeded 500 kDa.

7.2 Reaction Mechanism

The purified protein (Cl^- pump) had been subjected to phosphorylation within the proteoliposome and the reaction sequence and kinetics of the reaction sequence of the enzyme have been determined: Mg^{2+} causing phosphorylation, Cl^- causing dephosphorylation, and all in a time frame consistent with an aspartyl phosphate linkage [14, 43]. Hydroxylamine and high pH destabilise this phosphorylation confirming an acyl phosphate bond as an intermediate in the reaction sequence. Vanadate almost completely inhibited the Mg^{2+}-driven phosphorylation reaction, which corroborates the protein catalytic subunit molecular mass of 110 kDa and it also defines the protein as a P-type ATPase, because vanadate is a transition state inhibitor of phosphate [20]. More recently it has been shown that the calculated rate constant for E_1 -P formation was 26 s^{-1} [86]. This approximated E_1 -P rate constant values for other electrogenic P-type ATPases and therefore it was concluded from these results that the Cl^--ATPase phosphorylation kinetics did not differ greatly from P-type cation -ATPase phosphorylation kinetics. The calculated rate constant for disintegration of the phosphoprotein complex was 20 s^{-1} which, also, was not different than P-type cation ATPase dephosphosylation kinetics [87]. Figure 3 is an operational model of the reaction sequence of the Cl^- pump.

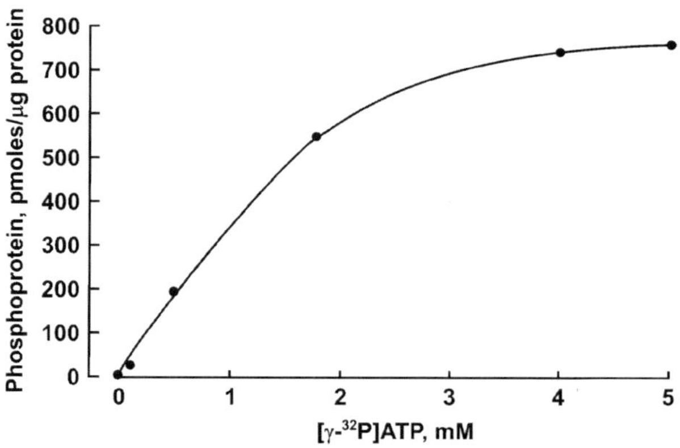

Fig. 3 An operational model of the reaction sequence of the $C1^-$ pump

7.3 Stoichiometry

The stoichiometry of ATP hydrolysed to Cl^- transported during a single cycle of the reaction sequence was ascertained through thermodynamic means [88]. Intracellular concentrations of ATP, ADP, and inorganic phosphate were determined and, coupled with an estimate of the standard free energy of hydrolysis for ATP, the operant free energy for ATP hydrolysis was calculated. Because the operating free energy of the Cl^- pump (electromotive force) was approximately one-half the energy (140 mV) obtained from the total free energy of ATP hydrolysis (270 mV), the only possible integral stoichiometries were one or, at the most, two Cl^- transported per cycle per ATP hydrolysed. Physiologically, the electrogenic Cl^- pump [68] most likely transports one Cl^- per ATP hydrolysed per reaction cycle. This increased electrochemical driving force created by the electrogenic nature of the pump could fuel secondary, electrophoretic (or electroneutral) transport processes such as the nutritional uptake of sugars and/or amino acids [68].

7.4 Phosphorylation

When the pure proteoliposomal preparation of Cl^--stimulated ATPase [14] was incubated with [γ-^{32}P]ATP, there was a fairly rapid formation of phosphoenzyme which remained stable for 4 min [89]. This and the finding that there is a curvilinear relationship between labelled ATP concentration and phosphoprotein levels suggests that at low ATP concentrations phosphoprotein formation is directly proportional to the ATP concentration; that is, the system follows Michaelis–Menten kinetics (Fig. 4). The relatively high K_m of ATP for the protein (1.25 mmol l^{-1}) suggests a relatively low affinity of the nucleotide for the enzyme. These findings are similar to what has been found for ATP affinity for fungal and plant P-type proton pumps [23, 90, 91].

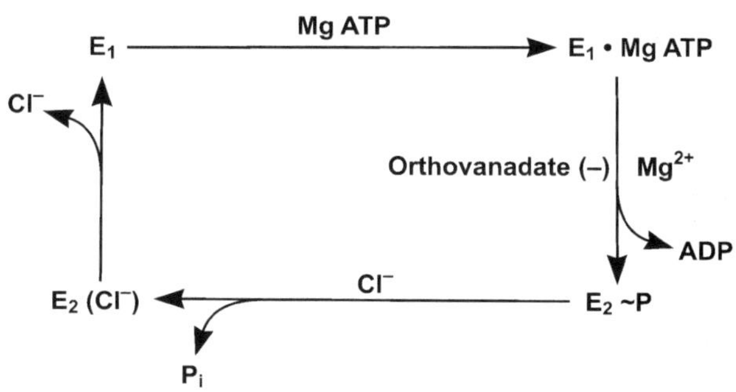

Fig. 4 Proteoliposomal phosphoprotein formation exhibits a curvilinear relationship with exogenous ATP concentrations

The finding of a millimolar K_m value was very surprising in view of well-established micromolar K_m values for ATP reactivity in P-type ATPases from various animal cells. During hydrolysis by brain or electroplax Na^+/K^+-ATPase, ATP reactivity to the enzyme exhibited a K_m below 1 mmol l^{-1} [92]. Phosphorylation studies on the Ca^{2+}-ATPase of rabbit sarcoplasmic reticulum have yielded ATP-K_m values of approximately 10 mmol l^{-1} [92]. The H^+/K^+-ATPase from pig gastric mucosa displays two kinetic components for ATP hydrolysis; one with a K_m near 2 mmol l^{-1}, and the other with a K_m near 50 mmol l^{-1} [93].

Despite some exaggeration due to methodological differences, it seems likely that the disparity of K_m values between the *A. californica* Cl^--ATPase, fungal and plant proton ATPases and other cation P-type ATPases is real. One possible interpretation is that the high ATP-K_m ATPases might reflect obligatory activation by nucleotide binding at a non-hydrolysing, low-affinity binding site, thereby masking the higher-affinity ATP at the catalytic site, such as has been proposed for the Ca^{2+}-ATPase of sarcoplasmic reticulum [27]. This suggestion is circumstantially supported by a slight sigmoidicity in plots of membrane potential, H^+-ATPase activity, and vesicular H^+ transport, against ATP concentration in *Neurospora crassa* [23, 94].

It was also found that adding large amounts of unlabeled ATP concentrations to the incubation medium increased the amount of enzyme dephosphorylation which suggests that there is at least one binding site for phosphate on the protein [89]. This premise was strengthened by the observation that the ATP molecule also needed to be hydrolysed in order for this phenomenon to occur (Fig. 5). These experiments also suggested that one of the rate-limiting factors for Cl^--ATPase phosphorylation-dephosphorylation is ambient ATP concentration as was found for plant P-type ATPases [91].

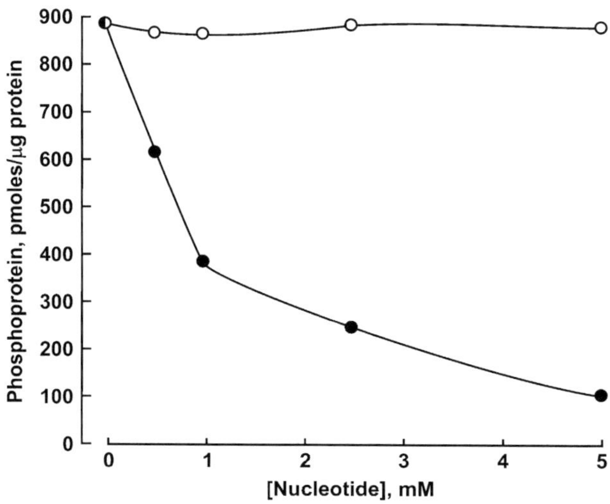

Fig. 5 Unlabeled ATP dephosphorylates $C1^-$-ATPase phosphoprotein in a proteoliposomal system

To summarise these phosphorylation experiments, it appears that the Cl^--stimulated ATPase is a P-type ATPase similar in characteristics and reaction scheme to those described for various cation-pumping P-type ATPases present in plasma membranes from a diversity of biological organisms [20, 23]. However, it appears that there are two types of this kind of P-type ATPase. One has a high affinity (micromolar) for ATP and is prevalent, for the most part, in plasma membranes of higher animals while the other group has a low affinity (millimolar) for ATP and is present in bacteria [23], fungi [90], plants [91], and in molluscs [14, 89].

7.5 Kinetics

Utilising a purified BLM vesicle preparation containing Cl^--ATPase from *A. californica* gut, it was demonstrated that ATP, and its subsequent hydrolysis, stimulated both intravesicular Cl^- accumulation and intravesicular negativity with almost identical kinetics [55, 56]. Similarly, the apparent K_ms of ATP for ATP-dependent Cl^- uptake, ATP-dependent membrane potential change and Cl^--stimulated ATPase in the proteoliposomal preparation were similar to each other [56]. They were also similar to the apparent K_ms, respectively, of Cl^- and ATP found for Cl^--ATPase in the BLM of *A. californica* [7] and for ATP-induced phosphorylation of Cl^--ATPase in the same proteoliposomal preparation of *A. californica* [14, 89]. Also, the proteoliposomal ATP-dependent Cl^- accumulation, ATP-dependent change in membrane potential and Cl^--stimulated ATPase activity showed identical pH optima and identical optimum Mg^{2+} concentrations [56]. These kinetic experiments demonstrate the correspondence between overall ATPase activity, Cl^--ATPase phosphorylation, ATP-dependent Cl^- transport, ATP-dependent membrane potential change, and Cl^--ATPase activity which are similar to those characteristics detected in cation-activated and cation-motive ATPases [20, 23].

These kinetics are uniquely significant not only because they are the first and only results obtained with an isolated discreet anion transporter ATPase but because they demonstrate the interrelationship, interchangability, and universality between both transport and catalysis by the Cl^- pump (Table 5).

7.6 Regulation of Cl^--ATPase Pump Activity

Utilising either a purified BLM preparation [55] or a proteoliposomal preparation [56] containing Cl^--ATPase activity from *Aplysia* foregut, it was demonstrated that intracellular ATP, Mg^{2+}, Cl^-, and pH regulated the activity of the ATPase and also its transport activity (Fig. 6). The catalytic and transport activity of the ATPase followed Michaelis–Menten kinetics as either intracellular ATP, Mg^{2+}, Cl^-, or pH was varied (Fig. 7).

When the solubilised Cl^--ATPase from rat brain was reconstituted into liposomes, it required phosphatidylinositol-4-monophosphate for its maximum activity, suggesting that in situ Cl^--ATPase activity is regulated by phosphatidylinositol metabolism [54, 83, 85]. Supporting this idea, Li^+, a potent uncompetitive inhibitor of inositol monophosphatase, reduced Cl^--ATPase activity and increased intracellular Cl^- concentration in cultured rat hippocampal neurons under conditions of stimulated phosphatidylinositol turnover [16, 83, 85].

8 Molecular Biology of the CL^- Pump

New bacterial rhodopsins of the cruxrhodopsin tribe were identified in a type strain *Haloarcula vallismortis* [95]. The genes encoding a halorhodopsin-like Cl^- pump were cloned and sequenced. The amino acid sequence of this photon-driven Cl^-

Table 5 Effect of inhibitors and/or reactants on ATP-dependent $\Delta\psi$, ATP-dependent Cl^- transport, and Cl^--ATPase activity

Inhibitors/reactants	$\Delta\psi$, mV	Inhibition, %	Cl^- transport, nmol/mg protein	Inhibition, %	Cl^--ATPase activity, μmol/ 15 min/mg protein	Inhibition, %
None: (Control, -ATP)	0.0 ± 2.5		82.5 ± 7.0			
ATP (5 x 10^{-3} M): (Control, +ATP)	−41.6 ± 2.8		270.6 ± 16.1		7.0 ± 0.7	
Orthovanadate (10^{-7} M)	−4.5 ± 0.8	89.2	92.5 ± 10.0	94.7	0.2 ± 0.2	97.1
Efrapeptin (10^{-6} M)	−44.2 ± 3.0	–	288.6 ± 18.1	–	7.6 ± 0.8	–
Furosemide (10^{-6} M)	−39.1 ± 1.6	6.1	283.1 ± 10.0	–	6.5 ± 1.0	7.1
SITS (10^{-5} M)	−35.9 ± 4.3	13.7	275.1 ± 17.9	–	5.8 ± 1.6	17.1
Ouabain (10^{-6} M)	−42.1 ± 2.0	–	274.6 ± 16.5	–	7.1 ± 2.3	–
DCCD (10^{-5} M)	−40.3 ± 4.1	3.2	288.1 ± 19.2	–	7.0 ± 1.7	–
Ruthenium Red (10^{-6} M)	−40.4 ± 3.7	2.9	281.3 ± 19.9	–	6.7 ± 1.0	4.3
Picrotoxin (10^{-6} M)	−38.1 ± 3.9	8.4	274.2 ± 16.5	–	6.8 ± 1.5	2.9
PCMBS (10^{-4} M)	−19.3 ± 3.5	53.6	155.1 ± 9.3	61.4	2.6 ± 1.6	62.9
DTT (10^{-4} M)	−40.9 ± 5.0	1.7	258.5 ± 16.3	6.4	7.3 ± 0.8	–
PCMBS (10^{-4} M) + DTT (10^{-4} M)	−41.5 ± 3.7	0.3	255.1 ± 18.1	8.2	7.2 ± 1.7	

Kinetics of $C1^-$-ATPase activity, ATP-dependent $C1^-$ transport and ATP-dependent membrane potential difference in a proteoliposomal preparation.

Fig. 6 Michaelis constants for ATP-dependent C1$^-$ transport, C1$^-$-stimulated ATPase activity and ATP-dependent membrane potential change in a proteoliposomal system

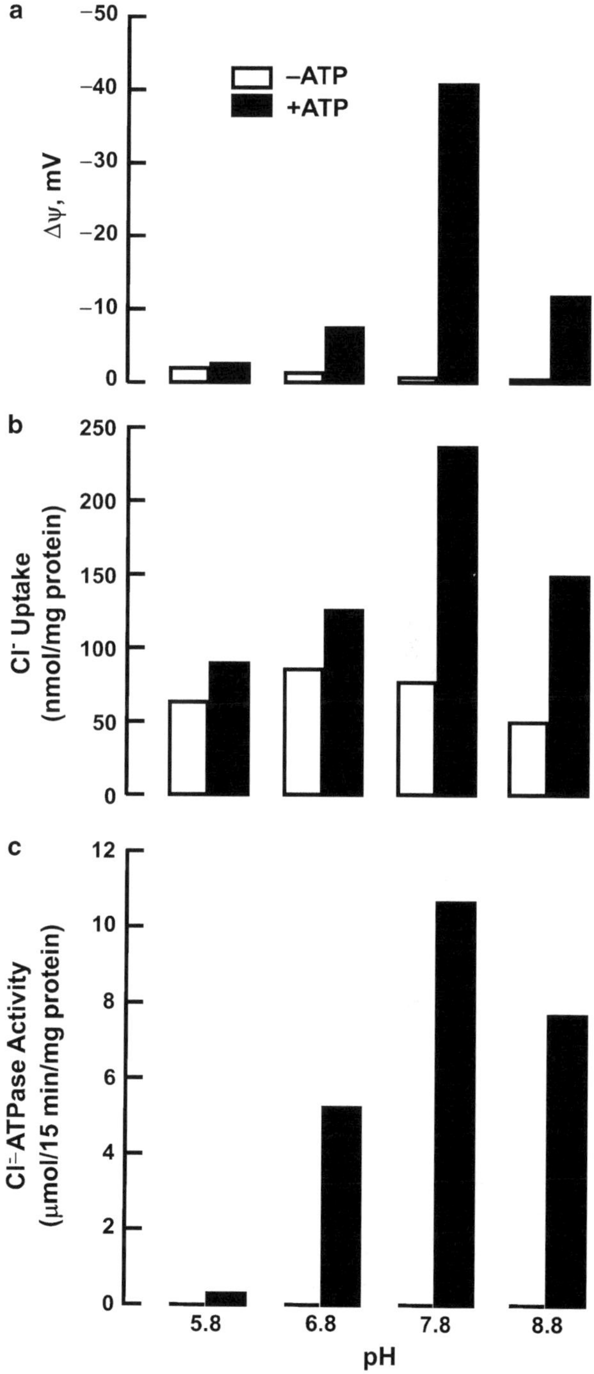

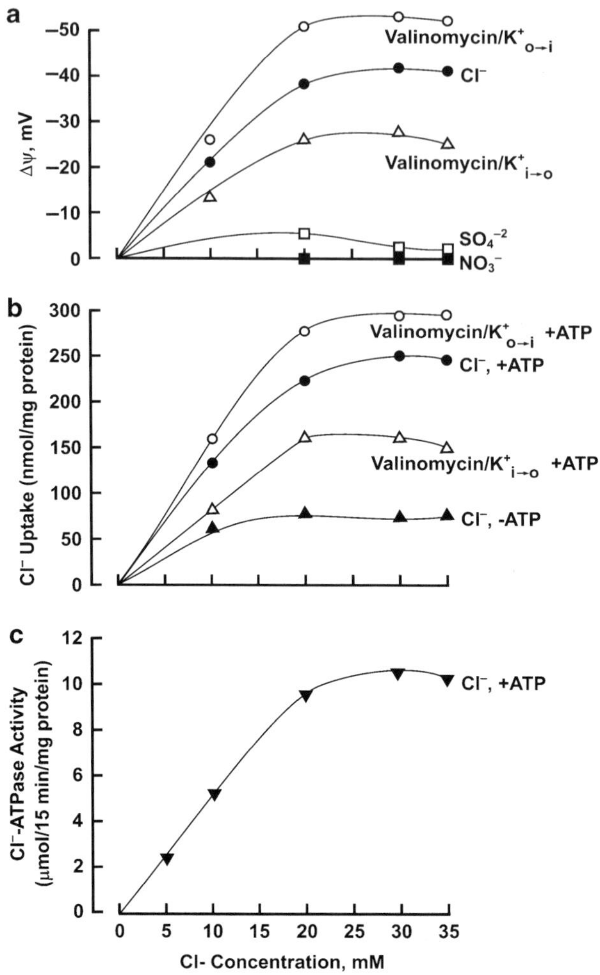

Fig. 7 Catalytic and transport activity of the $C1^-$ pump followed Michaelis–Menten kinetics as either ATP, Mg^{2+}, $C1^-$, or pH was varied

pump showed little homology with the ATP-driven Cl^- pumps [43], but showed a high degree of homology with other documented halorhodopsins [11].

The genes encoding the β subunit (50 kDa) of the Cl^- translocating ATPase of *A. acetabulum* were cloned from total RNA and from poly $(A)^+$ RNA and sequenced. The deduced amino acid sequence of the open reading frame consisted of 478 amino acids and showed high similarity to the β subunit of chloroplast F_1-ATPases [84]. Gene fragments encoding the putative β subunit of chloroplast F_1- (273 bp) and mitochondrial F_1-ATPases (332 bp) were also cloned from *A. acetabulum* and sequenced, respectively. The deduced amino acid sequence of the chloroplast F_1-ATPase showed 92.5% identity to the primary structure of the β subunit of the Cl^--translocating ATPase, while the nucleotide sequences were 79.9% identical. The deduced amino acid sequence was 79.9% identical. The deduced amino acid sequence of the latter was 77.3% identical to that of the β subunit of the Cl^--translocating ATPase and the nucleotide sequences were 67.5% identical.

Reverse transcriptase polymerase chain reaction was used to detect the mRNA of the *A. californica* foregut Cl^- pump. The Cl^- pump RNA had a high homology relative to mRNA's of subunits of Na^+/K^+-ATPase and other P-type ATPases. The cDNA sequence of the *A. californica* Cl^--ATPase was cloned and the resulting 200 base pair sequence was shown to be 76% identical to regions of alpha-subunits of P-type ATPases [96, 97].

Recently the DNA of a 55-kDa protein derived from a 520-kDa Cl^--ATPase complex was cloned from rat brain [83]. Sequences of nucleic acids in the cDNA and the deduced amino acids were not homologous with any known ion-translocating ATPases [20], and the application of its antisense oligonucleotides induced increases in Cl^- concentrations in primary cultured rat hippocampal neurons, suggesting that the 55-kDa protein acts as a catalytic subunit of the Cl^--ATPase pump.

9 Conclusions and Speculations

(1) In summary, it is quite apparent that in the past few years there has been an increasing number of convincing studies in a variety of biological tissues that have provided indirect, correlative evidence that active Cl^- transport is primary by nature. The active translocation of Cl^- by an enzyme that directly utilises the energy from ATP hydrolysis is not unlike that observed in plants [10, 15]. Indeed, the evidence for primary active Cl^- transport in these simple living things is almost as convincing as that presented for (Na^+ + K^+)-stimulated ATPase and (Ca^{2+} + Mg^{2+})-stimulated ATPase in their respective roles for actively transferring Na^+, K^+, and Ca^{2+} across animal plasma membranes. As alluded to by DePont and Bonting [18] future experimental steps aimed at confirming that an animal Cl^--stimulated ATPase is involved in primary Cl^- transmembrane movement should include the following: (a) a specific inhibitor for the enzyme should be found or synthesised (e.g. an antibody), and this inhibitor should be shown to inhibit the transport process; (b) the Cl^--stimulated ATPase should be biochemically isolated or purified and after its incorporation into liposomes should then be shown to support active Cl^- transport; and (c) the insertion of the Cl^- pump's mRNA into an expression system (e.g. an oocyte) focused by testing for Cl^- pump activity in this system. (2) Demonstrations of reconstitution and phosphorylation of Cl^--ATPase have been discussed in the present review which provides the first direct evidence for the existence of a new P-type ATPase: the Cl^- pump. However, it appears that V- and F-type Cl^--ATPases also exist as Cl^- pump mechanisms [83, 84] along with the photon-driven halorhodopsins [95]. (3) The variety and number of Cl^- pumps attests to the adaptability of a particular

organism existing in a particular environment and not to the non-existence of this primary active transporter in nature.

In contrast to most higher animals, the *A. californica* gut has evolved a double-salt pump system. It has a Na^+ pump (Na^+/K^+-ATPase) and a Cl^- pump (Cl^--ATPase) which, upon first view, suggests that there is inefficient salt absorption by these cells in terms of ATP utilisation. One can easily envision the Na^+ pump generating the intracellular electrochemical asymmetry needed for Na^+-coupled entry of energetic and structural equivalents. These equivalents (e.g. sugars, amino acids, etc.) are needed for the continued maintenance and survival of the organism. However, a Cl^- pump that utilises more ATP than the Na^+ pump, appears to be a frivolous vestigium of a gone-awry evolutionary process. Not necessarily so if one considers the environment that the *A. californica* gut cells are bathed in throughout their existence. These cells are surrounded by sea water, a medium so rich in Cl^- that the Cl^- concentration is, at least, 100 meq l higher than any other ion in this medium. The *A. californica* gut cells had to adapt to survive the onslaught of Cl^- rushing into their internal environment potentially causing their demise and therefore that of the animal itself. The Cl^- pump may represent a survival mechanism intended to prevent osmotic lysis of the *Aplysia* gut cells and this has helped maintain the existence of this species for millions of years.

The basic studies by Gerencser and Zhang [79, 80, 86, 94] and by Inagaki [85], have elucidated a mechanism that could be directly involved in convulsive brain activity. Inagaki [83, 85] has speculated on its role in Alzheimer's disease with convincing evidence. These studies, especially those of Inagaki [83] have laid bare a means to probe abnormal brain activity and, hopefully to resolve or partially resolve its manifestations.

Acknowledgments The author wishes to acknowledge his gratitude to his technologists, students, and collaborators for their able contributions to the studies reviewed and performed herein. These studies were supported by D.S.R. Seed Award (No. 229K15), Whitehall Foundation Grant (No. 78–156 ck-1), D.S.R. Award (No. 122101010), and the Eppley Foundation for Research, Inc.

References

1. DuBois-Reymond, E. *Untersuchungen uber tierische Elektrizitat*, Berlin. 1848.
2. Galeotti, G. Concerning the E.M.F. which is generated at the surface of animal membranes on contact with different electrolytes. *Zellforsh Phys Chem* 1904; 49: 542–562.
3. Ussing, HH, Zerahn, K. Active transport of sodium as the source of electric current in the short-circuited isolated frog skin. *Acta Physiologica Scandinavica* 1951; 23:110–127.
4. Skou, JC. Enzymatic basis for active transport of Na^+ and K^+ across cell membrane. *Physiol Rev* 1965; 45:596–617.
5. Frizzell, RA, Field, M and Schultz, SG. Sodium-coupled chloride transport by epithelial tissues. *Am J Physiol* 1979; 236: F1–F8.
6. Gerencser, GA. Invertebrate epithelial transport. *Am J Physiol* 1983b; 244: R127–R129.
7. Gerencser, GA, White, JF, Gradmann, D and Bonting SL. Is there a Cl^- Pump? *Am J Physiol* 1988; 255:R677–R692.
8. Hanrahan, JW and Phillips, JE. Cellular mechanisms and control of KCl absorption in insect hindgut. *J Exp Biol* 1983a; 106: 71–89.

9. Gradmann, D. Electrogenic Cl^- pump in the marine alga *Acetabularia*, in *Chloride Transport Coupling in Biological Membranes and Epithelia*, Gerencser, GA, Ed., Elsevier, Amsterdam, 1984; 13–62.
10. Hill, BS, Hanke, DE. Properties of the chloride ATPase from *Limonium* salt glands: activation by, and binding to, specific sugars. *J Membr Biol* 1979; 51: 185–194.
11. Zimanyi, L, Lanyi, J. Halorhodopsin: a light-driven active chloride transport system, in *Bicarbonate, Chloride and Proton Transport Systems*, Durham, J. and Hardy, M., Eds., Ann NY Acad Sci 1989; 11–19.
12. Gerencser, GA. Reconstitution of a chloride-translocating ATPase from *Aplysia californica* gut. *Biochimica et Biophysica Acta* 1990; 1030: 301–303.
13. Shiroya, T, Fukunaga, R, Akashi, K, Shimada, N, Takagi, Y, Nashino, T, Hara, M and Inagaki, C. An ATP-driven Cl^- pump in the brain. *J Biol Chem* 1989; 264:17416–17421.
14. Gerencser, GA and Zelezna, B. Reaction sequence and molecular mass of a Cl^--translocating P-type ATPase. *Proc Natl Acad Sci (USA)* 1993; 90: 7970–7974.
15. Ikeda, M and Oesterhelt, D. A Cl^--translocating adenosinetriphosphatase in *Acetabularia acetabulum*, 2, Reconstitution of the enzyme into liposomes and effect of net charges of liposomes on chloride permeability and reconstitution. *Biochemistry* 1990; 29(8): 2065–2070.
16. Zeng, X-T, Hara, M and Inagaki, C. Electrogenic and phosphatidylinositol-4-monophosphate-stimulated Cl^- transport by Cl^- pump in the rat brain. *Brain Res* 1994; 641:167–170.
17. Schultz, SG. Chloride transport by gastrointestinal epithelia: An overview, in *Mechanisms of Intestinal Secretion* Binder, HJ, Eds, New York: Alan R. Liss, Inc., 1979; 93–100.
18. DePont, JJHHM and Bonting, SL. Anion-sensitive ATPase and (K^+ + H^+)-ATPase, in *Membrane Transport*, Bonting, SL and DePont, JJHHM., Eds., Elsevier/North Holland Biomedical Press 1981; 209–222.
19. Kedem, O and Katchalsky, A. A physiological interpretation of the phenomenological coefficients of membrane permeability. *Am J Physiol* 1961; 45:143.
20. Pedersen, PL and Carafoli, E. Ion motive ATPases. I. Ubiquity, properties, and significance to cell function. *Trends Biochem Sci* 1987; 12:146–150.
21. Amzel, LM and Pedersen, PL. Proton ATPases: structure and mechanism. *Ann Rev Biochem* 1983; 52: 801–822.
22. Harvey, WR. Physiology of V-ATPases. *J Exp Biol* 1992; 172:1–17.
23. Slayman, CL and Zuckier, GR. Differential functional properties of a P-type ATPase/proton pump, in *Bicarbonate, Chloride and Proton Transport Systems*, Durham, J. and Hardy, M., Eds., Ann NY Acad Sci 1989; 574:233–245.
24. Cantley, LC, Jr, Josephson, L, Warner, R, Yanagisawa, M, Lechene, C and Guidotti, G. Vanadate is a potent Na,K-ATPase inhibitor found in ATP derived from muscle. *J Biol Chem* 1977; 252:7421–7424.
25. O'Neal, SG, Rhoads, DB and Racker, E. Vanadate inhibition of sarcoplasmic reticulum Ca^{2+}-ATPase and other ATPases. *Biochem Biophys Res Commun* 1979; 89:845–850.
26. Jorgensen, PL and Anderson, JP. Structural basis for E_1–E_2 conformational transitions in Na, K-pump and Ca-pump proteins. *J Membr Biol* 1988; 103: 95–103.
27. Inesi, G and Kirtley, MR. Structural features of cation transport ATPases. *J Bioenerg Biomembr* 1992; 24: 271–279.
28. Toyoshima, C, Nakasako, M, Nomura, H and Ogawa, H. Crystal structure of the calcium pump of sarcoplasmic reticulum at 2.6 A resolution. *Nature* (London) 2002; 405: 647–655.
29. Ogawa, H, Strokes, DL, Sasabe, H and Toyoshima, C. Structure of the Ca^{2+} pump of sarcoplasmic reticulum: a view along the lipid bilayer at 9-A resolution. *Biophy J* 1998; 75:41–52.
30. Zhang, P, Toyoshima, C, Yonekura, K, Green, N M and Strokes, DL. Structure of the calcium pump from sarcoplasmic reticulum at 8-A resolution. *Nature* (London) 1998; 392:835–839.
31. Auer, M, Scarborough, GA and Kuhlbrandt, W. Three-dimensional map of the plasma membrane H^+ -ATPase in the open conformation. *Nature* (London) 1998; 392:840–843.

32. Scarborough, GA. Structure and function of the P-type ATPases. *Curr Opin Cell Biol* 1999; 11:517–522.
33. Rice, WJ, Green, NM and MacLennan, DH. Site directed disulfide mapping of hellices M4 and M6 in the Ca^{2+} binding domain of SERCA1a, the Ca^{2+} ATPase of fast twitch skeletal muscle sarcoplasmic reticulum. *J Biol Chem* 1997; 272:31412–31419.
34. Jorgensen, PL, Nielsen, JM, Rasmussen, JH and Pedersen, PA. Structure-function relationships of E1–E2 transitions and cation binding in Na, K-pump protein. *Biochim Biophys Acta* 1998; 1365:65–70
35. Moller, JV, Juul, B and le Maire, M. Structural organization, ion transport, and energy transduction of P-type ATPases. *Biochim Biophys Acta* 1996; 1286:1–51.
36. Schuurmans, Stekhoven, F and Bonting, SL. Transport adenosine triphosphatases: Properties and functions. *Physiol Rev* 1981; 61:1–76.
37. Gerencser, GA and Lee, SH. Cl^--stimulated adenosine triphosphatase: existence, location and function. *J Exper Biol* 1983; 106: 142–161.
38. Gerencser, GA and Lee, SH. Cl^-/HCO_3^--stimulated ATPase in intestinal mucosa of *Aplysia*. *Am J Physiol* 1985b; 248: R241–R248.
39. Mircheff, AK, Sachs, G, Hanna, SD, Labiner, CS, Rabon, E, Douglas, AP, Walling, M W and Wright, EM. Highly purified basal lateral plasma membranes from rat duodenum: physical criteria for purity. *J Membr Biol* 1979; 50:343–363.
40. Bullough, DA, Jackson, CG, Henderson, PJ, Beechey, RB and Linnett, PE. The isolation and purification of the elvapeptins: A family of peptide inhibitors of mitochondrial ATPase activity. *FEBS Lett* 1982; 145: 258–262.
41. Gerencser, GA. Transport across the invertebrate intestine, *Transport Processes Iono- and Osmoregulation, Vol. II*, Gilles, R and Gilles-Baillien, M G, Eds., Springer, Berlin, 1985; 251–264.
42. Gerencser, GA. Primary electrogenic chloride transport across the *Aplysia* gut *Bicarbonate, Chloride and Proton Transport Systems*, Durham, J and Hardy, M, Eds., *Ann NY Acad Sci* 1989; 547:1–10.
43. Gerencser, GA and Zelezna, B. Existence of a chloride pump in molluscs, in *Electrogenic Cl^- transporters in biological membranes*, as part of *Advances in Comparative and Environmental Physiology* 19, Gerencser, GA, Ed., Springer-Verlag, Berlin 1994; 39–58.
44. DeRenzis, G and Bornancin, M. Cl^-/HCO_3^--ATPase in the gills of *Carassius auratus*: Its inhibition by thiocyanate. *Biochimica et Biophysica Acta* 1977; 467: 192–207.
45. Bornancin, M, DeRenzis, G and Maetz, J. Branchial Cl transport, anion-stimulated ATPase and acid-base balance in *Anguilla anguilla* adapted to freshwater: Effects of hyperoxia. *J Compar Physiol* 1977; 117:313–322.
46. Bornancin, M, DeRenzis, G and Naon, R. Cl^--HCO_3^--ATPase in gills of the rainbow trout: Evidence for its microsomal localization. *Am J Physiol* 1980; 238:R251–R259.
47. Hanrahan, J and Phillips, JE. Mechanism and control of salt absorption in locust rectum. *Am J Physiol* 1983b; 244:R131–R142.
48. Lechleitner, RA and Phillips, JE. Cl^-/HCO_3^--ATPase in locust rectum. *Can J Zool* 1988; 66(2): 431–438.
49. Phillips, JE, Wiens, C, Audsley, N, Jeffs, L, Bilgen, T and Meredith, J Nature and control of chloride transport in insect absorptive epithelia. *J Exp Zool* 1996; 275:292–299.
50. Wieczorek, H. The insect V-ATPase, a plasma membrane proton pump energizing secondary active transport: molecular analysis of electrogenic potassium transport in the tobacco hornworm midgut. *J Exp Biol* 1992; 172:335–343.
51. Inagaki, C, Tanaka, T, Hara, M and Ishiko, J. Novel microsomal anion-sensitive Mg^{2+}-ATPase activity in rat brain. *Biochem Pharmacol* 1985; 34:1705–1712.
52. Inagaki, C, Hara, M and Inoue, M. Transporting Cl^--ATPase in rat brain, in *Electrogenic Cl^- transporters in biological membranes*, as part of Advances in Comparative and Environmental Physiology 19, Gerencser, G.A., Ed., Springer-Verlag, Berlin, 1994; 59–79.

53. Graves, JS and Gutknecht, J. Current-voltage relationships and voltage sensitivity of the Cl^- pump in *Halicystis parvula*. *J Membr Biol* 1977; 36:83–99.
54. Moritani, C, Ohhashi, T, Satoh, S, Oesterhelt, D and Ikeda, M. Purification and characterization of a membrane-bound ATPase from Acetabularia cliftoni that corresponds to a Cl^--translocating ATPase in Acetabularia acetabulum. *Biosci Biotech Biochem* 1994; 58(11): 2087–2089.
55. Gerencser, GA and Purushotham, KR. A novel Cl^--pump: intracellular regulation of transport activity. *Biochem Biophys Res Commun* 1995; 215:994–1000.
56. Gerencser, GA and Purushotham, KR. Reconstituted Cl^- Pump Protein: a novel ion (Cl^-) - motive ATPase. *J Bioenerg Biomembr* 1996; 28:459–469.
57. Barry, RJC, Smyth, DH and Wright, EM. Short-circuit current and solute transfer by rat jejunum. *J Physiol* 1965; 181: 410–431.
58. Clarkson, TW, Cross, AC and Toole, SR. Electrical potentials across isolated small intestine of the rat. *Am J Physiol* 1961; 200:1233–1235.
59. Quay, JF and Armstrong, W McD. Sodium and chloride transport by isolated bullfrog small intestine. *Am J Physiol* 1969; 217:694–702.
60. Gerencser, GA. Electrical characteristics of isolated *Aplysia californica* intestine. *Comp Biochem Physiol* 1978; 61A: 209–212.
61. Gerencser, GA. Effects of amino acids on chloride transport in *Aplysia* intestine. *Am J Physiol* 1981; 240: R61–R69.
62. Gerencser, GA. Electrophysiology of chloride transport in *Aplysia* (mollusk) intestine. *Am J. Physiol* 1983a; 244: R143–R149.
63. Gerencser, GA and Hong, SK. Ion transport in *Aplysia juliana* intestine: stimulation by exogenous sugars. *Comp Biochem Physiol* 1977; 58A:275–280.
64. Gerencser, GA. Thiocyanate inhibition of active chloride absorption in *Aplysia* intestine *Biochimica et Biophysica Acta* 1984b; 775: 389–394.
65. Armstrong, W McD, Wojtkowski, W and Bixenman, WR. A new solid-state microelectrode for measuring intracellular chloride activities. *Biochimica et Biophysica Acta* 1977; 465: 165–170.
66. Armstrong, W McD, Bixenman, W R, Frey, K R, Garcia-Diaz, J F, O'Regan M G, and Owens, JH. Energetics of coupled Na^+ and Cl^- entry into epithelial cells of bullfrog small intestine. *Biochimica et Biophysica Acta* 1979; 551:207–212.
67. Gerencser, GA and White, JF. Membrane potentials and chloride activities in e epithelial cells of *Aplysia* intestine. *Am J Physiol* 1980; 239:R445–R449.
68. Gerencser, GA. Electrogenic ATP-dependent Cl^- transport by plasma membrane vesicles from *Aplysia* intestine. *Am J Physiol* 1988; 254:R127–R133.
69. Gerencser, GA. Transport energetics of the Cl^- pump in *Aplysia* gut. *Biochimica et Biophysica Acta* 1997; 1330:110–112.
70. Gerencser, GA. Electrogenic and electrically coupled chloride transport across molluscan intestine, *Chloride Transport Coupling in Biological Membranes and epithelia*, Gerencser, GA, Ed., Elsevier, Amsterdam, 1984a; 183–203.
71. Hill, BS. Metabolic coupling of chloride transport in higher plant cells, in *Chloride Transport Coupling in Biological Membranes and Epithelia*, Gerencser, G. A., Ed., Elsevier, Amsterdam, 1984; p. 1–11.
72. Akera, T, Temma, K and Takeda, K. Cardiac actions of vanadium. *Feder Proc* 1983; 42: 2984–2988
73. Gerencser, GA. Subcellular carbonic anhydrase profile in *Aplysia* gut. *J Exp Biol* 1991; 161:515–517.
74. Lee, SH. Salinity adaptation of HCO_3^--dependent ATPase activity in the gills of blue crab (*Callinectes sapidus*). *Biochimica et Biophysica Acta* 1982; 689: 143–154 .
75. Maren, TH. Use of inhibitors in physiological studies of carbonic anhydrase. *Am J Phys* 1977; 232:F291–F297.

76. White, JF. Bicarbonate-dependent chloride absorption in small gut: ion fluxes and intracellular chloride activities. *J Membr Biol* 1980; 53: 95–107.
77. Gerencser, GA. Properties and functions of Cl^--stimulated ATPase. *Trends Life Sci* 1986; 1: 1–18.
78. Gerencser, GA and Lee, SH. ATP-dependent chloride transport in plasma membrane vesicles from *Aplysia* intestine *Biochimica et Biophysica Acta* 1985a; 816:415–417.
79. Gerencser, GA and Zhang, J. Inhibition of a Cl^--transporting P-type ATPase in *Aplysia* gut. *Zool Sci* 2001a; 18:17–19.
80. Gerencser, GA and Zhang, J. The *Aplysia californica* Cl^- pump is a P-type ATPase: evidence through inhibition studies. *Can J Physiol Pharmacol* 2001c; 79:367–370.
81. Bowman, EJ, Siebers, A and Altendorf, K. Bafilomycins: A class of inhibitors of membrane ATPases from microorganisms, animal cells, and plant cells. *Proc Natl Acad Sci USA 1986;* 85:7972–7976.
82. Krogh, A. Osmotic regulation in the frog (*R. esculenta*) by active absorption of chloride ions. *Skandinavica Arch Physiol* 1937; 76:60–74.
83. Zeng, X-T, Higashida, T, Hara, M, Hattori, N, Kitagawa, K, Omori, K and Ingaki, C. Antiserum against Cl^- pump complex recognized 51 kDa protein, a possible catalytic unit in the rat brain. *Neurosci Lett* 1998; 258:85–88.
84. Moritani, C, Ohhashi, T, Kadowaki, H, Tagaya, M, Fukui, T, Lottspeich, F, Oesterhelt, D and Ikeda, M. The primary structure of the Cl^--translocating ATPase, b subunit of *Acetabularia acetabulm*, which belongs to F-type ATPase family. *Arch Biochem Biophys* 1997; 339: 115–124.
85. Inagaki, C, Hattori, N, Kitagawa, K, Zeng, X-T and Yagyu, K. Cl^--ATPase in rat brain and kidney. *J Exp Zool* 2001; 289:224–231.
86. Gerencser, GA and Zhang, J. Phosphorylation of chloride -ATPase reconstituted from *Aplysia* gut. *J Exp Zool* 2001b; 289:472–475.
87. Gerencser, GA, Zhang, J. Chloride ATPase: dephosphorylation in Aplysia gut. *J Exp Zool* 2002; 293:89–93.
88. Gerencser, GA. Stoichiometry of a Cl^--translocating ATPase. *Feder Eur Biochem Soc Lett* 1993b; 333:133–140.
89. Gerencser, GA. A novel P-type Cl^--stimulated ATPase: phosphorylation and specificity. *Biochem Biophys Res Commun* 1993a; 196:1188–1194.
90. Slayman, CL, Long, WS and Lu, CY. The relationship between ATP and an electrogenic pump in the plasma membrane of *Neurospora crassa. J Membr Biol* 1973; 14:305–338.
91. Vara, F. and Serrano, R. Phosphorylated intermediate of the ATPase of plant plasma membranes. *J Biol Chem* 1982; 257:12826–12830.
92. Fernandez-Belda, F, Kurzmack, M and Inesi ,G. A comparative study of calcium transients by isotopic tracer, metallochromic indicator, and intrinsic fluorescence in sarcoplasmic reticulum ATPase. *J Biol Chem* 1984; 259: 9687–9698.
93. Wallmark, B, Stewart, HB, Rabon, E, Saccomani, G and Sachs, G. The catalytic cycle of gastric (H^++K^+) ATPase. *J Biol Chem* 1980; 255:5313–5319.
94. Kitajima, T, Hirayama, J, Ihara, K, Sugiyama, Y, Kamo, N and Mukohata, Y. Novel bacterial rhodopsins from *Haloarcula vallismortis. Biochem Biophys Res Commun* 1996; 220:341–345.
95. Gerencser, GA and Zhang, J. CI-ATPases: Novel Primary Active Transporters in Biology. *J Exp Zool* 2001; 289:215–223.
96. Gerencser, GA and Zhang, J. The Cl^- pump in mollusks: a Cl^--translocating P-type ATPase. *Comp Biochem Physiol* 2000; 126A(1): S59.
97. Perlin, DS, Kasami, K, Brooker, RJ and Slayman, CW. Electrogenic H^+ translocation by the plasma membrane ATPase of *Neurospora. J Biolo Chem* 1984; 259:7884–7892.

Divalent Anion Transport in Crustacean and Molluscan Gastrointestinal Epithelia

George A. Gerencser and Gregory A. Ahearn

Abstract A novel invertebrate gastrointestinal transport mechanism has been shown to couple chloride/sulfate exchange in an electrogenic fashion. In the lobster, *Homarus americanus*, the hepatopancreas, or digestive gland, exists as an outpocketing of the digestive tract, representing a single cell layer separating the gut lumen and an open circulatory system comprised of hemolymph. Investigations utilizing independently prepared brush-border and basolateral membrane vesicles revealed discrete antiport systems which possess the capacity to bring about a transcellular secretion of sulfate. The luminal antiport system functions as a high affinity, one-to-one chloride/sulfate exchanger that is stimulated by an increase in luminal hydrogen ion concentration. Such a system would take advantage of the high chloride concentration of ingested seawater, as well as the high proton concentrations generated during digestion, which further suggests a potential regulation by resident sodium-proton exchangers. Exchange of one chloride for one divalent sulfate ion provides the driving force for electrogenic vectorial translocation. The basolateral antiport system was found to be electroneutral in nature, responsive to gradients of the dicarboxylic anion oxalate, while lacking in proton stimulation. No evidence of sodium/sulfate cotransport, commonly reported for the brush border of vertebrate renal and intestinal epithelia, was observed in either membrane preparation. The two antiporters together can account for the low hemolymph to seawater sulfate levels previously described in decapod crustaceans. A secretory pathway for sulfate based upon electrogenic chloride-antiport may appear among invertebrates partly in response to digestion taking place in a seawater environment.

Keywords Antiporter · Gastrointestinal epithelia · Divalent anion exchangers · Electrogenic · Nonelectrogenic · Mollusk · Crustacean

G.A. Gerencser (✉)
Department of Physiology and Functional Genomics, College of Medicine, University of Florida, Gainesville, FL 32610-0274, USA
e-mail: ggerencs@ufl.edu

G.A. Gerencser (ed.), *Epithelial Transport Physiology*,
DOI 10.1007/978-1-60327-229-2_2,

1 Introduction – Crustacean Sulfate Transport

The first half of this chapter describes a proposed transcellular secretory mechanism for the divalent anion sulfate which has been characterized in the hepatopancreas of a decapod crustacean, *Homarus americanus*. The driving force for the overall transcellular movement of sulfate is a sulfate/chloride antiporter present in the brush border membrane which operates in an electrogenic fashion, exhibiting a 1:1 coupling ratio. The brush border antiporter is regulated by external protons, a physiologically relevant characteristic due to the generally low luminal pH of the crustacean hepatopancreas during digestion/absorption functions. There are separate anion exchangers on the hepatopancreatic basolateral membrane which provides for sulfate entry into the epithelial cell from the hemolymph or, conversely, for the movement of intracellular sulfate across the basolateral membrane into the hemolymph. The different processes involved with sulfate secretion will be discussed with reference to the current models of sulfate transport in vertebrate tissues.

The chapter in this volume by Markavich will extensively cover sulfate transporters in vertebrate epithelial membranes; therefore, we will give a cursory presentation of this topic in an introductory format as a prelude to sulfate transporters in crustacean and molluscan epithelia.

2 Vertebrate Sulfate Transport

Most of our current knowledge of sulfate transport mechanisms has been accumulated from investigations among many vertebrate organisms, with insects being the only invertebrate group where sulfate regulation has been studied [1, 2]. Mosquito larvae, existing in sulfate-rich hyperosmotic lakes, appear to absorb sulfate through the gut wall and secrete the anion, via a saturable system against a concentration gradient, by the Malpighian tubules [1].

2.1 Intestinal and Hepatic Sulfate Transport

Sulfate is actively reabsorbed across the apical border of the small intestine of the rabbit [3–5], the rat [6], and the pig [7] by a Na-cotransport system. Intestinal brush border membrane vesicles have also been shown to possess pH-dependent sulfate uptake [8] and the anion exchange mechanism [9]. The only reported mechanism for sulfate transport across the intestinal basolateral membrane is anion exchange [4, 10–12].

Anion exchange of inorganic and organic anions with sulfate has also been described in liver sinusoidal and canalicular plasma membrane vesicles from the rat [13, 14] and sinusoidal basolateral membrane vesicles of the elasmobranch [15].

2.2 Renal Sulfate Transport

Sulfate reabsorption in the mammalian kidney has been observed by numerous investigations. Luminal uptake is predominately via the sodium-dependent pathway [16–18], with anion exchange exhibited in both perfused tubules [19] and membrane vesicles [20, 21]. The transport systems described for sulfate exit from renal epithelia involve a basolateral exchanger which can accept both inorganic and organic anions [10, 22–24].

Other vertebrate renal systems have been investigated with respect to sulfate regulation. In contrast to the mammalian reabsorption paradigm teleost fish exhibit a net secretory sulfate mechanism. The driving force for transepithelial sulfate movement appears to be proton-dependent sulfate cotransport at the basolateral membrane [25]. No effect of sodium on sulfate uptake was observed in basolateral or brush border membrane vesicles. The efflux of sulfate from flounder renal tubule occurred by the anion exchanger demonstrated on the brush border [26].

Sulfate transport in the chick renal tubule demonstrated characteristics of both the mammalian and teleost systems described above [27]. Brush border membrane vesicles were shown to maintain multiple pathways for sulfate transport. Sulfate uptake was stimulated by both sodium and proton cotransport and anion exchange with bicarbonate. Basolateral membrane vesicles did not respond to sodium or proton cotransport, but sulfate/bicarbonate exchange produced concentrative uptake. It appears that the avian renal tubule maintains the capacity for potential bidirectional sulfate transport.

3 Transmembrane SO_4/Cl Exchange in Vertebrate Epithelia

Among the many sulfate/anion exchange mechanisms described for epithelial cells, there exists a wide variety of specificities for acceptable substrates which can drive the antiporter. Sulfate exchangers are saturable with respect to external sulfate and internal anions. Most of the antiporters are suppressed by the classic disulfonic stilbene exchange inhibitors such as SITS (4-acetamido-4′-isothiocyano-2,2′-disulphonic stilbene) and DIDS (4,4′-diisothiocyanostilbene-2,2′-disulfonic acid). There are few studies which directly address both the coupling ratio of sulfate anion exchange and electrogenicity. Table 1 provides a comparison of apparent kinetic constants for sulfate antiport in brush border and basolateral membranes of several vertebrate species and the lobster.

3.1 Luminal Sulfate Transport

Talor and coworkers [21] demonstrated the presence of an anion antiporter at the luminal membrane of the beef kidney epithelium by combined methods. First, they showed [^{3}H]DIDS binding to the brush border membrane exhibiting kinetics

Table 1 Comparison of apparent kinetic characteristics of sulfate/anion antiport in various animal epithelial systems

Antiporter	Epithelia	K_{in}	J_{max}	Reference
SO_4/HCO_3^-	Rat kidney BBM	0.4	1.1	[20]
SO_4/OH	Rabbit ileal BBM	0.48	4.1	[8]
SO_4/Cl^-	Lobster hepatopancreas BBM	0.27	11.0	[35]
SO_4/OH	Rat liver BLM	16	12.2	[13]
SO_4/HCO_3^-	Rabbit ileal BLM	0.12	539	[4]
SO_4/Cl^-	Rabbit ileal BLM	0.30	1.6	[11]
SO_4/Oxalate	Lobster hepatopancreas BLM	6.0	3.3	[38]

which suggested a single class of binding sites. Second, they observed stimulated sulfate uptake in the presence of outwardly directed chloride, bicarbonate, and hydroxyl ion gradients which were inhibited by DIDS and furosemide. On the basis of potassium/valinomycin-induced diffusion potentials, they concluded that the exchange process was electroneutral. It was suggested that the brush border antiporter may be involved in the secretory flux of sulfate [21]. At the brush border membrane of the rat kidney epithelium, a sulfate/bicarbonate antiporter has been shown which accepted chloride, nitrate, and hydroxyl ions [20] and which was inhibitable by both DIDS and SITS. Flounder renal brush border membrane vesicles demonstrated a concentrative sulfate/anion antiporter which was stimulated strongly by bicarbonate and to a lesser extent by chloride, thiosulfate, and thiocyanate [26]. A significant inhibition of sulfate/bicarbonate exchange occurred when challenged by DIDS and probenecid. Because of the fact that this antiporter did not respond to the manipulation of membrane potential, the authors proposed an exchange of one sulfate to two monovalent anions.

3.2 Basolateral Sulfate Transport

The rat renal sulfate/anion exchanger in the basolateral membrane showed a very broad range of substrate specificity with respect to inorganic and organic anions [23]. Stimulated uptake of sulfate occurred in the presence of chloride, bicarbonate, phosphate, thiosulfate, formate, acetate, lactate, p-aminohippuric acid, and oxalate. Probenecid was able to block the effects of the above anions. Pritchard and Renfro [22], working with the same tissue also observed sulfate anion exchange; although the carrier did not utilize chloride, it was found to be electroneutral and probenecid-sensitive. The chick renal tubule basolateral membrane also exhibited SITS-sensitive sulfate/bicarbonate exchange which was not *cis*-inhibited by chloride [27].

The rabbit ileum exhibited a carrier mediated sulfate/chloride exchange in the basolateral domain [11]. Among the anions tested oxalate was the most effective in *cis*-inhibition of the antiporter. The same membrane was shown to contain a sulfate/bicarbonate antiporter which also transported oxalate [24]. This appeared to be an exchanger separate from the sulfate/chloride antiporter because chloride had no effect on sulfate/bicarbonate exchange [12].

Sulfate/oxalate exchange appeared to be a common mechanism in both sinusoidal and canalicular membrane vesicles of liver epithelia [13–15]. In rat sinusoidal membrane vesicles, sulfate uptake was stimulated by oxalate and succinate but not by chloride or bicarbonate [13]. In skate sinusoidal membrane vesicles, oxalate and sulfate *cis*-inhibited pH gradient stimulated sulfate uptake but chloride was ineffective [15]. In rat liver canalicular membrane vesicles, there was a sulfate/bicarbonate antiporter which was *trans*-stimulated by sulfate, oxalate, and thiosulfate, but not by chloride or other dicarboxylic acids [14]. All of the sulfate antiporters which accepted oxalate in liver membrane vesicles were suppressed by the anion exchange inhibitor DIDS.

4 Transmembrane Sulfate Antiport in Crustacean Hepatopancreatic Epithelium

The decapod hepatopancreas, also known as the digestive or midgut gland, is a multilobed structure located in the cephalothorax composed of many epithelial-lined blind-ended tubules. The gland exists as an outpocketing of the digestive tract in the pyloric region of the stomach and the anterior intestine [28–31]. Much research has been done which implicates the organ as a site for secretion of digestive enzymes, the site of final digestion of ingested materials, and the major site of nutrient absorption. Although the hepatopancreas has been largely known for its digestive and absorptive properties, it has been speculated that it may also play a role in excretion of ions or metabolic wastes [30, 32–34].

4.1 Electrogenic Sulfate/Chloride Exchange at the Luminal Membrane

Unlike sulfate transport mechanisms in vertebrate renal and intestinal brush border membranes, the lobster hepatopancreas was not shown to demonstrate a sodium/sulfate cotransport system [35]. However, recently, Gerencser and Levin [36] have demonstrated a nonelectrogenic Na^{+-}/SO_4^{2-} symporter in the foregut apical membrane of the invertebrate *Aplysia californica.* Sulfate exchange via antiporters in the luminal membrane of lobster hepatopancreas was stimulated by chloride and sulfate, but not by bicarbonate gradients [35]. Because of the orientation of brush border membrane vesicles, the translocation of sulfate into the vesicular interior represented net sulfate movement into the hepatopancreatic epithelium. There are several lines of evidence which suggest that the physiological role of this antiporter is secretion of cytoplasmic sulfate into the lumen of hepatopancreatic tubules. First, the antiporter was found to respond strongly to the imposition of a potassium/valinomycin-induced membrane potential. When the vesicular interior was positive, there was a stimulation of sulfate influx and an inhibition in the presence of a negative interior. On the basis of the negative transmembrane potential typical of intact epithelial cells [37], the movement of a net negative charge into an electronegative cytoplasm seems unlikely. A second line of evidence for the electrogenicity of this antiporter was provided when comparing chloride/sulfate exchange to sulfate/sulfate exchange under opposing membrane potentials. There was no effect of membrane potential on sulfate/sulfate exchange, while sulfate exchange with the monovalent anion reacted as described above. This suggested that the carrier protein operated in a 1:1 ratio regardless of substrate charge. Finally, radiolabeled chloride uptake into sulfate-loaded vesicles was stimulated in the presence of a negative interior when compared to short-circuited vesicles, which would imply the net movement of a negative charge out of the vesicle.

4.2 Modes of Chloride Transport

Previously described in the lobster hepatopancreatic epithelial basolateral membranes were a $SO_4{}^{2-}$/oxalate^{2-} antiporter [38], a $SO_4{}^{2-}/HCO_3{}^{-}$ antiporter [39], and a $Cl^-/HCO_3{}^-$ antiporter [40] (Fig. 1). In support of the hypothesis that Cl^- could utilize any of the three documented antiporters, it was shown in recent studies [39, 40] that Cl^- could be driven to accumulate in the basolateral membrane vesicles (BLMV) via favorable gradients of either $SO_4{}^{2-}$, oxalate $^{2-}$, or $HCO_3{}^-$. Therefore, it is logical to assume that Cl^- appears to be a universal exchangeable anion for any of these three antiporters; therefore, one, two, or all three antiporters could provide the transport means for Cl^- homeostasis in the lobster.

4.3 Electroneutral Sulfate/Oxalate Exchange at the Basolateral Membrane

In the lobster hepatopancreatic epithelial basolateral membrane, we observed another antiport mechanism which accepted the divalent carboxylic anion oxalate [38]. This exchange mechanism was electroneutral and transported one sulfate for one oxalate ion. There appeared to be a quite specific acceptance of substrates with respect to other carboxylic acids. Formate, succinate, oxaloacetate, α-ketoglutarate, and citrate did not stimulate sulfate uptake as they do in other sulfate exchange systems. The sulfate/oxalate antiport did not respond to the imposition of transmembrane pH gradients as observed in the brush border exchanger. The sulfate/oxalate exchanger was extremely sensitive to the anion antiport inhibitors SITS and DIDS [38].

It was also shown in a later study [41] that sulfate exhibited similar binding characteristics to the antiporter whether bound internally or externally. Similarly, oxalate had similar binding characteristics to the antiporter whether it was bound internally

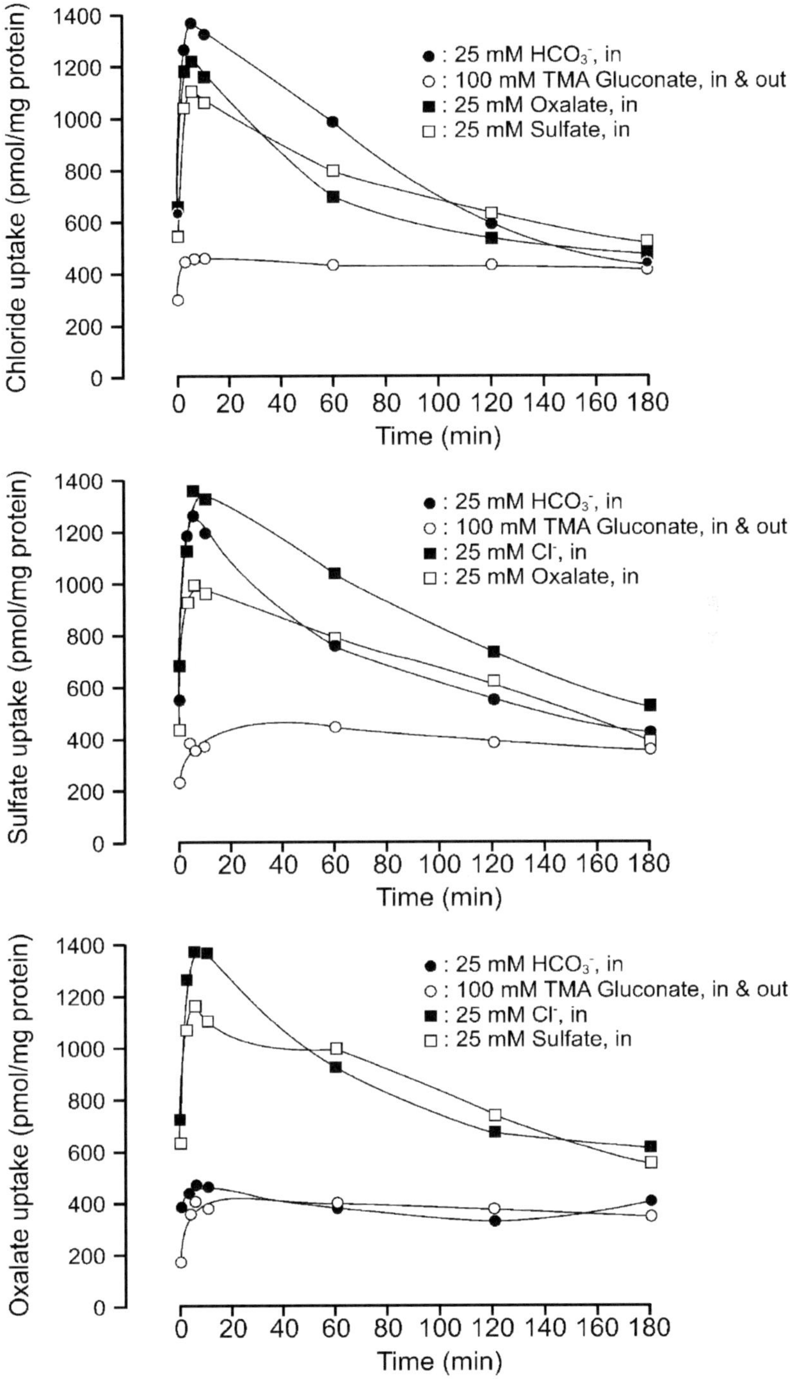

Fig. 1 Experimental protocols illustrating $C1^-$ as a ubiquitous substrate for either of three antiporters described in the BLM of lobster hepatopancreas

or externally. However, oxalate had a greater affinity for the antiporter than did sulfate. These results suggested that oxalate, not sulfate, regulates the antiporter transport rate because of its greater affinity to the transporter.

4.4 HCO_3^-/SO_4^{2-} Exchange in the Basolateral Membrane (BLM)

Recently, a HCO_3^-/SO_4^{2-} antiporter was described in BLMV of lobster hepatopancreatic epithelial cells [39]. The antiporter was electrogenic and shown to have a HCO_3^-/SO_4^{2-} transport stoichiometry of 1:1 (Fig. 2). This mechanism was also inhibited by disulfonic stilbenes; however, it was stimulated by external protons. Speculatively, this antiporter could provide a cellular means for cytoplasmic pH regulation which could be further modified by extracellular pH (Fig. 3). At the same time, this antiporter could also provide intracellular uptake of sulfate. This event would be beneficial for cellular viability of cellular metabolic reactions such as: (1) sulfur conjugation [42], and/or (2) complexing with heavy metals as concretions [38] in intracellular membrane-bound vesicles. The sulfate may be needed to detoxify these heavy metals before they are transferred to the blood where they could have deleterious effects in other parts of the crustacean's body. Additionally, sulfate movement from the cytoplasm to the hepatopancreatic lumen via the sulfate/chloride exchanger located in the brush border membrane [35] could execute a secretory function [38, 42]as a means for sulfate excretion [32, 33]. However, it is incumbent to include the possibility of sulfate absorption by this brush border sulfate/chloride exchanger since it is a bidirectional transport device. Then accumulated intracellular sulfate could facilitate the absorption of bicarbonate across

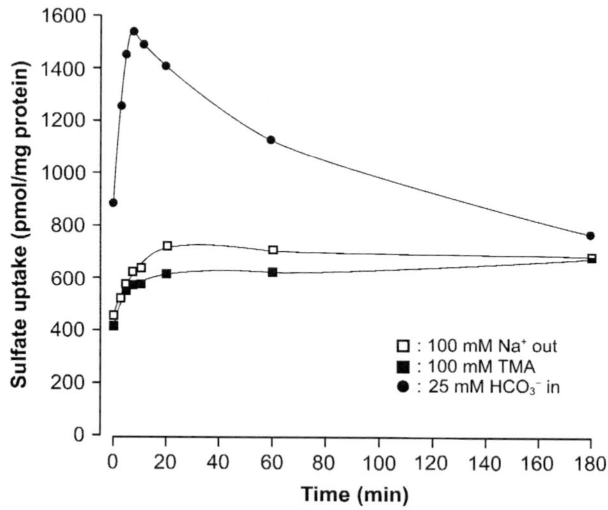

Fig. 2 Time course of bicarbonate-driven sulfate uptake by lobster hepatopancreatic BLMV

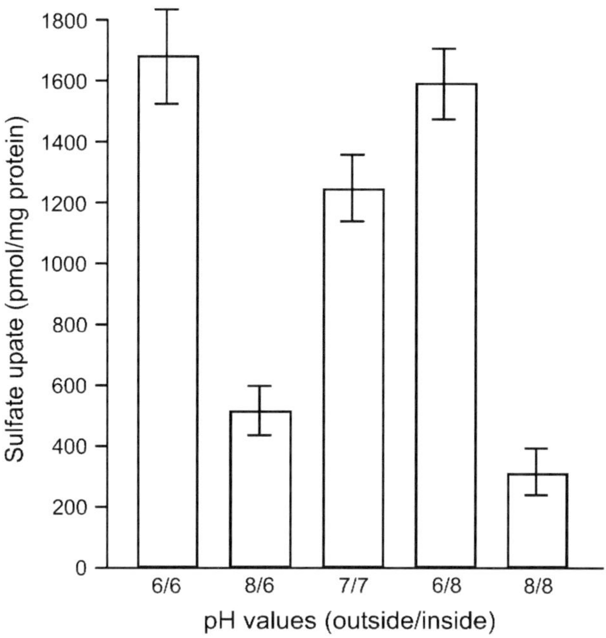

Fig. 3 Effect of various bilateral pH conditions and pH gradients on active uptake of sulfate into hepatopancreatic BLMV

the basolateral membrane via the sulfate/bicarbonate exchanger thereby providing a regulatory process for intracellular acid–base regulation.

4.5 Proton Regulation of Hepatopancreatic Chloride/Sulfate Exchange

As mentioned above, the luminal contents of the crustacean hepatopancreas have been recorded at a low pH value of 4 [30] due to the digestive processes of the organ. In the light of the reports of pH-dependent sulfate transport in different epithelial membranes [11, 25], and this natural proton gradient maintained across the hepatopancreatic brush border, it seemed likely that there might be a similar system in the lobster. Although no pH-dependent sulfate transport was shown in hepatopancreatic vesicles, there was a marked effect of low pH on chloride/sulfate exchange. In the presence of low extravesicular pH (high proton concentration), chloride/sulfate exchange was significantly enhanced. This effect was observed in the presence of both a transmembrane proton gradient and bilaterally equal pH conditions [42].

Regulation of membrane bound transport proteins by protons at either internal or external modifier sites has been described for several different cell types. Chloride/bicarbonate exchange in rabbit ileal brush border membranes was stimulated by a cytoplasmic pH modifier site which was presumed to be the effect of internal hydroxyl ions [43]. In lobster basolateral membrane vesicles, chloride/bicarbonate exchange was stimulated by protons [39, 40, 42], but due to the response of this antiporter to transmembrane chloride and pH gradients, it was suggested that the process involved simultaneous movements of a proton with chloride. However, it was shown in the previous section that a newly discovered SO_4^{2-}/HCO_3^- antiporter in the BLM of hepatopancreatic epithelial cells was stimulated by protons, but not by a proton gradient [39] (Fig. 3). Since this mechanism may be involved in both intracellular and extracellular pH regulation and the Cl^-/SO_4^{2-} antiporter in nonpH related functions, the mode of proton regulation for HCO_3^-/SO_4^{2-} and HCO_3^-/Cl^- antiporters would be expected to be different. Lymphocytes also demonstrated an internal pH regulatory site which adjusted chloride/bicarbonate exchange in response to sodium/proton exchange [44]. A kidney cell line, Vero cells, showed a pH regulation of chloride antiport (either chloride/bicarbonate or chloride/chloride). This investigation performed on intact cells suggested that in response to internal hydroxyl ions chloride uptake was significantly enhanced [19]. Recently we observed the luminal membrane of lobster hepatopancreatic epithelium exhibits a sulfate antiporter which accepts only chloride, oxalate, or sulfate as the counter ion and it is stimulated by protons [45]. The relative specificity of this antiporter and its response to external proton concentration significantly differs from the basolateral membrane antiporter for sulfate [39] (Figs. 4, 5, and 6).

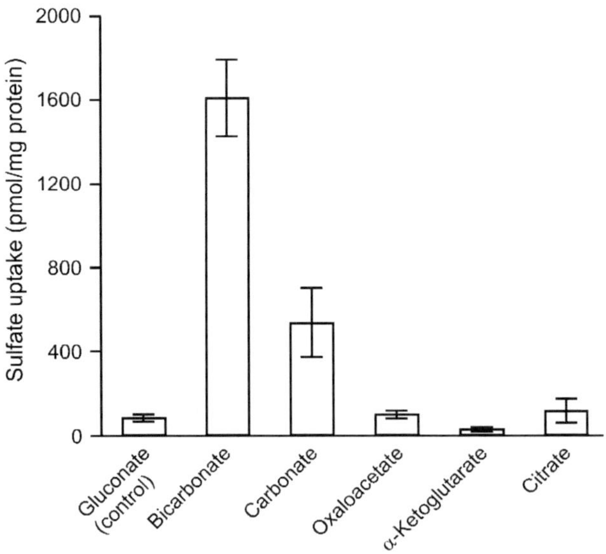

Fig. 4 Effect of internal organic anions on sulfate uptake into hepatopancreatic BLMV

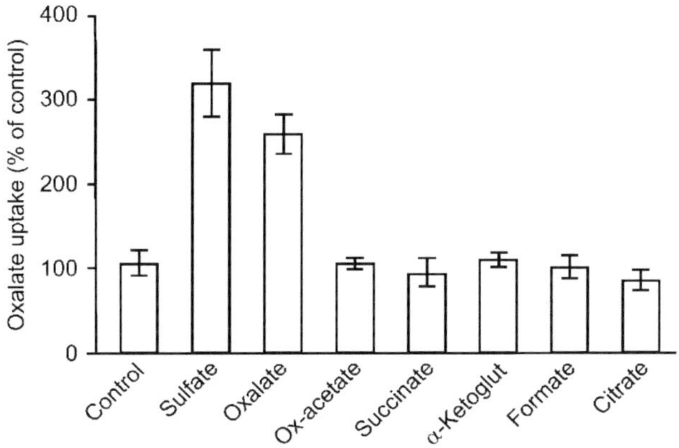

Fig. 5 Effect of internal organic anions on active oxalate uptake into hepatopancreatic Brush Border Membrane Vesicles (BBMV)

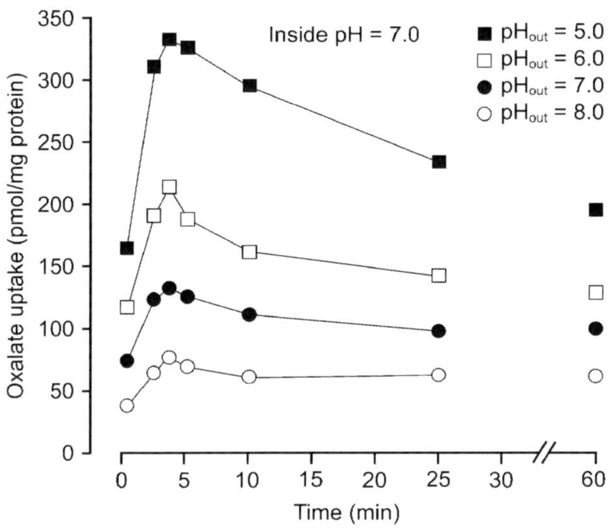

Fig. 6 Influence of external pH on active uptake of oxalate into hepatopancreatic BBMV

5 Transcellular Sulfate Transport in Lobster Hepatopancreas

The epithelial cells of the hepatopancreas operate in the secretion of digestive enzymes and the absorption of nutrients from the luminal contents [30]. There have been a variety of nutrient transporters described for the brush border of the lobster hepatopancreas which confirm the absorptive nature of the organ [45]. The lobster controls the concentration of sulfate in the hemolymph (18 mM) at a value lower than found in seawater which is 25 mM [46]. The lobster, as with other crustaceans, ingests seawater during both eating and drinking, which provides a high sulfate concentration in the hepatopancreas lumen [47, 48]. On the basis of our studies with

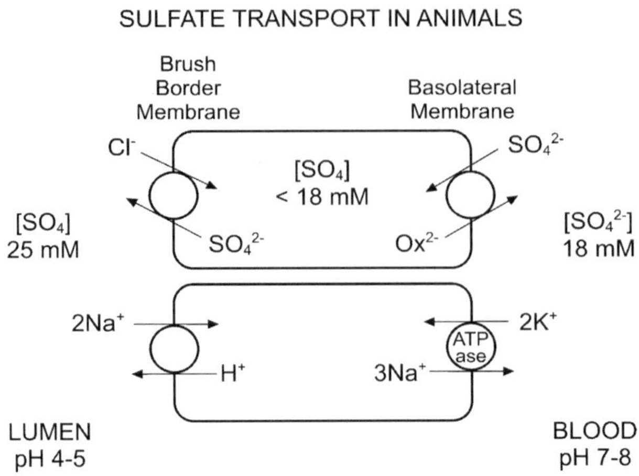

Fig. 7 Schematic diagram of hepatopancreatic epithelium illustrating the proposed model of sulfate secretion

sulfate transport in isolated membrane vesicles, we proposed a secretory pathway for this anion by the epithelial cells of the hepatopancreas (Fig. 7).

The initial step for sulfate movement from hemolymph to cytoplasm of the hepatopancreas cells occurs by an anion exchange system located in the basolateral membrane. The exchange of cytoplasmic oxalate for sulfate occurs by a saturable carrier-mediated mechanism which does not accept similar dicarboxylic acids. The cytoplasmic concentration of sulfate is unknown and assumed to be relatively low, so that the driving force for this antiporter would be the transmembrane sulfate gradient. The kinetic characteristics of the exchanger support this observation with the apparent binding constant for sulfate on the extracellular surface of the carrier being physiologically relevant ($Km = 6.0\,\text{mM}$) to the blood concentration. The apparent binding coefficient for cytoplasmic oxalate is relatively low ($Km = 0.\ 18\,\text{mM}$), suggesting a high affinity of the antiporter for oxalate [41]. The physiological role of oxalate in the lobster is uncertain; it is possible that the anion, in the form of calcium oxalate, is involved with the sequestrating of, or in the deposition, of, calcium that occurs prior to molting and the formation of the new exoskeleton [38].

The sulfate/oxalate antiporter is an electroneutral carrier which exchanges one sulfate for one oxalate ion. The basolateral carrier also exhibits self-exchange and appears to accept chloride as a substrate. There is no H^+ stimulation of the carrier due to manipulations of transmembrane pH gradients. The carrier is very sensitive to the anion exchange inhibitors DIDS and SITS, which provides support for an antiport system [38].

The movement of sulfate from the hepatopancreas cytoplasm into the tubular lumen occurs by a separate antiporter present in the brush border membrane of the epithelial cells [35]. This electrogenic chloride antiporter appears to be the overall driving force for the transcellular movement of sulfate by this epithelium.

The combined transmembrane chloride gradient and the extrusion of a net negative charge from the cytoplasm of the epithelial cell would drive the carrier. When a potassium/valinomycin-induced membrane potential was established across the vesicle wall, sulfate uptake was stimulated by a positive, and inhibited by a negative electrical, interior. In contrast, when measuring chloride uptake into sulfate-loaded vesicles, there was enhanced chloride uptake due to an inside-negative membrane potential. On the basis of this observation and the normal gradients established in the lumen and the cytoplasm of hepatopancreas cells, the evidence suggests that the predominant movement of sulfate is out of the cell. The luminal membrane of the hepatopancreas lacks the sodium/sulfate cotransport system typical of vertebrate renal [16, 46] and intestinal [17] epithelia. Along with the absence of a sodium cotransport system, the brush border does not exhibit proton (hydroxyl)-stimulated transport such as that of the rabbit ileal brush border [8] or flounder basolateral renal tubule [25]. This was observed in the presence of either inward- or outward-directed pH gradients and under equilibrated sulfate conditions with a simultaneous inward-directed proton gradient [35].

It was interesting to find that chloride/sulfate exchange was enhanced by elevated external protons; the lower the extravesicular pH, the greater the sulfate uptake [35]. This was observed under pH gradient or equilibrated pH conditions. This is important when considering other physiological properties of the crustacean hepatopancreas. The gastrointestinal contents are acidic and are continuous with the hepatopancreas lumen, thereby providing an environment relatively high in protons during digestion and absorption. Until recently, the mechanism of tubular acidification was unclear. Recently, a sodium/proton exchanger was described in both the lobster and freshwater prawn hepatopancreatic brush border which could adequately account for observed luminal proton concentrations [47]. This acid environment would allow for increased activity of the electrogenic chloride/sulfate antiporter and play an important role in the regulation of sulfate [48] (Fig. 1).

The net secretion of sulfate by the lobster hepatopancreas represents an example of sulfate handling very different from that of vertebrate sulfate handling. Although there are sulfate anion exchange mechanisms present on the brush border of renal tubules [20, 21], the potent sodium cotransport system prevails driving tubule reabsorption. The antiporters in the lobster hepatopancreas have many properties in common with mammalian epithelia, yet the overall function appears to differ markedly. Sulfate secretion, also driven by anion exchange at the brush border, has been observed in the flounder renal tubules which also shares attributes of the lobster antiporter [49]. It appears that the lobster hepatopancreas acts in a manner similar to the teleost kidney with respect to the regulation of sulfate homeostasis.

6 Sulfate and Phosphate Transport in Molluscan Epithelia

Gastrointestinal and renal transport of the divalent anion sulfate across epithelia has been investigated in various vertebrate groups including mammals [5, 20], teleost fish [25, 26], and the domestic chicken [27]. A number of mechanisms for brush-border carrier-mediated sulfate transport across epithelial membranes have been proposed and include sodium/sulfate cotransport [5, 16], anion exchange [21, 25], and pH gradient-dependent transfer [8]. These processes contribute to transepithelial regulation of sulfate levels, and may affect acid–base balance and plasma osmolarity [48].

However, there are very few studies of sulfate transport across epithelia of invertebrates. A proton-stimulated sulfate/chloride exchanger has recently been described in apical membranes of lobster (*H. americanus*) hepatopancreatic epithelial cells (this chapter).

6.1 Sulfate Transport in Aplysia Gut

However, it was not until 1979 [50] that active absorption of sulfate in a molluscan epithelium was demonstrated. It was further shown, more recently, that active sulfate absorption by the *A. californica* foregut was sodium dependent and this action dependence was also concentration dependent [36]. The symporter for sodium and sulfate resided in the luminal membrane, had a stoichiometry of 2 sodium: 1 sulfate, and was not inhibited by phosphate, but was competitively inhibited by thiosulfate.

6.2 Regulation of Sulfate Transport in Aplysia Gut

In contrast to the wealth of knowledge of renal and intestinal apical sodium/sulfate symport mechanisms very little information is available in the literature about possible regulatory control mechanisms for renal and intestinal apical sodium/sulfate symport. Glucocorticoids have been shown to be inhibitory for renal reabsorption of sulfate in the chicken [27] and thyroid hormone has been shown to stimulate sodium/sulfate symport activity in rat kidney [51]. Furthermore, the intestinal transport of sulfate in the rabbit seems to be regulated by some secretory stimuli, such as theophylline or the heat-stable enterotoxin [3–5].

It was recently demonstrated that the sodium sulfate symporter located in the luminal membrane of *A. californica* foregut was up-regulated by triiodothyronine [52]. The gastropod, *A. californica*, has been shown to have a hormone equivalent to triiodothyronine [35] which strengthens the physiological significance of this finding. Recently we reported that cyclic guanosine monophosphate stimulated the sodium/sulfate symporter in the apical membrane of *Aplysia* foregut absorptive cells [53]. Therefore, triiodothyronine could stimulate the production of cGMP which would act as the secondary messenger of triiodothyronine in stimulating the Na/SO_4 symporter similar to what has been reported in amphibian tissue [54]. Also, it appears that the triiodothyronine stimulation of the sodium/sulfate symporter via induced protein synthesis for actinomycin D, puromycin, or cycloheximide thwarted the triiodothyronine effects on active absorption of sulfate [55]. This transport event could be beneficial for viability of cellular metabolic reactions such as sulfur conjugation [20] complexing with heavy metals, as happens in lobster hepatopancreas [38]. Sulfate homeostasis in the *Aplysia* is, at least, partly maintained by this luminal T_3-regulated sodium/sulfate symporter [36, 52] which is mediated through de novo protein synthesis [55]. In contrast it was demonstrated that the sodium/sulfate symporter was down-regulated by glucocorticoids [56]. The physiological significance of glucocorticoid inhibition of the sodium/sulfate symporter in the *Aplysia* foregut may be related to glucocorticoid inhibition of glucose uptake and utilization by cells in general. Since there is less glucose utilization by cellular metabolism, there would be less need for cellular sulfate for sulfonation reactions such as protein synthesis. Therefore, glucocorticoid inhibition of cellular sulfate uptake would work in tandem with its inhibition of cellular glucose uptake and utilization.

6.3 Phosphate Transport in Aplysia Gut

Gastrointestinal and renal transport of the anion phosphate across epithelial apical membranes has been investigated in various vertebrate groups including: mammals such as rabbit [57] and rat [58]; avian such as chicken [59], and other lower vertebrates [60]. Studies with intact vertebrate tissue preparations have documented that transepithelial inorganic phosphate (P_i) transport against an electrochemical potential difference in the small intestine is dependent on the presence of sodium (Na^+) [61]. In vertebrates, this process can contribute to the transepithelial regulation of P_i levels, and may affect acid–base balance and plasma osmolarity.

However, there is a dearth of studies regarding P_i transport across epithelia of invertebrates. In view of this vacuum of P_i transport information in invertebrates, the present chapter was written to describe the nature of the P_i transporter in the mucosa of *Aplysia* gut. Recently, Gerencser and colleagues [62] used isolated foregut from *A. californica* to characterize a Na^+/P_i symporter that is located in the mucosal membrane of the gut cells, is inhibited by arsenate and ouabain, and has a stoichiometry of 2 Na^+:1 P_i.

6.4 Regulation of Phosphate Transport in Aplysia Gut

1. Hormonal regulation of P_i transport in vertebrate renal or gastrointestinal epithelia has shown an up-regulation of phosphate absorption by triiodothyronine [63] and a down-regulation of P_i absorption by glucocorticoids [64].
2. Similarly, the Na^+/P_i symporter in the invertebrate *Aplysia* foregut luminal membrane was up-regulated by T_3 [65] and down-regulated by glucocorticoids [66]. Additionally, it seems that the T_3 stimulation of Na^+/P_i symporter activity in the *Aplysia* foregut is via induced protein synthesis because actinomycin D, puromycin, or cycloheximide inhibited the T_3 effects on Na^+/P_i active absorption

[67]. Since the gastropod, *A. californica*, has been shown to possess a hormone equivalent to T_3 [35], based upon the presence and activity of iodinated thyronine-like compounds in the animal, these present observations strengthen the physiological significance of this finding. This event could be beneficial for cellular viability of cellular metabolic reactions such as phosphorylation. P_i homeostasis in *Aplysia*, is at least, partly maintained by T_3 regulation of the luminal Na^+/P_i symport transport mechanism [65].

6.5 Statement of Significance

The evolutionary ties of invertebrates to vertebrates intellectualizes the genetic bond between the two animal groups. This can facilitate, in the clinical setting, amelioration of protein-deficient diseases by drugs such as triiodothyronine which can aid in incorporation of peptide side-groups that contain sulfate and/or phosphate.

Acknowledgements This project was supported by the Eppley Foundation for Research, Inc. (G.A.G.) and by N.S.F. IBN 99-74569 (G.A.A.).

References

1. Maddrell, HP and Phillips, JE. Active transport of sulphate ions by the Malpighian tubules of larvae of the mosquito *Aedes campestris*. J. Exp. Biol., 1975; 62:367–378.
2. Maddrell, HP and Phillips, JE. Induction of sulphate transport and hormonal control of fluid secretion by Malpighian tubules of larvae of the mosquito *Aedes taeniorhynchus*. J. Exp. Biol., 1978; 72:181–202.
3. Smith, PL, Orella, SA and Field, M. Active sulfate absorption in rabbit ileum: dependence on sodium and chloride and effects of agents that alter chloride transport. J. Membr. Biol., 1981; 63:199–206.
4. Langridge-Smith, JE and Field, M. Sulfate transport in rabbit ileum: characterization of the serosal border anion exchange process. J. Membr. Biol., 1981; 63:207–214.
5. Langridge-Smith, JE, Sellin, JH and Field, M. Sodium influx across the rabbit ileal brush border membrane: sodium and proton dependence, and substrate specificities. J. Membr. Biol., 1983; 72:131–139.
6. Cardin, CJ, and Mason, J. Sulphate transport by rat ileum: effect of molybdate and other anions. Biochim. Biophys. Acta, 1975; 394:46–54.
7. Wolffram, S, Grenacher, B and Scharrer, E. Transport of selenate and sulphate across the intestinal brush-border membrane of pig jejunum by two common mechanisms. J. Exp. Physiol., 1988; 73:103–111.
8. Schron, CM, Knickelbein, RG, Aronson, PS, Della Puca, J and Dobbins, JW. pH gradient-stimulated sulfate transport by rabbit ileal brush-border membrane vesicles: evidence for SO_4-OH exchange. Am. J. Physiol., 1985; 249:G607–G613.
9. Knickelbein, RG., Aronson, PS and Dobbins, JW. Substrate and inhibitor specificity of anion exchangers on the brush border membrane of rabbit ileum. J. Member Biol., 1985; 88:199–204.
10. Hagenbuch, B., Stange, G and Murer, H. Transport of sulphate in rat jejunal and rat proximal tubular basolateral membrane vesicles. Pflugers Arch, 1985; 405:202–208.
11. Schron, CM, Knickelbein, RG, Aronson, PS and Dobbins, JW. Evidence for carrier-mediated $Cl\text{-}SO_4$ exchange in rabbit ileal basolateral membrane vesicles. Am. J. Physiol., 1987; 253:G404–G410.

12. Knickelbein, RG and Dobbins, JW. Sulfate and oxalate exchange for bicarbonate across the basolateral membrane of rabbit ileum. Am. J. Physiol., 1990; 259: G807–G813.
13. Hugentobler, G and Meier, PJ. Multispecific anion exchange in basolateral (sinusoidal) rat liver plasma membrane vesicles. Am. J. Physiol., 1986; 251:G656–G664.
14. Meier, PJ, Valantinas, J, Hugentobler, G and Rahm, I. Bicarbonate sulfate exchange in canalicular rat liver plasma membrane vesicles. Am. J. Physiol., 1987; 253: G461–G468.
15. Hugentobler, G, Fricker, G, Boyer, JL and Meier, PJ. Anion transport in basolateral (sinusoidal) liver plasma-membrane vesicles of the little skate (*Raja erinacea*). Biochem. J., 1987; 247:589–595.
16. Lucke, H, Stange, G and Murer, H. Sulphate-ion/sodium-ion co-transport by brush-border membrane vesicles isolated from rat kidney cortex. Biochem. J., 1979; 182:223–229.
17. Schneider, EG, Durham, JC and Sacktor, B. Sodium-dependent transport of inorganic sulfate by rabbit renal brush-border membrane vesicles. Effects of other ions. J. Biol., Chem., 1984; 259:14591–14599.
18. Turner, RJ. Sodium-dependent sulfate transport in renal outer cortical brush border membrane vesicles. Am. J. Physiol., 1984; 247:F793–F798.
19. Brazy, PC. and Dennis, VW. Sulfate transport in rabbit proximal convoluted tubules: presence of anion exchange. Am. J. Physiol., 1981; 241:F300–F310.
20. Pritchard, J B. Sulfate-bicarbonate exchange in brush-border membranes from rat renal cortex. Am. J. Physiol., 1987; 252:F346–F356.
21. Talor, Z, Gold, RM., Yang, WC and Arruda, JAL. Anion exchanger is present in both luminal and basolateral renal membranes. Eur. J. Biochem., 1987; 164:695–702.
22. Pritchard, JB. and Renfro, JL. Renal sulfate transport at the basolateral membrane is mediated by anion exchange. Proc. Natl. Acad. Sci. (USA), 1983; 80:2603–2607.
23. Low, I, Friedrich, T and Burckzhardt G. Properties of an anion exchanger in rat renal basolateral membrane vesicles. Am. J. Physiol., 1984; 246:F334–F342.
24. Kuo, SH and Aronson, PS. Oxalate transport via the sulfate/HCO_3 exchanger in rabbit renal basolateral membrane vesicles. J. Biol. Chem., 1988; 263:9710–9717.
25. Renfro, JL and Pritchard, JB. H^+-dependent sulfate secretion in the marine teleost renal tubule. Am. J. Physiol., 1982; 243:F150–F159.
26. Renfro, JL and Pritchard, JB. Sulfate transport by flounder renal tubule brush border: presence of anion exchange. Am. J. Physiol., 1983; 244:F488–F496.
27. Renfro, JL, Clark, NB, Metts, RE and Lynch, MA. Sulfate transport by chick renal tubule brush-border and basolateral membranes. Am. J. Physiol., 1987; 252: R85–R93.
28. Loizzi, RF. Interpretation of crayfish hepatopancreatic function based on fine structural analysis of epithelial cell lines and muscle network. Z. Zellforsch, 1971; 113:420–440.
29. Van Weel, PB. Hepatopancreas? Comp. Biochem. Physiol., 1974; A47: 1–9.
30. Gibson, R and Barker, PL. The decapod hepatopancreas. Oceanogr. Mar.Biol. Annu. Rev., 1979; 17:285–346.
31. Dall, W and Moriarty, DJW. Functional aspects of nutrition and digestion. In: Mantel LH, ed., The biology of crustacea. Internal anatomy and physiological regulation. Vol. 5, Academic Press, New York, 1983; 215–261.
32. Gifford, CA. Some aspects of osmotic and ionic regulation in the Blue crab *Callinectes sapidus*, and the Ghost crab, *Ocypode albicans*. Publ. Inst. Mar. Sci. Univ. Tex., 1962; 8:97–125.
33. Dall, W. Osmoregulation in the lobster *Hornarus americanus*. J. Fish Res. Board Can., 1970; 27:1123–1130.
34. Miller, DS and Hollida, CW. Organic cation secretion by *Cancer borealis* urinary bladder. Am. J. Physiol., 1987; 253:R153–R159.
35. Cattey, MA, Gerencser, GA and Ahearn GA. Electrogenic H^+-regulated sulfate-chloride exchange in lobster hepatopancreatic brush-border membrane vesicles. Am. J. Physiol., 1992; 262:R255–R262.

36. Gerencser, GA and Levin, R. Sodium-sulfate symport by *Aplysia california gut*. Zool. Sci., 2000; 17(5):27–32.
37. Gerencser, GA. Transport across the invertebrate intestine. In: R. Gilles, M. Gilles-Baillien, eds., Transport processes, iono- and osmoregulation. Springer, Berlin Heidelberg New York, 1985; 25–64.
38. Gerencser, GA, Cattey, MA and Ahearn, GA. Sulfate-oxalate exchange in lobster hepatopancreatic basolateral membrane vesicles. Am. J. Physiol., 1995; 38(3): R572–R577.
39. Gerencser, GA, Ahearn, GA and Cattey, MA. Sulfate/bicarbonate antiport by lobster hepatopancreatic basolateral membrane vesicles. J. Exp. Zool., 1999; 284:158–167.
40. Gerencser, GA, Ahearn, GA, Robbins, F and Cattey, MA. Chloride transport by lobster hepatopancreas in facilitated by several anion antiport mechanisms. Comp. Biochem. Physiol., 2000; 125A:223–228.
41. Gerencser, GA, Burgin, C, Robbins, F and Ahearn, GA. The oxalate/sulfate antiporter in lobster hepatopancreas: internal and external binding constants. J. Exp. Biol., 2000; 203:1497–1502.
42. Gerencser, GA, Ahearn, GA and Cattey, MA. Antiport-driven sulfate secretion in an invertebrate epithelium. J. Exp. Zool., 1996; 275:269–276.
43. Mugharbil, A, Knickelbein, RG, Aronson, PS and Dobbins, JW. Rabbit ileal brush-border membrane Cl-HCO_3 exchanger is activated by an internal pH-sensitive modifier site. Am. J. Physiol., 1990; 259:G666–G670.
44. Mason, MJ, Smith, JD, Garcia-Soto, JD and Grinstein, RJ. Grinstein Internal pH-sensitive site couples $Cl^{-}HC0_3{}^{-}$ exchange to Na^+ -H^+ antiport in lymphocytes. Am. J. Physiol., 1989; 256:C428–C433.
45. Ahearn, GA, Gerencser, GA, Thamotharan, M, Behnke, RD and Lemme, TH. Invertebrate gut diverticula are nutrient absorptive organs. Am. J. Physiol., 1992; 26: R472–R481.
46. Lucke, H, Stange, G and Murer, H. Sulfate-sodium cotransport by brush-border membrane vesicles isolated from rat ileum. Gastroenterology, 1981; 80:22–30.
47. Ahearn, GA, Franco, P and Clay, LP. Electrogenic 2 Na^+/1 H^+ exchange in crustaceans. J. Membr. Biol., 1990; 116:215–226.
48. Robertson, JD. Ionic regulation in some marine invertebrates. J. Exp. Biol., 1949; 26: 182–200.
49. Gerencser, GA, Robbins, F, Zhang, J and Ahearn, GA. Electrogenic proton regulated oxalate/chloride exchange by lobster hepatopancreatic brush border membrane vesicles. J. Exp. Biol., 2004; 207(4):571–578.
50. Gerencser, GA. Metabolic dependence of active sulfate transport in *Aplysia californica* intestine. Comp. Biochem. Physiol. A., 1979; 63:519–522.
51. Tenehouse, HS, Lee, J, and Harvey, N. Renal brush border membrane Na-sulfate cotransport: stimulation by thyroid hormone. Am J. Physiol., 1991; 261: F420–F4326.
52. Gerencser, GA, Levin, R, Zhang, J. Sulfate absorption in *Aplysia californica* gut: thyroid hormone stimulation. Can. J. Zool., 2002; 80(6):964–966.
53. Gerencser, GA, Levin, R, and Robbins, F. Sulfate absorption in *Aplysia californica gut*: intracellular regulation by cyclic guanosine monophosphate. Can. J. Zool., 2001; 79:1–3.
54. Sidlowski, H, Frieden, B. T_3 induces cGMP in tadpole. Biosci. Rep., 1982; 2:569–571.
55. Gerencser, GA, Loo, SY, Zhang, J. Thyroid hormone-induced sulfate absorption in *Aplysia californica* gut is mediated by protein synthesis. Can. J. Physiol. Pharm., 2003; 81(4):405–408.
56. Gerencser, GA, Zhang, J, Levin, R. Sulfate absorption in *Aplysia californica* gut:glucocorticoid inhibition. Can. J. Zool., 2002; 80:2037–2040.
57. Ahearn, GA and Murer, H. Functional roles of Na^+ and H^+ in $SO_2{}^{-4}$ transport by rabbit ileal brush border membrane vesicles. J. Membr. Biol., 1984; 78: 177–186.
58. Berner, WR, Kinne, R, Murer, H. Phosphate transport into brush border membrane vesicles isolated from small intestine. Biochem. J., 1976; 160: 467–474.

59. Sacktor, B, Cheng, L. Sodium gradient dependent phosphate transport in renal brush border membrane vesicles. Effect of an intravesicular to extravesicular proton gradient. J. biol. Chem., 1981; 256:8080–8084.
60. Danisi, G and Murer, H. Inorganic phosphate in small intestine. In Handbook of Physiology, Vol. 2 Edited by Field, M and Frizzell, R. New York Oxford University Press. 1991; 232–336.
61. Gerencser, GA. Effects of amino acids on chloride transport in *Aplysia* intestine. Am. J. Physiol., 1981; 240:R61–R69.
62. Gerencser, GA, Levin, R and Zhang J. Sodium phosphate symport by *Aplysia californica* gut. Zool. Sci., 2002; 19:163–166.
63. Tenenhouse, HS, Lee, J and Harvey, N. Renal brush border membrane Na – sulfate cotransport: Stimulation by thyroid hormone. Am. J. Physiol., 1991; 261:F420–F4326
64. Taylor, Z, Gold, RM, Yang, WC and Arruda, JAL Anion exchanger is present in both luminal and basolateral renal membranes. Eur. J. Biochem., 1987; 261: F420–F4326.
65. Gerencser, GA, Levin, R and Zhang, J. Phosphate absorption of *Aplysia californica* gut: thyroid hormone stimulation. Can. J. Physiol. Pharm., 2002; 80(6):604–607.
66. Gerencser, GA, Loo, SY, Robbins, FW and Zhang, J. Phosphate absorption in *Aplysia californica* gut is mediated by protein synthesis. Can. J. Physiol. Pharm., 2003; 81(4): 405–408.
67. Gerencser, GA, Cornette, KM and Zhang, J. Thyroid-hormone-induced phosphate absorption in *Aplysia californica* gut is mediated by protein synthesis. Can. J. Physiol. Pharm., 2003; 81(4):409–412.

Heavy Metal Transport and Detoxification by Crustacean Epithelial Lysosomes

Gregory A. Ahearn, Kenneth M. Sterling, Prabir K. Mandal, and Barbara Roggenbeck

Abstract Lysosomes are multi-functional organelles that aid in the disassembly of large organic molecules and store a variety of xenobiotics. Lysosomes, and vacuolar components of the endo-membrane system, play apparently ubiquitous sequestration and detoxification roles for heavy metals in cells of many organisms. X-ray microprobe analysis of metal-containing granules (concretions) in these organelles from many animal phyla suggest that monovalent, divalent, and trivalent metal cations can be stored in these compartments in conjunction with anionic elements such as phosphorus and sulfur. There is also evidence that thiol-containing compounds such as glutathione and metallothionein, which bind metals in the cytoplasm with high affinity, may also be translocated across lysosomal membranes for metal storage. Few studies have examined the nature of the sequestration and detoxification processes for heavy metals displayed by invertebrate lysosomes or other endo-membrane components. This review summarizes recent investigations focused on lysosomal function in crustacean hepatopancreatic absorptive epithelia. It describes the carrier-mediated transport processes that occur on lysosomal membranes for accumulating metals from the cytoplasm and how these metal transporters are linked with the uptake of multivalent anions that may precipitate concretions within the organelle at appropriate ion concentrations and pH conditions. In addition, preliminary data describing the potential role of the Organic Anion Transporter (OAT) in transporting glutathione with its associated metal load from cytoplasm to lysosomal interior are described. A model summarizing proposed coupling between cationic metal and polyvalent anion transports and how they might be linked with concretion formation and metal detoxification is presented.

Keywords Lysosomes · Heavy metal transport · Zinc · OAT1 · Organic anion exchanger · Detoxification · Concretion formation · Sequestration · *Homarus americanus* · Hepatopancreas

---

G.A. Ahearn (✉)
Department of Biology, University of North Florida, Jacksonville, Florida 32224
e-mail: gahearn@unf.edu

G.A. Gerencser (ed.), *Epithelial Transport Physiology*,
DOI 10.1007/978-1-60327-229-2_3,

1 Introduction: Heavy Metal Fates in Invertebrate Epithelial Cells

Epithelial cells of the digestive system, integument (including gills of aquatic species), and kidneys are cellular barriers between environmental toxic elements, such as heavy metals, and the blood of animals. High blood concentrations of many metals are known to lead to a variety of pathological conditions involving the nervous system, the muscles, and cellular genomic activity. Toxic exposure is unavoidable in aquatic species. To reduce the accumulation of metals in their bodies to concentrations that may prevent these conditions, aquatic invertebrates have developed a variety of epithelial cellular detoxification pathways that reduce the body burden of harmful metals while at the same time retaining any metals that have a physiological role (such as copper for enzymatic activity or for the synthesis of the arthropod blood pigment, hemocyanin). These pathways include: (1) physiological regulatory mechanisms balancing metal excretion with uptake rates; (2) intracellular sequestration processes for metals within organelles such as lysosomes that produce solid metallic phosphorous or sulfur concretions which may remain with the cell throughout its life cycle or undergo exocytosis for elimination; and (3) cytosolic binding to high affinity, low molecular weight, peptides such as metallothionein or glutathione, followed by vacuolar sequestration or elimination through the endomembrane system.

Potentially toxic heavy metals may enter animals across the integument [1, 2], by way of the gills [3–6], or through the digestive system after consumption of food [7, 8]. Elimination of metals from the animal may occur across the same tissues and by way of the kidneys. This review will focus on sequestration and detoxification mechanisms of crustacean gastrointestinal tract epithelial cells for zinc and other heavy metals, using the lobster (*Homarus americanus*) as a model organism.

In crustaceans, such as the lobster, epithelia of the hepatopancreas and intestine are the major site for absorption from dietary sources. Figure 1 illustrates possible fates of ingested monovalent, divalent, and trivalent metals (X) that pass across the brush border or luminal membrane of the lobster hepatopancreatic or intestinal epithelium. In the epithelial cell, metallic ions may (1) associate with glutathione or metallothionein [9]; (2) be transported into mitochondria [10, 11]; (3) undergo efflux across the basolateral cell membrane to the blood [7, 8]; (4) be accumulated by lysosomes [12, 13]; or (5) be transferred into the endoplasmic reticulum [14]. This review will focus on the role played by lysosomes as a heavy metal depot and detoxification center in invertebrate epithelial cells and the membrane transport processes that these organelles possess for the transfer of these elements from the cytoplasm to the storage site.

2 Lysosomal Concretion Formation as a Detoxification Process

Concretions containing heavy metals are found in animal cell lysosomes or vacuoles of virtually every invertebrate phylum [9, 15–17]. While these concretions or

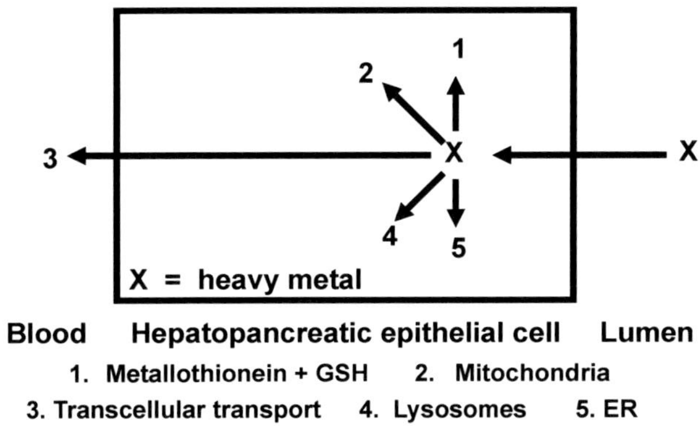

Fig. 1 Possible fates of luminal heavy metals in crustacean hepatopancreatic or intestinal epithelial cells. Luminal metal (X) may cross the brush border membrane into the cytoplasm by membrane carrier or channel proteins. In the cytoplasm the metal may complex with metallothionein or glutathione (GSH) (*1*), enter mitochondria (*2*), be transported across the basolateral membrane to the blood (*3*), be accumulated by lysosomes (*4*), or be transferred to the endoplasmic reticulum (ER) (*5*). Other fates for cytoplasmic metals are also possible. Glutathione (GSH) [43]

granules seem virtually ubiquitous in distribution among the invertebrates, the physiological/biochemical events responsible for their formation are largely unknown [9]. X-ray microprobe analysis of the metal composition of these granules suggests that they may contain either calcium or heavy metal cations such as zinc, copper, and iron, complexed with sulfur or phosphorus [7, 18–21]. As a result of these complex formation events with anions, the potentially toxic heavy metal cation is removed from the cytoplasm and sequestered within the lysosomal membrane in an insoluble, detoxified, form. Subsequent cellular exocytotic events may extrude the concretion from the cell followed by organismic excretory mechanisms that deposit the metal back into the environment.

Figure 2 illustrates the nature of several types of invertebrate intracellular concretions based on microprobe analysis [22]. As this figure suggests, concretions within lysosomes or vacuoles appear to sequester metals into three different intracellular types of membrane-enclosed compartments that may or may not include the thiol-rich peptide, metallothionein (M). Type A vacuoles (lysosomes) appear to accumulate zinc and phosphorus, and Type B appear to contain cadmium, copper, mercury, and silver along with sulfur. Type C vacuoles (lysosomes) appear restricted to iron, and the associated anion(s) in these structures are undefined. According to the model, there is a fourth type of concretion (Type D) that is extracellular and is formed between calcium and carbonate. These apparent distinguishing characteristics suggest that specific transport proteins may occur on the membranes of these different cellular compartments for both metals and anions. Until recently, there have been no experimental studies in invertebrates examining the nature of metal transport systems present in vacuolar (lysosomal) membranes that are responsible for the accumulation of metal cations within these membrane systems. Similarly,

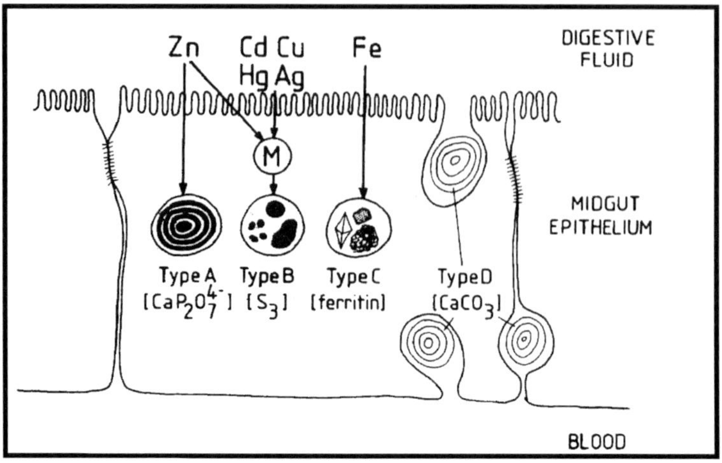

Fig. 2 Type and composition of cellular lysosomal- or vacuolar-enclosed concretions in invertebrate cells. Three types of intracellular (Types *A, B, C*) and a single type of extracellular (Type *D*) concretions involving both cations and anions have been described with electron microprobe elemental analysis. Cationic metals associate with concretions which appear to illustrate specificity toward both cations and anions. Metallothionein (M) [22]

there are no published studies describing the physiological processes responsible for the transfer of sulfur, phosphorus, or any other multivalent anions, such as oxalate, across concretion vacuolar (lysosomal) membranes.

3 Lysosomal Heavy Metal Transport Systems

Lysosomes are cellular organelles with an acidic interior where xenobiotics are disassembled and stored. Enzymatic digestion of large macromolecules, such as proteins, and subsequent storage and elimination of products of digestion (di and tripeptides) have been reviewed [23, 24]. Little is known about the mechanisms responsible for heavy metal transport and sequestration within these detoxification sites. Molecular studies of lysosomal transporters for copper accumulation in these organelles have reported both primary active transport systems (ATP7A and ATP7B; Menkes and Wilson's disease proteins) in mammals [25] and a facilitated diffusion process (CTR2) for this metal in lysosome-like vacuoles in yeast [26]. Similarly, a facilitated diffusion mechanism for zinc accumulation in mammalian cells (Znt2) has been described [27–30], but there are few described Zn-ATPases for eukaryotic organisms [31] other than plants [32] or yeast [33]. A heavy metal ATPase that transports both monovalent and divalent metals (Ag^+, Cu^{2+}, and Cd^{2+}) in rat liver has been described [34]. Chavez-Crooker et al. [12] used the fluorescent dye, Phen Green, to characterize the mechanisms responsible for copper uptake by lobster hepatopancreatic lysosomal membrane vesicles (LMV). This study suggested that copper uptake by this invertebrate organelle occurred by the combination of a copper ATPase and a facilitated diffusion Cu^{2+}/H^+ antiporter in the same membrane, the former powered by cytoplasmic ATP and the latter by an outwardly directed proton gradient.

Zinc, the second most abundant trace element in the body, plays an important role in catalytic functioning of many enzymes and in the structural stability of a variety of proteins involved in transcription and protein trafficking and, therefore, is a very important component of cell viability [35–38]. However, as with other heavy metals, too much zinc impairs cell function and, therefore, the cytosolic concentrations of this metal and others, has to be precisely controlled. Both ion channels and zinc transporters are responsible for controlling intracellular levels of this cation. Zinc homeostasis is maintained largely through the combined contributions from transmembrane transporters belonging to the ZIP and ZnT families. ZIP proteins are involved in cellular uptake of zinc, while ZnT proteins are responsible for the extrusion of the metal outside the cytoplasm (e.g., into the extracellular space or into intracellular organelles). Additional contributions to the control of cytosolic zinc and copper concentrations come from the high affinity binding roles of metallothioneins and glutathione, and from the use of calcium transport proteins, such as the ruthenium red-sensitive mitochondrial calcium uniporter [39] and the ER Sarco-Endoplasmic Reticulum Calcium ATPase (SERCA) [13]. The following section describes important new data for the role of epithelial lysosomes in transporting, sequestering, and detoxifying zinc and potentially other heavy metals in invertebrates.

4 Epithelial Lysosomal Zinc Transport in Invertebrates

Isolation and purification of lobster hepatopancreatic LMV followed the centrifugation methods described by Chavez-Crooker et al. [12] as modified by Mandal et al. [13]. Enzyme characterization of LMV produced by these methods resulted in purification factors of 12.25 ± 0.46, 0.12 ± 0.09, and 0.43 ± 0.12 (means $\pm$ SEM) for acid phosphatase, alkaline phosphatase, and NADPH-cytochrome c reductase, respectively. These data suggest that the LMV fractions used in the following study were relatively pure lysosomal membranes and had minimal membrane contamination from other parts of the cell such as the plasma membrane or the endoplasmic reticulum.

Lysosomal membrane vesicle preparations made with the methods described above were used in a series of experiments to define the nature of the transport processes responsible for accumulating heavy metals in these organelles. Characteristics of $^{65}Zn^{2+}$ transport by LMV were studied at room temperature (23°C) by diluting a small volume of vesicle suspension into a medium containing trace amounts of $^{65}ZnCl_2$ and unlabelled zinc sulfate. The composition of the final vesicle suspension solutions (inside vesicles) and incubation media were different for each experiment and are described in the attached figure legends. Uptake of radiolabeled substrate was initiated by rapidly mixing 20 μl of membrane suspension with 180 μl of transport medium and incubating this mixture for appropriate time intervals. Transport was terminated by addition of 2 ml (10-fold dilution) ice-cold stop solution and the suspension was immediately collected under vacuum on a Millipore filter (HAWP, 65-μm pore size), using the Millipore filtration technique developed by Hopfer et al. [40] Filters were then dissolved in liquid scintillation cocktail (Ecolume, Beckman, Fullerton, CA, USA) and the radioactivity counted in a Beckman LS 6500 multi-purpose scintillation counter. Uptake was expressed as pmol/mg protein x sec and the protein content of the vesicles was determined according to the Bradford technique (BioRad, Hercules, CA, USA), using bovine serum albumen as a standard. Displayed zinc activities were achieved using appropriate concentrations of zinc, NTA (nitriloacetic acid; *N, N*-bis[carboxymethyl]glycine), $MgCl_2$, (0.05 mM free Mg^{2+}), and EDTA using Winmax Chelator 2.0 software [41].

Isotope uptake into vesicles was corrected for nonspecific isotope binding by injecting an aliquot of vesicles and isotope directly into ice-cold stop solution without prior mixing. The resulting vesicle suspension was then filtered, rinsed, and counted as described previously. Resulting values for nonspecific isotope binding were subtracted from total isotope uptake in each experiment, providing an index of transmembrane transport of the radiolabeled cation. Data points are presented as means of 3–5 replicates and their associated standard errors Experiments were repeated at least twice with different animals. Curve fitting procedures were accomplished using Sigma Plot 7.101 software (Jandel) which provided an iterative best fit to experimental values.

The time course of $^{65}Zn^{2+}$ uptake by LMV was examined in the presence and absence of known inhibitors of mammalian lysosomes. As suggested by the results in Fig. 3, in the presence of 0.2 mM ATP, $^{65}Zn^{2+}$ uptake by hepatopancreatic LMV exhibited a maximal transport rate by 30 s and attained equilibrium by 300 s of incubation. Addition of 250 μM vanadate (Fig. 3) or 25 μM calcium or 25 μM copper [13] each reduced the uptake of zinc by approximately half of that under control conditions lacking the inhibitors, suggesting that the same transfer mechanism may have been similarly influenced by each of the substrates tested. These results suggest that hepatopancreatic LMV possessed a zinc transporter that was vanadate sensitive and likely shared by other metals.

The effects of ATP (1 mM), vanadate (250 μM), and thapsigargin (10 μM) on 10 s $^{65}Zn^{2+}$ influx kinetics into LMV were next evaluated. Figure 4 shows that under control conditions (0.2 mM ATP, no inhibitors) $^{65}Zn^{2+}$ influx (10 s uptakes) was a hyperbolic function of zinc concentration (apparent $K_m = 32.26 \pm 10.77$ μM; apparent $J_{max} = 20.65 \pm 2.62$ pmol/mg protein/s). Increasing ATP from 0.2 to 1.0 mM resulted in a significant ($P < 0.01$) increase in apparent J_{max} (31.94 ± 3.72 pmol/mg protein/s), without affecting ($P > 0.05$) the apparent K_m (35.89 ± 10.58 μM) of the zinc transport system. Addition of either 250 μM vanadate or 10 μM thapsigargin significantly ($P < 0.05$) reduced the influx of $^{65}Zn^{2+}$ from control levels. These data suggest that zinc transport by LMV occurred by a carrier-mediated process that was stimulated by ATP and inhibited by the P-ATPase inhibitor, vanadate, and the SERCA inhibitor, thapsigargin.

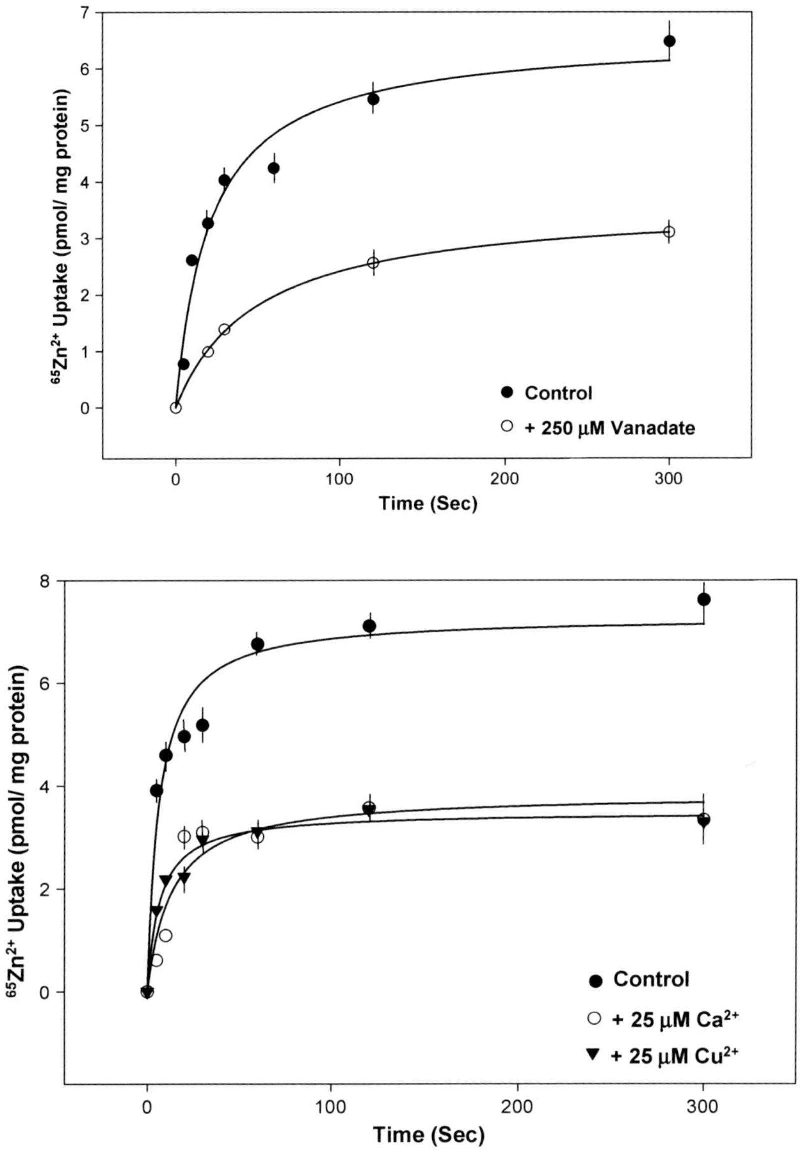

Fig. 3 Effects of 250 μM vanadate (*upper panel*), 25 μM Ca^{2+} and 25 μM Cu^{2+} (*lower panel*) on the time course of 25-μM $^{65}Zn^{2+}$ uptake by lobster hepatopancreatic LMV. Vesicles were loaded with 150 mM KCl and 20 mM Hepes/Tris at pH 7.2 and were incubated in a medium containing 25 μM $^{65}Zn^{2+}$-sulfate, 0.5 NTA, 20 mM Hepes/Tris, 150 mM KCl, and 0.2 mM ATP at pH 7.2. Values are means ± SEM (n = 3–5 replicates/mean) [13]

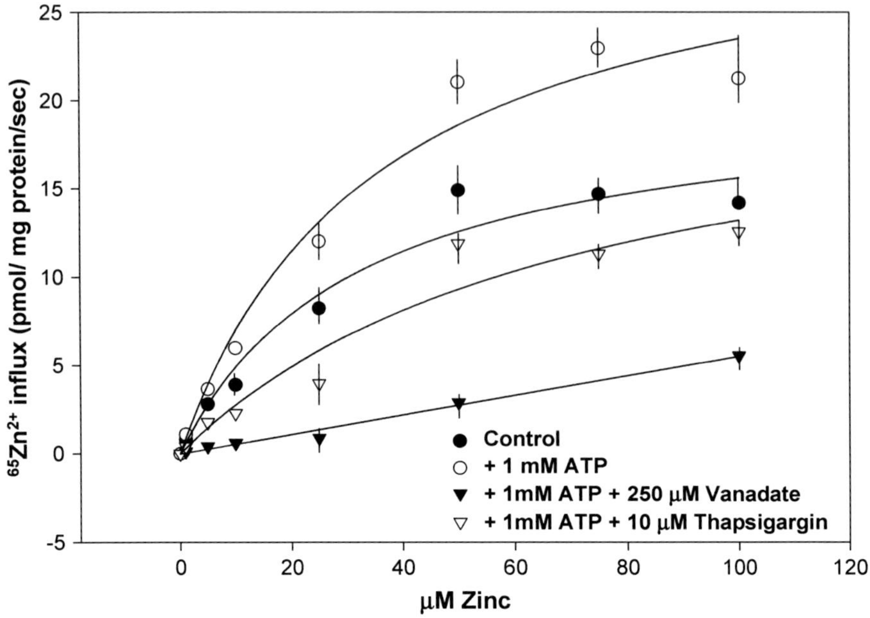

Fig. 4 Effects of 1 mM ATP, 250 μM vanadate, and 10 μM thapsigargin on the influx kinetics of $^{65}Zn^{2+}$by lobster hepatopancreatic LMV. Vesicles were loaded with 150 mM KCl and 20 mM Hepes/Tris and pH 7.2 and were incubated for 10 s in a medium containing $^{65}Zn^{2+}$-sulfate concentrations ranging from 0 to 100 μM, 0.5 mM NTA, 20 mM Hepes/Tris, 150 mM KCl, and 0.2 mM ATP at pH 7.2. The other three treatment groups had the ATP concentration raised to 1.0 mM. Values are means ± SEM (n = 3–5 replicates/mean) [13]

The effects of transmembrane pH gradients on lysosomal vesicle $^{65}Zn^{2+}$ influxes were next examined. Figure 5 describes the effects of equal pH values and transmembrane pH gradients on 10 s 25-μM $^{65}Zn^{2+}$ influxes into lobster LMV. No ATP was added to the incubation medium in this experiment. No significant differences ($P > 0.05$) were found for zinc influx K_m or J_{max} when bilateral pH values were at 7.0 or 5.0. However, an outwardly directed proton gradient ($pH_i < pH_o$) significantly ($P < 0.01$) increased $^{65}Zn^{2+}$ influx compared to vesicles with equal pH, and an inwardly directed proton gradient ($pH_i > pH_o$) significantly ($P < 0.01$) reduced metal uptake compared to the controls. The effect of these treatments was on the K_m of the transport system without significant influence on the J_{max} (Table 1). The results of this experiment suggest that carrier-mediated zinc transport by hepatopancreatic LMV likely occurs by exchange with protons.

To investigate the effects of other heavy metals on 5 and 25-μM $^{65}Zn^{2+}$ influxes into hepatopancreatic LMV in the presence of 0.2 mM ATP, experiments were performed measuring 10 s zinc uptakes in buffers containing either Cd^{2+} or Cu^{2+} activities ranging from 0 to 250 μM. Increasing the cadmium or copper concentrations over this range significantly ($P < 0.01$) reduced $^{65}Zn^{2+}$ influxes into LMV at both radiolabeled zinc concentrations (Figs. 6, 7). For both metal inhibitors,

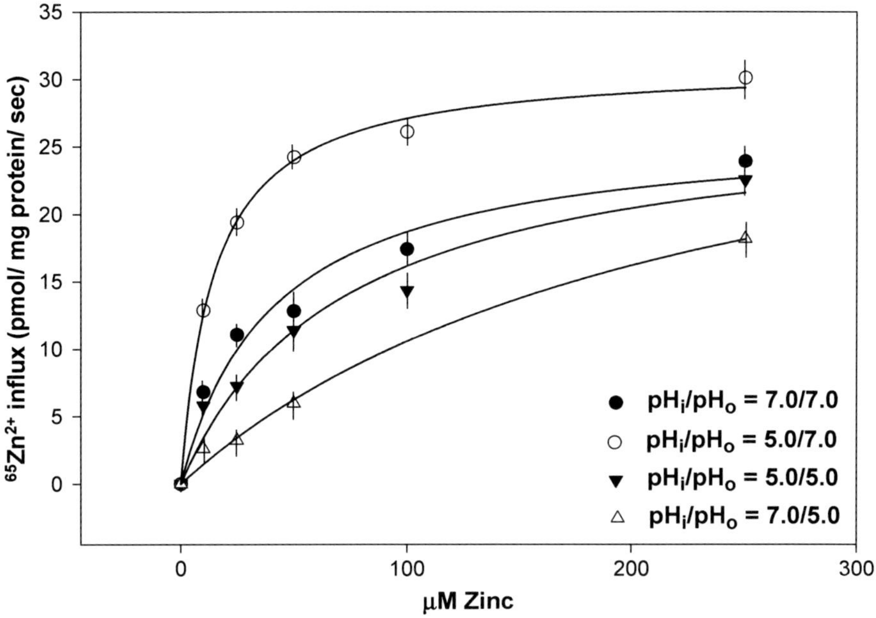

Fig. 5 Effects of variable pH on 25-µM $^{65}Zn^{2+}$ influx in lobster hepatopancreatic LMV. Vesicles were loaded with 150 mM KCl and 20 mM Hepes/Tris and were incubated for 10 s in a medium containing 25 µM $^{65}Zn^{2+}$-sulfate, 0.5 mM NTA, 20 mM Hepes/Tris or 20 mM Mes/Tris, and 150 mM KCl. No ATP was added to any treatment group in this experiment. Values are means ± SEM (n = 3–5 replicates/mean) [13]

Table 1 Effect of pH on the kinetics of $^{65}Zn^{2+}$influx into lobster hepatopancreatic lysosomal membrane vesicles

pH conditions		K_m (µM)	J_{max} (pmol/mg protein x sec)
Equal pH	$pH_i/pH_o = 7.0/7.0$	41.3 ± 11.0	26.4 ± 2.4
Outward Gradient	$pH_i/pH_o = 5.0/7.0$	14.8 ± 1.3	31.1 ± 0.6
Equal pH	$pH_i/pH_o = 5.0/5.0$	72.0 ± 21.8	27.8 ± 3.4
Inward Gradient	$pH_i/pH_o = 7.0/5.0$	228.8 ± 63.9	34.7 ± 5.3

Values were determined using Sigma Plot software for best fit to Michaelis–Menten relationships between $^{65}Zn^{2+}$influx vs. [Zn] for each pH condition. Triplicate samples were used at each [Zn] [13].

hyperbolic decay curves were fitted to the data using Sigma Plot software, which yielded asymptotic decreases in zinc influx values at each zinc concentration. These asymptotic values were subtracted from $^{65}Zn^{2+}$ influxes at each cadmium or copper concentration and the differences were plotted in Dixon plots to yield the component of $^{65}Zn^{2+}$ influx that was inhibited by each metal. The Dixon plots provided

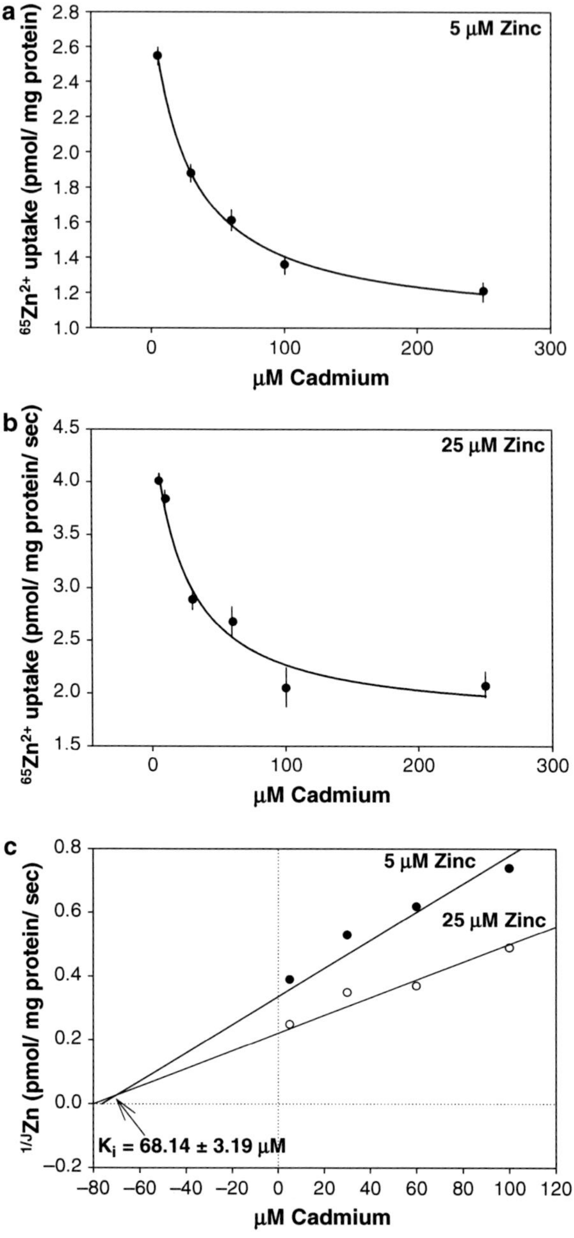

Fig. 6 Effects of variable cadmium concentrations on 5-μM (panel*A*) and 25-μM (panel *B*) $^{65}Zn^{2+}$ influx by lobster hepatopancreatic LMV. Vesicles were loaded with 150 mM KCl and 20 mM Hepes/Tris at pH 7.2 and were incubated in a medium containing cadmium sulfate concentrations from 0 to 250 μM. External incubation medium contained 0.2 mM ATP. Values are means ± SEM (n = 3–5 replicates/mean). Dixon plot of data from panels (*A*) and (*B*). The inhibitory constant (K_i) was obtained from the Dixon plot by drawing regression lines through the data using Jandel Sigma Plot software [13]

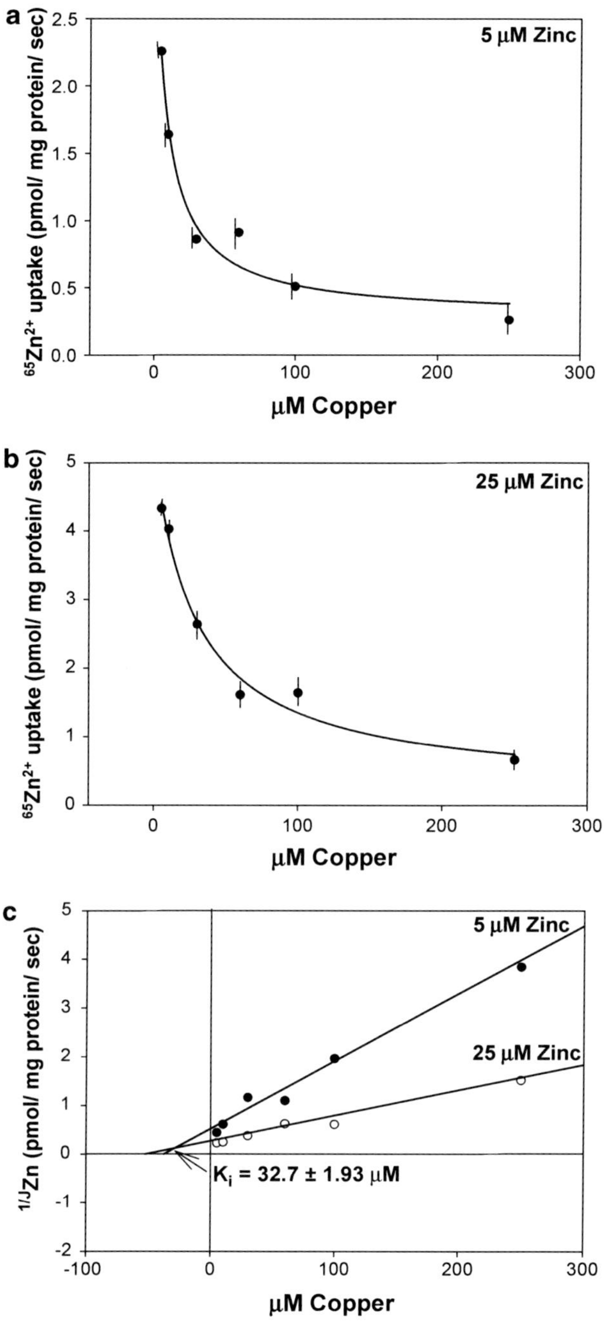

Fig. 7 Experiment as in Fig. 6 except that copper sulfate concentrations from 0 to 250 μM were used instead of cadmium sulfate [13]

an index of the apparent binding affinities of each competitive inhibitor (e.g., K_i) for the shared binding site with the substrate (e.g., $^{65}Zn^{2+}$). As displayed on the figures, the K_i for cadmium was $68.1 \pm 3.2\,\mu M$, while that for copper was $32.7 \pm 1.93\,\mu M$. The results of Figs. 6 and 7 clearly indicate that zinc, cadmium, and copper are competitive inhibitors of each other's transport and likely share a common carrier process on hepatopancreatic LMV.

5 Effect of Anions on Lysosomal Heavy Metal Sequestration and Detoxification

Detoxification of heavy metals in organelles such as lysosomes requires the presence of a mechanism that sequesters the cation in a form that is relatively immobile and cannot be transferred back into the cytoplasm. Electron microprobe analyses of a variety of invertebrate epithelial cells indicate that metals are sequestered together with sulfur and phosphorus in solid concretions within membrane-bound organelles [17]. While the composition of these concretions has been evaluated in a wide variety of invertebrate species, the mechanisms whereby complexing anions are transferred into these sites of detoxification are unknown. The following experiments examine the nature of anion transport by lobster hepatopancreatic lysosomal membranes and suggest a mechanism by which metal and polyvalent anions are brought together to form detoxifying concretions in these cells [42].

Several monovalent and divalent anions were loaded into LMV to see if the nature of the anion had an effect on the uptake and equilibrium attained by $^{65}Zn^{2+}$ during its accumulation by these preparations. Lysosomal membrane vesicles were preloaded with SO_4^{2-}, PO_4^{3-}, Cl^-, Oxalate^{2-}, and mannitol and the rate of 25-μM $^{65}Zn^{2+}$ uptake and final metal equilibrium level achieved by each were compared. As shown in Fig. 8A, vesicles preloaded with mannitol at pH 7.0 exhibited a slow uptake rate and low level of metal equilibrium. In contrast, when vesicles were preloaded with either SO_4^{2-} or PO_4^{3-} at pH 7.0, the $^{65}Zn^{2+}$ uptake rate was twice as fast as in mannitol-loaded vesicles and the metal equilibrium value was twice that of vesicles without preloaded anions. Figure 8B indicates that preloading vesicles with the monovalent anion, Cl^-, or the divalent organic anion, oxalate^{2-}, resulted in $^{65}Zn^{2+}$ uptake rates and equilibrium values that were similar to those of mannitol.

Because SO_4^{2-} was one of the polyvalent inorganic anions that led to enhanced vesicle accumulation of $^{65}Zn^{2+}$, an experiment was conducted to see what effect a wide range of intravesicular SO_4^{2-} concentrations would have on metal uptake. Figure 9 shows that increasing intravesicular $SO_4^{2-}$from 25 to 1,000 μM resulted in a stepwise increase in uptake rate and equilibrium of $^{65}Zn^{2+}$ by hepatopancreatic LMV with a maximum of both parameters attained at about 500 μM SO_4^{2-}.

Figure 5 showed that $^{65}Zn^{2+}$ uptake was stimulated by an acidic LMV interior ($pH_i = 5.0$; $pH_o = 7.0$) which suggested that metal transport occurred by way of a Zn^{2+}/H^+ antiporter, but the exchange stoichiometry of this exchanger was unclear. In order to clarify the stoichiometric nature of this transporter an experiment was

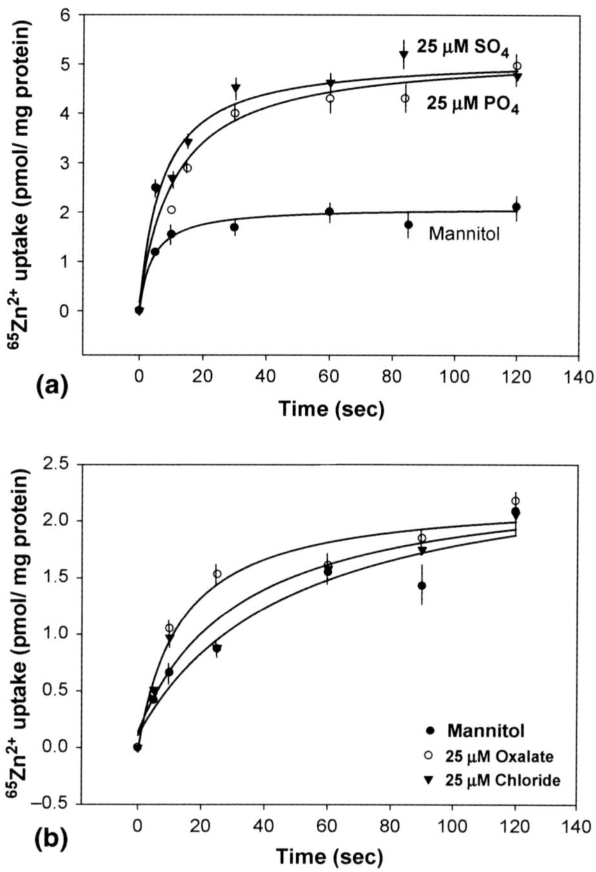

Fig. 8 (**A**) Effects of intravesicular polyvalent inorganic anions on the time course of 25-μM $^{65}Zn^{2+}$ uptake by hepatopancreatic LMV. Vesicles were loaded with 200 mM mannitol, 20 mM Hepes/Tris, pH 7.0 and 25 μM K_2SO_4, K_3PO_4, or mannitol, and were then incubated in a medium containing 200 mM mannitol, 25 μM $^{65}Zn^{2+}$-sulfate, 0.5 mM NTA, 0.2 mM ATP and 20 mM Hepes/Tris, pH 7.0. (**B**) Effects of intravesicular monovalent inorganic anions (Cl^-), polyvalent organic anions (oxalate^{2-}), and mannitol at pH 7.0 on the time course of 25 μM $^{65}Zn^{2+}$ uptake by LMV. Uptake conditions as in **A** except that in this instance vesicles were loaded with 25 μM oxalic acid, NaCl, or mannitol. Experiments were conducted in triplicate; values are means ± SEM at each time point [42]

conducted using valinomycin (potassium ionophore) and different concentrations of K^+ across the vesicle membrane. As displayed in Fig. 10 the influx of $^{65}Zn^{2+}$ was stimulated by a decrease in intravesicular pH and by the presence of external ATP. Furthermore, in the presence of ATP, metal transport was enhanced by an intravesicular positive electrical potential difference (e.g., $K^+_i < K^+_o$) compared to when the vesicles were either short-circuited (e.g., $K^+_i = K^+_o$) or contained an intravesicular negative electrical potential difference (e.g., $K^+_i > K^+_o$). These results suggest that because an electrically positive intravesicular potential difference increased the uptake of $^{65}Zn^{2+}$, more positive charge must have been transferred out of the vesicles than into the vesicles during the exchange of zinc and protons. The minimum transport stoichiometry that could account for this exchange would be 1 $Zn^{2+}/3H^+$.

Because intravesicular polyvalent inorganic anions like SO_4^{2-} and PO_4^{3-} enhanced the equilibrium accumulation of $^{65}Zn^{2+}$ within LMV (Fig. 8A), the nature of the uptake process facilitating the transfer of these anions from the cytoplasm was investigated. Two groups of vesicles were prepared. The first group was loaded

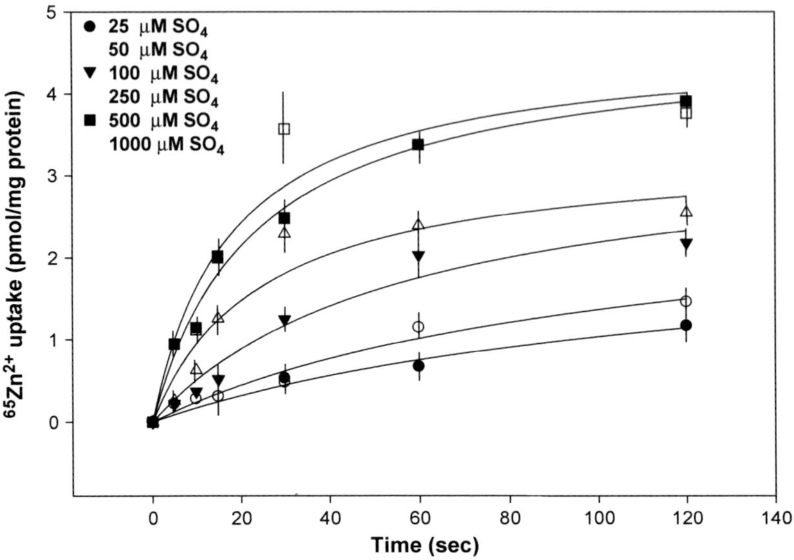

Fig. 9 Effects of varying intravesicular SO_4^{2-} concentration on the time course of 25-μM $^{65}Zn^{2+}$uptake by LMV. Vesicles were loaded with 200 mM mannitol, 20 mM Hepes/Tris, pH 7.0, and K_2SO_4 concentrations of 25, 50, 100, 500 or 1,000 μM and were then incubated in a medium containing 200 mM mannitol, 25 μM $^{65}Zn^{2+}$-sulfate, 0.5 mM NTA, 0.2 mM ATP and 20 mM Hepes/Tris, pH 7.0. Experiments were conducted in triplicate. Values are means ± SEM at each time point [42]

with mannitol and Hepes/Tris at pH 7.0, while the second group had mannitol and intravesicular chloride at pH 7.0. These two groups were then incubated in media containing $^{35}SO_4^{2-}$ or ^{14}C-oxalate^{2-} and the time course of isotope uptake into both membrane groups was observed. Figure 11A,B show that both $^{35}SO_4^{2-}$ and ^{14}C-oxalate^{2-} uptakes were stimulated in vesicles containing intravesicular Cl^- compared to those shown by mannitol-loaded vesicles. These results suggest the presence of an anion countertransport process in LMV, but do not define any of its properties except that it exchanges anions.

A series of experiments were conducted to examine the nature of the LMV anion exchange process. Vesicles were loaded with mannitol at pH 7.0, 8.0, and 9.0 and were incubated in media containing $^{35}SO_4^{2-}$ at concentrations from 2.5 to 50 mM. Under these conditions OH^- were the only anions inside the vesicles that were available for exchange with external $^{35}SO_4^{2-}$. Sigmoidal influx kinetics, as described by Eq. (1) below, were obtained for SO_4^{2-} influx as a function of external $[SO_4^{2-}]$ at each of the chosen internal pH conditions (Fig. 12).

$$J_{SO4} = J_{max}\,[SO_4]^n \,/\, \left(K_m^n + [SO_4]^n\right) \tag{1}$$

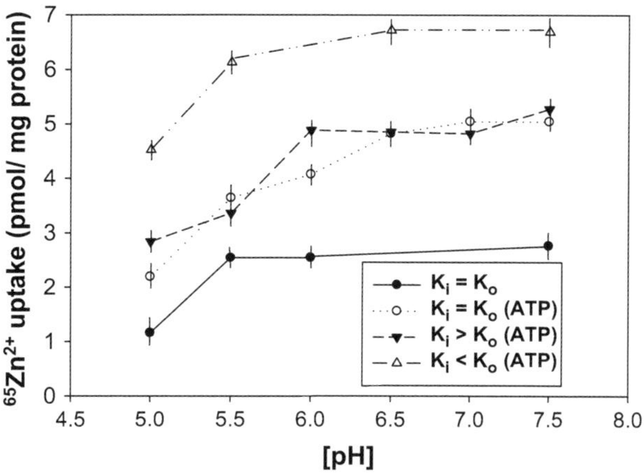

Fig. 10 Effects of intravesicular pH (pH 7.5–5.0), extravesicular ATP (1 mM), and induced membrane potential on the influx (10 s uptake) of 25 μM $^{65}Zn^{2+}$. A transmembrane electrical potential was induced by equilibrating the vesicles with 50 μM valinomycin, varying the extravesicular K^+ concentration [0 mM ($K_i > K_o$), 100 mM ($K_i = K_o$), or 200 mM ($K_i < K_o$)], and maintaining a constant intravesicular K^+ concentration (100 mM). Vesicles were loaded with 100 mM K_2SO_4, either 20 mM Hepes/Tris (pH 6.5–7.5) or Mes/Tris (pH 5.0–6.0), 50 μM valinomycin, and appropriate mannitol to maintain osmolarity. Loaded vesicles were then incubated in media containing 25 μM $^{65}Zn^{2+}$-sulfate, 0, 100 or 200 mM K_2SO_4, 0.2 or 1.0 mM ATP (+ATP), 0.5 mM NTA, 20 mM Hepes/Tris, pH 7.0 and mannitol to maintain osmolarity. The experiment was conducted in triplicate and values are means ± SEM at each pH value [42]

Where J_{SO4} is sulfate influx at any given [SO_4], J_{max} is apparent maximal sulfate influx rate (nmol/mg protein × s), $K_m{}^n$ (mM) is apparent binding affinity of the transporter for the anion corrected for multi-site cooperative interactions, $[SO_4]^n$ is sulfate concentration with multi-site corrections (mM), and n is the apparent Hill coefficient, which is an estimate of the number of $SO_4{}^{2-}$ ions transported during each transport cycle. Results indicated that both sulfate influx $K_m{}^n$ and J_{max} increased significantly (P < 0.01) from pH 7.0 to 9.0, but a greater change was observed in the J_{max} parameter than in the apparent affinity values. A small increase (P < 0.02) was also observed in the magnitude of the Hill coefficient, n, over the pH range examined, but all values approximated 2.0.

An experiment was conducted to clarify the exchange stoichiometry between $^{35}SO_4{}^{2-}$ influx into and OH^- efflux from lysosomal vesicles. $^{35}SO_4{}^{2-}$ influx was measured into vesicles loaded at several internal pH values and incubated in media containing either 5 or 10 mM $^{35}SO_4{}^{2-}$. $^{35}SO_4{}^{2-}$ influx was a hyperbolic function of internal [OH^-] (nM) in Fig. 13 and followed the modified Michaelis–Menten Eq. (2) below:

$$J_{SO4} = J_{max}\left[OH^-\right] / \left(K_m + \left[OH^-\right]\right) \qquad (2)$$

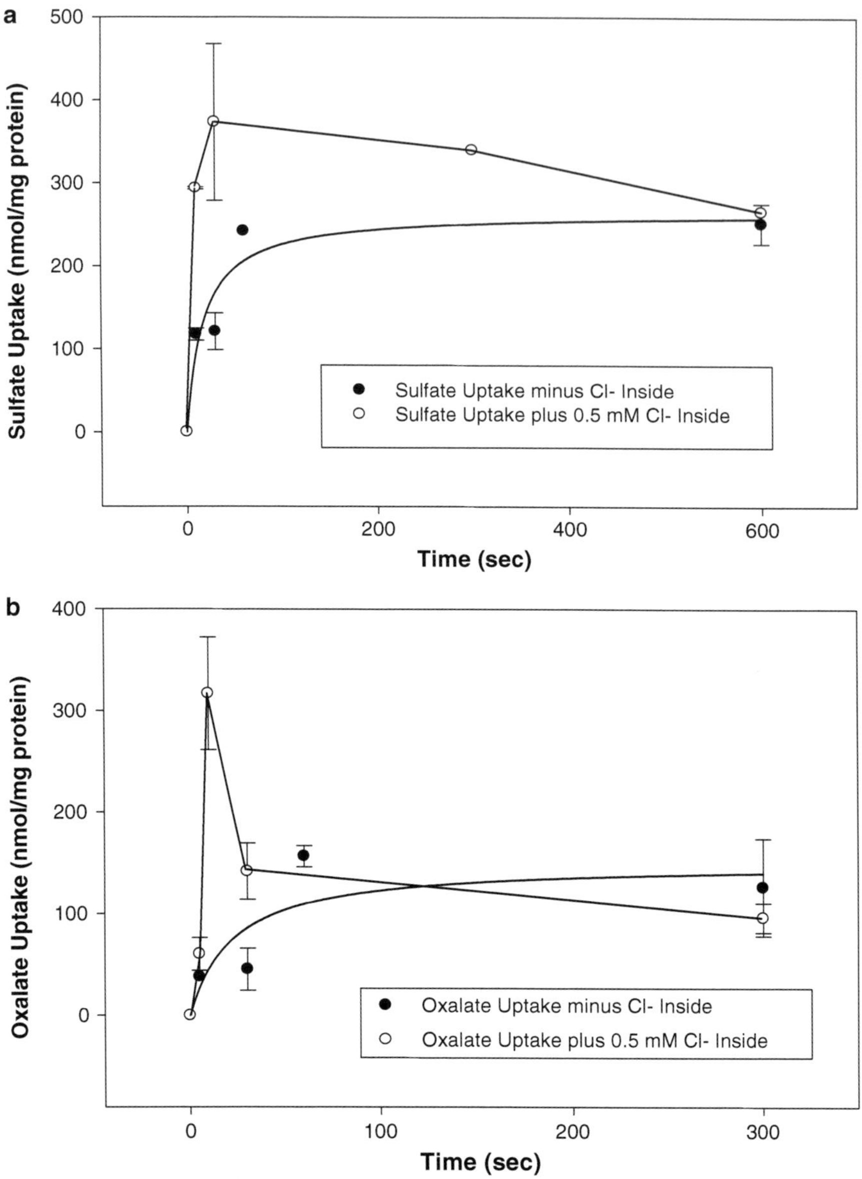

Fig. 11 Effect of preloaded Cl^- on the time course of $^{35}SO_4{}^{2-}$ and ^{14}C-oxalate^{2-} uptake by hepatopancreatic LMV. (**A**) Vesicles were loaded with 200 mM mannitol, 20 mM Hepes/Tris, pH 7.0, with or without 0.5 mM KCl, and were incubated in media containing 200 mM mannitol, 20 mM Hepes/Tris, pH 7.0, with or without 0.5 mM potassium gluconate and 5 mM $K_2{}^{35}SO_4$. (**B**) Vesicles were loaded with 200 mM mannitol, 20 mM Hepes/Tris, pH 7.0, with or without 0.5 mM KCl, and were incubated in media containing 200 mM mannitol, 20 mM Hepes/Tris, pH 7.0, with or without 0.5 mM potassium gluconate, and 5 mM ^{14}C-oxalate^{2-}. Experiments were in triplicate and values are means ± SEM at each time point [42]

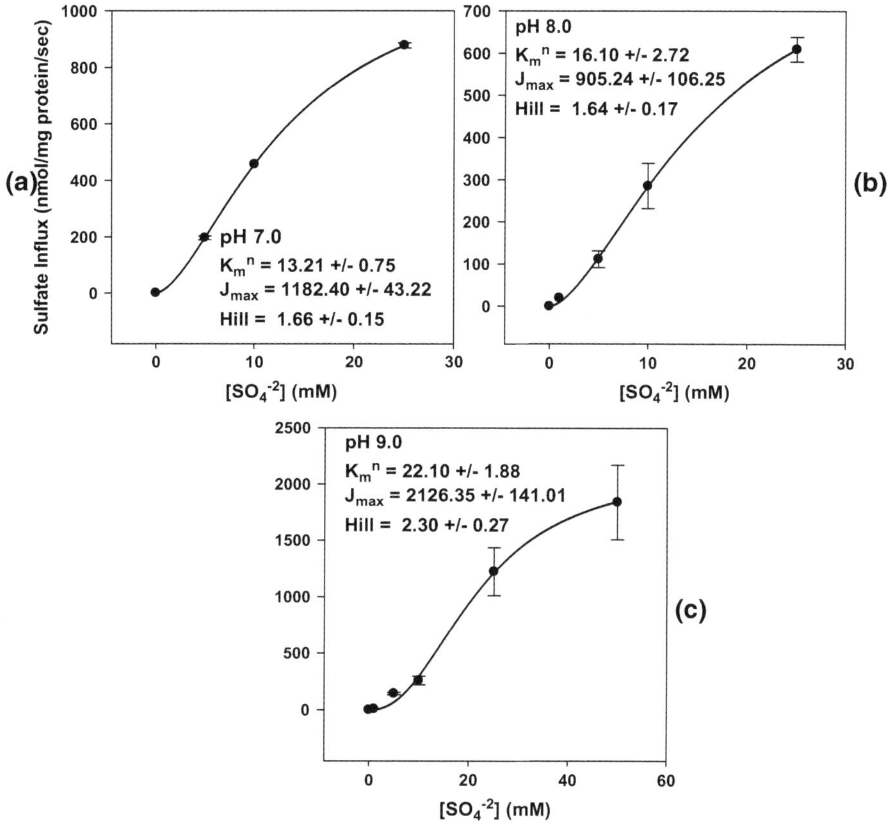

Fig. 12 Effect of intravesicular pH on $^{35}SO_4^{2-}$ influx kinetics in hepatopancreatic LMV. In all cases extravesicular pH was held at pH 7.0. Vesicles were loaded with 200 mM mannitol, 20 mM Hepes//Tris at pH 7.0. (**A**), 8.0 (**B**), and 9.0 (**C**). Loaded vesicles were then incubated in media containing 2.5, 5, 10, 25 and 50 mM $K_2{}^{35}SO_4$, 20 mM Hepes/Tris, pH 7.0 and mannitol to maintain osmolarity. The experiment was repeated in triplicate; values are means ± SEM. The sigmoid curves were drawn through the data using Sigma Plot software (Jandal) [42]

Where J_{SO4} is sulfate influx by exchange with intravesicular OH^- (nmol/mg protein × s), J_{max} is maximal sulfate influx, K_m is an apparent binding affinity constant of the exchanger for OH^- ions (nM), and $[OH^-]$ is the intravesicular concentration of hydroxyl ions (nM). Increasing external $[SO_4]$ from 5 to 10 mM resulted in apparent K_m values for OH^- binding that were not significantly different ($P > 0.05$). In contrast, sulfate influx J_{max} values were significantly ($P < 0.01$) greater at 10 mM SO_4^{2-}. The hyperbolic nature of the influx curve in this experiment, and the sigmoidal curves displayed in Fig. 12, suggests an electrogenic exchange stoichiometry between the polyvalent and monovalent anions of 2 SO_4^{2-} / 1 OH^-.

A working model of the role of polyvalent anions in hepatopancreatic lysosomal heavy metal sequestration and detoxification is shown in Fig. 14. The model

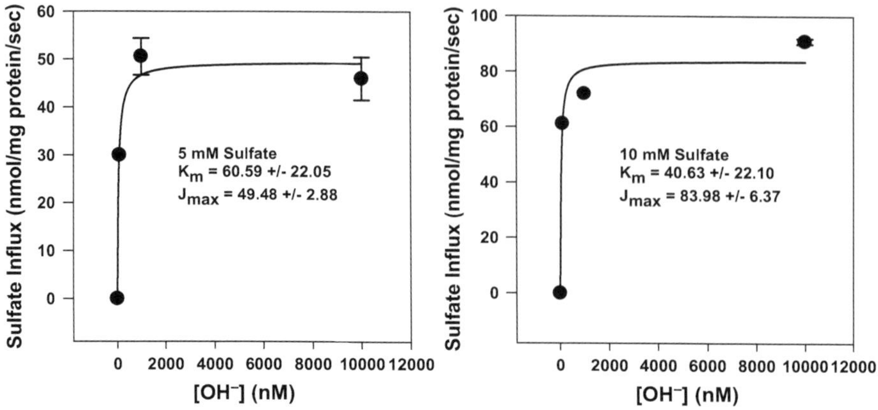

Fig. 13 Effect of intravesicular OH^- concentration on $^{35}SO_4^{2-}$ influx at 5 mM (**A**) and 10 mM SO_4^{2-} (**B**) (5 s uptakes) in hepatopancreatic LMV. In both cases the extravesicular pH was held at 7.0. Vesicles were loaded with 200 mM mannitol and either 20 mM Hepes/Tris or Mes/Tris pH 7.0, 8.0, and 9.0. Loaded vesicles were then incubated in media containing either 5 or 10 mM $K_2{}^{35}SO_4$, 20 mM Hepes/Tris, pH 6.0 and mannitol to maintain osmolarity. The experiment was repeated in triplicate and values are means ± SEM. Hyperbolic curves through the data were drawn with Sigma Plot software (Jandal) [42]

incorporates results that are presented in this review and from other previous publications [12, 13, 42, 43]. While the model describes the processes that are likely responsible for the accumulation and detoxification of zinc, other divalent cations such as calcium, copper, and cadmium appear to share this carrier [13] and may be sequestered within the organelle as well by this mechanism. Mandal et al. [13] presented evidence that $^{65}Zn^{2+}$ uptake by hepatopancreatic lysosomes was ATP-dependent and inhibited by both vanadate (ATPase inhibitor) and thapsigargin (SERCA – Sarco/Endoplasmic reticulum Ca^{2+}-ATPase – inhibitor). Because enzyme marker measurements indicated that the LMV preparation in this study was highly purified with minimal endoplasmic reticulum contamination, the inhibitory effect of thapsigargin on lysosomal $^{65}Zn^{2+}$ uptake was proposed to occur by an ATP-dependent lysosomal-specific SERCA isoform. Other experiments using the V-ATPase inhibitor, bafilomycin, confirmed the presence of an ATP-dependent H^+-ATPase in these membranes that acidified the vesicle interior and that appeared to be at least partially linked to metal accumulation [12].

Figure 14 suggests that the hepatopancreatic V-ATPase may power two electrogenic ion exchangers through the use of ATP and the induced membrane potential and bring about the accumulation of metals and polyvalent anions within the organelles. As this figure indicates, the asymmetric exchange stoichiometries for metal (e.g., Zn^{2+}, Cu^{2+}, Cd^{2+}, etc.) and polyvalent anion (e.g., SO_4^{2-}, PO_4^{3-}, etc.) uptakes lead to the accumulation of both ions within the organelle. High lysosomal concentrations of metals and polyvalent anions may condense into a solid concretion with a change in pH brought about by alterations in V-ATPase activity.

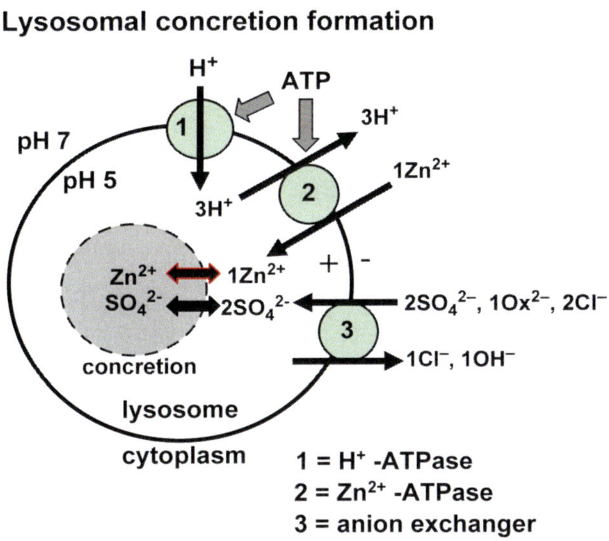

Fig. 14 Working model of the role of polyvalent anions in hepatopancreatic lysosomal heavy metal sequestration and detoxification. Protein 1 (ATP-dependent, V-ATPase) transfers protons into the vesicle interior creating a decrease in pH, an accumulation of hydrogen ions, and an inside-positive membrane potential. The outwardly directed proton gradient and positive vesicular interior provide the driving force for the asymmetric exchange of cytosolic divalent metals for intravesicular hydrogen ions by an electrogenic, ATP-dependent, Zn^{2+}-ATPase or $3H^+/1Zn^{2+}$ exchanger (Protein 2). Polyvalent inorganic anions such as sulfate^{2-} or phosphate^{3-} exchange with intravesicular monovalent inorganic anions such as Cl^- or OH^- by a second asymmetric antiporter (Protein 3), which uses the membrane potential as a driving force for exchange. Both divalent metals and polyvalent anions increase in concentration inside vesicles at acidic pH and are retained because they cannot be accommodated on the intravesicular monovalent binding sites of the exchangers. Divalent metals and polyvalent anions form precipitates (concretions) as the V-ATPase decreases in activity and the intravesicular pH rises [42]

6 Does the Organic Anion Transporter (OAT) Occur on Lysosomal Membranes and Facilitate the Accumulation of Zinc and Glutathione Within the Organelles?

As shown in Fig. 1, heavy metals entering the cytoplasm of invertebrate cells have a variety of possible fates including sequestration in organelles such as mitochondria, endoplasmic reticulum, and lysosomes. In addition, specific metal-binding compounds such as metallothionein and glutathione (GSH) are upregulated in the cytoplasm in response to elevated cytosolic concentrations of these cations [44, 45]. These compounds contain cysteine residues with thiol groups that bind metals avidly and thereby reduce the cytoplasmic free concentrations of the metals considerably. The question arises as to whether these metal-binding compounds remain in the cytoplasm and build in concentration as cytoplasmic metals accumulate from the

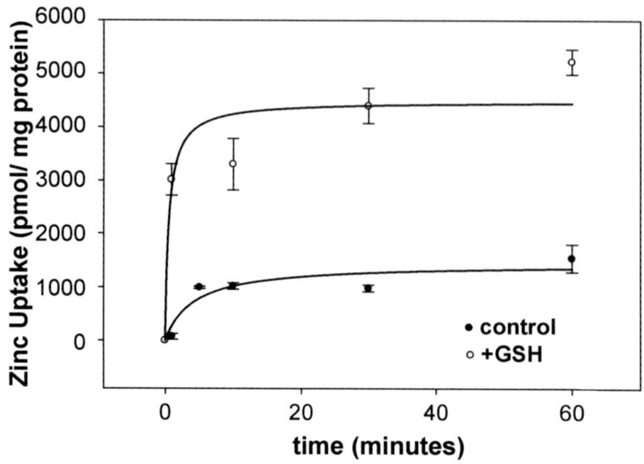

Fig. 15 Effect of 1 mM glutathione (GSH) on the time course of uptake of 25 μM $^{65}Zn^{2+}$ by LMV of lobster hepatopancreas. Vesicles were loaded with 150 mM KCl, 20 mM Hepes/Tris, pH 7.0, 1 mM α-ketoglutarate and were incubated in a medium containing 150 mM KCl, 20 mM Hepes/Tris, pH 7.0, 0.5 mM NTA, 2.0 mM DTT (dithiothreitol), 1 mM glutathione, and 25 μM $^{65}Zn^{2+}$-sulfate. Symbols are means ± SEM of 3–5 replicates/mean

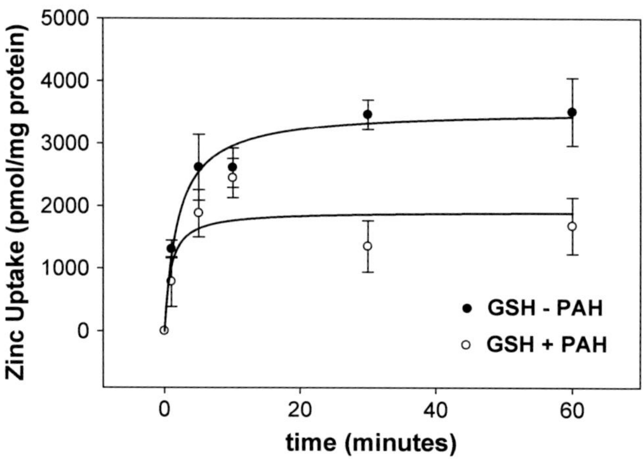

Fig. 16 Effect of 1 mM PAH (para-aminohippuric acid) on the time course of 1-mM glutathione-stimulated 25-μM $^{65}Zn^{2+}$ uptake by LMV of lobster hepatopancreas. Vesicles were loaded and incubated in the same internal and external media as in Fig. 15 with the addition of 1 mM PAH to the test condition in the external medium. Symbols are means ± SEM of 3–5 replicates / mean

environment, or are they sequestered within organelles after complexing with the metals?

The Organic Anion Transporter (OAT1) of the mammalian renal epithelium exchanges α-ketoglutarate (metabolically produced) for organic anions complexed with heavy metals as a means of clearing the blood of contaminating metals such as mercury [46]. The transporter is inhibited by p-aminohippurate (PAH) and probenicid and appears equally capable of transporting mercury–thiol conjugates of cysteine (e.g., Cys–Hg–Cys) or metal conjugates of GSH (e.g., GSH–Hg–GSH). While excretion of metals by way of exchange via the OAT1 system by renal tissue is clearly a means of reducing circulating concentrations of metals, long-term sequestration of metal–thiol conjugates in epithelial organelles such as lysosomes would produce the same result.

Experiments were performed with lobster hepatopancreatic LMV to assess whether these organelles were capable of transporting GSH and, if so, whether this transport in the presence of zinc would lead to the uptake and sequestration of the metal. Figure 15 illustrates the time course of 25 μM $^{65}Zn^{2+}$ uptake by LMV in the presence and absence of 1 mM GSH. Zinc uptake was increased by a factor of 4 or 5 in the presence of GSH. Figure 16 shows the effect of 1 mM PAH on the GSH-stimulated uptake of 25 μM $^{65}Zn^{2+}$. In this instance the GSH-stimulated uptake of the metal was decreased by half in the presence of the inhibitor, PAH. These combined results suggest that LMV from lobster hepatopancreas may display OAT1-like transporters on their membranes which serve to accumulate both GSH and associated zinc as GSH–Zn–GSH thiol conjugates. Because GSH is a tripeptide composed of cysteine, glycine, and glutamate residues, this compound may be recycled to the cytoplasm from the lysosome by transport via the proton-dependent, lysosomal dipeptide transporter (PEPT1; [47]) after dissociation of the metal within the organelle. Free lysosomal zinc would then be available to contribute to concretion formation by the process described in Fig. 14.

The preliminary data described here for a possible role of the OAT-1 transport system in heavy metal accumulation by epithelial lysosomes are only one form of evidence for the existence and functioning of this carrier system in hepatopancreatic cells. Future studies using molecular methods such as Western blotting and cloning techniques may support the physiological evidence presented here and provide an alternative process for metal accumulation and detoxification within these important organelles.

Acknowledgments Work reported herein from the author's laboratory was supported by National Science Foundation grant number IBN04-21986.

References

1. Ahearn HRH, Ahearn GA, Gomme J. Effects of cadmium on integumentary uptake of ^{3}H-L-histidine in the marine polychaete worm, *Nereis succinea*. J. Exp. Biol. 2000; 203: 2877–2885.
2. Peppler JE, Ahearn GA. Effect of heavy metals on the uptake of ^{3}H-L-histidine by the polychaete *Nereis succinea*. Comp. Biochem. Physiol. Part C 2003; 136(2):181–189.
3. Verbost PM, Van Rooij J, Flik G, Lock RAC, Bonga SEW. The movement of cadmium through freshwater trout branchial epithelium and its interference with calcium transport. J. Exp. Biol. 1989; 145: 185–197.
4. Bury NR, Grosell M, Grover AK, Wood CM. ATP-dependent silver transport across the basolateral membrane of rainbow trout gills. Toxicol. Appl. Pharmacol. 1999; 159: 1–8.
5. Bury NR, Walker PA, Glover CN. Nutritive metal uptake in teleost fish. J. Exp. Biol. 2003; 206: 11–23.

6. Grosell M, Wood CM. Copper uptake across rainbow trout gills: mechanisms of apical entry. J. Exp. Biol. 2002; 205: 1179–1188.
7. Conrad EM, Ahearn GA. ^{3}H-L-histidine and $^{65}Zn^{2+}$ are co-transported by a dipeptide transport system in lobster (*Homarus americanus*) intestine. J. Exp. Biol. 2005; 208: 287–296.
8. Conrad EM, Ahearn GA. Transepithelial transport of zinc and L-histidine across perfused intestine of American lobster, *Homarus americanus*. J. Comp. Physiol. B. 2007; 177: 297–307.
9. Viarengo A. Heavy metals in marine invertebrates: Mechanisms of regulation and toxicity at the cellular level. Rev. Aquatic Sci. 1989; 1: 295–317.
10. Klein MJ, Ahearn GA. Calcium transport processes of lobster hepatopancreatic mitochondria. J. Exp. Zool. 1999; 283: 147–159.
11. Chavez-Crooker P, Garrido N, Ahearn GA. Copper transport by lobster hepatopancreatic epithelial cells separated by centrifugal elutriation: measurements with the fluorescent dye, Phen Green. J. Exp. Biol. 2001; 204: 1433–1444.
12. Chavez-Crooker P, Garrido N, Pozo P, Ahearn GA. Copper transport by lobster (*Homarus americanus*) hepatopancreatic lysosomes. Comp. Biochem. Physiol. Part C. 2003; 135(2): 107–118.
13. Mandal PK, Mandal A, Ahearn GA $^{65}Zn^{2+}$ transport by lobster hepatopancreatic lysosomal membrane vesicles. J. Exp. Zool. 2006; 305A: 203–214.
14. Mandal PK, Mandal A, Ahearn GA. Physiological characterization of $^{45}Ca^{2+}$ and $^{65}Zn^{2+}$ transport by lobster hepatopancreatic endoplasmic reticulum. J. Exp. Zool. 2005; 303A: 515–526.
15. Brown BE. The form and function of metal-containing "granules" in invertebrate tissues. Biol. Rev. 1982; 57: 621–667.
16. Mason AZ, Simkiss K. Sites of mineral deposition in metal-accumulating cells. Exp. Cell Res. 1982; 139:383–391.
17. Al-Mohanna SY, Nott JA. The accumulation of metals in the hepatopancreas of the shrimp, *Penaeus semisulcatus* de Haan (Crustacea: Decapoda) during the moult cycle. Proc. Confr. Mar. Environ. Pollution, 1985; pp. 195–209.
18. Coombs T, George SG. Mechanisms of immobilization and detoxification of metals in marine organisms. In: Physiology and behaviour of marine organisms (McLusky, D. S. and Berry, A. J., eds.), 1978; pp. 179–187, Pergamon Press, New York.
19. George SG, Pirie BJS, Cheyne AR, Coombs TL, Grant PT. Detoxification of metals by marine bivalves: an ultrastructural study on the compartmentation of copper and zinc in the oyster *Ostrea edulis*. Mar. Biol. 1978; 45: 147–156.
20. Mauri M, Orlando E. Experimental study on renal concretions in the wedge shell *Donax trunculus* L. J. Exp. Mar. Biol. UK. 1982; 63: 47–57.
21. Mason AZ, Simkiss K, Ryan KP. The ultrastructural localization of metals in specimens of *Littorina littorea* collected from clean and polluted sites. J. Exp. Mar. Biol. UK. 1984; 64: 699–720.
22. Hopkin SP. Ecophysiology of metals in terrestrial invertebrates. 1989; London, Elsevier Applied Science.
23. Pisoni RL, Thoene JG. The transport systems of mammalian lysosomes. Biochim. Biophys. Acta 1991; 1071: 351–373.
24. Chou HF, Vadgama J, Jonas AJ. Lysosomal transport of small molecules. Biochem. Med. Metab. Biol. 1992; 48: 179–193.
25. Suzuki M, Gitlin JD. Intracellular localization of the Menkes and Wilson's disease proteins and their role in intracellular copper transport. Pediatrics Int. 1999; 41: 436–442.
26. Rees EM, Lee J, Thiele DJ. Mobilization of intracellular copper stores by the Ctr2 vacuolar copper transporter. J. Biol. Chem. 2004; 279(52) 54221–54229.
27. Palmiter RD, Cole TB, Findley SD. ZnT-2, a mammalian protein that confers resistance to zinc by facilitating vesicular sequestration. EMBO J. 1996a; 15:1784–1791.

28. Palmiter RD, Cole TB, Findley SD. ZnT-3, a putative transporter of zinc into synaptic vesicles. Proc. Natl. Acad. Sci. USA 1996b; 93: 14934–14939.
29. Gaither LA, Eide DJ. Eukaryotic zinc transporters and their regulation. Biometals. 2001; 14: 251–270.
30. Liuzzi JP, Cousins RJ. Mammalian zinc transporters. Ann. Rev. Nutr. 2004; 24: 151–172.
31. Wang A, Cortax N, Edelman IS. Mg-dependent, Zn-ATPase: Enzymatic characteristics, ion specificities, and tissue distribution. J. Membr. Biol. 2001; 181: 11–20.
32. Hussain D, Haydon MJ, Wang Y, Wong E, Sherson SM, Young J, Camakaris J, Harper JF, Cobbett CS. P-type ATPase heavy metal transporters with roles in essential zinc homeostasis in *Arabidopsis*. The Plant Cell. 2004; 16: 1327–1339.
33. MacDiarmid CW, Milanick MA, Eide DJ. Biochemical properties of vacuolar zinc transport systems of *Saccharomyces cerevisiae*. J. Biol. Chem. 2002; 277: 39187–39194.
34. Havelaar AC, de Gast IL, Snijders S, Beerens CEMT, Mancini GMS, Verheijen FW. Characterization of a heavy metal ion transporter in the lysosomal membrane. FEBS Lett. 1998; 436: 223–227.
35. Truong-Tran AQ, Ruffin RE, Zalewski PD. Visualization of labile zinc in apoptosis of primary airway epithelial cells and cell lines. Am. J. Phyisol. Lung Cell Mol. Physiol. 2000; 279: L1172–L1183.
36. Truong-Tran AQ, Carter J, Ruffin R, Zalewski PD. New insights into the role of zinc in the respiratory epithelium. Immun. Cell Biol. 2001; 70: 170–177.
37. Beyersmann D, Haase H. Functions of zinc in signaling, proliferation and differentiation of mammalian cells. Biometals. 2001; 14: 331–341.
38. Maret W. Cellular zinc and redox states converge in the metallothionein/thionein pair. J. Nutr. 2003; 133: 1460S–1462S.
39. Chavez-Crooker P, Garrido N, Ahearn GA. Copper transport by lobster (*Homarus americanus*) hepatopancreatic mitochondria. J. Exp. Biol. 2002; 205: 405–413.
40. Hopfer U, Nelson K, Perrotto J, Isselbacher KJ. Glucose transport in isolated brush border membrane from rat small intestine. J. Biol. Chem. 1973; 248: 25–32.
41. Bers DM, Patton CW, Nuccitelli R. A practical guide to the preparation of Ca^{2+} buffers. Methods Cell Biol. 1994; 40: 3–29.
42. Sterling KM, Mandal, PK, Roggenbeck, BA, Ahearn, GA, Gerencser, GA, and Ahearn, GA. Heavy metal detoxification in crustacean epithelial lysosomes: role of anions in the compartmentalization process. J. Exp. Biol. 2007; 210: 3484–3493.
43. Ahearn GA, Mandal PK, Mandal A. Mechanisms of heavy metal sequestration and detoxification in crustaceans: a review. J. Comp. Physiol. B 2004; 174: 439–452.
44. Brouwer M, Schlenk D, Ringwood AH, Brouwer-Hoexum T. Metal-specific induction of metallothionein isoforms in the blue crab *Callinectes sapidus* in response to single- and mixed-metal exposure. Arch. Biochem. Physiol. 1992; 294(2): 461–468.
45. Syring RA, Brouwer-Hoexum T, Brouwer M. Cloning and sequencing of cDNAs encoding for a novel copper-specific metallothionein and two cadmium-inducible metallothioneins from the blue crab *Callinectes sapidus*. Comp. Biochem. Physiol. C Toxicol. Pharmacol. 2000; 125: 325–332
46. Zalups R, Barfuss, DW. Renal organic anion transport system: A mechanism for the basolateral uptake of mercury–thiol conjugates along the pars recta of the proximal tubule. Toxicol. Appl. Pharmacol. 2002; 182: 234–243.
47. Thamotharan M, Lombardo YB, Bawani SZ, Adibi SA. An active mechanism for completion of the final stage of protein degradation in the liver, lysosomal transport of dipeptides. J. Biol. Chem. 1997; 272:11786–11790.

Epithelial Calcium Transport in Crustaceans: Adaptation to Intrinsic and Extrinsic Stressors

Michele G. Wheatly, Yongping Gao, and Christopher M. Gillen

Abstract Since the classical studies of Ussing employing a nonmammalian isolated epithelium (frog skin) to explore the basic principles of ion transport, physiologists have adopted increasingly reductionist approaches to dissect the biophysical mechanisms undergirding biological transport. In vitro characterization has employed isolated perfused organs, isolated epithelia, and reconstituted vesicle studies. Depth of resolution has been further enhanced by the emerging molecular revolution. Following years of deconstruction, physiologists are now engaging in reconstruction, namely putting the genes back into the organism. This contribution attempts such an integrative approach for a single electrolyte, calcium (Ca^{2+}), in a nonmammalian epithelium, the crayfish antennal gland (kidney). Two collaborating laboratories have archived an inventory of Ca^{2+} associated proteins believed to play a role in transcellular Ca^{2+} movement. Using the basic building blocks (expression profiles of key Ca^{2+} associated proteins and their regulators), the authors attempt to reconstruct a whole cell model for Ca^{2+} regulation in transporting epithelium (compared with a nonepithelial tissue) under stressors that perturb Ca^{2+} homeostasis which originate either intrinsically (the postmolt stage of the molting cycle) or extrinsically (unanticipated cold acclimation). Through horizontal integration of expression profiles of seven target Ca^{2+} associated proteins in epithelial and nonepithelial tissue under two contrasting experimental conditions, emergent themes inform the physiological complexity of Ca^{2+} homeostasis. Integration at the next level will require placing the epithelium in the context of organismic Ca^{2+} balance. The unique Ca^{2+} handling capabilities of the freshwater crayfish make it an excellent nonmammalian model for those studies.

Keywords Crayfish · Calcium · Antennal gland · Muscle · Posmolt · Cold · ECaC · PMCA · NCX · SERCA · SCP · CaM · eEF1Bγ

M.G. Wheatly (✉)
Department of Biological Sciences, College of Science and Mathematics, Wright State University, Dayton, OH 45435, USA
e-mail: michele.wheatly@wright.edu

G.A. Gerencser (ed.), *Epithelial Transport Physiology*,
DOI 10.1007/978-1-60327-229-2_4,

1 Introduction

Over the past decade several labs have developed invertebrate models for vectorial Ca^{2+} movement across epithelia [1–3]. The Wheatly lab employs a freshwater crustacean model (the crayfish *Procambarus clarkii*) and contrasts periods of impressive net unidirectional Ca^{2+} influx (postmolt stage of the molting cycle) with the control state of organismal calcium balance (intermolt stage). The naturally occurring molting cycle has provided the opportunity to study acute changes in whole organism Ca balance and the role played by key transporting epithelia including renal, branchial, and digestive epithelia [1, 4, 5]. The lab has documented the interdependent activity and regulation of a suite of proteins that are either directly or indirectly involved in cellular Ca^{2+} homeostasis including transmembrane proteins on apical, basolateral, and endomembranes as well as proteins that bind Ca^{2+} or serve a regulatory role [6, 7].

The arthropod molting cycle is orchestrated through a combination of intrinsic (hormonal) and extrinsic factors (daylength, lunar cycle, temperature) that conspire to precipitate ecdysis. The periodicity of this event may be as frequent as several times in the first growing season to once yearly for adults; certain species undergo a terminal ecdysis that heralds adulthood. Life expectancy of this crayfish species is about 18 months in the wild (4 years in captivity); throughout their lifetime it is estimated that a crayfish undergoes about 11 molts to achieve sexual maturity [8, 9]. Thus, in the life of an individual organism ecdysis could be viewed as an *extraordinary* or "special event" (by analogy, the periodicity of molting for a crayfish would be approximately equivalent to the number of times an average American would change automobiles over a lifetime of driving).

With a better understanding of cellular Ca^{2+} homeostasis in transporting epithelia during the molting cycle, in the past few years, and in collaboration with the Gillen lab at Kenyon College, we have sought to study Ca^{2+} homeostasis through the more generic lens of adaptation to *ordinary* everyday environmental stressors (e.g. temperature, salinity, endocrine disruptors). While special events mark milestones in growth and development, more meaningful to the organism is its ability to adapt to environmental assault on a day-to-day basis (by analogy an owner acquires understanding of automobile performance through routine maintenance). The freshwater crayfish has been long recognized for its ability to adapt to environmental extremes (temperature, salinity, aerial and acid exposure), that can occur on a routine basis and without the benefit of advanced warning [10]. Inland environments are inherently less stable than oceanic environments in their physical and chemical properties. Our research has led us to question how epithelial Ca^{2+} transport adjusts to everyday unanticipated environmental challenges as opposed to the occasional, highly conspicuous and well-orchestrated molting cycle that permits incremental growth.

In this paper we have employed low temperature acclimation (28 days at 4°C compared with acclimation to room temperature 23°C) as an "everyday" environmental stressor through which to compare reactive adaptation to unanticipated Ca^{2+} dyshomeostasis versus the anticipated and hormonally mediated alterations in Ca^{2+} balance associated with postmolt remineralization following the "special event" of ecdysis (postmolt as compared with intermolt). Our studies have compared Ca^{2+} homeostasis in epithelial tissue (antennal gland analogous to kidney) with a nonepithelial tissue (muscle) in order to determine which of the changes are systemic and which are specific to epithelial transport. After summarizing the comparison across target proteins within a given treatment (postmolt, cold acclimation) and tissue (epithelial, nonepithelial), this article will attempt to summarize two emergent integrative themes: comparison across treatments within a tissue (postmolt vs. cold); and comparison across tissues within a treatment (antennal gland vs. muscle).

The data presented in this review compare responses of seven target proteins to postmolt mineralization or cold acclimation and are grouped by experimental approach: quantification of mRNAs (Table 1, Fig. 1), localization of mRNAs

Table 1 Real-time PCR assay for the expression of mRNA of seven target Ca^{2+} associated proteins (ECaC, PMCA, NCX, SERCA, SCP, CaM, EF1Bγ) in control (intermolt at 23°C, left), postmolt at 23°C (center), and cold acclimation (intermolt at 4°C right) in antennal gland (upper) and axial (A) or cardiac (C) muscle (lower) of crayfish *Procambarus clarkii*. Values are presented as relative quantification value (RQ), mean ± STDV from three different samples with four to five crayfish in each sample. Quantitative real-time PCR reaction was performed in a 96-well microtiter plate using the relative quantification ΔΔCt method. The threshold cycle (Ct) represents the PCR cycle at which an increase in SYBR Green fluorescence above a baseline signal can first be detected. Relative quantification (RQ) was performed by normalizing Ct values of each sample with Ct values of the endogenous control 18s rRNA gene (ΔCt) and finally calculated using ΔCt of the control (intermolt at 23°C) as calibrator. ΔCt corresponds to the difference between the ΔCt of the gene of interest and the ΔCt of the endogenous control 18s rRNA

Protein	Control 23°C	Postmolt 23°C	Cold 4°C
A. Relative quantitation mRNA *Antennal gland*			
ECaC	1	23.9 ± 3.7	8.6 ± 1.3
PMCA	1	22.2 ± 3.4	4.0 ± 0.7
NCX	1	11.9 ± 1.7	5.1 ± 0.7
SERCA	1	6.0 ± 1.3	1.3 ± 0.2
SCP	1	0.06 ± 0.02	1.6 ± 0.5
CaM	1	5.5 ± 1.6	1.6 ± 0.8
EFBγ	1	0.9 ± 0.6	0.7 ± .1
B. Relative quantitation mRNA *Muscle (Axial or Cardiac)*			
ECaC	Nd	Nd	Nd
PMCA	1	6.5 ± 2.6	2.1 ± 0.7 (A)
NCX	1	4.5 ± 1.8 (A)	3.3 ± 0.6 (A)
		6.1 ± 2.0 (C)	
SERCA	1	11.3 ± 3.8 (A)	1.2 ± 0.3 (A)
		8.1 ± 1.5 (C)	1.1 ± 0.3 (C)
SCP	1	0.084 ± 0.014 (A)	0.8 ± 0.1 (A)
	1	0.151 ± 0.046 (C)	0.2 ± 0.1 (C)
CaM	1	4.1 ± 0.7 (A)	1.0 ± 0.6 (A)
	1	7.2 ± 2.8 (C)	1.1 ±0.8 (C)
EFBγ	1	11.9 (A)	0.7 ± .1 (A)
	1	2.0 (C)	0.8 ± .2 (C)

Nd: Not detectable.

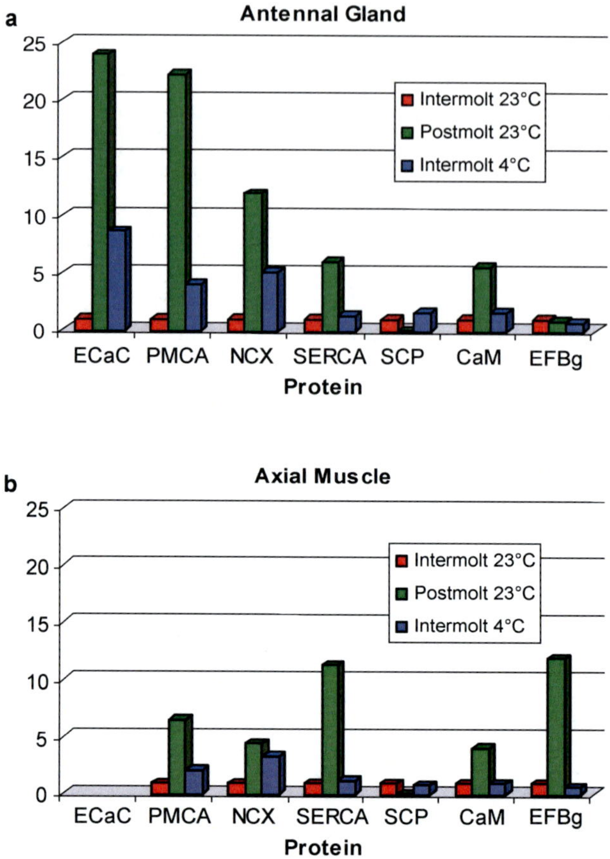

Fig. 1 Graphical representation of the relative quantification of mRNA of seven target Ca^{2+} associated proteins (ECaC, PMCA, NCX, SERCA, SCP, CaM, EF1Bγ) in control (intermolt at 23°C, *red*), postmolt at 23°C (*center, green*), and cold acclimation (intermolt at 4°C *right, blue*) in antennal gland (**A**) and axial muscle (**B**) of crayfish *Procambarus clarkii*. Data are taken from Table 1

(Figs. 2 and 3), quantification of protein (Figs. 4 and 5) and localization of protein (Figs. 6 and 7). The dataset, while broad ranging, is not comprehensive. As our interest in Ca^{2+} associated proteins has evolved over time, development of the necessary molecular probes has lagged behind. Thus in a nutshell, while this review touches upon two treatments, in two tissues, using four different methodological approaches to study a suite of seven Ca^{2+} associated proteins, we do not have the complete 112 experimental data sets in hand!

Briefly, in our primary research publications, a range of routine molecular techniques have been employed to determine the temporal and spatial expression of mRNAs and proteins that are believed to be involved in cellular Ca^{2+} homeostasis. Temporal mRNA expression has been documented using real-time PCR [11]. Fold

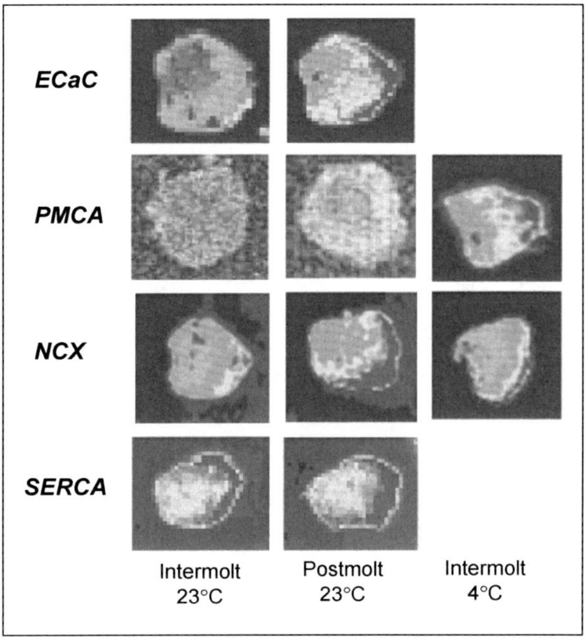

Fig. 2 Localization and expression of ECaC, PMCA, NCX, and SERCA mRNA in crayfish antennal gland sections in intermolt at 23°C (control, *left*), postmolt at 23°C (*center*), and intermolt at 4°C (*right*) using in situ hybridization of an antisense probe. Shown are representative digitized computer images where abundance of mRNA shows up as yellow to orange coloration when scanned on Fuji film with Image Gauge software. The specificity of the probe was confirmed by running comparable sections with a sense probe (control, not shown). Scale: diameter of antennal gland is approximately 5 mm

changes in expression of each target protein are reported compared with the control condition. In situ hybridization has been employed to localize and assess relative abundance of mRNA for the proteins of interest in tissue sections [12]. Western blot analysis has been employed to quantify protein expression for target proteins [13] against which we have previously generated homologous polyclonal antibodies [14]. These same antibodies have been used to localize the proteins of interest in tissue sections using immunohistochemistry [14]. Subcellular localization of targeted proteins has been determined through immunocytochemistry with laser scanning confocal visualization [15].

2 Postmolt Mineralization as an Anticipated "Special Event" Challenging Ca^{2+} Transport

Over the past decade our group has documented the changes in expression profile for a series of proteins that are of strategic importance in effecting transcellular Ca^{2+} flux, specifically comparing periods of net Ca^{2+} influx (postmolt, apical to basolateral vectorial transport) with periods of Ca^{2+} balance (intermolt) using the cyclical "special event" of postmolt remineralization following ecdysis in crustaceans [7]. Our work has attempted to clarify how epithelial cells are able to maintain low cytosolic Ca^{2+} in the face of vectorial Ca^{2+} transit, through comparing temporal and

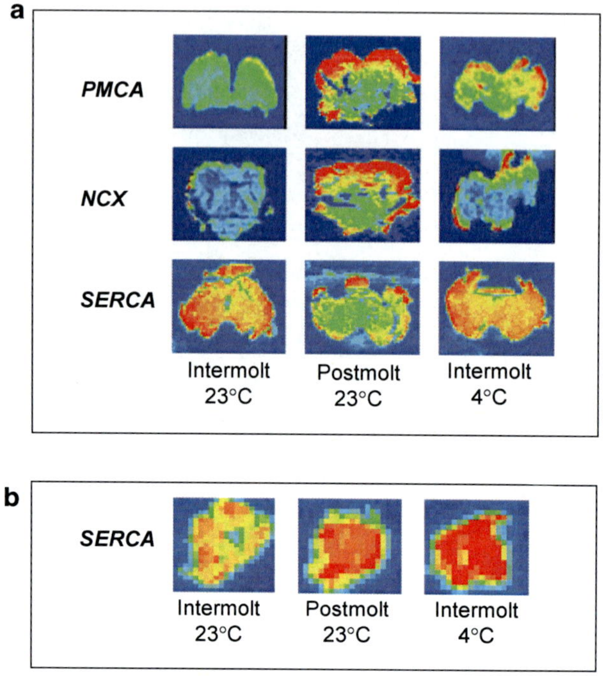

Fig. 3 Localization and expression of PMCA, NCX, and SERCA mRNA in crayfish axial (tail) muscle sections (**A**) and SERCA expression in cardiac muscle sections (**B**) in intermolt at 23°C (control, *left*), postmolt at 23°C (*center*), and intermolt at 4°C (*right*) using in situ hybridization of an antisense probe. Shown are representative digitized computer images where abundance of mRNA shows up as yellow to orange coloration when scanned on Fuji film with Image Gauge software. The specificity of the probe was confirmed by running comparable sections with a sense probe (control, not shown). Scale: diameter of axial muscle is approximately 20 mm

spatial expression patterns with those of nonepithelial tissues (such as axial abdominal and cardiac muscle). The antennal gland, which is the crustacean equivalent of the kidney, has emerged as the epithelium of choice for our experimental purposes. Freshwater crayfish exhibit the somewhat rare ability to produce a dilute urine through active reabsorption of 97% of filtered electrolytes [16]. The unidirectional Ca^{2+} influx rate (postfiltrational reabsorption) measured at the renal epithelium in intermolt exceeds branchial and digestive uptake; in immediate postmolt renal Ca^{2+} reabsorption increases further still as the crayfish attempts to correct a hemodilution [17] resulting from uptake of external fresh water to precipitate ecdysis (shedding). So it transpires that the antennal gland exhibits the dynamic upregulation of unidirectional Ca^{2+} influx in postmolt (compared with intermolt) that would be experimentally difficult to study in the gills or the hypodermis, which are both less accessible for experimental purposes.

Our journey into understanding the molecular basis of epithelial Ca^{2+} uptake has led us from the "prime suspects" namely basolateral Ca^{2+} motive pumps and

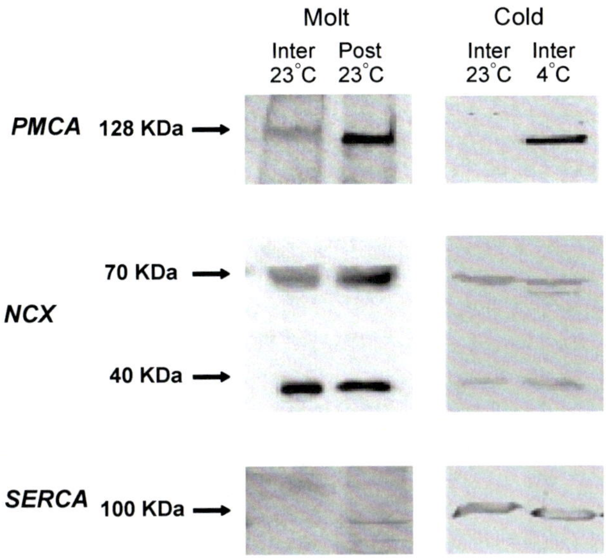

Fig. 4 Western blot analysis showing the effect of either postmolt (*left panels*) or cold acclimation (*right panels*) on protein expression of PMCA, NCX, and SERCA in crayfish antennal gland. In each treatment the left lane is intermolt at 23°C (control). The right lane is either postmolt at 23°C (*left panels*) or intermolt at 4°C (*right panels*). Total membrane protein (30 μg) was loaded in each lane and the membrane was hybridized to the appropriate polyclonal homologous antisera for crayfish target proteins raised in rabbits

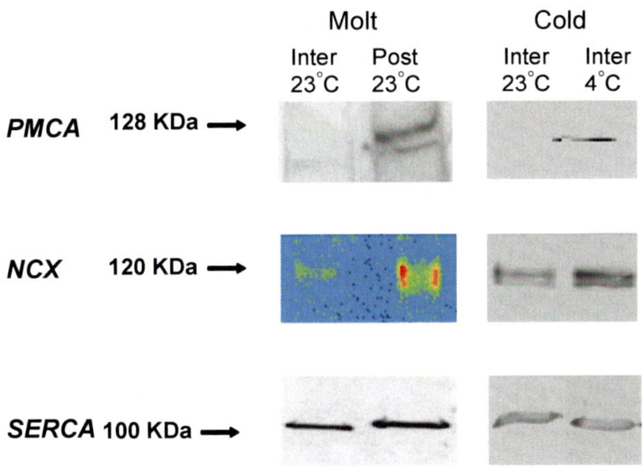

Fig. 5 Western blot analysis showing the effect of either postmolt (*left panels*) or cold acclimation (*right panels*) on protein expression of PMCA, NCX, and SERCA in crayfish axial muscle. In each treatment the left lane is intermolt at 23°C (control). The right lane is either postmolt at 23°C (*left panels*) or intermolt at 4°C (*right panels*). Total membrane protein (30 μg) was loaded in each lane and the membrane was hybridized to the appropriate polyclonal homologous antisera for crayfish target proteins raised in rabbits

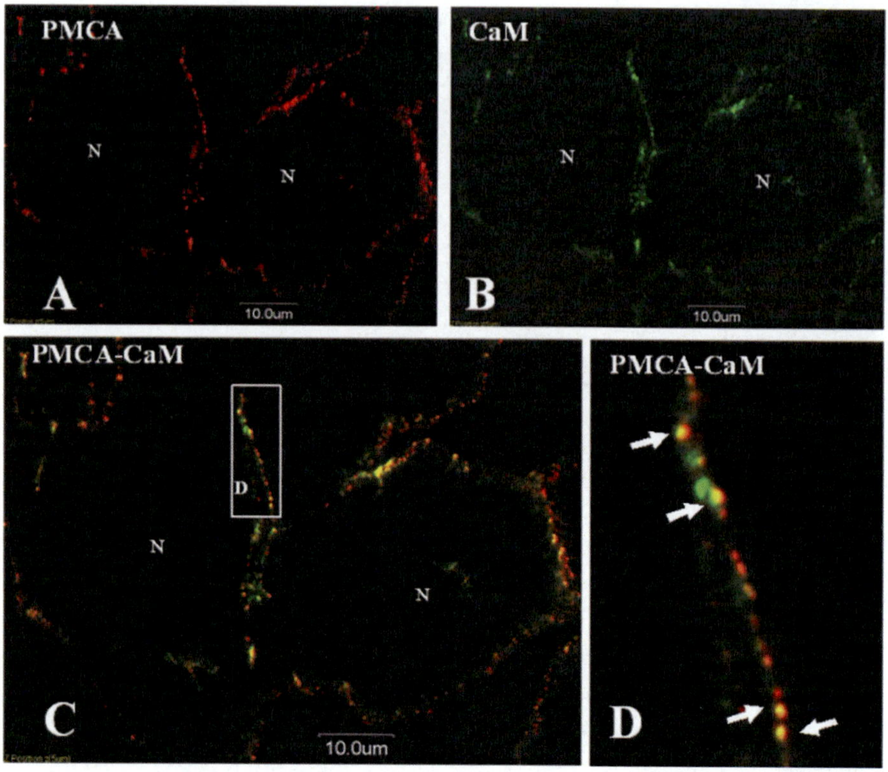

Fig. 6 Projected stacks of optical sections through crayfish tail muscle showing colocalization of PMCA and CaM on the plasma membrane. The immunoreactivity of CaM was detected with Cy3-conjugated secondary antibodies and the PMCA immunoreactivity was detected with Fluorescein (FITC)-conjugated secondary antibodies directed against the PMCA and CaM primary antibody species. Sections were mounted and coverslipped under Vectashield mounting medium and examined with confocal microscopy (Olympus Fluoview; BX50 microscope with 60X oil immersion objective, N.A. 1.4). Images were analyzed with ImagePro Plus (version 4.1; Media Cybernetics), and final figure composition and labeling were performed in CorelDraw (version 11). Double staining for PMCA (*Panel A, red*) and CaM (*Panel B, green*). *Panel C* is the merged image of *Panels A and B*. *Panel D* is an enlarged image of *insert D* from *Panel C*. The immunoreactivity of PMCA-CaM complex clusters (*yellow*) are indicated by *white arrows*

exchangers that have been linked to active Ca^{2+} uptake through in vitro characterization [18, 19] to a more integrative approach that attempts to explain Ca^{2+} homeostasis from a cellular perspective through studying a suite of Ca^{2+} associated proteins that are dynamically regulated in interdependent ways. Target proteins for study have been selected because of their putative role in key stages in transcellular Ca^{2+} transfer: import across the apical membrane (epithelial Ca^{2+} channel, ECaC, [20]); transfer across sarco- or endoplasmic reticulum endomembranes for storage in the Intracellular Ca^{2+} Store (ICS, Sarco/Endoplasmic Reticulum Ca^{2+} ATPase, SERCA, [21–23]); buffering of cytosolic Ca^{2+} (Sarcoplasmic Ca^{2+} binding Protein,

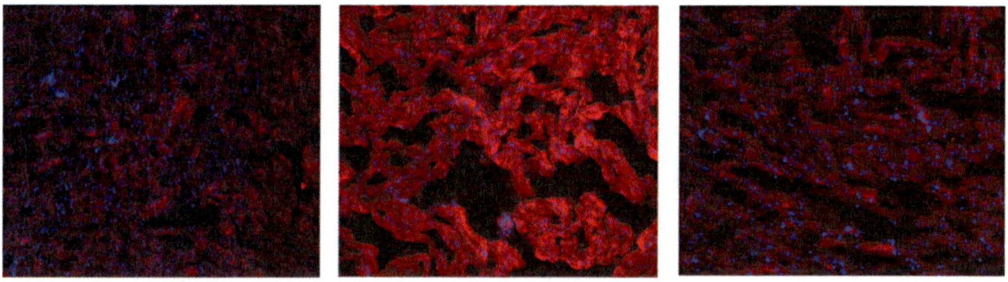

Fig. 7 Immunohistochemistry for SERCA in cardiac muscle tissue sections in intermolt at 23°C (control, *left*), postmolt at 23°C (*center*), and intermolt at 4°C (*right*). SERCA antibody was used as the primary antibody and Cy-3 as the secondary antibody. Images shown are the positive controls where fluorescence is proportional to the amount of SERCA expressed. Dapi was used to stain the DNA (*blue dots*) which represent nucleoli or nuclei heterochromatin

SCP; [11]); calcium binding proteins known to regulate Ca^{2+} motive enzymes (CalModulin, CaM, Gao and Wheatly, personal communication); and export from the cell across the basolateral membrane (Plasma Membrane Ca^{2+} ATPase, PMCA, [20]; Na^+/Ca^{2+} exchanger, NCX, [24]). Additionally we have recently investigated regulatory factor eEF1Bγ (an elongation factor subunit that regulates the recruitment of amino acyl-tRNAs to the ribosome during protein synthesis in eukaryotes) that may participate in sensing stress and regulation of membrane protein synthesis and may also be a Ca^{2+} dependent membrane binding protein [25–27].

3 Event Planning: Expression of Ca^{2+} Associated Proteins in Postmolt vs. Intermolt

In epithelial tissue impressive increases (greater than 10-fold) in mRNA expression were measured in postmolt compared with intermolt for primary apical import (ECaC) and basolateral export pathways (PMCA and NCX) as one would expect based on the significant increase in unidirectional Ca^{2+} influx predicted to occur across the renal epithelium at this time (Table 1, Fig. 1A). Unexpectedly the increase in PMCA expression was double that observed for NCX even though NCX has historically been viewed as the "workhorse" for basal Ca^{2+} efflux based on vesicle studies [19]. In epithelial tissue SERCA also exhibited increased expression in postmolt, although the magnitude of the increase was half that observed for NCX suggesting that intracellular sequestration of Ca^{2+} is less important than basolateral efflux during mass transcellular conveyance. Meantime a decrease in expression of SCP, the sarcoplasmic Ca^{2+} binding protein, suggests that IC buffering of Ca^{2+} is less important in cellular Ca^{2+} homeostasis while export processes are upregulated. An alternative explanation would be that reduced buffering enhances basolateral transport or sequestration in IC organelles by increasing the availability of free

Ca^{2+} which has been shown to kinetically stimulate transport in pre- and postmolt hepatopancreatic cells of shrimp [28]. Calmodulin (CaM), the calcium binding protein known to regulate Ca^{2+} motive enzymes, increased in expression as one might predict, although not to the same extent as the Ca^{2+} motive protein it is known to regulate (PMCA). Expression of the elongation factor eEF1Bγ however was unchanged.

The mRNAs for target proteins (ECaC, PMCA, NCX, and SERCA) were localized in antennal gland sections using in situ hybridization (Fig. 2). Three proteins (ECaC, PMCA, and NCX) were expressed ubiquitously in all regions of the antennal gland but were most abundant in the peripheral regions (labyrinth and nephridial canal) believed to be associated with ion reabsorption. Expression in postmolt increased significantly compared with intermolt. SERCA expression was more diffuse throughout the antennal gland sections and was more abundant relative to ECaC, PMCA, or NCX. Expression increased in postmolt but was less pronounced than the increases observed in the major import/export mechanisms.

Postmolt protein expression in antennal gland was quantified for proteins against which we have previously generated antibodies (PMCA, NCX, and SERCA, Fig. 4). These data were largely confirmatory of the real-time PCR data. PMCA (128 kDa) showed a significant increase in postmolt compared with intermolt. While both bands of NCX (70 and 40 kDa) showed increased binding in postmolt, the increase was less dramatic than exhibited by PMCA. Consistent with relative changes observed in real-time PCR, SERCA protein expression (100 kDa) was also increased in postmolt but not to the same extent as observed in the primary efflux mechanisms located on basolateral plasma membranes.

The mRNA expression profiles for the proteins of interest in nonepithelial tissue (axial abdominal and cardiac muscle) provide some interesting similarities and comparisons, enabling us to discern which changes were systemic during postmolt and which are specific to or more exaggerated in epithelial compared with nonepithelial (control) tissues. First, as a Ca^{2+} channel specific to epithelial tissues, ECaC is not expressed in muscle. The primary export mechanisms (PMCA and NCX) showed a modest (about 5-fold) increase in expression in postmolt compared to intermolt, whereas SERCA, a mechanism known to be more abundant in muscle, showed a 10-fold increase (Table 1, Fig. 1B). Thus, the relative roles of the proteins involved in exporting Ca^{2+} from the cell tend to be reversed in nonepithelial cells (where sequestration in the ICS predominates) compared with epithelial cells (where export into extracellular fluid effects transepithelial flux).

The eEFBγ increased significantly in abdominal muscle in about the same proportion as SERCA (about 11-fold). This was the only significant change in eEF1Bγ observed in either tissue or either treatment. The eEF1Bγ subunit is believed to act as a scaffold independently binding the α and β subunits of the multisubunit complex that is tethered to the ER through interactions with ER membrane protein. The finding that eEF1Bγ mRNA levels only increased during postmolt in axial muscle suggests that transcriptional regulation may correlate with synthesis rates of muscle proteins. Prior work has shown that fractional rates of protein synthesis increase 10-fold in leg muscles of the shore crab, *Carcinus*, during premolt in preparation for ecdysis and that they remain elevated in postmolt [29]. Similarly in abdominal muscle, protein synthesis rates increase in late premolt and early postmolt [30, 31]. In a related crayfish, *Austropotamobius pallipes* [32] an increase in actin mRNA was demonstrated during pre- and postmolt; however a more detailed study in lobster suggested that the increase in protein synthesis was translationally regulated, which is consistent with our finding of increased expression of the elongation factor subunit eEF1Bγ. Interestingly, the corresponding trend towards increased eEF1Bγ in cardiac muscle was of a smaller magnitude than axial muscle and not significantly different from intermolt, suggesting that the muscle growth that is symptomatic of somatic muscle in the postmolt period is missing or less extensive in the heart. The effect of molting cycle on muscle has primarily focused on crustacean striated muscle particularly claw, walking leg, and abdominal muscle. After ecdysis, there is a rapid increase in body volume due to high rates of water uptake. This stretches the somatic muscle creating one of the triggers for muscle elongation (in addition to neural and hormonal stimuli) by addition or insertion of sarcomeres or transverse sarcomere splitting. To accommodate incremental growth, the skeletal muscle fibers must elongate rapidly (within several hours of ecdysis) through rapid synthesis of sarcomeric proteins that begins in premolt [33–35]. Behavioral studies have shown characteristic differences in the escape response over the molt cycle in juvenile lobsters that are attributable to molt-related changes in nerve and muscle physiology in the tail [36]. To the best of our knowledge, the effect of the molting cycle on cardiac muscle morphology has not yet been determined. While the heart would obviously need to enlarge over time to perfuse a growing organism, since the organ is suspended in the body cavity and not encased in cuticle, the growth patterns would not need to be harnessed to the cyclicity of the molting cycle.

In situ hybridization of PMCA and NCX in tail muscle sections (Fig. 3A) revealed modest increases in mRNA expression during postmolt compared to intermolt. The hybridization increase was greater in the dorsal deep lateral and medial extensor muscles (that effect the recovery stroke from tail flip) compared with the ventral deep flexor muscle. Unexpectedly SERCA in tail sections showed decreased mRNA expression which is inconsistent with the real time quantification. Western analysis (Fig. 5) of PMCA showed increased protein expression (3.5-fold increase of the 128 kDa band) in postmolt compared with intermolt. NCX protein expression was also significantly upregulated. SERCA protein expression however showed less marked increase which would be consistent with the in situ hybridization results. In tail muscle PMCA and CaM have been colocalized using immunoreactivity and visualization under confocal microscopy (Fig. 6). This has revealed PMCA-CaM clusters on the plasma membrane of crayfish muscle cells.

In cardiac muscle SERCA mRNA expression has been assessed through in-situ hybridization and SERCA protein expression using immunofluorescence (Jamal, personal communication). In postmolt cardiac muscle we observed a dynamic upregulation of SERCA mRNA compared with intermolt with hybridization spread throughout the entire section (Fig. 3B). Immunofluorescence revealed a corresponding dramatic increase in protein expression in heart tissue sections (Fig. 7).

4 Temperature Acclimation as an Unplanned "Everyday Event" Challenging Ca^{2+} Transport

For aquatic poikilotherms, temperature can influence geographical distribution, growth, metabolism, reproduction and life history. Several studies have been conducted on the thermal tolerance and preferenda of crayfish [37, 38]. As ectotherms and residents of temperate fresh water, the thermal tolerance of *P. clarkii* has been documented to extend from around freezing to 35°C [39]. Temperate crayfish species are often exposed to a 3 month "winter" with ambient temperatures below 5°C [40]. Thermal acclimation in crayfish physiological systems has primarily focused on metabolism, respiratory and cardiovascular physiology [41]. In addition to retarding growth and molting in crustaceans, acclimation to low temperature (both artificially and in the wild) has been shown to affect a variety of physiological processes via adjustments at the molecular and membrane level [42, 43]. Germaine to the present study, in species as diverse as rye (*Secale cereale*, [44]) and goldfish (*Carassius auratus,* [45]) exposure to low ambient temperature has been shown to elicit influx of Ca^{2+} into cells primarily through increased opening of Ca^{2+} channels [46, 47]. To compensate for cytotoxicity under these conditions cells export Ca^{2+}. A series of experimental studies in our laboratory has addressed the effect of 28 days of exposure to 4°C on Ca^{2+} homeostasis ([24], Gao, personal communication). At this lower temperature crayfish are very quiescent although not torpid or dormant per se.

5 Chilling Out: Expression of Ca^{2+} Associated Proteins in Low Temperature Exposure

Laboratory exposure to low ambient temperature in crayfish resulted in upregulation within 5 days of several proteins known to be associated with cellular Ca^{2+} homeostasis (Table 1, Fig. 1; Gao, personal communication), confirming that, in invertebrates, Ca^{2+} entry into cells as a result of cold exposure was compensated by export of Ca^{2+} either out of the cell or into the ICS. In antennal gland (Table 1, Fig. 1A) we detected an 8.6-fold increase in expression of ECaC, the epithelial Ca^{2+} channel after 28 days of cold exposure. There is debate over whether Ca^{2+} entry on cold acclimation is a secondary response, or whether Ca^{2+} entry from the extracellular fluid (ECF) into the cytosol serves as an early signaling event in eliciting cold signal transductions [48]. Changes in IC Ca^{2+} during cold stress are sensed by Ca^{2+} sensor proteins such as calmodulin and Ca^{2+}-dependent protein kinases and are transduced through changes in IP_3 levels into upregulation of expression of low temperature-inducible genes [49]. Our group has recently isolated the heat shock protein Hsp40 from crayfish antennal gland and is working on expression patterns. The fact that ECaC expression increased in antennal gland strongly suggests that proteins involved in import are upregulated and that the Ca^{2+} entry is intentional and thus a primary effect.

The mRNAs of primary exporter proteins (NCX and PMCA) increased by an equivalent amount (4–5-fold) in antennal gland. CaM expression doubled for purposes of regulating the Ca^{2+} motive enzymes. SCP expression doubled suggesting an increased role for Ca^{2+} buffering in the cold. Elongation factor expression, however, was unchanged, as was SERCA.

In situ hybridization confirmed in antennal gland a distinct increase in mRNA expression of both PMCA and NCX after 28 days cold acclimation (Fig. 2). Binding was stronger around the periphery (labyrinth and nephridial canal) as compared to the central region of the tissue section. Western analysis (Fig. 4) confirmed impressive upregulation (4-fold) of PMCA protein in cold acclimation from levels that were barely detectable at 23°C. By contrast the two NCX bands evident in Western analysis (70 and 50 kDa) only exhibited a 1.4-fold increase following cold exposure. Likewise SERCA protein expression did not change appreciably as a result of cold acclimation, confirming the real-time PCR data.

In muscle, cold exposure elicited comparable changes in expression patterns of Ca^{2+} associated proteins and their mRNAs, though not quite as pronounced (Table 1, Fig. 1B). However, since ECaC is not expressed in muscle tissue, the entry of Ca^{2+} must be attributed to other types of Ca^{2+} channels. Expression of the mRNAs for the primary export mechanisms (PMCA and NCX) increased by only 2–3-fold. As with antennal gland, in muscle, the NCX showed higher mRNA expression change than PMCA suggesting that the NCX is the primary export mechanism in response to cold acclimation. No changes in SERCA expression were observed during cold acclimation.

In axial abdominal muscle increased expression of both PMCA and NCX mRNA was observed using in situ hybridization (Fig. 3A). The hybridization increase was again greater in the dorsal deep lateral and medial extensor muscles compared with the ventral deep flexor muscle. SERCA expression however was unchanged. Western analysis (Fig. 5) revealed a 6-fold increase in expression of PMCA3 protein (128 kDa) in cold acclimated axial muscle. As with antennal gland, PMCA protein expression in axial muscle exceeded the change in mRNA suggesting increased protein stability of PMCA protein in addition to increased transcription. NCX protein expression (120 kDa band) similarly increased by 1.6-fold in cold acclimation. SERCA protein expression was virtually unchanged confirming mRNA expression trends. Interpretation of mRNA/protein expression data after cold acclimation is somewhat complicated by physical effects of reduced temperature on biochemical processes including transport per se but also reduced rates of transcription and translation. Increased expression of transport proteins upon cold acclimation could alternatively be traced to reduced transport capacity/efficiency at reduced temperature. Likewise increased mRNA expression could be compensating for direct effects of temperature on rates of protein synthesis, mRNA degradation or transcription. Confirming this was a study on a related temperate crayfish *Austropotamobius pallipes*, where cold acclimation (12 to 1°C) resulted in a decrease in the rate of protein synthesis in skeletal muscle that was attributed to a decrease in RNA activity [50].

In cold acclimated cardiac muscle there appeared to be an increased expression of mRNA for SERCA throughout the entire myocardial section by in situ hybridization (Fig. 3B); however immunofluorescence did not show any appreciable change in SERCA protein expression in cardiac sections (Fig. 7). Exposure of goldfish *Carassius auratus* to a similar temperature regime [45] was associated with a 23-fold increase in transcription of SERCA and parvalbumin β, a Ca^{2+} buffering protein associated with SR (the vertebrate equivalent of SCP). In crayfish SCP in cardiac muscle decreased on cold acclimation.

6 Emergent Integrative Themes

6.1 Comparison Across Treatments Within a Tissue (Postmolt vs. Cold Acclimation)

For the majority of target proteins examined (ECaC, PMCA, NCX, SERCA, CaM), the two experimental treatments (postmolt compared with intermolt at 23°C; 28 days acclimation to 4°C compared with acclimation to 23°C) elicited similar responses, namely an increase in expression of mRNAs and proteins in both epithelial and nonepithelial tissues. Within a tissue the upregulation of a given protein elicited during postmolt was far greater (more than double) than that observed after 28 days of cold acclimation. This suggests that the Ca^{2+} dyshomeostasis resulting from the preplanned events following ecdysis far exceed the imbalance resulting from unanticipated exposure of intermolt crayfish to low temperature.

The only exception to this theme was SCP which showed a significant decrease in postmolt, yet was either maintained (muscle) or moderately increased (epithelial) in cold acclimation. This suggests that the two experimental treatments created differing demands for IC Ca^{2+} buffering. In postmolt, enhanced Ca^{2+} export capacity may decrease the necessity for IC buffering; however in cold acclimation sequestration of Ca^{2+} in ICS is the primary strategy to lower cytosolic levels. A second exception was eEF1Bγ which exhibited a tissue-specific response, namely a significant increase in expression in postmolt axial muscle that was not apparent in either antennal gland or cardiac muscle. These two exceptions highlight the finding that transmembrane Ca^{2+} motive proteins and their regulators (ECaC, PMCA, NCX, SERCA, CaM) demonstrate a common expression pattern that is different from the genes with other functions (SCP, eEF1Bγ).

6.2 Comparison Across Tissues Within a Treatment (Epithelial vs. Nonepithelial Response to Either Postmolt Mineralization or Cold Acclimation)

Within a treatment, it was apparent that epithelial and nonepithelial tissues exhibited different strategies to correct Ca^{2+} imbalance. In general, within an experimental treatment, epithelial tissue (antennal gland) exhibited a more pronounced upregulation of both "importers" (ECaC) and "exporters" on the plasma membrane (PMCA,

NCX). This is consistent with the role of renal epithelia in vectorial Ca^{2+} transport in postmolt; it also suggests that the antennal gland exhibits an organ level response to cold acclimation. Since cellular dehydration is known to result from exposure to near-freezing temperatures, as the primary organ for volume regulation, the antennal gland would be expected to show a compensatory response. By contrast, nonepithelial muscle cells showed greater upregulation of "exporters" on endomembranes (SERCA) suggesting that sequestration of Ca^{2+} in the ICS is the preferred response to elevated IC Ca^{2+}, at least in postmolt. For two of the target proteins (SCP and CaM) virtually the same response was observed in both tissues under either experimental treatment. This would suggest that the responses are systemic to all tissues and not specific to epithelial cells.

Expression of the elongation factor eEF1Bγ was only significantly elevated in postmolt axial muscle possibly associated with the protein synthesis effecting extensive myofibrillar remodeling at this stage. Thus, while the induction of transmembrane Ca^{2+} motive protein expression (PMCA, SERCA, NCX) during postmolt is greater in epithelial tissues than in muscle, this pattern is reversed for eEF1Bγ. These findings argue against a role for eEF1Bγ as a regulatory factor for the induction of transmembrane Ca^{2+} motive proteins either during postmolt or cold acclimation although it remains possible that post-translational modification may influence eEF1Bγ activity during either treatment.

7 Future Directions in Physiological Complexity

The cold acclimation employed in this study was unanticipated and did not occur through natural means. In the wild, many temperate species undergo a natural "wintering" cold water period that is linked to molting, maturation, and fecundity [51]. For example in Norway winter lasts from September to June during which no growth or molting occurs and water temperature ranges from 2 to 14°C. Studies on these animals show that wild adult noble crayfish (*Astacus astacus*) seem to be adapting to winter conditions by late summer/early fall and cannot be "cheated" into molting by manipulating temperature. It should be appreciated that the life cycle of the crayfish is a continuum and that, even though ecdysis is a "special event" the intermolt is the main functional phase of the crayfish life during which it feeds, reproduces and undergoes a series of preparatory metabolic and molecular processes in preparation for shedding. In the same way that the intermolt (preparation for ecdysis) is a several month process, so is natural acclimation to lowered temperature in the wild. A nice complement to the lab studies summarized in this review would be sampling of target proteins in field populations of crayfish over the course of a natural wintering event. There is every expectation that cold acclimation in natural populations would be better anticipated and planned. In fact, in natural populations, the molting cycle is very much interlinked with the annual thermal cycle. In temperate species pre- and postmolt are usually restricted to the summer months when conditions are most favorable for growth (increased temperature and food availability) whilst crayfish overwinter in the intermolt stage [52]. If cold acclimation and pre- and postmolt are ordinarily mutually exclusive conditions then the similar pattern of upregulation of Ca^{2+} motive proteins in these conditions is unlikely due to factors other than Ca^{2+} dyshomeostasis common to these conditions. Thus, while this study has reported on the independent effects of ecdysis and cold acclimation on Ca^{2+} homeostasis, ambient temperature and molting cyclicity are inextricably linked.

Acknowledgments The research reported from the authors' laboratories at Wright State University and Kenyon College has been supported by National Science Foundation grant number IBN 0445202 (to M.W., Y.G., and C.G.). The original data on cold acclimation resulted from the thesis research of two master's students, Daniel Whalen and Zeenat Jamal (assisted by undergraduates and interns Francieli Vigo and Ashkahn Golshani).

References

1. Wheatly MG. Calcium homeostasis in crustaceans: the evolving role of branchial, renal, digestive and hypodermal epithelia. J Exp Zool 1999; 283:620–640.
2. Ahearn GA, Mandal PK, Mandal A. Calcium regulation in crustaceans during the molt cycle: a review and update. Comp Biochem Physiol 2004; 137(A):247–257.
3. Ziegler A, Hagedorn M, Ahearn GA, Carefoot TH. Calcium translocations during the molt cycle of the semiterrestrial isopod *Ligia hawaiiensis* (Oniscidea, Crustacea). J Comp Physiol 2007; 177(B):99–108.
4. Zanotto FP, Wheatly MG. Calcium balance in crustaceans: nutritional aspects of physiological regulation. Comp Biochem Physiol 2003; 133A:645–660.
5. Ahearn GA, Mandal PK, Mandal A. Mechanisms of heavy metal sequestration and detoxification in crustaceans: a review. J Comp Physiol B 2004; 174:439–452.
6. Wheatly MG, Zanotto FP, Hubbard MG. Calcium homeostasis in crustaceans: subcellular Ca dynamics. Comp Biochem Physiol 2002; 132 (B):163–178.
7. Wheatly MG, Gao Y, Gillen CM. Paradox of epithelial cell calcium homeostasis during vectorial transfer in crayfish kidney. Gen Comp Endocrinol 2007; 152:267–272.
8. Huner JV. *Procambarus* in North America and elsewhere. In: Holdich DM, Lowery RS, eds. Freshwater crayfish: biology, management and exploitation. London: Chapman Hall, 1988; 239–261.
9. Huner JV. *Procambarus*. In: Holdich DM, ed. Biology of freshwater crayfish. Oxford: Blackwell Science Ltd, 2002; 541–584.
10. McMahon BR. (2002). Physiological adaptation to environment. In: Holdich DM, ed. Biology of freshwater crayfish. Oxford: Blackwell Science Ltd, 2002; 327–376.
11. Gao Y, Gillen CM, Wheatly MG. Molecular characterization of the sarcoplasmic calcium-binding protein (SCP) from crayfish *Procambarus clarkii*. Comp Biochem Physiol 2006; 144(B):478–487.
12. Wheatly MG, Gao Y, Nade M. Integrative aspects of renal epithelial calcium transport in crayfish: temporal and spatial regulation of PMCA. Int Congr Ser 2004; 1275: 96–103.
13. Gao Y, Wheatly MG. Characterization and expression of plasma membrane Ca^{2+} ATPase (PMCA3) in crayfish, *Procambarus clarkii*, antennal gland during moulting. J Exp Biol 2004; 207:2991–3002.
14. Wheatly MG, Zhang Z, Weil JR, et al. Novel subcellular and molecular tools to study Ca transport mechanisms during the elusive moulting stages of crustaceans: flow cytometry and polyclonal antibodies. J Exp Biol 2001; 204:959–966.
15. Muennich EAL, Fyffe REW. Focal aggregation of voltage-gated, Kv2.1 subunit-containing, potassium channels at synaptic sites in rat spinal motoneurones. J Physiol 2003; 554: 673–685.

16. Wheatly MG, Toop T. The mechanisms of acid-base and ionoregulation in the freshwater crayfish during environmental hyperoxia and recovery. II. The role of the antennal gland. J Exp Biol 1989; 143:53–70.
17. Wheatly MG, Ignaszewski LA. Electrolyte and gas exchange during the molting cycle of a freshwater crayfish. J Exp Biol 1990; 151:469–483.
18. Wheatly MG, Pence RC, Weil JR. ATP-dependent calcium uptake into basolateral vesicles from transporting epithelia of intermolt crayfish. Am J Physiol 1999; 276:R566–R574.
19. Wheatly MG, Hubbard MG, Corbett AM. Physiological characterization of the Na^{+}/Ca^{2+} exchanger (NCX) in hepatopancreatic and antennal gland basolateral membrane vesicles isolated from the freshwater crayfish *Procambarus clarkii*. Comp Biochem Physiol 2002; 131:343–361.
20. Gao Y, Wheatly MG. Molecular cloning of an epithelial Ca^{2+} channel (ECaC) like gene from crayfish *Procambarus clarkii*. J Exp Biol 2007; 210:1813–1824.
21. Zhang Z, Chen D, Wheatly MG. Cloning and characterization of sarco/endoplasmic reticulum Ca^{2+} ATPase (SERCA) from crayfish axial muscle. J Exp Biol 2000; 203:3411–3423.
22. Chen D, Zhang Z, Wheatly MG et al. Cloning and characterization of the heart muscle isoform of sarco/endoplasmic reticulum Ca^{2+} ATPase (SERCA) from crayfish. J Exp Biol 2002; 205:2677–2686.
23. Mandal PK, Mandal A, Ahearn GA. Physiological characterization of $^{45}Ca^{2+}$ and $^{65}Zn^{2+}$ transport by lobster hepatopancreatic endoplasmic reticulum. J Exp Zool 2005; 303A: 515–526.
24. Wheatly MG, Gao Y, Stiner LM, et al. Roles of NCX and PMCA in basolateral calcium export associated with mineralization cycles and cold acclimation in crayfish. NY Ann Sci 2007; 1099:190–192.
25. Creutz CE, Snyder SL, Kambouris NG. Calcium-dependent secretory vesicle-binding and lipid-binding proteins of *Saccharomyces cerevisiae*. Yeast 1991; 7(3):229–244.
26. Le Sourd F, Boulben S, Le Bouffant R, et al. eEF1B: at the dawn of the 21st century. Biochim Biophys Acta 2006; 1759:13–31.
27. Gillen CM, Gao Y, Niehaus-Sauter MM et al. Elongation factor 1Bγ (eEF1Bγ) expression during the molting cycle and cold acclimation in the crayfish *Procambarus clarkii*. Comp Biochem Physiol 2008; 150(2):170-176 (in press).
28. Zilli L, Schiavone R, Storelli C, et al. Analysis of calcium concentration fluctuations in hepatopancreatic R cells of *Marsupenaeus japonicus* during the molting cycle. Biol Bull 2007; 212:161–168.
29. El Haj AJ, Houlihan DF. *In vitro* and *in vivo* protein synthesis rates in a crustacean muscle during the moult cycle. J Exp Biol 1987; 127:413–426.
30. Whiteley NM, El Haj AJ. Regulation of muscle gene expression over the moult in crustacea. Comp Biochem Physiol 1997; 117B:323–331.
31. El Haj AJ. Regulation of muscle growth and sarcomeric protein gene expression over the intermolt cycle. Am Zool 1999; 39(3):570–579.
32. El Haj AJ, Whiteley NM, Harrison P. Molecular regulation of muscle growth over the crustacean moult cycle. In: El Haj AJ, ed. Molecular biology of muscle (SEB Seminar Series 46). Cambridge: Cambridge University Press, 1992; 151–165.
33. El Haj AJ, Govind CK, Houlihan DF. Growth of lobster leg muscle fibers over intermolt and molt. J Crust Biol 1984; 4:536–545.
34. El Haj AJ, Clarke SR, Harrison DF et al. *In vivo* muscle protein synthesis rates in the American lobster *Homarus americanus* during the moult cycle and in response to 20-hydroxyecdysone. J Exp Biol 1996; 199:579–585.
35. West JM. Ultrastructural and contractile activation properties of crustacean muscle fibres over the molt cycle. Comp Biochem Physiol 1997; 117B:333–345.
36. Cromarty SI, Cobb JS, Kass-Simon G. Behavioral analysis of the escape response in the juvenile lobster *Homarus americanus* over the molt cycle. J Exp Biol 1991; 158: 565–581.

37. Becker CD, Genoway, RD. Evaluation of the critical thermal maximum for determining thermal tolerance of freshwater fish. Environ Biol Fishes 1979; 4(3):245–256.
38. Firkins I, Holdich DM. Thermal studies with three species of freshwater crayfish. In: Holdich DM, Warner GF, eds. Freshwater crayfish. IX. Lafayette, LA: University of Southwestern Louisiana Press, 1993; 241–245.
39. Huner JW, Barr JE. Red swamp crayfish: biology and exploitation 3rd edition. Baton Rouge: Louisiana Sea Grant College Program, Center for Wetland Resources, Louisiana State University 1991.
40. Whiteley NM, Taylor EW, deSouza SCR, et al. Seasonal and moult related changes in haemolymph oxygen and acid-base levels in a wild population of *Austropotamobius pallipes* (Lereboullet). In: Holdich DM, Warner GF, eds. Freshwater crayfish. IX. Lafayette, LA: University of Southwestern Louisiana Press, 1993; 189–199.
41. Rutledge PR. Circulation and oxygen transport during activity in the crayfish,*Pacifastacus leniusculus*. Am J Physiol 1981; 240:99–105.
42. Whiteley NM, Taylor EW. The effects of seasonal variations in temperature on extracellular acid-base status in a wild population of the crayfish *Austropotamobius pallipes*. J Exp Biol 1993; 181:295–311.
43. Portner HO. Physiological basis of temperature-dependent biogeography: trade-offs in muscle design and performance in polar ectotherms. J Exp Biol 2002; 205:2217–2230.
44. Puhakainen T, Pihakaski-Maunsbach K, Widell S, et al. Cold acclimation enhances the activity of plasma membrane Ca^{2+} ATPase in winter rye leaves. Plant Physiol Biochem 1999; 37: 231–239.
45. Nelson T, McEachron D, Freedman W, et al. Cold acclimation increases gene transcription of two calcium transport molecules, calcium transporting ATPase and parvalbumin beta, in *Carassius auratus* lateral musculature. J Therm Biol 2003; 28:227–234.
46. Ushio H, Oshima T, Koizumi C, et al. Effect of dietary fatty acids on Ca^{2+} ATPase activity of the sarcoplasmic reticulum of rainbow trout skeletal muscle. Comp Biochem Physiol 1997; 118:681–691.
47. Shiels H, Vornanen M, Farrell AP. Temperature dependence of cardiac sarcoplasmic reticulum function in rainbow trout myocytes. J Exp Biol 2002; 205:3631–3639.
48. Plieth C, Hansen UP, Knight H, et al. Temperature sensing by plants: the primary characteristics of signal perception and calcium response. Plant J 1999; 18:491–497.
49. Chinnusamy V, Zhu JK. Molecular genetic analysis of cold-regulated gene transcription. Philos Trans R Soc Lond Biol Sci 2002; 357B:877–886.
50. Whiteley NM, Taylor EW, El Haj AJ. Seasonal and latitudinal adaptations to temperature in crustaceans. J Therm Biol 1997; 22:419–427.
51. Taugbol T, Skurdal J. The significance of a cold water (winter) period for molting, maturation and fecundity in wild-caught adult noble crayfish *Astacus astacus* in Norway. In: Romaire R, ed. Freshwater crayfish. VIII. Baton Rouge: Louisiana State University Press, 1995; 148–156.
52. Lowery, RS. Growth, moulting and reproduction. In: Holdich DM, Lowery RS, eds. Freshwater crayfish: biology, management and exploitation. London: Chapman and Hall, 1988; 83–113.

The Cellular Basis of Extreme Alkali Secretion in Insects: A Tale of Two Tissues

David F. Moffett and Horst Onken

Abstract Dipteran and lepidopteran insect larvae generate gut pH values that may be as high as 12; the highest values known in animal biology. This chapter reviews the cellular mechanisms hypothesized to be involved in such extreme alkalinization, and compares them to the mechanisms responsible for alkali secretion in two other well-studied tissues, the beta cells of the vertebrate renal tubule, a tissue that secretes alkali but does not achieve extreme pH values, and the dipteran Malpighian tubule, a tissue that does not exhibit significant alkalinization. The common feature in all of the tissues is a central role of the plasma membrane vacuolar-type H^+ ATPase (V-ATPase) in energizing secondary transport of mineral ions and organics. Whether alkali secretion, acid secretion, or no net acid–base secretion occurs appears not to be a direct result of the primary H^+ secretion, but instead relates to the magnitude and polarity of the "strong ion difference" resulting from transport processes secondary to the H^+ transport. The energy demands of extreme alkalinization are calculated to approach, or even exceed, the energetic limits of the V-ATPase, suggesting that additional energy sources remain to be identified. Although the V-ATPase energizes alkali secretion in both dipteran and lepidopteran larvae, the cellular arrangement of the transport systems is so different that extreme alkalinization must have evolved separately in the lineages of the two orders.

Keywords Mosquito · Lepidopteran · Insect · Midgut · Ion transport · Alkali secretion · Electrogenic proton transport · V-ATPase · Renal tubule · Malpighian tubule · Carbonic anhydrase · Anion exchanger · Sodium/hydrogen exchanger · Potassium/hydrogen exchanger · Chloride transport · Carbonate transport · Potassium transport · Transepithelial potential · Pump electromotive force

D.F. Moffett (✉)
School of Biological Sciences, Washington State University, Pullman, WA, USA 99164-4236
e-mail: dmoffett@wsu.edu

G.A. Gerencser (ed.), *Epithelial Transport Physiology*,
DOI 10.1007/978-1-60327-229-2_5,

1 Introduction

Within the animal kingdom insects represent a group that outnumbers all others in relation to species and individuals. Insects first appeared in the fossil record about 400 million years ago, and have since radiated to fill niches in every terrestrial and freshwater aquatic habitat. In the process, they have diversified physiologically to such an extent that current textbooks of insect physiology are, to a large extent, simply eclectic collections representing the small number of example species or systems that have been more or less well-studied. For several reasons, many biologists consider insects to be considerably different from other animals. Although it does not hold for all insects, one of these differences is that, unlike all other animals, certain larval insects evolved a midgut compartment with extremely alkaline pH values which is believed to dissociate tannin-protein complexes and result in a more efficient assimilation and utilization of dietary protein [1]. It should not surprise us that in their diversification, insects have evolved at least two cellular mechanisms for achieving the extreme alkalinization mentioned above. The cells involved are as different from each other as London and Paris in the Dickens novel, *A Tale of Two Cities,* parodied in our title. One cell type has been studied most thoroughly in the gut of the mosquito larva. The other type reflects a cellular adaptation that may well have evolved after the lepidopteran branch separated from the insect trunk, and thus may never be found in any other order of insects.

The purpose of this chapter is to provide a critical review of our current understanding of how the two main characters in our story can generate the highest pH values known in biology. However great the differences between these two cell types, in both of them the vacuolar-type proton ATPase ("V-ATPase") is the linchpin of the process, central both to its cellular organization and providing most, if perhaps not all of its energization. For comparative purposes it will be helpful to consider two other tissues that also express the plasma membrane V-ATPase but achieve, at best, only a moderate level of alkalinization: these are the vertebrate renal tubule and the insect Malpighian tubule. Since the plasma membrane V-ATPase is common to all of these examples, we must first address the nature of the V-ATPase and its possible relationship to the general physiology of acids and bases.

2 The V-ATPase and Alkali Secretion

In Stewart's [2] explanatory scheme of acid–base chemistry in physiological solutions (hereafter referred to as the "SS"), the dominance of alkali or acid, as reflected by the ratio of $[H^+]$ to $[OH^-]$, is the result of a "strong ion difference" ([SID]). The [SID] itself is not a novel concept, because it relates closely to the Gamblegram, a bar graph showing on one side the total strong cation concentration of an extracellular body fluid (usually approximated by the sum of $[Na^+]$ and $[K^+]$) and on the other side the total strong anion (usually approximated simply by the $[Cl^-]$). An alkaline solution is thus one in which the concentration of "strong" cation

(Na^+, K^+, etc.) exceeds the concentration of "strong" anion (Cl^-). The resulting "anion gap" is filled by weak anions, of which the main one in extracellular fluids is usually bicarbonate. A consequence of the positive [SID] is an excess of the "weak" anion OH^- relative to the "weak ion" H^+, a condition traditionally regarded as definitive for an alkaline solution. Conversely, if there is a change in the [H^+] of a solution, there must also have been a change in the [SID]. Thus, a corollary of the SS is that handling of strong ions by an epithelium is essential for affecting the alkalinity or acidity of extracellular compartments. By implication, we might expect that alkalinizing or acidifying epithelia will have to be provided with cellular mechanisms that carry out active transport of Na^+, K^+, and Cl^-, because transport of solely H^+, OH^-, or HCO_3^-, while potentially helpful, will not be sufficient. For example, the SS asserts that to acidify the vertebrate stomach, it would be sufficient merely for the gastric mucosa to actively secrete Cl^-, creating a negative [SID], in which strong anion dominates strong cation. The ions HCO_3^-, CO_3^{2-}, and NH_4^+, characterized by Stewart as "add-on" acids or bases, are important insofar as replacement of them with strong ions is a potential mechanism of changing the [SID].

This way of understanding the behavior of acids and bases has enjoyed acceptance in some quarters, but hardly seems to have penetrated into the mainstream canon of physiology. For example, a recent medical physiology text, apparently up to date in every way, presents a cartoon of cellular mechanisms of renal tubular acidification in which H^+ transport is implicitly equated with acidification, and with never a strong ion to be seen [3]. So, do we really need the SS to understand the physiology of acids and bases? The multiple roles of the V-ATPase in insect tissues provide an interesting test of its usefulness. Ultimately we can use the SS's insight to understand how different tissues might be able to use the V-ATPase to drive alkali secretion either by a primary secretion of H^+ or by a primary absorption of H^+, whereas to a conventional mind that equates absorption of H^+ with alkalinization and its secretion with acidification, this would seem a counterintuitive proposition. Also, we can understand how a primary secretion of H^+ can drive other transport pathways while resulting in neither acidification nor alkalization.

3 The Vertebrate Renal Beta Cell – A Reigning Paradigm for Moderate Alkalinization

The mammalian renal distal tubule and its analogue, the urinary bladder in amphibians and reptiles, are major sites of homeostasis of whole-body extracellular fixed acid and base. The cells involved have many common features that cut across taxonomic lines. To consider only one of the several well-studied systems, in the mammalian cortical distal tubule, secretion of fixed acid is carried out by alpha intercalated cells, whereas that of fixed base is by beta intercalated cells. Both the acid-secreting (alpha) and base-secreting (beta) cell subtypes are characterized by numerous mitochondria and abundant cytoplasmic carbonic anhydrase (CA), so, depending on the biochemical perspectives of individual investigators, they have

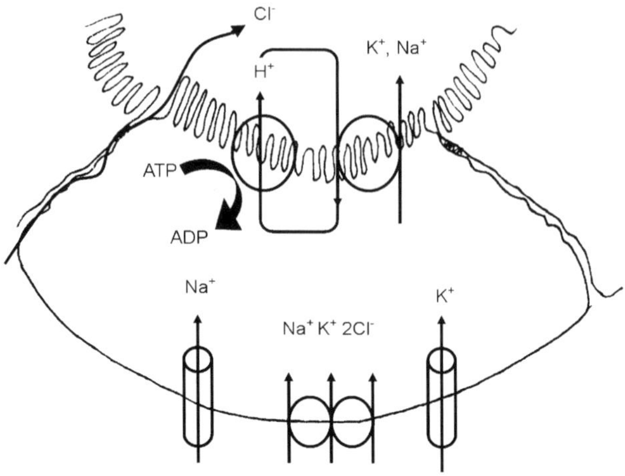

Fig. 1 Mammalian renal beta cell showing transport processes relevant to alkali secretion, including apical anion exchanger and basolateral V-ATPase with parallel Cl^- channels

been termed "mitochondria-rich cells" or "carbonic anhydrase-rich cells." The beta cell is characterized by abundant V-ATPase expressed on the basolateral membrane and by an apically located Cl^-/HCO_3^- exchanger (AE1 [4, 5]; see Fig. 1). Although beta cells sometimes have been termed "bicarbonate-secreting cells," in the conceptional framework of the SS the key process is the Cl^- absorption, which, in the absence of simultaneous absorption of a strong cation, would lead to a positive [SID] and result in alkalinization of the tubular fluid. The CAII expressed by both beta and alpha cells is a cytoplasmic form of the enzyme that is generally believed to have the role of rapidly supplying the V-ATPase and the AE1 with H^+ and HCO_3^- from metabolic CO_2.

4 The Mosquito Malpighian Tubule – Secreting H^+, but Not Secreting Either Acid or Alkali

The insect Malpighian tubule represents one form of secretory kidney, in which solute secretion drives the formation of primary urine. Depending on the species and the circumstances, insect primary urine is a KCl- and/or NaCl-rich solution that is isosmotic with plasma. Figure 2 shows the model configuration for a principal cell, the type believed to be responsible for primary ionic secretion in Malpighian tubules of mosquitoes and other dipterans. In these cells, an apical V-ATPase [6] is coupled to a cation exchanger. In the Malpighian tubules of the mosquito *Aedes*the cation exchanger has recently been shown to be NHE8, a relatively nonselective member of the epithelial Na^+/H^+ exchanger family [7]. The relative nonselectivity for alkali metals of this exchanger would allow the preference of the system for Na^+ versus K^+ to be controlled by basolateral entry mechanisms, including both

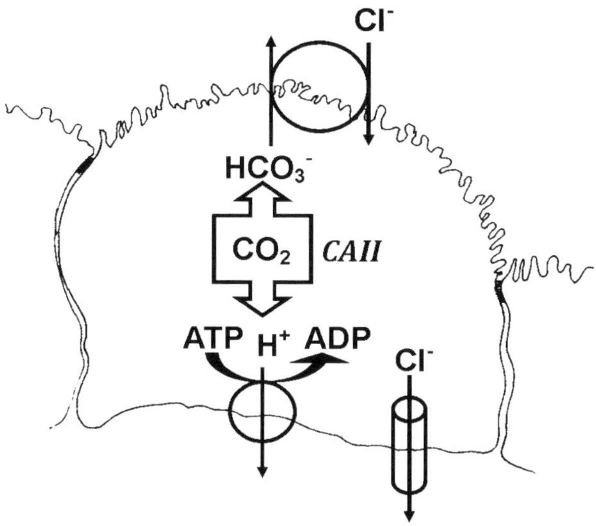

Fig. 2 Malpighian tubule principal cell showing transport processes believed to drive ion secretion. Basolateral processes, including the NK2C cotransporter and ion channels, deliver Na^+ and K^+ to the V-ATPase-coupled alkali metal exchanger in the apical membrane and control the relative rates of secretion of the two ions. Chloride transport is passive and is by routes that do not pass through the principal cell

ion channels and the NK2C cotransporter, and indeed, after a blood meal the adult mosquito has the ability to excrete first the immediate Na^+ load resulting from the ingestion of blood plasma, and later the K^+ load that results when red blood cells are lysed [8]. Interestingly, in the mosquito Malpighian tubule the exchanger apparently operates in the reverse mode, absorbing H^+ and secreting Na^+, whereas in vertebrate systems it is a mechanism for absorbing Na^+ that is destined to enter a transepithelial transport pathway [9]. This exchanger is a member of the subset of the SLC9 gene family that is characteristically expressed to intracellular organelles, rather than the plasma membrane [9]. It may not be just a coincidence that the "vacuolar-type" H^+-ATPase was also once so characterized.

Although a vigorous proton secretion is the primary active transport process in solute secretion in the Malpighian tubule, the pH of its primary secretion is generally not decidedly acidic; for example, in the adult mosquito a pH of 7.2 is typical for tubules at rest or in diuresis [10]. This outcome is not intuitive, but can be understood as the outcome of two features of the stimulated Malpighian tubule. First, at least in the Malpighian tubule of the adult mosquito, there appears to be a 1/1 exchange of H^+ for Na^+ or K^+ by the apical alkali metal ion exchanger, so the net effect is secretion of a strong cation. However, the dominant anion permeability under conditions of diuresis, is to Cl^- [8]. The net effect is that each strong cation secreted is paired with a strong anion, Cl^-, and thus a [SID] does not develop.

5 The Larval Mosquito Midgut: Secreting Alkali by Absorbing H^+

The anterior midgut of mosquito larvae (or "anterior stomach"; unfortunately the terms are used interchangeably) generates an alkaline lumen that may approach pH 12 [11]. A similar situation has been recently demonstrated in the posterior midgut of larval *Drosophila* [12]. Three main experimental approaches have been applied to the study of alkalinization in the mosquito larva. These are: the intact, free-swimming larva, where alkalinization may be estimated through color change of ingested indicator dye (cf. [11]) and inhibitors can be added to the medium with the expectation that they will be ingested; an in-situ preparation, in which the immobilized larva is opened to expose the gut [13], and a fully isolated, perfused preparation, first used by Clark et al. [14]. Each of these preparations has advantages and disadvantages. In the free-swimming and in situ approaches, experimental manipulation of the hemolymph composition and luminal contents and measurement of the tissue's electrical responses, are limited or impossible, and drugs applied to the whole animal may have consequences on gut transport that are secondary to their effects elsewhere in the body. In the isolated, perfused preparation, transepithelial potential [14, 15] and even short-circuit current [16] can be readily measured, and solution composition can be manipulated, but alkali secretion must be re-stimulated by addition of serotonin [14, 17], and the likely interaction of processes in the anterior midgut with other body structures (the peritrophic membrane, CNS, other parts of the gut, etc.) is of course eliminated.

In the mosquito anterior midgut and the *Drosophila* posterior midgut, conventional thinking finds it reassuringly sensible that, as in the vertebrate beta cell model, the V-ATPase is located on the basal border of the cells [12, 18], directing a net flux of H^+ away from the gut lumen [13] , and expects that this be accompanied by some apical process or processes that have the net effect of absorbing H^+ and Cl^- and/or of secreting Na^+ and $CO_3{}^{2-}$ in order to produce the extremely alkaline lumen. The SS simply demands that there be Cl^-absorption without corresponding absorption of strong cation and/or secretion of a strong cation without corresponding secretion of chloride. An apparent Cl^- flux from the cells to the hemolymph has been detected in the in situ preparation [13]. A pharmacological search for an apical anion exchanger or exchangers proved disappointing [17], even though several candidate anion exchangers are detectable in the *Aedes* and *Anopheles* genomes. We have found that in the perfused tissue, luminal alkalinization can be observed after perfusion stop if the lumen is provided with an unbuffered solution containing only NaCl and the indicator dye m-cresol purple ([17]; see also Fig. 3). According to the SS, this alkalinization must be accompanied by a positive [SID]. This can be accomplished by one, or some combination of, net Na^+ secretion, net Cl^- absorption, or secretion of some other "strong" cation (K^+ would be the only plausible possibility), but not, according to the SS, by unaccompanied transport of "weak" cation or anion. Therefore, a simple combination of the basal V-ATPase and some apical H^+ channels could not alkalinize the lumen, and indeed, experiments designed to be diagnostic for the presence of apical H^+ channels returned negative results [19]. The hypothesis that the apical membrane must contain one or both of a NHE-type cation/H^+ exchanger and a $Cl^-/HCO_3{}^-$ exchanger lies squarely in our path. These possibilities are especially appealing, both because they mediate an exchange of strong ion for weak ion, and because they could impart extra energy from a cation gradient or a Cl^- gradient to the task of alkali secretion. Unfortunately, both of these

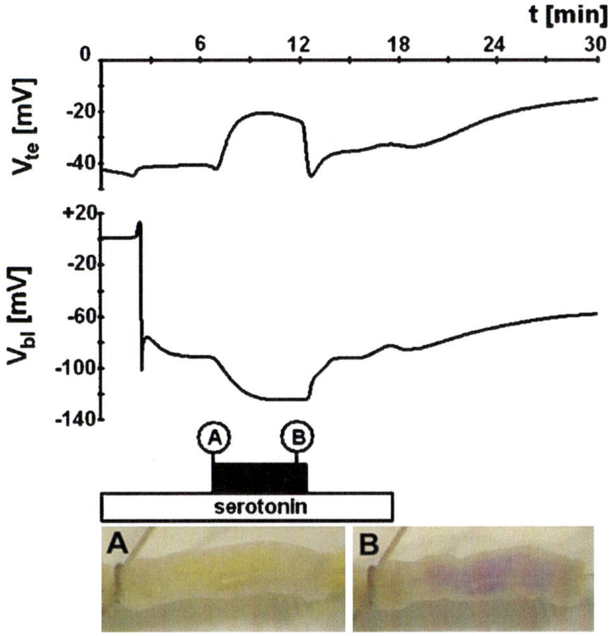

Fig. 3 Alkalinization and cell voltages in mosquito anterior midgut cells. Experiment with an isolated and perfused anterior midgut of larval (4th instar) *Aedes aegypti* perfused with 100 mM NaCl containing 0.04% m-cresol purple and bathed in mosquito saline containing 0.2 μM serotonin, measuring transepithelial (V_{te}) and transbasal voltage (V_{bl}). After stabilization of V_{te} at approximately –40 mV (lumen negative), an epithelial cell in the anterior portion of the preparation is impaled with a microelectrode (see *lower right* in photographs A and B). V_{bl} stabilized at –90 mV (cell negative). From these values the transapical voltage (V_{api}) can be calculated to –50 mV (cell negative). Approximately 5 min after successful impalement a photograph of the preparation is taken (A, showing yellow perfusate of neutral pH) and the perfusion pump is stopped for the time period indicated by the *black bar*. Within approximately 3 min V_{te} drops to approximately –20 mV, whereas V_{bl} hyperpolarizes to over –120 mV (V_{api} calculates under these conditions to –103 mV), and the m-cresol purple indicates an alkaline midgut lumen (photo B). Restarting the luminal perfusion results in a return of the voltages to their original values. Subsequent washout of hemolymph-side serotonin reduced the lumen negative V_{te} to –15 mV and the membrane voltages V_{bl} and V_{api} decreased to –57 and –42 mV, respectively

helpful possibilities have, so far, proven ephemeral. Neither luminal infusion of the anion exchange inhibitor DIDS, nor luminal amiloride, nor elevation of luminal [K^+] (a maneuver intended to abolish the transapical K^+ gradient and thus suppress the activity of an apical H^+/K^+ exchanger), affect alkalinization, and alkalinization was even observed in the absence of chloride on both sides of the tissue [17].

Two reports of cation/proton exchangers in ion-transporting tissues of mosquitoes have appeared recently. In the first case, an isoform of the NHE family of electroneutral sodium/hydrogen exchangers (NHE3) was found to be expressed in various ion-transporting tissues of *Aedes aegypti* larvae and adults [20]. This exchanger was immunolocated mainly in the basal membranes of all of the relevant larval tissues, including all midgut regions and the Malpighian tubules. This result is disappointing, but still potentially relevant, for even though it suggests that NHE3 probably plays mainly a housekeeping role in [H^+] homeostasis of the epithelial cells, such a role might be significant in balancing cellular homeostasis with secretory function. NHE3 is described as only partly sensitive to amiloride at high concentrations [20], so its presence probably cannot account for our finding of a substantial decrease in transepithelial potential and luminal alkalinization of the isolated anterior midgut with application of amiloride or Na^+-free saline to the hemolymphal side of the tissue [15]. However, the latter finding suggests that transcellular Na^+ secretion is involved in the overall process.

In the second case, Rheault et al. [21] reported cloning and localization of an electrophoretic sodium/proton exchanger (AgNHA1) from *Anopheles gambiae*. This gene product is expressed widely throughout the larval body, including the central nervous system and all parts of the gut. Expression levels were highest in the rectum and gastric caeca. Expression levels in the anterior midgut were modest, and in sectioned material it was clear that immunolocalization was almost entirely confined to intracellular vesicles, whereas in the posterior midgut significant immunostaining appeared on the apical membrane. This expression pattern would not seem consistent with a role for this exchanger in alkali secretion, but is consistent with a role in the reacidification of midgut contents that occurs in the posterior midgut. However, it could be that the anterior midgut tissue used for these studies reflected an unstimulated condition, and with serotonin stimulation, vesicle fusion might deliver the enzyme to a plasma membrane domain.

Since we have so far found no apical mechanism that could serve as an energetic booster for alkali secretion at the apical membrane, we must consider whether it is possible that the entire energetic burden of alkali secretion is borne by the basal V-ATPase. Is this mechanism strong enough to do the job? For the purposes of energetic analysis, the maximum power output of the V-ATPase can be measured by determining its pump electromotive force, or the maximum potential energy it can generate expressed in electrical terms, a measurement analogous to measuring the brake horsepower of an automobile engine. The analogy is not unreasonable since the V-ATPase is a molecular engine, with parts that rotate relative to one another. As the ATPase is faced with an increasingly unfavorable chemical and/or electrical gradient, electromechanical models suggest that slippage would occur in its mechanism, so that the rate of H^+ transport would fall and ultimately cease as the gradient approaches the pump EMF. Models of the V-ATPase suggest that this would occur at voltages in the range of 140–220 mV, depending on model assumptions [22, 23].

We can approach the question from a second point of view, by asking how much chemical potential energy could be delivered to the V-ATPase by ATP. If the V-ATPase transports $2H^+$/ATP [22], ATP hydrolysis would provide about 6 Kcal/mole H^+, using a generally accepted estimate of the energy yield of a mole of ATP under cytoplasmic conditions [24]. Faraday's Number indicates that a volt of electrical energy is the equivalent of slightly more than 23 Kcal/mole H^+. This would predict a theoretical maximum pump EMF of about 260 mV. In the physiological literature, a value of 240 mV has repeatedly been asserted for the EMF of the V-ATPase, but frequently without citation of actual measurements or calculations (cf. [25–27]). Since the ATPase cannot be expected to convert phosphorylation potential to electrochemical potential with 100% efficiency, this analysis suggests that the value of 240 mV may be somewhat of an overstatement. A current-voltage analysis estimated values of 150–190 mV for the pump acting in concert with an electrochemically coupled cation exchanger in lepidopteran midgut ([28]; see also below).

Now let us estimate the energy expressed by the transport system under an extreme, but not unrealistic scenario. Working in round numbers, if one accepts a hemolymph pH of 7 as approximately the normal value, achieving a luminal pH of 12, the maximum value reported for the mosquito, requires a pump EMF of at least 275 mV at 55 mV/decade of chemical gradient assuming a negligible contribution from the transepithelial potential (see Fig. 3). To move H^+ against an aggregate gradient of this size would require about 6.3 Kcal/mole, which is clearly at the upper range of theoretical maximums. Although there is no doubt that the V-ATPase is an important prime mover of protons in this system, it is necessary to look for additional cellular energy sources that can assist the V-ATPase in the maintenance of such large proton gradients. One candidate is the Na^+/K^+-ATPase which has been immunolocalized to the basolateral membrane of the very anterior portion of the anterior midgut [29]. Surprisingly, it was found in the apical membrane of the rest of the anterior midgut. Basolateral ouabain did indeed slightly reduce the transepithelial voltage [15], but luminal perfusion with the drug for up to 60 min did not affect voltage or alkalinization [17]. Although the latter findings seem to rule out that the Na^+/K^+-pump is a major player in the alkalinization game in larval mosquitoes, it is likely that it supports V-ATPase by maintaining gradients for Na^+ and K^+ across apical and basolateral membranes. These seem to be of importance for the transepithelial voltage and for alkalinization as the effects of hemolymph-side amiloride or Na^+-free medium suggest.

How are the transepithelial chemical and electrical gradients partitioned between apical and basolateral membranes? In recent studies [19] we measured the cytoplasmic pH of mosquito anterior midgut cells using the fluorescent indicator BCECF. Under control conditions, with buffered mosquito saline (pH 7.0) for both sides of the epithelium the cytoplasmic pH fell to the 6.8–7.2 range. An increase into the range of 7.4–7.8 occurred after the addition of serotonin which stimulates luminal alkalinization. If the luminal pH was then raised from 7 to 10, the cytoplasmic pH rose further, into the 8.0–8.5 range. Luminal pH values of 10 were recently recorded during experiments including a stop of the luminal perfusion [17] and under such conditions, the transapical potential rises into the range of 80–100 mV (see Fig. 3). Thus, the transapical H^+ gradient is at least reasonably close to equilibrium with the transapical electrical potential. The intracellular pH changes are surprising from the point of view of intracellular homeostasis, but such changes might have the beneficial effect of making free CO_3^{2-} available for an apical anion exchanger that could accept it.

In summary, we are so far unable to draw a complete model of strong alkalinization in the midgut of larval mosquitoes. However, some key components have been identified:

- V-ATPase has been localized in the basolateral membrane [18] and results of the in situ approach indicated that protons and chloride ions move from the cells to the hemolymph [13], suggesting a transcellular absorption of HCl.
- Luminal alkalinization depends on hemolymph-side Na^+ (Onken et al., 2008) and the transepithelial voltage decreases (instead of increasing) in the absence of hemolymph-side Na^+ [15].Together with the high intracellular pH values observed [19] this suggests transcellular secretion of Na_2CO_3.

Although the findings are not yet conclusive when it comes to individual transport proteins, they readily explain strong luminal alkalinization and are consistent with the demands of SS. Even more important, alkalinization can actually be measured with the isolated tissue which is very promising for future explorations of the involved transporters. For the tissue addressed in the next section of this chapter it is exactly the other way around: A model mechanism is well established, but any significant alkalinization immediately ceases as soon as the tissue is isolated.

6 The Larval Lepidopteran Midgut – Secreting Alkali by Secreting H^+?

The midgut of lepidopteran insect larvae (the caterpillar form of moths and butterflies) mediates a vigorous secretion of K^+ – up to $2\,\mu eq\,cm^{-2}\,min^{-1}$ under short-circuit conditions in vitro. The lumen-positive transepithelial potential may exceed 100 mV immediately after the tissue is isolated and may remain well above 50 mV for up to several hours [30]; similar values can be recorded by implanted electrodes in the intact animal [31]. The K^+ transport is fully rheogenic – that is, the net isotopic K^+ flux and the short circuit current closely approximate one another [32], with small differences attributable to much smaller transports of Ca^{2+}[33], Mg^{2+}[34], and Cl^- [35]. The studies of this tissue were pioneered by William Harvey and Karl Zerahn and their students and associates, starting in 1963 with a paper by William Harvey and Signe Nedergaard [36]. These studies established that this tissue radically departed from the model for solute transport originated shortly before that by Hans Ussing, in which transbasal active transport of Na^+ and K^+ by the Na^+/K^+ ATPase energizes both transepithelial Na^+transport and a variety of secondary transapical solute transport mechanisms, such as intestinal sugar and amino acid uptake. To summarize the differences, in the lepidopteran gut model, Na^+/K^+ ATPase is absent [37], the lumen-positive transepithelial potential appears to arise directly from net K^+transport, with the complete absence of the Na^+ ion being of no consequence [30], and the energizing force for this process resides at the apical surface of the tissue rather than at the basolateral face [38].

Further studies in the Harvey laboratory implicated the "goblet" cells [39], a cell type that accounts for roughly one-quarter of the total cell mass of the epithelium, as the agents of K^+ secretion. The goblet cells are characterized by an apical crypt or cavity that is filled with a gel-like matrix [40]. This cavity is both electrically and chemically isolated from the free luminal solution by a valve-like structure [41, 42]. Paradoxically, we (unpublished) have found goblet cells almost indistinguishable from those in the lepidopteran posterior midgut in the mosquito anterior midgut, but they are far too few to have any major role in the latter's transport activities.

The path to confirm that the goblet cells are the cell type responsible for net K^+ transport in the lepidopteran midgut was a rocky one. An indirect approach based on measurements of the K^+ transport pool size generated disparate results in two laboratories, summarized in [32]. An initial microelectrode approach apparently conflated penetrations that were probably not valid, as attributable to goblet cells [43]; the situation was ultimately clarified by iontophoretic marking of penetrated cells for microscopic identification [44, 45]. However, convincing evidence was ultimately provided by a combination of electrophysiological, biochemical, and histochemical approaches that implicated the V-ATPase as the driver of K^+ transport in this tissue and the goblet cell apical membrane (GCAM) as the driver's seat. These experiments are described below.

As a feature of insect epithelial physiology, or indeed, the plasma membrane of any animal cell, the V-ATPase made its first public appearance in a report by Wieczorek et al. [46] in which it was demonstrated that inside-out apical membrane vesicles from the GCAM of the posterior midgut of the tobacco hornworm *Manduca sexta* contained an electrogenic proton pump (ultimately found to be the V-ATPase). In the presence of ATP and the absence of external K^+, the vesicles could develop both an inside-positive electrical potential and an interior pH of 2–3 units lower than that of the external solution. This proton electrochemical gradient was dissipated by the addition of K^+, consistent with the presence of a H^+/K^+ antiporter [47], and the effect of K^+ was blockable by amiloride, suggesting the antiporter belongs to the NHE family of alkali metal exchangers. The voltage gradient was not detected when external Cl^- was elevated, consistent with the presence of a leak pathway for Cl^-.

These findings led to the hypothesis that K^+transport by the GCAM, as well as in Malpighian tubules (outlined above) and potentially in a number of other insect epithelia, is the outcome of a primary secretion of H^+ linked to a secondary exchange of H^+ for K^+ (the "Wieczorek–Harvey Hypothesis", Fig. 4A). A series of studies with double-barreled H^+ and K^+ intracellular electrodes [38] showed that the chemical activity of K^+ in the goblet cavity is well above its calculated equilibrium distribution relative to both the cytoplasm and the gut lumen, confirming active secretion of K^+ into the goblet cavity. Surprisingly, the goblet cavity $[H^+]$ was found to be slightly more alkaline than the cytoplasmic $[H^+]$. This fact demands that the H^+/K^+ exchanger be electrogenic, since only an electrogenic exchanger could draw on the V-ATPase's electrical driving force in the absence of an H^+ gradient. Furthermore, in the intact midgut, a nonelectrogenic exchange of K^+ for H^+ can be ruled out upon energetic grounds: measurements of the electrochemical gradients of

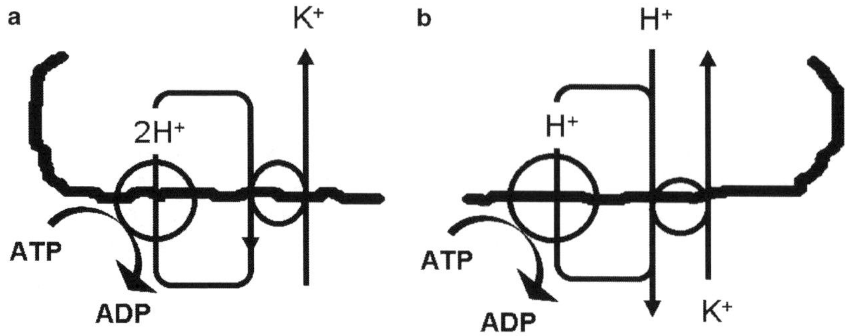

Fig. 4 Two hypotheses for alkali metal secretion, and alkalinization coupled to V-ATPase, in lepidopteran goblet cell apical membrane. **A**: The Wieczorek–Harvey Hypothesis. For each pump cycle of the V-ATPase, 1 ATP energizes secretion of 2 H^+ into the goblet cavity – these return to the cytoplasm, driven mainly by the transapical electrical potential, driving secretion of a K^+ ion. **B**: The Azuma–Harvey–Wieczorek Hypothesis. A transGCAM electrical potential is established by the V-ATPase, which drives both return of H^+ from the goblet cavity and uptake of H^+ from the gut lumen, in exchange for secreted K^+

H^+ and K^+ across the goblet cell apical membrane under short-circuit showed that at least $2H^+$ would be necessary to energize the movement of 1 K^+ [38]. Thus, the exchanger in this system cannot be the same NHE8 that is apparently operative in the mosquito Malpighian tubule as discussed above, because the latter exchanger is electroneutral. An electrogenic exchanger ($2H^+/K^+$) is now an established part of the hypothesis, as shown in Fig. 4B. Initial attempts to isolate the K^+/H^+ exchanger by affinity chromatograph using putative binding to concanavalin A, by immunochemical approaches, or by RT-PCR using degenerate primers derived from NHE1 were reported to have failed [47]. A genomic search for this exchanger could well be undertaken now, since in the interim the complete genome of the Oriental silkmoth *Bombyx mori* and substantial parts of the *M. sexta* genome have become available.

In a report cited in the previous section, Rheault et al. [21] cloned an electrophoretic NHA exchanger from the *Anopheles* genome. They conducted an extensive phylogenetic analysis that included primate, canine, rodent, nematode, and bacterial paralogues as well as ones from coleopteran, dipteran, and hymenopteran (but not lepidopteran) insects. They then suggested that this exchanger could be the long-sought $K^+/2H^+$ antiporter of the Wieczorek–Harvey Hypothesis. Clearly, ultimate confirmation of this suggestion awaits definitive studies in lepidopteran tissue.

Historically, interest in the extreme alkalinity of the gut lumen in lepidopteran larvae developed much later than that for the mechanism of K^+ transport. For those researchers now curious about alkali secretion by the lepidopteran midgut, it is an inconvenient fact that much of the earlier work on K^+ transport by this tissue utilized specifically the posterior midgut. Indeed, this segment of the gut was the source of the GCAM vesicles utilized by Wieczorek et al. [46] to discover the V-ATPase–cation exchanger couple. However, the posterior midgut is apparently not the part of the gut most responsible for alkalinization, and, as in the mosquito larva, is more likely to mediate alkali recovery, because the pH of luminal contents reaches a maximum as the contents reach the end of the anterior midgut, and the transtissue H^+gradient diminishes to an insignificant level as the contents pass along the posterior midgut [48].

Azuma et al. [49] have proposed a hypothesis (the "Azuma–Harvey–Wieczorek Hypothesis") under which K^+ secretion and gut alkalinization might involve a common mechanism. The simplicity of a single mechanism that accomplishes multiple ends is seductive, but the nature of the putative coupling between K^+ secretion and alkalinization, like most seductresses, has been ultimately elusive. In this hypothesis (Fig. 4B) we are asked to imagine that for each cycle of the mechanism, the V-ATPase secretes an H^+ and the antiporter exchanges this H^+, together with another H^+ taken from the gut lumen, for 1 K^+, an aspect of the hypothesis that has a Maxwell's-Demonic air to it. Note that, in the aggregate, the process is nonrheogenic – the current of H^+ exactly cancels the current of K^+ – so thus this cannot be the mechanism that also operates to transport K^+ in vitro. Perhaps, though, when the gut is removed from the animal the Demon takes leave, for in our hands, at least, gut alkalinization, as measured by pH change in a minimally buffered luminal saline, apparently ceases within minutes after the gut is mounted [50], whereas the large, lumen-positive transepithelial potential, short-circuit current, and net isotopic K^+ transport closely approximating the short-circuit current, continue for some hours after mounting [32, 50, 51]. These results rule out an obligatory exchange of an H^+ from the gut lumen for each K^+ secreted by the tissue, and thus would seem to obviate the hypothesis shown in Fig. 4B, at least for the conditions chosen for these experiments. It is of course possible that a tissue, given appropriate conditions of metabolic substrate and neurochemical stimulation, can turn on alkalinization by engaging a different coupling mechanism.

The Azuma–Harvey–Wieczorek Hypothesis for the lepidopteran anterior midgut is susceptible to an energetic analysis similar to that presented above for the mosquito system. The energy demands are also substantial in this system, because in vivo the midgut sustains simultaneously a $[H^+]$ gradient of up to five orders of magnitude [48] and a $[K^+]$ gradient of one order of magnitude [52]. Because the transepithelial potential of about 100 mV is lumen-positive, it reduces the energy needed to account for transepithelial transport of protons (or their equivalents) to about 175 mV, but increases the energy needed to transport K^+ to about 155 mV. Thus, the sum of the energy required to sustain both the proton and the K^+ gradients is about 330 mV. So in the lepidopteran midgut, as well as the mosquito midgut, we must conclude that an additional, unidentified energy source is required to support alkalinization to the levels reported.

In our hands, acid–base transport by both anterior and posterior midguts isolated in minimal saline has been disappointing, although we found that continued alkalinization could be supported by partly dissected, open-circuited, in situ preparations in which nervous and respiratory supplies to the gut were maintained [50]. However, in other hands, some alkali secretion is reported for the fully isolated, short-circuited gut, with the use of more complex saline-containing substrates [53], but even so it is far from approaching the magnitude of secretion estimated for the gut in vivo.

We will now turn to the conundrum posed by the fact that all parts of the gut secrete K^+ [54], but only the anterior segment also alkalinizes [48]. We are somewhat handicapped from the start, for the published data that we have for intracellular and intragoblet ion concentrations and membrane potentials largely represent the posterior midgut, and in any case are derived from tissues that were not secreting alkali at the time. Instead, we are compelled to address this question through a comparison of the cells of the two gut segments. Both segments contain goblet cells, but there are some morphological and histochemical differences between them [54]. The anterior midgut goblet cells have deep, basally enlarged cavities. Each microvillus projecting into the cavity contains a mitochondrion, inserted as a finger is into a glove. In contrast, the cavities of posterior midgut goblet cells are apically displaced, and their microvilli do not contain mitochondria. In the anterior midgut, carbonic anhydrase activity, as measured by Hanson's method, is abundant on the surface of the microvilli; in the posterior midgut, it is almost absent from goblet cells and instead, is expressed at the apical brush border of the columnar cells [55]. The V-ATPase is expressed at the apical membrane of goblet cells in both segments [56].

The close positioning of mitochondria to the GCAM in the anterior midgut is provocative. Such an arrangement has also been found in the principal cells of Malpighian tubules during diuresis [57], so at first one is inclined to see this juxtaposition of mitochondria and V-ATPase as an adaptation that provides ready turnover of ATP for the V-ATPase, and to discount the possibility that it might be an adaptation for alkali secretion. However, in the posterior midgut, the V-ATPase seems to operate perfectly well in microvilli with no inserted mitochondria [54]. Furthermore, in Malpighian tubules the bulk of carbonic anhydrase activity appears to be cytoplasmic [58]. It may be that in the anterior midgut the placement of the mitochondrion insures that most of the CO_2 it produces will immediately encounter a screen of membrane-bound CA, ensuring that it rapidly equilibrates with its hydration products. If the hydration reaction mostly takes place on the cavity side of the GCAM, protons released by the reaction would be available for reabsorption through the H^+/K^+ exchanger. If the exchanger were sufficiently avid for protons and the transapical voltage were sufficiently great, it could, in concept, raise the goblet cavity pH high enough to strip protons from HCO_3^-, resulting in an ultimate secretion of K_2CO_3.

In the posterior midgut, K^+-coupled absorption of amino acids is driven by the transmembrane K^+ electrochemical gradient. Expression of the K^+-coupled amino acid transporters is apparently restricted to columnar cells [26], giving them a potential role in a complete cycle of K^+ movement back and forth between hemolymph and gut lumen. As K^+ and amino acids are absorbed across the columnar cell apical membrane (CCAM), a local high concentration of CO_3^{2-}(and/or HCO_3^-, depending on the position along the length of the posterior midgut segment) would occur in the unstirred layer at the membrane surface. Logically, the CA located on this surface can have but one purpose: to catalyze conversion of luminal CO_3^{2-}/HCO_3^- to CO_2, which could be recovered by diffusion as it is through proximal tubular urinary acidification in the mammalian kidney. If we did not know apical H^+ transport was coupled to K^+ secretion in the posterior midgut as in the anterior midgut, we might be justified in feeling that we understood posterior midgut acidification and recovery of secreted alkali fairly well.

This set of hypotheses predicts a very alkaline goblet cavity for anterior midgut engaged in alkali secretion, and a more acidic one for the posterior midgut. It also demands a fairly alkaline cytoplasm for anterior midgut goblet cells actually engaged in alkalinization. It does not address the issue of why there is a goblet cavity, or how solutes enter and leave the goblet cavity from the gut lumen.

7 Some Persistent Questions Common to Both Tissues

7.1 How Are Anions Managed?

An important message of the SS is that management of anion movements is critical to attaining alkali or acid secretion, even when the anions are not, themselves, subject to primary active transport. This point is emphasized by the fact that in the lepidopteran midgut, alkali metal secretion does not equate to alkalinization; the anterior midgut alkalinizes whereas the posterior midgut does not, although both secrete K^+. From a black-box point of view, both the mosquito midgut and the lepidopteran midgut must secrete either a carbonate salt or other strong base in order to achieve the extreme pH values measured. Secretion of a bicarbonate salt would give maximum values only in the 8–9 range, as is the case with the mammalian kidney, which can achieve a urinary pH no higher than about 8.2. Although it is certainly conceptually possible for a HCO_3^- secretion coupled to an avid H^+ absorption to achieve a net secretion of carbonate, it would seem convenient to use instead a transporter that could accept CO_3^{2-}. The intracellular pH values observed in larval mosquitoes at a luminal pH of only 10 were already in a range in which availability of free CO_3^{2-} becomes significant (see above).

As a phenomenon, carbonate transport has been little studied, but is thought to be a function of some members of the SLC4 family of anion transporters [59]. The tissues described here are ideal places to look for them. At the time of this writing, transporters of the SLC4 family have not been positively identified in either of the example tissues, although at least two members of the family are apparently present in the *Aedes* genome. Luminal alkalinization in the isolated mosquito anterior midgut is insensitive to DIDS [17], a fairly wide-spectrum inhibitor of SLC4 transporters, indicating that if such a transporter is present, it must be resistant to DIDS. The failure of luminal chloride substitution to affect alkalinization observed in the midgut of larval mosquitoes [17] seems to rule out the hypothesis that apical anion exchange is indispensible for alkalinization in this tissue. On one hand, this may result in a helpful limitation of the candidates from the SLC4 family to those that are Na^+/HCO_3^- (or CO_3^{2-}) symporters. On the other hand this finding and our interpretation are inconsistent with the observation that chloride ions exit the cells across the basolateral membrane [60]. We are left with the question of how these chloride ions enter the cells.

In the lepidopteran midgut handling of anions is no better understood than in the mosquito. In our studies of Cl^- transport by *M. sexta* posterior midgut [35], the I_{sc} was not significantly affected by bilateral substitution of gluconate or isethionate for Cl^-. These results indicate that K^+ transport does not have a Cl^--dependent component, and that Cl^- transport itself is electrically silent. However, quite different results were reported for similar experiments by another group [61], which showed approximately 4X inhibition of the maximal K^+ transport rate with substitution of gluconate for chloride. The apparent stimulatory effect of Cl^- was specific to the hemolymphal (or K^+ uptake) side of the tissue. Further studies [62] using the noise analysis technique suggested that the Cl^- effect was due to an interaction of Cl^- with barium-inhibitable K^+ channels in the basolateral membrane Possible differences in experimental technique that might account for the conflicting results do not come readily to mind. First, in all of these studies, the validity of I_{sc} as a measure of net K^+ transport was not confirmed; this becomes a more important issue in the case that ion substitutions affect the I_{sc} than if they do not. Tissue oxygenation is achieved differently in different laboratories – in our laboratory bathing solutions are oxygenated by a vigorous current of 100% O_2, whereas in some laboratories the solutions are bubbled with compressed air. In our experience, oxygenation with air results in a substantially lower I_{sc}, a difference that presumably results from some level of cellular hypoxia. Bubbling with air might also be expected to increase tissue retention of metabolic CO_2. Both of these effects could decrease intracellular pH and possibly expose a latent Cl^-/HCO_3^- exchanger. Also, hypoxia has been shown to favor an electrically silent, active basolateral K^+ uptake mechanism [63], which could interact with Cl^-.

7.2 What is the Role of Carbonic Anhydrase?

The enzyme carbonic anhydrase (CA) is almost universally found in situations in which acid or alkali secretion occur [64], and its presence seems almost diagnostic for such processes. Depending on the particular process and its cellular location, carbonic anhydrase may do one of the following:

1. If located in the cellular interior, rapidly hydrate metabolic CO_2, providing a ready supply of H^+ and HCO_3^- for primary and secondary membrane transport processes for these ions
2. If expressed to the cell surface, catalyze the conversion of extracellular H^+ and HCO_3^- to CO_2, which may then move between physiological compartments as a highly diffusible gas

In the anterior midgut of lepidopterans, the abundant carbonic anhydrase activity is largely confined to the GCAM, in close apposition to both the mitochondria that occupy the microvilli, and to the apical V-ATPase, whereas in the posterior midgut it is associated mainly with the apical membrane of the columnar cells [55]. The histochemical assay does not tell us which face of the anterior midgut GCAM sees the active site of the enzyme. If the CA is active versus its substrates in the cytoplasm, the enzyme could readily fulfill role 1 above, minimizing the extent to which metabolic CO_2 escapes into the goblet cavity as a diffusing gas, and supplying H^+ for the V-ATPase and HCO_3^- for a putative Cl^-/HCO_3^- exchanger or a putative HCO_3^- or CO_3^{2-} transporter. Early studies with sulfonamide inhibitors of CA in the silkmoth (*Hyalophora cecropia*) larval midgut supported a role for CA in K^+ transport [65].

In the case of the mosquito gut, pharmacological evidence collected from both the freely swimming model and the in situ model supported an important role of CA in luminal alkalinization [60, 66, 67], and this evidence was adduced to support a scheme of alkalinization based on Cl^-/HCO_3^- exchange. In evaluating this evidence, it is important to note that inhibition of CA would potentially affect any transporter that carried HCO_3^- or H^+, not just anion exchangers. Furthermore, the strength of the conclusion is threatened by two pieces of contradictory evidence. First, alkalinization by the isolated, perfused gut is insensitive to the inhibitors of CA that are effective in the freely swimming and in situ approaches [17], and second, CA activity has been found to be considerably higher in the caeca and posterior midgut than in the anterior midgut, where levels in the epithelium itself are so low as to be almost undetectable by biochemical means [68]. It would be possible to make too much of the differences between the intact animal and isolated tissues, since a negative result in the isolated preparation might only mean that the hydration reactions of CO_2 are not limiting for the rate of alkalinization achieved by the isolated tissue. However, with the evidence in hand presently it is difficult to assert a common thread between a possible role of carbonic anhydrase in alkali secretion in the lepidopteran gut and its putative role in the mosquito anterior midgut epithelial cells.

Several investigators [69] have proposed that SLC4 proteins may be physically linked to CAII through a putative binding motif in the cytoplasmic domain of SLC4. This arrangement could form a metabolon that would facilitate the transport of HCO_3^- and/or CO_3^{2-}. This hypothesis has been disputed by Piermarini et al. [69], partly on the basis that there is at least one example of effective transport of HCO_3^- / CO_3^{2-} by an exchanger not linked to CAII. The alkali-transporting insect tissues offer multiple models to test the role of membrane-bound CA in anion transport. The lepidopteran anterior midgut GCAM and the lepidopteran posterior midgut CCAM are both very likely sites of anion exchange, and in both cases the membranes may be extracted in vesicle form, so they are amenable to reductive experimentation. The mosquito anterior midgut apparently accomplishes the same goal as the lepidopteran anterior midgut, but without membrane-bound CA. The opportunities for comparative studies are obvious.

8 Conclusions

The deuterostome and protostome branches of the animal kingdom that led to the vertebrates and insects, respectively, diverged from one another at least 0.5 billion years ago, before the basic forms of either the vertebrate kidney or the insect gut had appeared. Thus, it is probable that the similar roles of the V-ATPase in the two tissues are the result of convergent evolution and not an evolutionarily conservative process. It is becoming clear that achieving truly high pH values the insect systems are capable of requires some revolutionary adaptations in the overall makeup of the cells involved. These are not necessarily visible in electron micrographs; for example, far-reaching adaptations of every cytoplasmic enzyme must be required to yield cellular machinery that could routinely function at pH values higher than 8.

The studies of the lepidopteran midgut by several laboratories over several decades opened a window that led to understanding of a critical alkali metal transport mechanism also used by other important insect tissues. This window did indeed provide an early look into what we now know as a world of transport systems energized by the V-ATPase in animal cells. To borrow from the famous opening line of *A Tale of Two Cities*, were these the best of times or the worst of times? To the good, the V-ATPase can now stand beside the Na^+/K^+ ATPase as a ubiquitous prime mover of transepithelial solute and water transport. To the bad, too many questions about the cellular basis of K^+ transport in the lepidopteran system remain unanswered or incompletely answered. For example, in our current hypothesis about how goblet cells work, K^+ and H^+ are assumed to freely traverse the goblet valve, even though a surrogate for them, tetramethylammonium, apparently does not [42], while Cl^- apparently can [35]. Even worse, though, almost every cellular aspect of alkali transport by this tissue remains not only unanswered, but barely addressed experimentally. This deficit of attention may even be a sign of the success of this tissue model. Knowledge gained from it was quickly translated into a greater understanding of other systems. Perhaps, with the prospect to do this in view, it was all too easy to conclude, in the words of an anonymous NSF review panel member, that with the apparent experimental validation of the Wieczorek–Harvey Hypothesis, "all of the important questions about this system have already been settled." The results obtained with the larval midgut of *A. aegypti* in the last decade underline our lack of knowledge about the mechanisms of strong alkalinization in insect midguts and their possible diversity. With regard to the extraordinary diversification of insects mentioned in the beginning of this article, future generations of physiologists may remain pretty busy studying the subject of insect midgut alkalinization.

For insect control measures, the potential payoff of pursuing these studies is not trivial. The gut is arguably the most vulnerable interface between the animal and the environment. It is the extreme alkalinity of the gut that protects the larval insect from many potential pathogens [70, 71], and, paradoxically, what makes it susceptible to the effects of *Bacillis thuringiensis* endotoxin [72], one of the most successful and environmentally benign biological insecticides. A chemical agent that interfered with gut alkalinization could be effective simply by rendering the larvae susceptible to opportunistic viral or bacteriological infection. Reasonable targets for such agents are numerous: the operation of the V-ATPase itself, the process that assembles the V-ATPase at the apical or basolateral membrane, a mechanism that might, in the lepidopteran system, couple and uncouple K^+ secretion to alkalinization, the as-yet-uncharacterized ion exchangers, and the neural and endocrine control systems that sustain alkalinization and coordinate it with other functional and homeostatic body systems. With relatively complete genomic information for *Aedes* and *Anopheles* mosquitoes and for at least one lepidopteran, *B. mori*, the arsenal of serotonin pharmacology provided by the pharmaceutical industry, and the tools of rational drug design, ultimate success in this endeavor seems to require only the will to pursue it.

Acknowledgments The work of the authors cited herein was supported by the NSF (DCB 8315739, DCB 8811354, and IBN 0091208) and the NIH (GM 26287 and AI 063463).

References

1. Berenbaum M. Adaptive significance of midgut pH in larval Lepidoptera. Am Nat 1980;115:138–146.
2. Stewart PA. How To Understand Acid-Base – A Quantitative Acid-Base Primer for Biology and Medicine. New York, USA; Elsevier 1981.
3. Boron WF, Boulpaep, EL. Medical Physiology – A cellular and molecular approach. Philadelphia, U.S.A. Saunders 2003: Figure 38-2.
4. Stetson DL, Steinmetz PR. Alpha and beta types of carbonic anhydrase-rich cells in turtle bladder. Am J Physiol 1985; 249: F553–F565.
5. Brown D, Sabolic I, Gluck S. Polarized targeting of V-ATPase in kidney epithelial cells. J Exp Biol 1992;172: 231–243.
6. Weng X-H, Huss M, Wieczorek H, Beyenbach KW. The V-type ATPase in Malpighian tubules of *Aedes aegypti*: localization and activity. J Exp Biol 2003; 206: 2211–2219.
7. Kang'ethe W, Aimanova KG, Pullikuth AK, Gill SS. NE8 mediates amiloride-sensitive Na^+/H^+ exchange across mosquito Malpighian tubules and catalyzes Na^+ and K^+ transport in reconstituted proteoliposomes. Am J Physiol Renal Physiol 2007; 292: F1501–F1512.
8. Beyenbach, K. Transport mechanisms of diuresis in Malpighian tubules of insects. J Exp Biol 2003; 206: 3845–3856.
9. Orlowski J, Grinstein S. Diversity of the mammalian sodium/proton exchanger SLC9 gene family. Pflugers Arch 2004; 447: 549–565.
10. Petzel DH, Pirotte PT, Van Kerkhove E. (1999) Intracellular and luminal pH measurements of Malpighian tubules of the mosquito *Aedes aegypti*: the effects of cAMP. J Insect Physiol 1999; 45: 937–982.
11. Dadd RH. Alkalinity within the midgut of mosquito larvae with alkaline-active digestive enzymes. J Insect Physiol 1975; 21: 1847–1853.
12. Shanbhag S, Tripathi S. Electrogenic H+ transport and pH gradients generated by a V-H+-ATPase in the isolated perfused larval *Drosophila* midgut. J Membrane Biol 2005; 206: 61–72.
13. Boudko DY, Moroz LL, Linser PJ, Trimarchi JR, Smith PJS, Harvey WR. In situ analysis of pH gradients in mosquito larvae using non-invasive, self-referencing pH-sensitive microelectrodes. J Exp Biol 2001; 204: 691–699.
14. Clark TM, Koch A, Moffett DF. The anterior and posterior stomach regions of larval Aedes aegypti midgut: regional specialization of ion transport and stimulation by 5-hydroxytryptamine. J Exp Biol 1999; 202: 247–252.
15. Onken H, Moffett SB, Moffett DF. The transepithelial voltage of the isolated anterior stomach of larval mosquitoes (*Aedes aegypti*): pharmacological characterization of the serotonin-stimulated cells. J Exp Biol 2004; 207: 1779–1787.

16. Onken H, Moffett DF, Moffett SB. The isolated anterior stomach of larval mosquitoes (*Aedes aegypti*): voltage clamp measurements with a tubular epithelium. Comp Biochem Physiol Part A 2006; 143: 24–34.
17. Onken H, Moffett SB, Moffett DF. Alkalinization in the isolated and perfused anterior stomach of larval *Aedes aegypti*. J. Insect Sci 2008; 8:46–66.
18. Zhuang Z, Linser PJ, Harvey WR. Antibody to H^+ V-ATPase subunit E colocalizes with portasomes in alkaline larval midgut of a freshwater mosquito (*Aedes aegypti* L.) J Exp Biol 1999; 202: 2449–2460.
19. Onken H, Parks SK, Goss GG, Moffett DF, unpublished.
20. Pullikuth AK, Aimanova K, Kang'ethe W, Sanders HR, Gill SS. Molecular characterization of sodium/proton exchanger 3 (NHE3) from the yellow fever vector, *Aedes aegypti*. J. Exp. Biol. 2006; 209: 3529–3544.
21. Rheault MR, Okech BA, Keen SB, Miller MM, Meleshkevitch EA, Linser PJ, Boudko DY, Harvey WR. Molecular cloning, phylogeny and localization of AgNHA1: the first Na^+/H^+ antiporter (NHA) from a metazoan, *Anopheles gambiae*. J Exp Biol 2007; 210: 3848–3861.
22. Grabe M, Wang H, Oster G. The mechanochemistry of V-ATPase proton pumps. Biophys J 2000; 78: 2798–2813.
23. Luo C, Clark JW Jr, Heming TA, Bidani A. A simplified model for V-ATPase H^+ extrusion. IEEE Trans Nanobiosci 2004; 3: 257–264.
24. Voet D, Voet JG. Biochemistry. 2nd ed. New York, USA: Wiley, 1995; 430.
25. Martin FG, Harvey WR. Ionic circuits of insect midgut vesicles. J Exp Biol 1994; 196: 77–92.
26. Castagna M, Shayakul C, Trotti D, Franca Sacchi, V, Harvey WR, Hediger MA. Cloning and characterization of a potassium-coupled amino acid transporter. PNAS USA 1998; 95: 5395–5400.
27. Beyenbach KW. Energizing epithelial transport with the vacuolar H^+-ATPase. News in Physiol Sci 2001; 16: 145–151.
28. Moffett DF. Voltage-current relation and K^+ transport in tobacco hornworm (*Manduca sexta*) midgut. J Membrane Biol 1980; 54: 213–219.
29. Patrick ML, Aimanova K, Saunders HR, Gill SS. P-Type Na^+/K^+-ATPase and V-Type H^+-ATPase expression patterns in the osmoregulatory organs of larval and adult mosquito *Aedes aegypti*. J Exp Biol 2006; 209: 4638–4651.
30. Harvey WR, Haskell JA, Nedergaard S. Active transport by the Cecropia midgut III. Midgut potential generated directly by active transport. J Exp Biol 1968; 48:1–12.
31. Moffett DF, Cummings SA. Transepithelial potential and alkalinization in an in situ preparation of tobacco hornworm (*Manduca sexta*) midgut. J. Exp. Biol. 1994; 194: 341–345.
32. Blankemeyer JT, Harvey WR. Insect midgut as a model epithelium In: Jungreis AM, Hodges TK, Kleinzeller A, Schultz SG, eds. Water Relations in Membrane Transport in Plants and Animals, New York, USA: Academic Press, 1977.
33. Wood JL, Harvey WR. Active transport of calcium across the isolated midgut of *Hyalophora cecropia*. J Exp Biol 1976; 65: 313–320.
34. Wood JL, Jungreis AM, Harvey WR. Active transport of magnesium across the isolated midgut of *Hyalophora cecropia*. J Exp Biol 1975; 63: 347–360.
35. Chao AC, Koch AR, Moffett DF. Active chloride transport in isolated posterior midgut of tobacco hornworm (*Manduca sexta*). Am J Physiol 1989; 257: R752–R761.
36. Harvey WR, Nedergaard S. Sodium-independent active transport of potassium in the isolated midgut of the Cecropia silkworm. PNAS USA 1964; 51: 757–765.
37. Jungreis AM, Vaughn GL. Insensitivity of lepidopteran tissues to ouabain: absence of ouabain binding of Na^+/K^+ATPases in larval and adult midgut. J Insect Physiol 1977; 23: 503–509.
38. Moffett DF, Koch A. Driving forces and pathways for H^+ and K^+ transport in insect midgut goblet cells. J Exp Biol 1992; 172: 403–415.
39. Anderson E, Harvey WR. Active transport by the Cecropia midgut. II. Fine structure of the midgut epithelium. J Cell Biol 1966; 31:107–134.

40. Schultz TW, Lozano G, Cajina-Quezada M. Histochemical analysis of the goblet cell matrix in the larval midgut of *Manduca sexta*. Trans Am Micros Soc 1981; 100: 204–209.
41. Dow JAT, Peacock JM. Microelectrode evidence for the electrical isolation of goblet cell cavities in *Manduca sexta* middle midgut. J Exp Biol 1989; 143:101–114.
42. Moffett DF, Koch A, Woods R. Electrophysiology of K^+ transport by midgut epithelium of lepidopteran larvae III. Goblet valve patency. J Exp Biol 1995; 198: 2103–2113.
43. Blankemeyer JT, Harvey WR. Identification of active cell in potassium transporting epithelium. J Exp Biol 1978; 77: 1–13.
44. Moffett DF, Hudson RL, Moffett SB, Ridgway RL. (1982) Intracellular K^+ activities and cell membrane potentials in a K^+-transporting epithelium, the midgut of tobacco hornworm (*Manduca sexta*). J Membrane Biol 1982; 70: 59–68.
45. Thomas MV, May TE. Active potassium transport across the caterpillar midgut II. Intracellular microelectrode studies. J Exp Biol 1984; 108; 293–304.
46. Wieczorek H, Putzenlechner M, Zeiske W, Klein U. A vacuolar-type proton pump energizes H^+/K^+ antiport in an animal plasma membrane. J Biol Chem 1991; 266: 15340–15347.
47. Lepier A, Azuma M, Harvey WR, Wieczorek H. K^+/H^+ antiport in the tobacco hornworm midgut: the K^+-transporting component of the K^+ pump. J Exp Biol 1994; 196: 361–373.
48. Dow JAT. Extremely high pH in biological systems: a model for carbonate transport. Am J Physiol 1984; 246; R633–R635.
49. Azuma M, Harvey WR, Wieczorek H. Stoichiometry of K^+/H^+ antiport helps to explain extracellular pH 11 in a model epithelium. FEBS Lett 1995; 361: 153–156.
50. Clark TM, Koch A, Moffett DF. Alkalinization by *Manduca sexta*anterior midgut *in vitro*: requirements and characteristics. Comp Biochem Physiol A 1998; 121: 181–187.
51. Harvey WR, Zerahn K. Active transport of potassium and other alkali metals by the isolated midgut of the silkworm. In: Bronner F, Kleinzeller A, eds. Current Topics in Membranes and Transport, Vol. III New York USA, Academic Press 1972.
52. Harvey WR, Wood JL, Quatrale, RP, Jungreis AM. Cation distributions across the larval and pupal midgut of the lepidopteran, *Hyalophora cecropia, in vivo*. J Exp Biol 1975; 63: 321–330.
53. Chamberlin ME. Luminal alkalinization in the isolated midgut of the tobacco hornworm (*Manduca sexta*). J Exp Biol 1990; 150: 467–71.
54. Cioffi M, Harvey WR. Comparison of potassium transport in three structurally distinct regions of the insect midgut. J Exp Biol 1981; 91: 103–116.
55. Ridgway RL, Moffett DF. Regional differences in the histochemical localization of carbonic anhydrase in the midgut of tobacco hornworm (*Manduca sexta*). J Exp Zool 1986; 237: 407–412.
56. Klein U, Loffelmann G, Wieczorek H. The midgut as a model system for insect K^+ transporting epithelia – Immunocytochemical localization of a vacuolar-type H^+ pump. J Exp Biol 1991; 161: 61–75.
57. Bradley TJ, Stuart AM, Satir P. The ultrastructure of the larval Malpighian tubules of a saline-water mosquito. Tissue & Cell 1982; 14: 759–773.
58. Wessing A, Zierold K, Bertram G. Carbonic anhydrase supports electrolyte transport in Drosophila Malpighian tubules. Evidence by X-ray microanalysis of cryosections. J Insect Physiol 1997; 43: 17–28.
59. Romero MF, Fulton CM, Boron WF. The SLC4 family of HCO_3^- transporters. Pflugers Arch 2004; 447: 495–509.
60. Boudko DY, Moroz LL, Harvey WR, Linser, PJ. Alkalinization by chloride-bicarbonate pathway in larval mosquito midgut. PNAS USA 2001; 98: 15354–15359.
61. Zeiske W, Schrőder H, Alpert G. K^+ current stimulation by Cl^- in the midgut epithelium of tobacco hornworm (*Manduca sexta*) I. kinetics and effect of Cl^- -site-specific elements. J Comp Physiol B, 1992; 162: 331–339.
62. Zeiske W, Marin H. K^+ current stimulation by Cl^- in the midgut epithelium of the tobacco hornworm (*Manduca sexta*). J Comp Physiol B 1992; 162: 340–344.

63. Chao AC, Koch AR, Moffett DF. Basal membrane uptake in potassium-secreting cells of midgut of tobacco hornworm (*Manduca sexta*). Am J Physiol 1990; 258: R112–R119.
64. Maren TH. Carbonic anhydrase: chemistry, physiology and inhibition. Physiol Rev 1967; 47: 595–781.
65. Haskell JA, Clemons RD, Harvey WR. Active transport by the Cecropia midgut I. inhibitors, stimulants and K^+ transport. J Cell Comp Physiol 1965; 98: 45–56.
66. Del Pilar Corena M, Seron TJ, Lehman HK, Ochrietor JD, Kohn A, Tu C, Linser PJ. Carbonic anhydrase in the midgut of larval *Aedes aegypti*: cloning, localization and inhibition. J Exp Biol 2002; 205: 591–602.
67. Del Pilar Corena M, Fiedler MM, Van Ekeris L, Tu C, Silverman DN, Linser PJ. Alkalinization of larval mosquito midgut and the role of carbonic anhydrase in different species of mosquitoes. Comp Biochem Physiol Part A; 2004; 207–225.
68. Seron TJ, Hill J, Linser PJ. A GPI-linked carbonic anhydrase expressed in the larval mosquito midgut. J Exp Biol 2004; 207: 4559–4572.
69. Piermarini PM, Kim EY, Boron WF. Evidence against a direct interaction between intracellular carbonic anhydrase II and pure C-terminal domains of SLC4 bicarbonate transporters. J Biol Chem 2007; 284: 1409–1421.
70. Mulligan F, Schaefer C, Wilder W. Efficacy and persistence of *Bacillis sphaericus* and *Bacillis thuringiensis israelensis* against mosquitoes under laboratory and field conditions. J Econ Entomol 1980; 73(5): 688–694.
71. Stiles B, Paschke JD. Midgut pH in different instars of three *Aedes* mosquito species and the relation between pH and susceptibility of larvae to a nuclear polyhedrosis virus. J Invert Pathol 1980; 35: 58–64.
72. Petrantonio PV, Gill SS. *Bacillus thuringiensis* endotoxins: action on the insect midgut. In: Lehane MJ, Billingsley PF, eds. Biology of the Insect Midgut. London, England: Chapman and Hall, 1996; 345–372.

H^+, Na^+, K^+, and Amino Acid Transport in Caterpillar and Larval Mosquito Alimentary Canal

William R. Harvey and Bernard A. Okech

Abstract Two principal strategies are used to energize membranes in living organisms, a Na^+ strategy and a voltage strategy. In the Na^+ strategy a primary Na^+/K^+ ATPase imposes both Na^+ and K^+ concentration gradients across cell membranes with Na^+ high outside and K^+ high inside the cells. The Na^+ gradient, $\Delta[Na^+]$ is used to drive diverse secondary transporters. For example, in many animal cells $\Delta[Na^+]$ drives Na^+ inwardly coupled to H^+ outwardly, mediated by Na^+/H^+ exchangers (NHEs). They provide the principal means by which metabolically produced acids are ejected from mammalian cells [70]. In the voltage strategy the electron transport system of prokaryotes or H^+ V-ATPases of eukaryotes, impose a voltage gradient, $\Delta\Psi$, across biological membranes with the outside positive. The $\Delta\Psi$ drives secondary (Na^+ or K^+)/nH^+ antiport that is mediated by Na^+/H^+ antiporters (NHAs). The stoichiometry of NHEs is $1Na^+$ to $1H^+$ so they are independent of the membrane potential and are said to be electroneutral. The stoichiometry of NHAs is $1Na^+$ or K^+ to more than $1H^+$ so they are driven both by the ion gradients and the membrane potential and are said to be electrophoretic. NHAs operate in the opposite direction from NHEs, moving nH^+ inwardly and Na^+ or K^+ outwardly. $\Delta\Psi$ also drives Na^+- or K^+-coupled nutrient amino acid uptake that is mediated by electrophoretic (Na^+ or K^+) amino acid symporters (NATs) [11]. In eukaryotic cells the primary sources of voltage gradients across plasma membranes have classically been considered to be K^+, Na^+, or other ionic diffusion potentials. Thus, K^+ diffusion potentials dominate the resting potential and Na^+ diffusion potentials dominate the action potential in squid axon and many other nerves. Only recently are $\Delta\Psi$s generated by H^+ V-ATPases becoming recognized as the energy source for electrophoretic transporters in animal cells [35, 65, 90]. The H^+ V-ATPases translocate H^+ outwardly across the cell membrane leaving their partner anion (gegenion) behind. Thus, they charge the capacitance of the membrane resulting in a transmembrane voltage, with the outside positive. The translocated H^+s exchange with

W.R. Harvey (✉)
Whitney Laboratory for Marine Bioscience, University of Florida, St. Augustine, FL 32080, USA; Department of Epidemiology and Biostatistics, Emerging Pathogens Institute, Gainesville, FL 32610, USA
e-mail: wharvey@whitney.ufl.edu

G.A. Gerencser (ed.), *Epithelial Transport Physiology*,
DOI 10.1007/978-1-60327-229-2_6,

more numerous Na^+s or K^+s in the outside bulk solution, transforming the H^+ electrochemical gradient to a Na^+ or K^+ electrochemical gradient which in turn drives Na^+- or K^+-coupled amino acid symport via a NAT into the cells. Membrane energization by H^+ V-ATPases is accomplished by a five-phase system consisting of (1) the bulk solution inside the cells, (2) the inside solution/membrane interface, (3) the membrane, (4) the outside solution/membrane interface, and (5) the outside bulk solution [36, 49, 50].

The chapter is divided into five parts: (1) voltage-driven transporters and their terminology, (2) a summary of progress from the concept of "active K^+ transport" through the discovery of portasomes and their role in the isolation of the so-called K^+ pump to the cloning of its component H^+ V-ATPase and $K^+/2H^+$ antiporter, (3) the cloning and localization of components of the H^+ V-ATPase-Na^+/H^+ antiporter-NAT system of mosquito larval alimentary canal (AC), with emphasis on the cloning of the first, putatively electrophoretic, Na^+/nH^+ antiporter from *Anopheles gambiae* (AgNHA1), (4) attempts to characterize NHEs and NHAs heterologously in *Xenopus* oocytes, and (5) the incorporation of existing data into a qualitative model of the mosquito system for taking up amino acids while recycling H^+, Na^+, and K^+ between lumen, cells, and hemolymph as well as generating longitudinal pH gradients in the absence of barriers along the AC of mosquito larvae.

Keywords *Anopheles gambiae* · African malaria mosquito · *Xenopus laevis* oocytes · NHA · NHE · NHE_{VNAT} · pH

1 Voltage-Driven Transporters and Their Terminology

The hypothesis that animal plasma membranes are energized not only by Na^+-gradient – generating Na^+/K^+ P-ATPases but also by voltage-gradient-generating H^+ V-ATPases is becoming accepted, especially in fresh water organisms. However, the hypothesis that electrophoretic Na^+/H^+ antiporters (NHAs) use the resulting voltage gradient to drive H^+ into cells while expelling K^+ or Na^+ has not been evaluated seriously. We will argue that voltage-driven NHAs and NATs are so widespread as to require an expansion of the animal membrane energization/utilization paradigm. We will introduce and discuss novel NHE_{VNAT}s which move H^+ out of cells via H^+ V-ATPases and Na^+ into the cells by Na^+-coupled amino acid symporters. They are called NHE_{VNAT}s because these Na^+ and H^+ movements are oriented in the same direction as those mediated by NHEs. We will analyze the way in which H^+, Na^+, K^+, Cl^-, and HCO_3^- fluxes interact during the uptake of the 12 essential amino acids (AAs) and their distribution throughout the larval AC of *A. gambiae* and *Aedes aegypti* – the principal vectors of malaria, yellow fever, dengue, and other tropical diseases.

The terminology of Na^+/H^+ transporters is confusing. These transporters move Na^+ in one direction across a membrane coupled to the movement of H^+ in the opposite direction and are called "exchangers" by mammalian scientists but are called "antiporters" by bacterial and insect scientists. These synonyms were given separate meanings based on phylogeny by Brett et al. [12] who showed that NHEs (Na^+/H^+ exchangers) are genetically separated from NHAs (Na^+/H^+ antiporters). We postulate that the two classes of transporter have separate functional properties as well, with NHEs being electroneutral, $\Delta[Na^+]$-driven exchangers whereas NHAs are electrophoretic, $\Delta\Psi$-driven, antiporters. Insect Na^+/H^+ transporters often antiport either Na^+ or K^+ for H^+ leading to the awkward expression "(Na^+ or K^+) transporters"; for brevity we will call them NHAs understanding that K^+ is often substituted for Na^+. Furthermore, two different systems for naming these transporters have been published. Table 1 enables translation between the human-based terminology [72] and the phylogeny-based terminology [12, 13] for NHAs and NHEs; the Brett terminology will be used in this chapter.

Table 1. Nomenclature of Na^+/H^+ transporters (NHTs) in *A. gambiae* alimentary canal

Brett et al. [12] *A. gambiae*	Pullikuth et al. [73] *A. aegypti*	GenBank Accession No.	AA. sequence Length	CPA Gene Family [12]
NHA1	NHE10	XP_320946 (predicted) ABJ91581 (Cloned)	647 (predicted) 647 (cloned)	CPA2/NHA
NHA2	NHE9	XP_312647 (predicted) ACM47586 (cloned)	453 (predicted) 568 (cloned)	CPA2/NHA
NHE1	NHE8	XP_307859 (predicted)	650 (predicted)	CPA1/NHE IC NHE8-like
NHE2	NHE3	XP_319788 (predicted) AAO34131 (cloned?)	948*(predicted) 1,221 (cloned)	CPA1/NHE PM Recycling
NHE3	NHE6	XP_314826 (predicted)	733 (predicted)	CPA1/NHE IC Endo/TGN

Throughout this chapter we will use the term "electrogenic" to describe primary transporters that generate transmembrane electrical potentials by transducing energy directly from exergonic processes such as ATP hydrolysis. Thus, the H^+ V-ATPase is an electrogenic "pump" that generates a transmembrane potential by coupling ATP hydrolysis to H^+ translocation across the membrane bilayer (dielectric). We will use the term "electrophoretic" to describe secondary transporters that use the voltage to drive ion translocations across the energized membrane. For example EcNhA1 is an electrophoretic antiporter which uses the membrane potential generated by the electron transport system to drive $Na^+/2H^+$ exchange [81]. The voltage gradients across biomembranes are enormous. For example, in the caterpillar midgut the phosphorylation potential is ~240 mV and voltages approximately equal to this amount were reported by Dow and Peacock [29]. 240 mV is equivalent to a 10,000-fold concentration gradient across a membrane for a monovalent ion such as Na^+. Thus, the enormous voltage gradients involved in H^+ V-ATPase membrane energization are far too large to be ignored.

2 From Active K^+ Transport to H^+ V-ATPase and NHA

2.1 A. K^+ Pumps, Portasomes, and H^+ V-ATPases

H^+ V-ATPases were identified, isolated, cloned, and characterized, more or less independently, from several sources. One of the first sources was identified by Arthur Ramsay who showed that K^+ concentrations are higher in urine than blood of seven insect species including the blood-sucking *Rhodnius prolixus* and concluded that active K^+ transport by what was later called the K^+ pump in Malpighian tubules (MTs) of these insects is responsible [74]. The insect K^+ pump turned out to be present in many insect epithelia, for example rectal papilla of blowflies, midgut of caterpillars, sensory sensilla of flies, and MTs of several insects (Gupta, and Berridge 1966[41, 54, 57] and other reviews; Berridge et al. [6]; Harvey et al. [38]; Maddrell [56]; Thurm and Wessel [83]). In 1966 Berridge and Gupta first identified a coat of repeating particles on the apical membrane in rectal papillae of the blowfly [34, 5]. On a visit to Harvey's laboratory in Amherst, Berridge pointed out similar particles on the goblet cell apical membranes (GCAMs) in electron micrographs of caterpillar midgut. Following this lead Anderson and Harvey [2] suggested that the particles might be mediating the active K^+ transport that had been demonstrated earlier in isolated caterpillar midguts bathed in Na^+-free medium [41]. They noted that "The organization of the entire unit reminds one of the construction of the "elementary particles" found on the cristae of mitochondria" [31]. Comparing transport thermodynamics and physiology of the caterpillar goblet cell apical membrane (GCAM) with cristae of the mitochondrial inner membrane Harvey et al. [39] postulated that the spike and sphere images on caterpillar GCAMs were functionally equivalent to F_1 ATP synthases and suggested that both images be called portasomes (transport bodies).

Fast-forwarding several years, studies with Signe Nedergård, Jack Wood, Julian Dow, Brig Gupta, and others convinced Harvey that the insect K^+-pump was on the apical membrane of the goblet cells. Using portasomes as an assay, Harvey's group separated the GCAMs from columnar cell apical membranes, lateral membranes, and basal membranes [22, 38]. But they could prepare only enough pure GCAMs for one or two ATPase assays after a laborious 2-day membrane preparation. They were able to demonstrate ATPase activity in a highly purified GCAM fraction and showed that it consisted of but a few proteins among which 64 kDa and 55 kDa proteins (now known to be V-ATPase subunits A and B) were most prominent. Miraculously Helmut Wieczorek appeared at Harvey's lab and a visit was arranged for him and his technician. Wieczorek and his associates had been trying to isolate the K^+ ATPase from the particle-studded apical membranes of blowfly sensory sensilla and had perfected a sensitive ATPase micro-assay. Combining forces with Wieczorek and using his micro-assay on their pure GCAM preparations the group showed that the K^+ ATPase, the putative biochemical equivalent to the physiological K^+-pump, was highly enriched on the GCAM fraction [92]. Wieczorek's group then solubilized the GCAM fraction, separated its proteins on gels and discovered that the GCAM ATPase was in fact a vacuolar-type ATPase [91] similar to that which had been discovered by Nathan Nelson's group in chromaffin granules [21] and Yasuhiro Anraku's group in yeast [85]. At about the same time Stephen Gluck and Dennis Brown found the V-ATPase in kidney tubules [16, 32] and others found it in other cells, for example in osteoclasts [19, 64]. A frenzy of molecular biology enabled Nelson, Wieczorek, Harvey, and others to clone all six V_o subunits and all eight V_1 subunits of the V-ATPase. Now regulation [87, 43, 33] including the crystal structure of a prokaryotic V-ATPase has been determined [63], and details of the Na^+ V-ATPase of *Enterococcus hirae* are known [62]. More relevant to the midgut physiology story, Grüber and associates [73] showed that the isolated portasomes are indeed V_1 ATPase complexes, a point worthy of expansion because it is virtually ignored in much of the literature, for example Clements [23].

2.2 Proof That Portasomes Are V_1 ATPase Sectors

The H^+ V-ATPase consists of a peripheral, V_1 sector and a membrane-bound V_o sector. The V_1-ATPase purified from starving *Manduca sexta* larvae consists of eight subunits with the stoichiometry thought to be $A_3B_3CDEFG_3H$ (Wieczorek et al., 1999) as deduced from SDS (sodium dodecyl sulfate) gels of the cytoplasmic V_1 complex. The purified V_1 ATPase accounts for up to 2% of the total cytosolic protein and can be obtained in milligram amounts (Nelson and Harvey, 1999). The high yield and purity of the V_1 preparation enabled Gerhard Grüber to produce the first V_1 ATPase crystals (Nelson and Harvey, 1999).

Electron micrographs taken by Moira Cioffi revealed that portasomes stud the basal plasma membranes in anterior midgut of *A. aegypti* larvae (Fig. 1a; Zhuang et al. [94]). Images of isolated, negatively stained portasomes from *M. sexta* are shown in Fig. 1b. In averaged, enlarged images portasomes appear as V_1 ATPase sectors in which the subunit A_3B_3 stoichiometry is clearly visible (Fig. 1c). The enzyme has a barrel-like structure that is 11 nm in height and 13.5 nm in diameter. The six large A and B subunits are inter-digitated for most of their length (9 nm; Fig. 1c). These averaged electron micrographs of V_1 particles (Fig. 1c) along with small-angle X-ray scattering analysis of the solubilized V_1 domain, and the SDS PAGE patterns showing all eight V_1 subunits [33], together provide overwhelming proof that the portasomes seen in electron micrographs are indeed V_1 ATPase domains.

That portasomes are V_1 ATPase sectors had been demonstrated clearly four years earlier by Brown and Breton [15] who showed that portasomes are labeled by immunogold-antibody against an H^+ V-ATPase subunit. First they showed a conventional electron micrograph image from a typical portasome-studded plasma membrane in rat kidney collecting duct intercalated cells. Then they showed a similar image stained with immunogold-antibody against a H^+ V-ATPase subunit. Finally, rapid-freeze, deep-etch micrographs and a freeze-fracture electron micrograph showed tightly packed hexagonal images of portasomes. This proof from caterpillars and rats is especially relevant for our studies on mosquito larvae, where

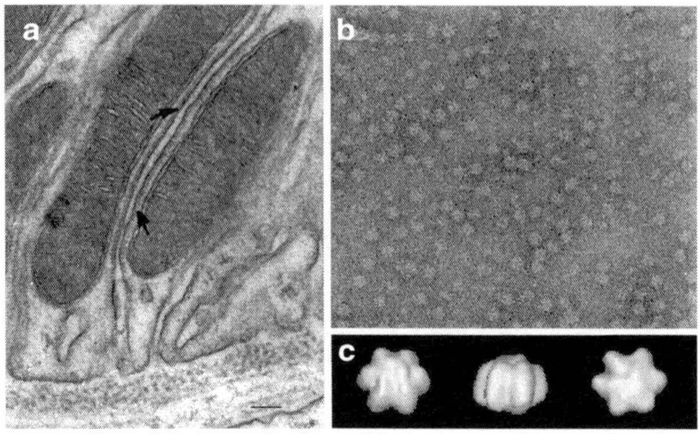

Fig. 1 Proof that portasomes are H^+ V_1-ATPase sectors (**a**. Zhuang et al. [94]; **b.** and **c**. Grüber et al. [33])

images of portasomes enabled Zhuang et al. to localize the V-ATPase on infoldings of the basal plasma membrane of anterior midgut cells (Fig. 1a; Zhuang et al. [94]). Micrographs of portasomes in many insect transporting epithelia [23] are located where Okech et al. [66] have immunolocalized H^+ V-ATPase.

2.3 Na^+ Concentration Gradients Versus Voltage Gradients As Membrane Energizers

Earlier R. B. Dean had introduced the concept of the sodium ion pump; he wrote that "the muscle can actively move potassium and sodium against concentration gradients. . .this requires work. Therefore there must be some sort of a pump possibly located in the fiber membrane, which can pump out sodium or, what is equivalent, pump in potassium" [25]. When Skou isolated and characterized the N^+/K^+ P-ATPase [see Nobel Lecture [78]] the Na^+ pump hypothesis became dogma and the Na^+ pump was said to energize all animal cell membranes. For example, the Na^+ pump was said to be the prime membrane energizer for Na^+ uptake in the skin of fresh water frogs and the frog skin became the model epithelium for alkali metal ion transport [51, 86]. The frog skin model held that a basolateral Na^+/K^+ ATPase moves Na^+ "actively" into the blood from the cells where it is replaced by passive diffusion of Na^+ from the "pond-side" bathing solution while the K^+ recycles between cell and extracellular fluid. Although this model worked well under laboratory conditions with high $[Na^+]$ on the outside, it faced a problem when applied to frogs living in ponds. How could Na^+ diffuse passively into frog epidermal cells when its concentration is only 0.1 mM in pond water and between 10- and 100-fold higher in the cells? The answer is that the apical (pond-facing) membrane is electrically charged, the interior being ~30 mM negative to the pond-side. But where does the voltage come from? Ehrenfeld and associates made a giant step toward solving the frog skin dilemma by showing that the voltage driving Na^+ into the frog skin is generated by an H^+ V-ATPase [30, 40], although the transporter or channel that mediates the Na^+ uptake remains controversial.

2.4 The Antiporter Story

The new challenge, broadly stated, is to determine the relative roles of the "new" electrogenic, H^+ V-ATPase of metazoans which acts via voltage gradients and the well-known electroneutral Na^+/K^+ P-ATPase which acts via $[Na^+]$ gradients. As noted above an important role of the Na^+ gradient generated by the Na^+/K^+ P-ATPase is to eject metabolic acid by an electroneutral Na^+/H^+ exchanger (NHE). This mechanism works well in marine organisms and their terrestrial descendents whose blood and extracellular fluid is always rich in Na^+ (~140 mM). But, it does not work for organisms that live or breed in freshwater where the Na^+ concentration is ~0.1 mM. Instead fresh water dwellers appear to use the outside-positive voltage generated by a H^+ V-ATPase to drive Na^+ into the cells while ejecting metabolic H^+, perhaps by the $2Na^+/1H^+$ transporter identified by Ahearn et al. [1], antibodies of which label many apical membranes in transporting cells. In the absence of any other mechanism by which freshwater organisms can accumulate Na^+, the evolution of H^+ V-ATPases which generate voltages that drive electrophoretic $2Na^+/H^+$ antiport would provide a mechanism that explains how organisms could invade and thrive in fresh water.

Having established that the postulated caterpillar GCAM K^+-ATPase is actually an H^+ V-ATPase the next question is – how can a V-ATPase, which transports only H^+, drive K^+ transport across the caterpillar midgut epithelium? And how can a H^+ V-ATPase that is said to acidify the output compartment render the goblet cavity slightly alkaline at pH 7.3 [18] and the lumen pH as high as 12 [27]? Wieczorek solved the problem by showing that a K^+/H^+ antiporter is present in GCAM vesicles along with the H^+ V-ATPase and that it has a stoichiometry of 1 K^+ to $2H^+$ [4, 90, 55] which implies that it is electrophoretic. Since there are insufficient Na^+ or K^+ chemical gradients across caterpillar midgut membranes to drive the antiport this finding suggests that the antiport is driven by the voltages, as high as –269 mV [29] that are generated by the H^+ V-ATPase. Although the suggestion that a voltage created by H^+ translocation in one direction could drive $2H^+$ translocation across the same membrane encountered some skepticism from metazoan physiologists, (Na^+ or K^+)/$2H^+$ antiport is well understood in bacteria. Thus, the electron transport system in alkalophilic bacteria generates a voltage across the cell membrane (outside positive) that drives an electrophoretic $Na^+/2H^+$ antiporter (NHA) which transports H^+ back into the cells and ejects the toxic Na^+ that leaks in from the alkaline environment [53]. The bacterial NHA has been studied extensively, especially by Padan and associates [44] who worked out its crystal structure and reaction mechanism [3].

The missing link in the metazoan Na^{+}/H^{+} exchange story and especially in the insect $K^{+}/2H^{+}$ antiporter story had been the failure to clone a metazoan, electrophoretic NHA or KHA. Recently, AgNHA1 was cloned from *A. gambiae* [76] and antibodies to it and a V-ATPase subunit both label the membranes across which antiport activity has been deduced [66]. Simultaneously, Lao and associates [93] cloned NHA2 from humans and a little later Dow and associates [24] cloned both NHA1 and NHA2 from *Drosophila melanogaster*. To describe this work we shift attention from the caterpillar midgut to the larval mosquito AC.

3 Localization of Pumps and Transporters in larval AC

Two decisions by the National Institute of Allergy and Infectious Diseases (NIAID) led to a shift in many laboratories from caterpillars to mosquito larvae; (1) NIAID commissioned the *A. gambiae* genome project and (2) they decided to support research on disease vector insects rather than model agricultural pests. The shift was fortuitous. The large size of the caterpillar midgut had made it possible to obtain sufficient material for protein biochemistry which led to the isolation of the K^{+}-pump-containing GCAM and eventually to the structure of the entire H^{+} V-ATPase macromolecule. The small size of mosquito larvae makes it possible to examine the entire AC by molecular techniques such as in situ hybridization, quantitative polymerase chain reaction (qPCR) and by immunolocalization of membrane proteins in a single whole mount. The sequence information in the *A. gambiae* genome makes it possible to generate antibodies to specific membrane proteins and the qPCR technique makes it possible to determine the regional localization of transcripts. Neither antibody production nor qPCR requires that the genes be cloned or that cRNA be generated; the genomic sequence information is all that is required. Studies on solute transport across biological membranes have the added advantage that the localization of pumps and transporters alone makes model building possible because in many cases the direction of ion movements and ion stoichiometry are fixed and transport direction can be deduced directly from immunolocalization patterns.

3.1 Function and Structure in Larval Mosquito Alimentary Canal

Mosquito larvae accomplish four remarkable feats. They absorb enough amino acids to grow ~1,000-fold in mass; they maintain an anterior-posterior pH gradient along their AC that increases from 6.8 to 10.5 then returns to 6.8 in the absence of diffusion barriers (Fig. 2) as reported by Arthur Ramsey long ago [75]; they defecate rather than exhale CO_2; and they conserve Na^{+} which they take up from an environment in which its concentration is ~0.0001 M. These feats are accomplished by at least eight components that interact synergistically between cytoplasm, plasma membranes, and extracellular fluid, forming a bicarbonate-carbonate buffer system in the AC lumen. The components are carbonic anhydrase (CA), bicarbonate/chloride

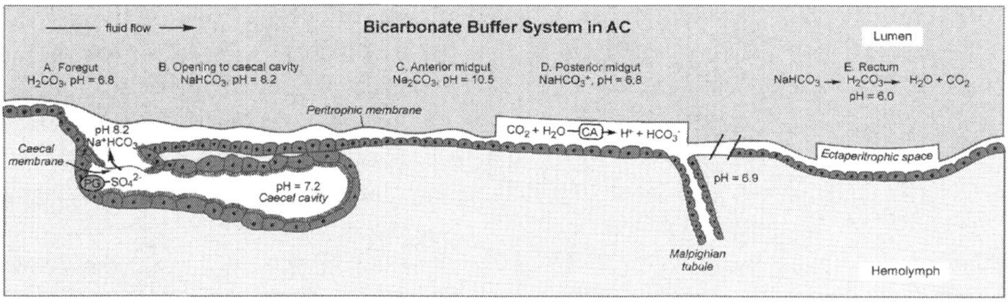

Fig. 2 Longitudinal pH gradients in the absence of barriers in *A. gambiae* AC

exchanger (AE), H^+ V-ATPase (V), Na^+/K^+ P-ATPase (Na/K), Na^+/H^+ Exchanger (NHE), Na^+ or K^+ Antiporter (NHA), Na^+ or K^+ Nutrient Amino Acid Transporter (NAT), and several functional NHEs comprising H^+ V-ATPases electrically coupled to NATs (NHE_{VNATs}). We will call the H^+ Amino Acid Transport complex HAAT for short. Immunolocalization of these HAAT proteins helps to answer some persistent questions about pH, alkali metal cation and amino acid homeostasis in larval mosquito AC (Fig. 2).

Are the principal ions in the lumen of the foregut Na^+, K^+, HCO_3^-, and Cl^-? Is the pH nearly neutral? Is the pH of the gastric caecum held at ~7.2 by a mixture of $NaHCO_3$ and H_2SO_4? Are there fixed SO_4^{2-} groups bound to proteoglycans projecting into the lumen of the caecal cavity from the apical membrane of the caecal epithelial cells; do they form an H^+ wire from V-ATPase-pumped protons adhering to the plasma membrane inside the unstirred layer? Does the pH of fluid that passes from the caecal cavity to the anterior midgut proper rise abruptly to pH 8.2 because the fixed sulfate groups are no longer present there? Does the pH of anterior midgut increase to 10.5 because Na^+ is added to the luminal contents while H^+ is removed with the result that Na^+ HCO_3^- is replaced by $2Na^+$ CO_3^{2-}? As luminal contents pass to the posterior midgut is Na^+ removed by Na^+ amino acid symport while H^+ is added by an apical H^+ V-ATPase so that the luminal contents pH drops back toward neutrality. Finally, is all of the luminal Na^+ in the rectum reabsorbed so that H_2CO_3 or $KHCO_3$ is defecated? These questions are tentatively answered in the affirmative and the events are summarized in a tentative model (Fig. 2). Our job is to determine if these answers are correct.

When the *A. gambiae* and *A. aegypti* genomes were released it became possible using qPCR to localize mRNA transcription patterns and using specific antibodies it also became possible to localize each of the HAAT proteins to specific regions of the midgut, even to specific plasma membrane sectors. For example, knowing that AgCA9 is abundant in the extra-peritrophic space all along the midgut allows one to deduce that metabolic CO_2 is transformed to HCO_3^- there. Since HCO_3^- is produced within cells one can predict that a HCO_3^-/Cl^- exchanger will move HCO_3^- outward and Cl^- inward. Since ATP is present within cells and translocates H^+ with no gegenion one can predict that H^+ V-ATPase will move H^+ outward and hyperpolarize the membrane. Na^+/K^+ ATPases always move $3Na^+$ out of cells while moving 2 K^+ inwardly. All published NHEs move Na^+ into cells and H^+ out. Eukaryotic NHAs are largely unknown but by analogy with their prokaryotic cousins, EcNhAs, and in the absence of sufficient Na^+ gradients they are expected to move Na^+s or K^+s out of cells in exchange for nH^+s driven in (and therefore NHAs are electrophoretic). All NATs that have been localized to date move Na^+ or K^+ into cells along with the symported amino acid. Finally, when H^+ V-ATPase that is moving H^+ outwardly is located in the same membrane where a symporter is moving Na^+ inwardly along with an amino acid, an NHE_{VNAT} is formed. These observations make it possible to formulate tentative, qualitative models such as Fig. 20 to guide future research.

3.2 Mosquito Membrane Thermodynamics

Activity of Na^+, K^+, and Cl^- in hemolymph of mosquito larvae: O'Donnell and others have calculated electrochemical gradients of specific ions across the epithelium of MTs in *R. prolixus* from measurements of V_t and V_b along with luminal and intracellular ion activities [45, 46]. They also carried out simultaneous measurements of pH and V_b in single MT cells using double-barreled, ion-selective microelectrodes [46]. Measurements of activities of Na^+, K^+, and Cl^- in the hemolymph of *A. gambiae* 4th instar larvae were carried out by Rheault and Harvey (unpublished results, Fig. 3). Values for *A. aegypti* larvae [26] are included in the bar graph for comparison. Na^+ and Cl^- values were approximately the same in both mosquitoes whereas K^+ was greater than 70 mmol l^{-1} in *A. gambiae* but only 9 mmol l^{-1} in *A. aegypti*. The 7-fold difference in K^+ levels in the hemolymph of the two mosquitoes implies that their transport physiologies are different. Thus, the yellow fever vector *A. aegypti* is not a suitable model for the physiology of the malaria vector *A. gambiae*.

Concentration of anions measured by capillary zone electrophoresis Boudko et al. [10] determined concentrations of several small anions in samples from 4th instar *A. aegypti* larvae by capillary zone electrophoresis (CZE) using a contact

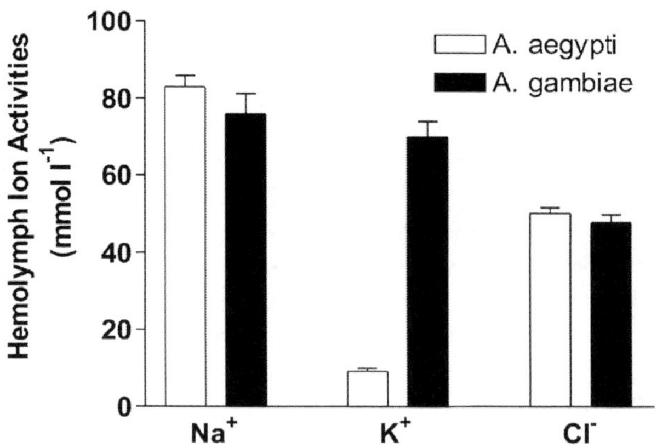

Fig. 3 Activity of Na^+, K^+, and Cl^- in hemolymph of mosquito larvae

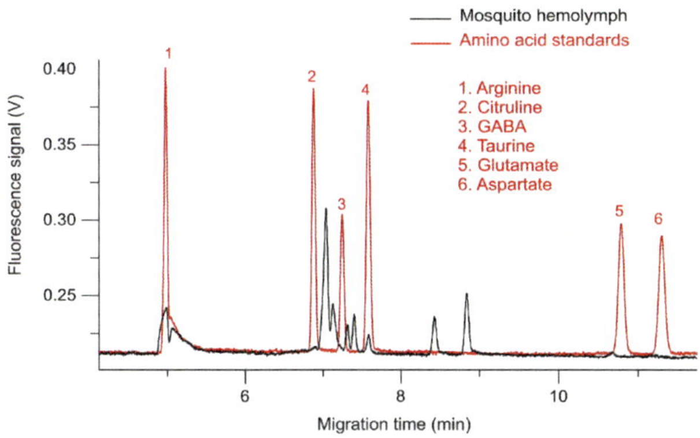

Fig. 4 Mean concentrations of anions in anterior midgut of *A. aegypti* larvae measured by capillary zone electrophoresis: HEM, hemolymph; TIS anterior midgut tissue, LUM, ectoperitrophic fluid/lumen, REM, rearing medium. The *z*-axis is the estimated ion concentration in undiluted samples with the value written on the top of each bar [9]

conductivity detector (Fig. 4). The CZE data suggested that chloride/bicarbonate exchange takes place at the apical membrane of anterior midgut cells adjacent to the gastric caeca. The exchange appeared to occur against chemical gradients of both ionic species, implying that it is stoichiometrically asymmetric and is electrophoretically driven. Parallel experiments with the Scanning Ion Electrode Technique (SIET) a.k.a. Self Referencing Ion Selective (SERIS) electrodes and inhibitors of anion exchange and carbonic anhydrase (DIDS and acetazolamide) confirmed these results [9].

3.3 Phylogeny of CPA Transporters in the SLC9 Superfamily

Although NHA genes have been identified in all phyla for which genomes are available, no NHA had been cloned previously from a metazoan organism nor had the encoded proteins been localized or characterized. Using a PCR strategy with sequence information from the genome, Harvey's group cloned a cDNA encoding a new ion transporter from the AC of the African malaria mosquito, *A. gambiae* Giles *sensu stricto* [76]. A phylogenetic tree was constructed and is shown as a maximum likelihood (ML)-distance tree (Fig. 5). It was deduced from an alignment of 65 sequences of prokaryotic and eukaryotic CPA genes and is arbitrarily rooted with the *Escherichia coli* NhaA sequence. Maximum likelihood support values are shown at branch nodes. Background highlighting: CPA1 genes, purple; CPA2 genes, yellow; further CPA2 gene highlighting: bacteria, red; vertebrate, green; nematode, orange; insect, blue. The scale bar indicates the estimated number of amino acid substitutions per site. The phylogenetic tree places the gene encoding the novel ion transporter in a group designated NHA by Brett et al. [12]; so we called it AgNHA1.

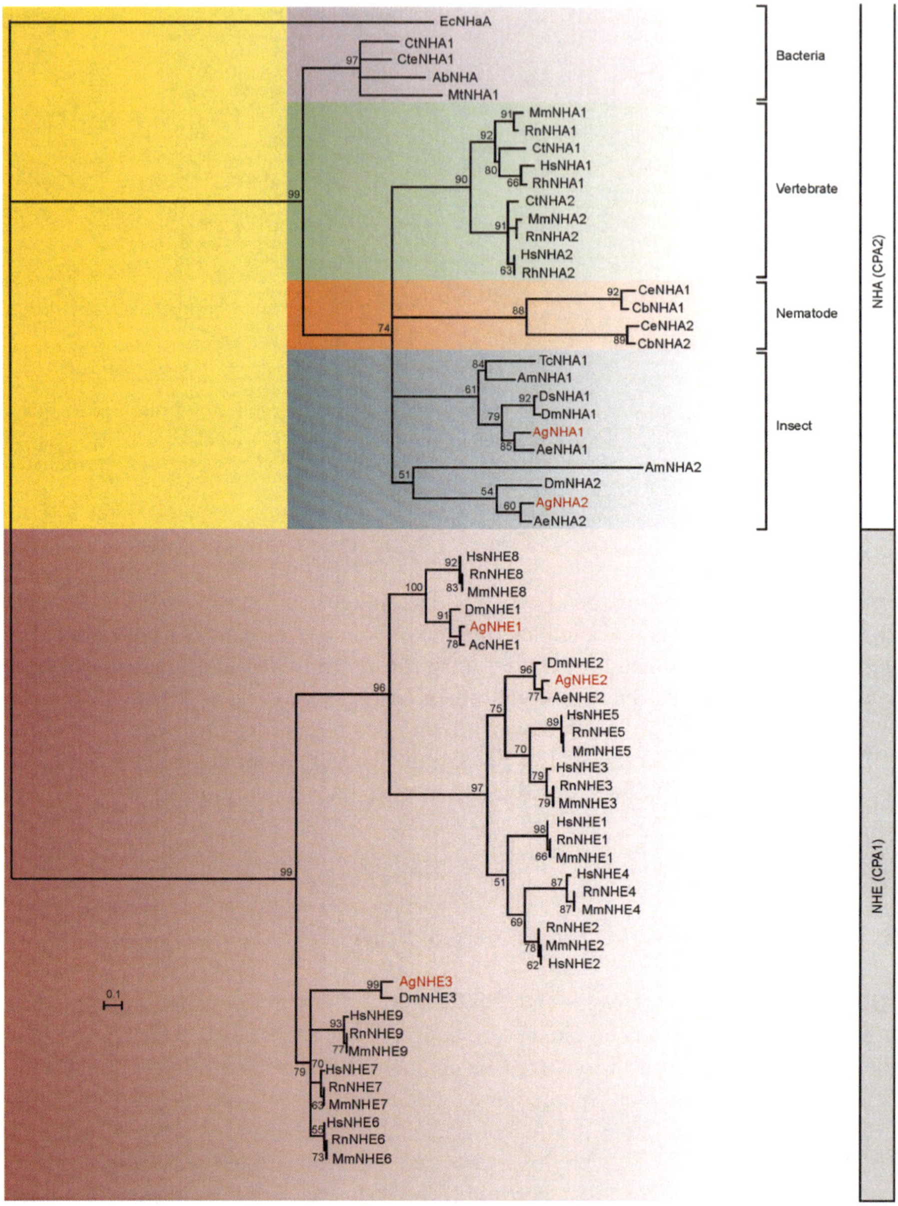

Fig. 5 Phylogeny of CPA transporters in the SLC9 superfamily [77]

The annotation of current insect genomes shows that both AgNHA1 and a close relative, AgNHA2, belong to the Cation Proton Antiporter 2 (CPA2) subfamily and cluster in an exclusive insect clade of highly identical genes from *A. aegypti*, *D. melanogaster*, *D. pseudoobscura*, *Apis mellifera*, and *Tribolium castaneum*.

3.4 Relative Transcriptions of AgNHA1 in A. gambiae *Tissues*

The transcription of AgNHA1 was investigated in 4th instar *A. gambiae* larvae using quantitative real-time PCR (qPCR) on preparations of various tissues (Fig. 6). Data were presented as the mean of three replicates of at least two independent experiments plus standard error. AgNHA transcript levels were normalized to the levels of the carcass at a value of 1. Preparations were studied from ventral nerve cord, VNC; salivary glands and cardia, SG/cardia; gastric caeca, GC; anterior midgut, AMG; posterior midgut, PG, and Malpighian tubules, MTs. AgNHA1 message was detected in gastric caeca and rectum with much weaker transcription in other parts of the AC. However, expression of transporter protein, as revealed by immunolabeling procedures, was prominent in other regions as described below.

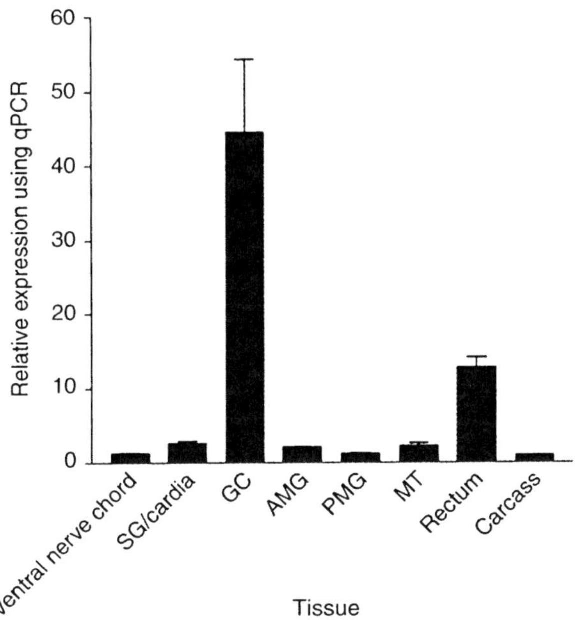

Fig. 6 Relative transcriptions of AgNHAs and AgNHEs in *A. gambiae* tissues [76]

3.5 Immunolabeling of Whole Mounts of A. gambiae *Alimentary Canal*

The antibody to AgNHA1 labeled cardia, gastric caeca, anterior midgut, posterior midgut, proximal MTs and rectum as well as the subesophageal and abdominal ganglia. Its relationships to other transport proteins and their deduced roles in larval AC biology are discussed below.

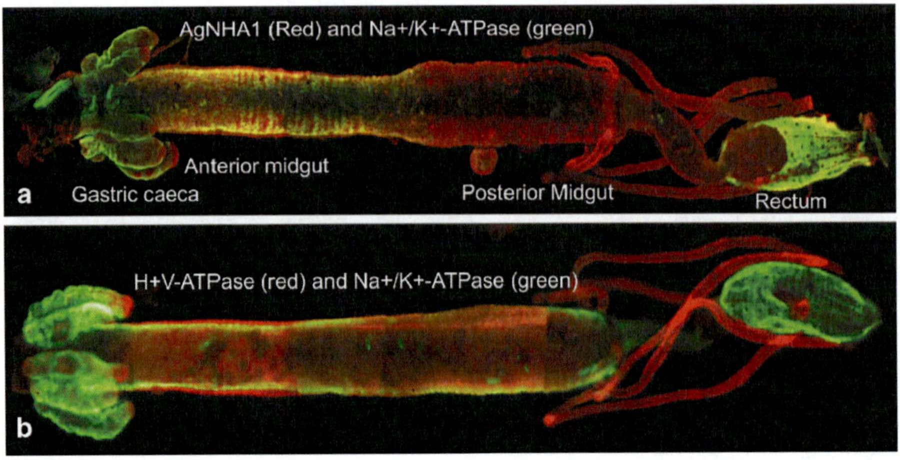

Fig. 7 Co-expression of AgNHA1 (A dorsal view) and H^+ V-ATPase (B ventral view) on the apical membrane of anterior midgut cells in whole mounts of *A. gambiae* larval AC [66]

1. *Co-expression of H^+ V-ATPase with AgNHA1 in whole counts of A. gambiae larval AC*: Co-localization of the voltage-generating H^+ V-ATPase and its voltage-driven companion, the Na^+/H^+ antiporter was deduced in studies on *A. gambiae* [66]. Strong labeling by both AgNHA1 (Fig. 7a) and H^+ V-ATPase (Fig. 7b) is seen at the distal tips of gastric caeca and in anterior midgut where AgNHA1 and H^+-V-ATPase are both at the apical border. Its co-expression with H^+ V-ATPase suggests strongly that AgNHA1 is electrophoretic. Moreover both antibodies labeled the MTs.

   A similar close relationship between H^+ V-ATPase and an NHA was also found by Day et al. [24] in *D. melanogaster.* However, co-expression is not necessary and the electrogenic H^+ V-ATPase and electrophoretic Na^+/nH^+ need not reside in the same membrane. Klaus Beyenbach has shown in an electrical analysis that the voltage generated across the basal plasma membrane can hyperpolarize a cell and the electrical coupling – "Beyenbach Coupling" – results in a somewhat attenuated but still substantial voltage across the apical membrane [7]. Neither is an H^+ V-ATPase absolutely necessary; any electrogenic process, such as Na^+/K^+ ATPase activity that generates a voltage across a membrane can energize an electrophoretic NHA.
2. *Co-expression of H^+ V-ATPase with AgNHA1 in apical membranes of gastric caecal cells*: The co-expression of the electrogenic ATPase and the electrophoretic antiporter is seen in more detail in cross sections where antibodies to H^+ V-ATPase (Fig. 8) and AgNHA1 (Fig. 10b) both clearly label the apical membrane of epithelial cells in the gastric caeca.
3. *Driving $C^+/2H^+$ Antiport under bulk alkaline conditions:* A persistent issue with the hypothesis that H^+ V-ATPase drives (K^+ or Na^+)/$2H^+$ antiport is the scarcity of hydrogen ions in a compartment with pH = 10. If the Na^+ concentration is

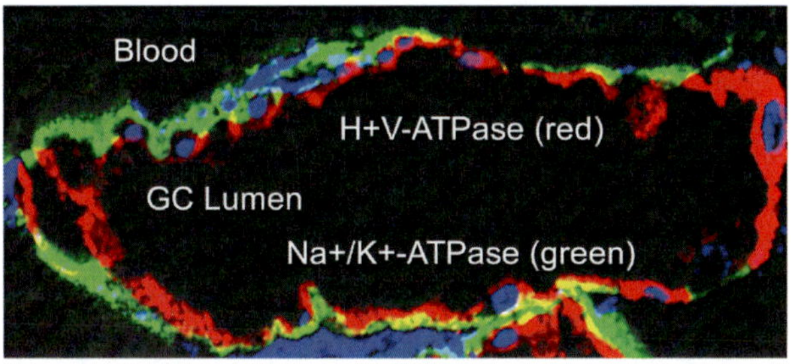

Fig. 8 Expression of H^+ V-ATPase in apical membranes and Na^+/K^+ P-ATPase in basal membranes of gastric caecal cells in *A. gambiae* larvae [66]

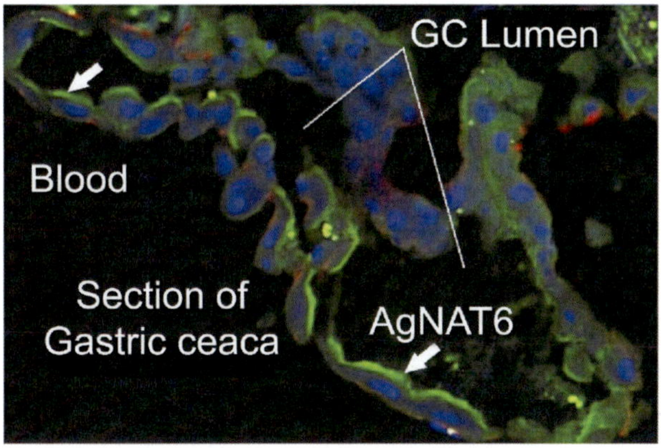

Fig. 9 Co-expression of H^+ V-ATPase and AgNAT6 forms NHE_{VNAT} in gastric caecal apical membranes [59]

0.050 M and the pH is 10 (0.0000000001 M) where are the H^+ to be antiported? The problem is made even worse by the strong signal for $3Na^+/2\ K^+$ P-ATPase on the basal membrane (Fig. 8) which removes one more strong cation (Na^+) than it adds (K^+) tending to make the H^+ concentration in the cells even higher and the antiport more difficult. In caterpillar midgut goblet cavity the sulfur concentration measured by X-ray microanalysis is high [28] and $SO_4{}^{2-}$ groups are postulated to project into the cavity and provide a local strong anion as counter ion for H^+ leaving the V-ATPase at the apical membrane. If $SO_4{}^{2-}$ projects into the caecal cavity then the local pH could be much lower than the bulk pH and could provide the required H^+. The entire topic of higher H^+ concentrations at the membrane fluid interface than in the bulk fluid outside membranes energized by H^+ V-ATPases is reviewed by Harvey [36].

4. *Co-expression of H^+ V-ATPase with AgNAT6 forms NHE_{VNAT} in gastric caecal apical membranes.* In the gastric caeca H^+ V ATPase localizes on the apical membrane (Fig. 8) with AgNAT6 (Na^+ coupled neutral amino acid$^{\pm}$ symporter 6 [59] (Fig. 9)) forming a NHE_{VNAT} [66]. NHE_{VNAT} is shown even more clearly in the longitudinal sections of posterior midgut (Fig. 13b and d).

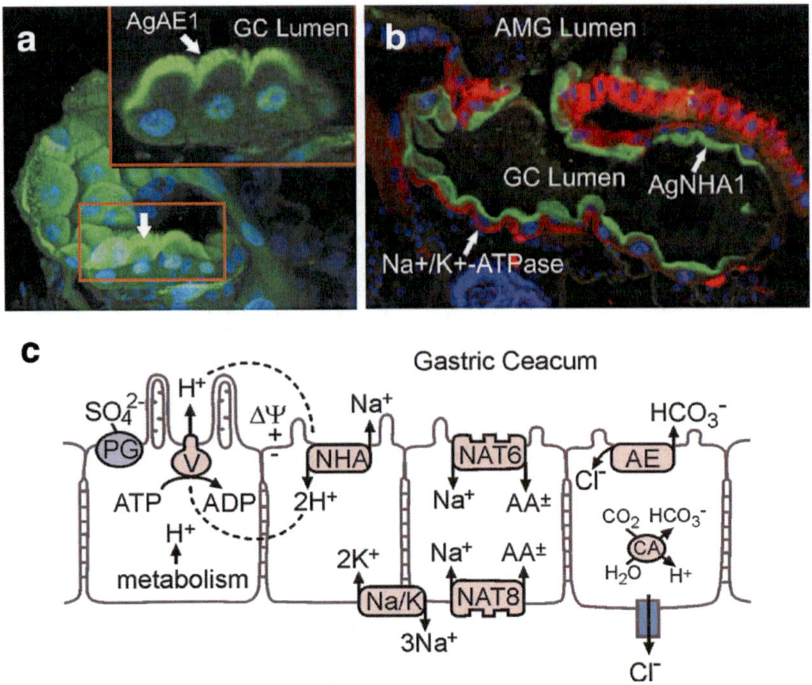

Fig. 10 Hydrogen ion and sodium ion cycles in gastric caeca of *A. gambiae* larva as well as the cooperation between carbonic anhydrase and the bicarbonate/chloride ion exchanger (**a**, courtesy Paul J. Linser; **b**, [66])

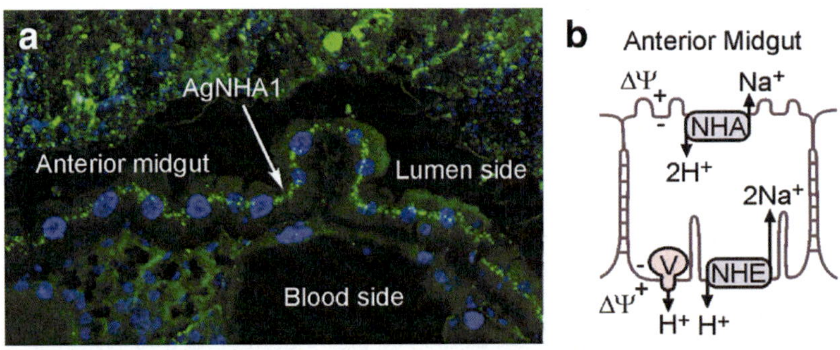

Fig. 11 Apical AgNHA1 and basal H^+ V-ATPase (not shown) remove H^+ and add Na^+ to lumen of anterior midgut in *A. gambiae* larvae (Okech, et al. 2008)

The apical location of AgNAT6 in gastric caeca is consistent with the work of Walter Terra who showed the final digestion of dipeptides to amino acids occurs there [82]. AgNAT8 [58] is located on the basal membrane (image not shown) which is consistent with the observation that many amino acids are taken up from the hemolymph into the gastric caeca.

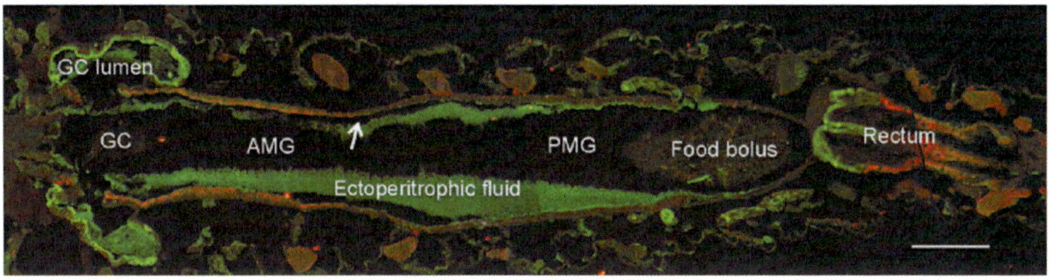

Fig. 12 Carbonic anhydrase in ectoperitrophic space traps metabolic CO_2 as HCO_3^- all along the AC [79]

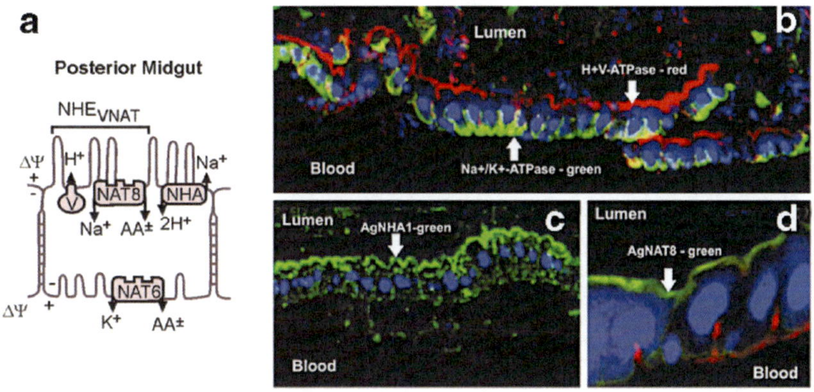

Fig. 13 Co-expression of H^+ V-ATPase and AgNAT8 (NHE_{VNAT}) with AgNHA1 in posterior midgut [66]

5. *Replacement of luminal Cl^- by cellular HCO_3^- across the apical membrane of gastric caecal cells* Since the pH in anterior midgut is 10.5 or higher [23], the Cl^- concentration is low and the HCO_3^- concentration is high (Fig. 4) [9] the question arises – how is Cl^- removed and HCO_3^- added? Dr. Linser has provided an important clue by showing that an antibody to AgAE1 (AE for anion exchanger) is present on the apical membrane of gastric caecal cells where it can remove Cl^- from the lumen and replace it by HCO_3^- (Fig. 10a).
6. *H^+ and Na^+ cycles in gastric caeca.* Na^+ that enters the caecal lumen by both apical AgNAT6 (Fig. 9) and basal AgNAT8 (not shown) is recycled both to the lumen by AgNHA1 on the apical membrane (Fig. 10b) and to the hemolymph by Na^+/K^+ P-ATPase on the basal membrane (Fig. 10b).

   The recycling of H^+ from the cells to the lumen by the V-ATPase and back to the cells via an NHA (e.g. AgNHA1) in the gastric caeca is illustrated by the cartoon (Fig. 10c.). The cycling of Na^+ to the lumen by NHA and back to the cells by a NAT such as AgNAT6 and a second cycle of Na^+ out of the cells to the hemolymph by basal Na^+/K^+ ATPase and back into the cells by AgNAT8 are also shown along with the efflux of cellular HCO_3^- to the lumen and influx of Cl^- to the cells by the anion exchanger (Fig. 10c.).

7. *Anterior midgut alkalinization*: AgNHA1 (Fig. 7a – red labeled antibody and Fig. 11a –green labeled antibody) and H^+ V-ATPase – red antibody (Fig. 7b) are positioned in anterior midgut so as to remove H^+ from the lumen and replace it by Na^+ [66]. The basal V-ATPase moves H^+ from cell to hemolymph while the apical AgNHA replaces the cell H^+ by removing it from the lumen and replacing it by Na^+. This appears to be the mechanism for the stripping of H^+ from bicarbonate to make carbonate that was proposed two decades ago by Dow [27]. However, AgNHA1 is located in vesicles near the apical membrane (arrow in Fig 11) not on the membrane. A precedent for fusing of vesicles with membranes occurs in kidney collecting ducts where fusion of V-ATPase-containing vesicles with the plasma membrane during acid stress is well known [14, 15].

3.6 Carbonic Anhydrase in Ectoperitrophic Space Traps Metabolic CO_2 As HCO_3^- All Along the AC

Just as bicarbonate is the principal anion in caterpillar midgut [84] it is also the principal anion in larval mosquito midgut (Fig. 4) [9]. If lumen Cl^- were exchanged for cell HCO_3^- and the H^+ were removed, lumen HCO_3^- would be transformed into $2Na^+\ CO_3^{2-}$ and account for the high alkalinity (Fig. 2). HCO_3^- enters the lumen in GC by exchange for Cl^-, mediated by AgAE1 (Fig. 10a). This removal of the strong Cl^- anion is essential for alkalinizing the lumen. The mechanism for introducing bicarbonate into the lumen was discovered by Kristin Smith et al. [79] who found that an antibody to CA9 labels the entire length of the ectoperitrophic space (Fig. 12). Metabolic CO_2 that leaves cells toward the hemolymph equilibrates with cell CO_2. But CO_2 that leaves cells to the ectoperitrophic fluid is converted by CA9 to HCO_3^- which is polar, electrically charged, and cannot return to the cells; so there is a continuous large gradient driving CO_2 from cells to ectoperitrophic space and driving bicarbonate from space to the lumen with no ATPase involved – an alternate mechanism for "active" transport.

3.7 Co-expression of H^+ V-ATPase and AgNAT8 (NHE_{VNAT}) with AgNHA1 in Posterior Midgut

A remarkable co-localization of H^+ V-ATPase (Fig. 13b) and AgNAT8 (Fig. 13d) on the apical membrane in posterior midgut cells constitutes the functional equivalent of an NHE (NHE_{VNAT}) the activity of which lowers the pH in posterior midgut lumen. The H^+ V-ATPase is clearly present on the apical membrane (Fig. 13b) where it pumps H^+ toward the lumen. AgNAT8 is also clearly present on the same membrane (Fig. 13d). It is widely recognized that the voltage generated by the ATPase drives Na^+ symport of neutral amino acids from lumen to cell, see [36]. Thus, the H^+ V-ATPase moves H^+ from cell to lumen while AgNAT8 moves Na^+ from lumen to cell. The combined movements are precisely what a classical NHE does – so we appear to have discovered a functional NHE and have called it NHE_{VNAT} [66]; there are seven NATs in the *A. gambiae* genome, implying seven NHE_{VNAT}s. These movements of H^+ and Na^+ would soon acidify the lumen and lower its Na^+ concentration to the point where Na^+:AA symport would stop. However, the presence of AgNHA1 (Fig. 13c) provides a pathway for local cycling of H^+ back to the cells and Na^+ back to the lumen. Together the three components on the apical membrane could produce a balanced ionic steady state in the poster midgut cells. In an unpublished study K. Sterling, B.A. Okech, and W.R. Harvey have isolated brush border membrane vesicles (likely apical plasma membranes of gastric caeca and posterior midgut) and shown by western blots, using antibodies to the V-ATPase, AgNAT8, and AgNHA1, that all three components are present. The BBMV may become a valuable tool in larval midgut physiology.

3.8 Is AgNHA1 the Mosquito Equivalent of the Long-Sought $K^+/2H^+$ Antiporter?

In 1991 Wieczorek and associates proposed the paradigm-changing hypothesis that the insect K^+ pump is composed of an H^+ V-ATPase acting in conjunction with a K^+/nH^+ antiporter [90]. This hypothesis implies that both proteins reside on the same membrane where the membrane potential generated by the H^+ V-ATPase is postulated to drive the electrophoretic antiport. Indeed, polyclonal antibodies against AgNHA1 labeled the apical membranes lining the goblet cavity in *M. sexta* larval midgut at the sites where antibodies to V-ATPase label these membranes (Drs. Olga Vitavska and Helmut Wieczorek, personal communication). The antibodies also labeled the tips of the apical brush border in adjacent columnar cells suggesting that K^+ is recycled to the lumen via an NHA1 antiporter after it enters the cells via K^+-coupled amino acid symport [17]. Unfortunately, unlike the single band at the proper molecular mass in western blots of AgNHA1 from *A. gambiae* membranes [66, 76] the blot of *M. sexta* membranes showed several bands including one at the right mass.

4 Current/Voltage Analysis of AgNHA1 in *Xenopus* Oocytes

The close relations of insect NHEs (CPA1) with mammalian NHEs (Fig. 14 right) and the isolation of insect NHAs (CPA2) from mammalian NHAs (Fig. 14 left) suggest that insect NHEs like all other metazoan NHEs may be Na^+ gradient driven, electroneutral Na^+/H^+ exchangers whereas insect NHAs like bacterial NHAs may

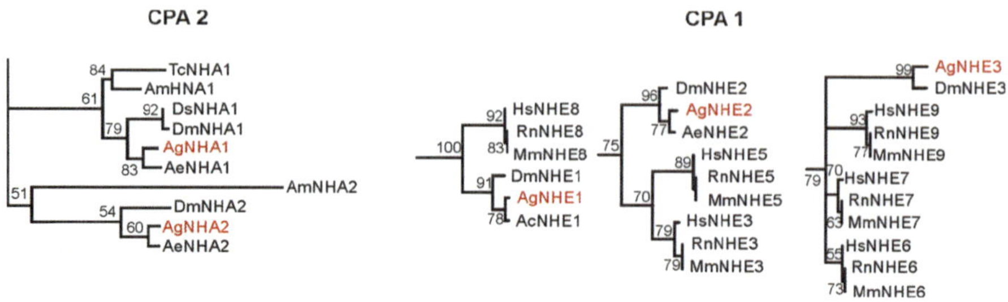

Fig. 14 Phylogeny of AgNHAs and AgNHEs in the CPA2 and CPA1 subfamilies (from Fig. 5)

be voltage gradient driven, electrophoretic Na^+/nH^+ antiporters. Moreover, the isolation of insect NHAs suggests that they might be different from mammalian NHAs and be good targets for the development of safe mosquitocides whereas the close proximity of insect NHEs to mammalian NHEs suggests that they would be poor targets for mosquitocide development. These immunolocalizations of AgNHA1 suggest that it plays a key role in maintaining the characteristic longitudinal pH gradient in the lumen of the AC of *A. gambiae* larvae [66]. To evaluate this hypothesis cRNA encoding AgNHA1 was injected into *Xenopus laevis* oocytes and current/voltage (I/V) measurements were made. Modified excerpts from that study are presented in detail below because, like the first heterologous characterization of an NHE in *Xenopus* oocytes [71], they represent the first attempt to characterize an NHA in oocytes; however, the results raise more problems than they solve and are best viewed as background for future investigations.

Heterologous expression. AgNHA1 cRNA was obtained by in vitro transcription of *Pme*I-linearized pXOOM-AgNHA1 plasmids carrying the gene for green fluorescent protein. It was injected into collagenase-treated stage V-VI *Xenopus* oocytes which were incubated in sterile N98 oocyte medium which contains (in mmol l^{-1}) NaCl 98.0, KCl 2.0, $MgCl_2$ 0.5, $CaCl_2$ 0.5, Hepes 10 (adjusted to pH 7.2 with NaOH), which was supplemented with 2.5 mmol l^{-1} sodium pyruvate, 100 units ml^{-1} penicillin, 0.1 mg ml^{-1} streptomycin, and 2.5% horse serum.

AgNHA1 is a membrane protein. The hydropathy plots of AgNHA1 reveal the 12-transmembrane domains that are typical of SLC9 transporters residing in membranes [76]. Their immunolocalization on plasma membranes throughout the AC demonstrates that they reside in mosquito plasma membranes [66]. AgNHA1 cRNA was also expressed in mammalian plasma membranes. Thus Sergio Grinstein and associates injected cRNA encoding AgNHA1-green fluorescent protein (AgNHA1-GFP) into Chinese hamster ovary (CHO) mutant cells that lack an endogenous Na^+/H^+ exchanger. They found prominent labeling in the central region, in scattered organelles and, most significantly, in the plasma membrane (Fig. 15, left). It is clear that AgNHA1-GFP is targeted to the plasma membrane in CHO cells. However, despite intensive efforts no NHA activity was detected in assays using fluorescent dyes, perhaps due to incompatibility of CHO boundary lipids with the mosquito boundary lipids to which AgNHA1 is adapted (Sergio Grinstein, personal

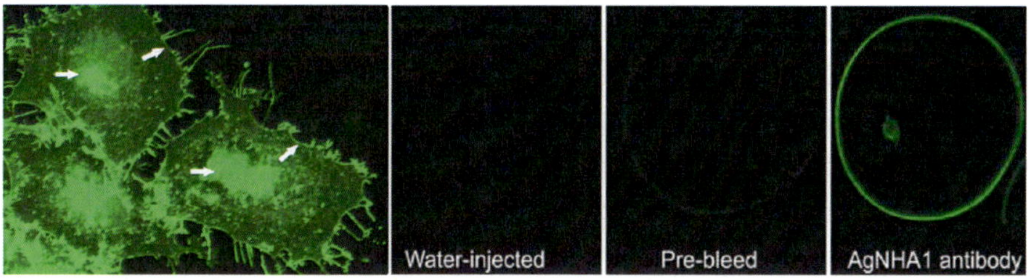

Fig. 15 Plasma membrane localization of AgNHA1 in Chinese hamster ovary cells and *Xenopus* oocytes (*Left*, courtesy of Sergio Grinstein; *right*)

communication). The plasma membrane of whole mounts of AgNHA1 RNA-GFP-injected *X. laevis* oocytes also fluoresced (data not shown) and was strongly labeled by antibody to AgNHA1 (Fig. 15, right). Neither oocytes injected with pre-immune serum or with water (Fig. 15, central) were labeled. The GFP and antibody localizations together demonstrate that AgNHA1 is targeted to the plasma membrane in two heterologous expression systems, CHO cells and *Xenopus* oocytes, as it is in the *A. gambiae* AC.

Electrophysiological measurements in AgNHA1-RNA-injected oocytes. X. laevis oocytes were continuously perfused at a rate of 2 ml min^{-1} with N98 solution at room temperature. The chamber solution was grounded via two, 2 mol l^{-1} KCl/agar bridges that connected through Ag/AgCl reference electrodes to a virtual ground of a current monitor head stage. Oocytes were voltage-clamped at a holding potential, V_h, of –50 mV using a standard two-electrode voltage-clamp device (OC 725A, Warner, New Haven, Conn., USA). Microelectrodes were made from 1.2-mm borosilicate glass capillary tubing and filled with 1 mol l^{-1} KCl, (0.5–1 MΩ tip resistance). To obtain current/voltage (I/V) relationships, 19 voltage steps from +60 to –120 mV with 10 mV intervals and 0.8 s duration were applied. Currents were recorded during the last 0.2 s of the voltage step when they were stable with no relaxation. Experiments were performed on several oocytes from different donors under each condition. Values depicted in graphs represent the mean ± s.e.m. Since the current was invariably linear from 0 to ∼ –60 mV the conductance was calculated as the slope of the IV relationship in this range and the current that was measured at –60 mV (I_{60}) was used as a representative value for each experiment.

AgNHA1-RNA injection results in electrophoretic transport. I/V curves from 28 AgNHA1-cRNA injected oocytes and 18 water-injected oocytes were measured. The raw data for currents exhibited large variations but when the values were normalized by dividing by the current values at +10 mV the curves were nearly congruent. In all 46 experiments the currents were invariably inward with reversal potentials of ∼10 mV. The average current and conductance in AgNHA1-expressing oocytes was ∼2-times that in controls at a confidence level <5%. The currents were rather small to be carried by a typical channel. The results suggest that the currents are carried by ion fluxes that are mediated by a transporter in the AgNHA1-RNA-injected oocytes that is driven by the voltage, i.e. a transporter that is electrophoretic.

Whether the transporter is an endogenous one activated by the RNA injection or is AgNHA1 itself is open to question.

AgNHA1-injected oocytes are more acidic than controls. The next question is whether AgNHA1 actually exchanges H^+ for Na^+. Strong evidence for inward H^+ translocation is the difference in initial intracellular pH in AgNHA1-injected oocytes compared to controls. The average pH measured with an ion-selective microelectrode in 10 AgNHA1-injected oocytes was 0.3 units less than controls, corresponding to a 2-fold increase in acidity in antiporter-expressing oocytes over controls [37].

Currents and conductance in AgNHA1 oocytes are not pH dependent. There was little difference in inward current in three sets of paired experiments at pH 6, 7, or 8. The average inward current at –60 mV potential for AgNHA1-injected oocytes was 157 nA compared to 57 nA in water-injected controls.

I/V curves in AgNHA1-injected oocytes are Na^+ dependent. When Na^+ was replaced by N-methyl-D-glucamine ($NMDG^+$) in NHA-injected oocytes the current at –60 mV dropped from 100 to 40 nA compared to a decrease from 60 to 50 nA in controls. Although the I/V trace in AgNHA1-injected oocytes is Na^+ dependent it is not clear whether the foreign transporter or an endogenous source is responsible.

I/V curves in AgNHA1-injected oocytes are K^+ dependent. AgNHA1-injected oocytes were measured with Na^+ replaced by K^+. The average values in Na^+ were: $V_r = -20$ mV, $I_{60} = 50$ nA, and $G_{0\text{-}60} = 1.4\,\mu S$ whereas in K^+ the average current had increased to 100 nA and the conductance to 1.7 μS then returned to the original values when Na^+ was restored. In the control oocytes the current and conductance were both less than in experimental oocytes and were unaffected by the change from Na^+ to K^+. These results show that the current in AgNHA1-injected oocytes can be carried by K^+ as well as Na^+ but again the carrier remains uncertain.

Accelerating increase in current and conductance at very negative membrane potentials. In all of the experiments the inward current and conductance from 0 to –60 mV in NHA-injected oocytes were much greater than in controls and in nearly all of the experiments the currents increased sharply at voltages more negative than –70 mV. Typical results are shown in Fig. 16. NHA-injected oocytes have a current of ~140 nA and a conductance of ~1.6 μS at a potential of –60 mV whereas water-injected controls have a current of ~40 nA and a conductance of ~1.1 μS at both pH 6 and 8. There is little difference in current or conductance between pH 8 and pH 6 in either AgNHA-injected oocytes or controls in Fig. 16. However, in virtually all of the experiments on AgNHA1-injected oocytes and to a lesser extent in the water-injected controls the conductance increased at an accelerating rate at potentials more negative than –70 mV. For example in Fig. 16 the average current at pH 6 and 8 in AgNHA1-cRNA-injected oocytes was ~600 nA at a potential of –120 mV. The conductance measured from –100 to –120 mV was 10 μS at pH 8 and more than 20 μS at pH 6.

The significance of conductances in *Xenopus* oocytes injected with foreign cRNA encoding foreign transporters (but not channels that have much larger conductances) is brought into question by the presence of endogenous cation channels and/or transporters [88]. In particular I/V relationships in oocytes injected with cRNA encoding

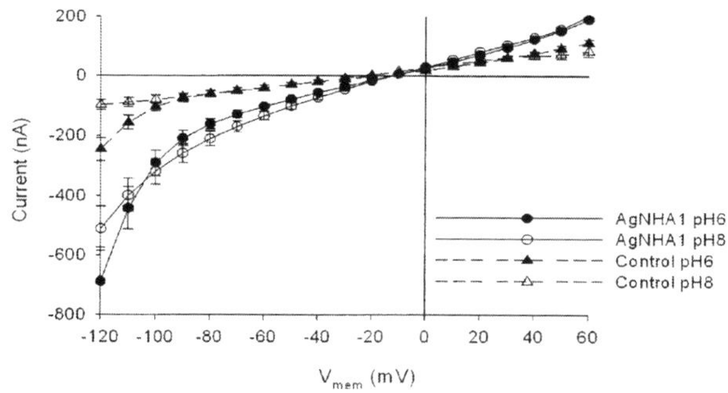

Fig. 16 Large increases in current and conductance at very negative V_m in *Xenopus* oocytes [37]

Plasmodium falciparum chloroquine resistant transporter (PfCRT) [65] and those encoding *A. aegypti* Na^+/H^+ exchanger 8 (AeNHE8; Brett's AeNHE1) [71] are very similar to those in oocytes expressing AgNHA1 [37].

I/V plot from oocytes expressing PfCRT. The *P. falciparum* chloroquine resistance (PfCRT) gene has been the subject of many publications. Oocytes injected with the corresponding cRNA have a higher intracellular pH and a lower membrane potential than water-injected controls. Nessler et al. [65] propose that the PfCRT cRNA is activating two conductances, one due to an endogenous NHE and the other due to activation of an endogenous, nonspecific cation channel. Although the currents are smaller and the reversal potentials more negative and variable, the trace for PfCRT-injected oocytes in regular Ringer solution with high 96 mM Na^+ (Fig. 17) are similar to those for AgNHA1 in 98 mM Na^+ in Fig. 16. Although Nessler et al. suggested that the endogenous NHE is electroneutral the strong voltage-dependence in Fig. 17 suggests that it is electrophoretic.

I/V plot from oocytes expressing AeNHE1. The plot for oocytes expressing AeNHE1 (Fig. 18) [71] is very similar to that for oocytes expressing AgNHA1 (Fig. 16). The current at –60 mV is ~75 nA and the conductance is linear from 0 to –60 at ~2.7 μS. The current increases at an accelerating rate beyond –120 mV.

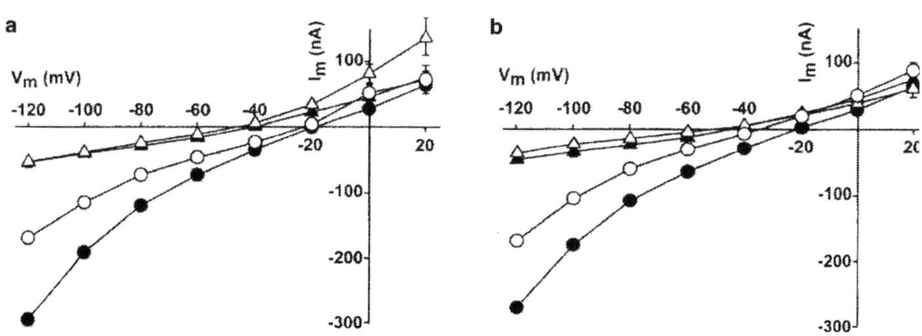

Fig. 17 I/V plot from oocytes expressing PfCRT; *triangles*, controls; *circles* PfCRT; open A, Cl^- free; open B Na^+ free ([65], their Figs. 5b, c)

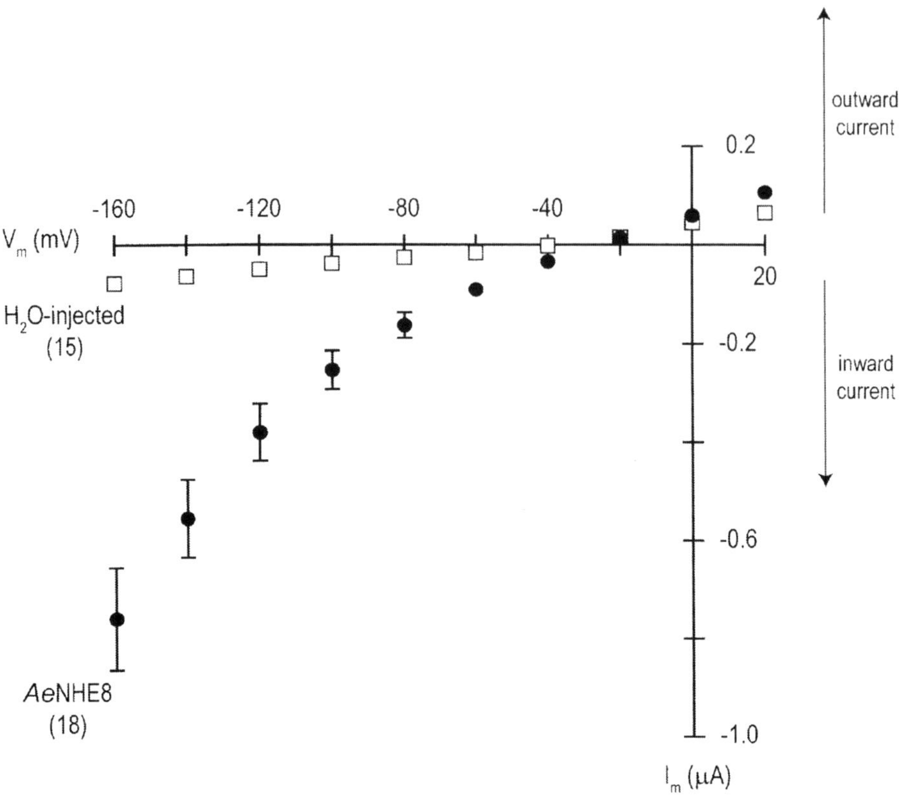

Fig. 18 I/V plot from *Xenopus* oocytes injected with water (*open squares*) or cRNA encoding AeNHE8 (*closed circles*); Piermarini et al. [71]

Piermarini et al. attribute the I/V trace to an endogenous, cation-selective channel or transporter. They used ion-selective electrodes to study the properties of the expressed AeNHE1 (8). However, if their injected cRNA is activating an endogenous NHE as suggested for PfCRT by Nessler et al. then the pHi would be expected to change. Therefore, pHi values measured with intracellular microelectrodes are not free from influence by the endogenous transporter/channel and again the results must be interpreted with caution.

I/V plot from oocytes expressing MsKAAT1 and AeAAT1. The I/V traces for oocytes injected with cRNA encoding MsKAAT1 in the presence of K^+ (Fig. 19 left) or with cRNA encoding AeAAT1 in the presence of Na^+ (Fig. 19 right) are very similar to those of oocytes injected with AgNHA1, PfCRT, AeNHE1 (8), and cRNA from other sources [88]. In the case of MsKAAT1 [17], AgNAT6 [59], and the three other characterized insect NATs the inward current depends upon the co-transported amino acid and the analysis is valid.

Futility of oocytes for characterizing exogenous NHEs and NHAs. Although many channels, with currents in the microampere range, have been characterized

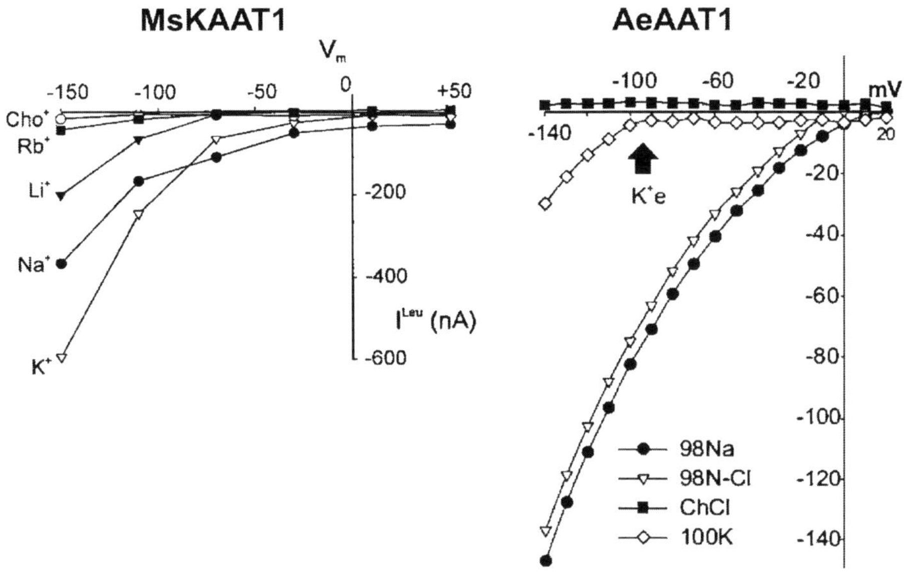

Fig. 19 I/V plot from oocytes injected with cRNA from MsKAAT1 and AeAAT1 (ordinate is nA) [8]

by two-electrode voltage clamp measurements after expression in *Xenopus* oocytes, the characterization of transporters with currents in the nanoamperes range are compromised by endogenous conductances in this same range. Therefore, all of the published attempts to characterize NHEs and NHAs by two-electrode voltage clamps must be viewed with caution. They all suffer from the complication of activation of endogenous cation-selective channels or transporters, which Nessler et al. [65] have shown to be separate membrane proteins. Recognizing this problem Gill's group [48] and Dow's group [24] have done their heterologous characterizations in yeast or other systems. Nevertheless, to determine whether or not AgNHA1 or any other transporter is electrophoretic (voltage-driven) over-expression in oocytes still remains essential. The best hope at present appears to be to clone and characterize the endogenous transporter(s) and channel(s) from *X. laevis* and/or to find inhibitors to block these membrane proteins.

5 Proton Paradigm and Mosquito Midgut Model

5.1 Interpretation of Data from Heterologous cRNA Expression in Xenopus *Oocytes*

Among many explanations of the large accelerating conductances shown in Figs. 16, 17, 18 and 19 from oocytes are the following:

1. Thermodynamics are flawed because reversal potentials are pH independent
2. Currents are too large for transporters and are artifacts due to leaky oocytes
3. Inward currents are carried by cations via nonspecific endogenous channels
4. Inward currents are carried by cations via endogenous Na^+ or K^+ channels or transporters
5. Inward currents are carried by an endogenous, electrophoretic N^+ nH^+ antiporter (XeNHA)
6. Inward currents are carried by a XeNHA component and a larger AgNHA1 component

Explanations 1–3 seem unlikely for the reasons discussed above. Explanations 4–6 all invoke transporters, whether endogenous or expressed via the injected cRNAs, which in all cases appear to be electrophoretic. The existence of electrophoretic cation exchangers augments the earlier reports of electrophoretic nutrient amino acid transporters and provides molecular support for expanding the membrane energization paradigm in metazoan animals. In the existing paradigm Na^+/K^+ P-ATPase energizes membranes by establishing an outward K^+ gradient (ΔK^+) and an inward Na^+ gradient (ΔNa^+) that drives electroneutral secondary transporters such as Na^+/H^+ exchangers (NHEs). In the expanded paradigm H^+ V-ATPase energizes membranes by establishing an inside negative voltage gradient ($\Delta\Psi$) that drives electrophoretic secondary transporters such as Na^+-coupled amino acid symporters (NATs) and the postulated Na^+ or K^+ antiporters (NHAs). In brief, the Na^+ gradient paradigm is expanded by a voltage-gradient component. We have reviewed evidence which suggests but does not prove that AgNHA1 is an electrophoretic, Na^+/H^+ antiporter that is H^+-, Na^+- and K^+-dependent. Together with its immunolocalization in specific membranes [66] these results suggest that AgNHA1 is like EcNhA1 and other bacterial antiporters in that it uses $\Delta\Psi$ to drive nH^+ into cells and Na^+ or K^+ out of cells. However, the $\Delta\Psi$ is generated by an H^+ V-ATPase in eukaryotic cells rather than by the electron transport system, as in prokaryotic cells, thereby simplifying the analysis.

5.2 An Electrophoretic AgNHA1 and the Membrane Potential Energization Paradigm

Peter Mitchell's chemiosmotic hypothesis that ATP is synthesized by using the energy from a proton electrochemical gradient ($\Delta\Psi + \Delta pH$) across the mitochondrial inner membrane to synthesize ATP via the F-ATP synthase was greeted with skepticism if not outright derision when it was first proposed. Now we know that chemiosmotic ATP synthesis by oxidative phosphorylation is far more efficient than substrate-level ATP synthesis by glycolysis and that chemiosmotic ATP synthesis is found in all aerobic organisms. The use of the electron transport system to create an electrical potential and pH difference that can drive a variety of energy-requiring reactions, such as sugar and amino acid transport in bacteria and other prokaryotes is also widely accepted [52, 53]. The use of the H^+ V-ATPase to create a potential difference to drive membrane transport in yeast and in endomembranes of all eukaryotic animals is also widely accepted [64]. However, many scientists still maintain that H^+ V-ATPases mainly operate in vacuoles (hence their name) and energize plasma membranes only in some animals or in "specialized" animal cells [77]. Nevertheless, the list of metazoans with plasma membrane H^+ V-ATPases grows daily [47]. The new question is: can the voltage generated by H^+ V-ATPases drive secondary transport of solutes such as amino acid uptake and cation exchange. Regarding amino acid uptake the answer is yes. As discussed above caterpillar and mosquito larval midguts possess several Na^+-coupled nutrient amino acid transporters (NATs) which use the voltage generated by a plasma membrane H^+ V-ATPase to drive Na^+ or K^+ coupled amino acid uptake by midgut cells [11, 58]. But can voltages drive cation exchange? All currently characterized NHEs use the inwardly directed Na^+ gradient established by the Na^+/K^+ P-ATPase to drive Na^+ into cells and expel metabolically produced H^+ [69].

However, in many phyla but not those of metazoans NHAs are known to be electrophoretic transporters. The best example is the EcNhA1 from *E. coli* which couples the electron transport system-generated voltage to drive $2H^+$ into their cells and expel the Na^+ that has leaked in from the caustic medium [53, 70]. These bacterial NHAs differ from metazoan NHEs not only in being electrophoretic rather than electroneutral but in moving H^+ into cells and Na^+ out: moreover they are present in every genome yet published, including the human genome. However, unlike metazoan NHEs which have been cloned and characterized from many cells [69], metazoan NHAs have been cloned and partially characterized only recently – AgNHA1 from *A. gambiae* larval midgut [66, 76], HsNHA1 from *Homo sapiens* [93], and DmNHAs 1 and 2 from *D. melanogaster* [23]. The evidence reported in those publications suggests that AgNHA1 is indeed an electrophoretic antiporter that is K^+, Na^+, and H^+ dependent. It suggests that, like prokaryotic NHAs, which use $\Delta\Psi$ from H^+ F-ATPase to move $1Na^+$ out of cells and $2H^+$ in [81], AgNHA1 uses $\Delta\Psi$ from H^+ V-ATPase to eject $1Na^+$ or K^+ and take $2H^+$ in.

5.3 AgNHA1 and Native XeNHA Both May Have Voltage-Activated Channel Components?

The current in AgNHA1-injected oocytes had increased to ~500 nA at –120 mV and the conductance to >20 μS in the range from –100 to –120 mV (Fig 16). This increased current and conductance of AgNHA1-injected oocytes might suggest that an antiporter is present but has voltage-activated channel properties. Moreover, the conductance in the controls at pH 6 had also increased to 10 μS at –100 to –120 mV. Recall that NHAs are present in all genomes so *X. laevis* must possess an endogenous NHA (XeNHA). The parallel properties of AgNHA1-cRNA-injected and water-injected oocytes imply that either or both of the heterologously expressed

AgNHA1 or the endogenous XeNHA are electrophoretic, possibly with a voltage-activated channel component. It may be permissible to speculate that, XeNHA, along with AgNHA1, HsNHA2, DmNHA1, and DmNHA2 are all putative electrophoretic antiporters. Perhaps all metazoan NHAs are electrophoretic. Although this speculation is attractive, the accelerating increase in current and conductance must be treated with caution because as discussed above they could be due to an endogenous cation channel in the oocyte membrane that is activated by injected AgNHA1 cRNA but is not associated with any NHA. Alternatively, these anomalous results may be due to incompatibility of AgNHA1 boundary lipids with the xenic boundary lipids in *Xenopus* oocytes. Long ago the loss of Ca^{2+} transport activity but the retention of ATPase activity was demonstrated when boundary lipids were removed from the endoplasmic reticulum Ca^{2+} P-ATPase [42]. Finally, these large values may simply reflect increasing leakiness of oocytes that are maintained in abnormally high Na^+ concentrations at extremely negative voltages, neither of which are encountered by *Xenopus* oocytes under natural conditions.

5.4 AgNHA1-like Bacterial NHA is Oriented to Absorb H^+ and Secrete Na^+

The H^+ V-ATPase always transports H^+ outwardly and renders the outside of the plasma membrane electrically positive to the inside. This voltage polarity would drive H^+ inwardly and Na^+ outwardly only if the stoichiometry of the antiporter is 1 K^+ or 1Na^+ exchanging for 2 or more H^+ as it is in the caterpillar midgut [4]. That AgNHA1 moves H^+ inwardly and K^+ or Na^+ outwardly was deduced from its immunolocalization with respect to the H^+ V-ATPase in mosquito larvae [66]. This transport orientation is like that in alkalophilic bacteria in which the positive outside voltage generated by the electron transport system provides the energy by which bacterial NHAs transport 2H^+ inwardly and 1Na^+ outwardly [81].

5.5 Larval Mosquito Alimentary Canal Model

A. *Rationale for model.* Even if the electrical data on AgNHA1 in oocytes is discounted, previously published work on transporters in the *A. gambiae* larval midgut, especially Okech et al. [66], Rheault et al. [76], and Smith et al. [79] can be integrated into a working model (Fig. 20). As discussed above, this integration is possible because Na/K P-ATPase always moves 3Na^+ out of cells and 2 K^+ in, H^+ V-ATPase always moves H^+ out of cells with no counter ion and we assume that AgNHA1 always moves Na^+ or K^+ out of cells and 2H^+ in, using the H^+ V-ATPase-generated hyper-polarization of both apical and basal membranes of AC cells.
B. *Opening to gastric caecum of anterior midgut.* H^+ is removed from the gastric caecal cavity by exchange for cellular Na^+ across the apical membrane of the

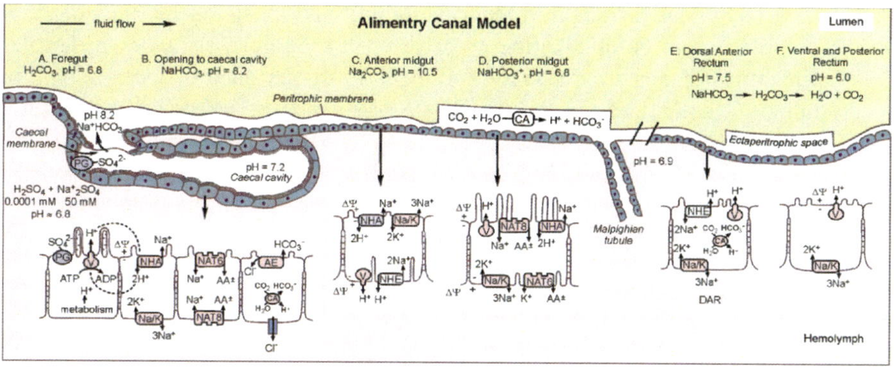

Fig. 20 Model of pH regulation and amino acid uptake in larval mosquito AC

epithelial cells mediated by AgNHA1 driven by the outside voltage generated by the H^+ V-ATPase in the same membrane [76]. We speculate that this process is aided by fixed, deprotonated, $SO_4{}^{2-}$ groups that protrude from the extracellular surface of the epithelial cells and provide a strong anion, resulting in a local, unstirred layer of $H^+{}_2\ SO_4{}^{2-}$, analogous to that deduced by Dow et al. [27] from X-ray microanalysis of fixed sulfate in caterpillar goblet cavity. The fixed sulfate serves as gegenion for H^+ which AgNHA1 drives back into the cells while ejecting Na^+ or K^+ to the lumen. The resulting Na^+ or K^+ $HCO_3{}^-$ diffuses across the caecal membrane into the anterior midgut lumen where $SO_4{}^{2-}$ is absent and the pH rises to 8.2 (the pKa of $NaHCO_3$ or $KHCO_3$ buffer).

C. *Anterior midgut AgNHA1 in the apical membrane exchanges cellular Na^+ for luminal H^+* which is expelled to the blood across the basal membrane by the H^+ V-ATPase. The result is high lumen $2Na^+$ or $2\ K^+$ and $CO_3{}^{2-}$ which brings the pH to 10.5 (the pKa of Na_2CO_3 or K_2CO_3 buffer) in anterior midgut. The AgNHA1 was actually observed in vesicles between the nucleus and apical membrane [66] and one must assume that under appropriate conditions the vesicles fuse with the apical membrane, thereby inserting the NHA into the membrane. Such a fusion is well documented in the case of H^+ V-ATPase-containing vesicles that fuse with the apical plasma membrane of kidney collecting duct cells under acid stress [14].
D. *Posterior midgut* AgNHA1 plays a key role in recycling Na^+ to the lumen after it has been symported into the cells along with essential amino acids by AgNAT8 [58]. The AgNHA is localized on the apical membrane along with the H^+ V-ATPase and AgNAT8; it would have to be electrophoretic to drive $2H^+$ in and Na^+ back out. The presence of electrogenic H^+ V-ATPase and the electrophoretic Na^+: amino acid$^\pm$ symporter (AgNAT8) in the same membrane constitutes a new kind of NHE that we have called NHE_{VNAT}. Starting in middle midgut and continuing through posterior midgut, we postulate that NHE_{VNAT} drives H^+ back to the lumen and Na^+ into the cells thereby restoring lumen pH to neutrality, where it persists from posterior midgut to hindgut. Horst Onken and David Moffett

[67, 68] have new insight into anterior midgut alkalinization by showing that the cytoplasm can be as alkaline as pH 8 or 9; they have also shown that de-alkalinization of posterior midgut continues in the absence of amino acids in the lumen. This finding implies that NHE_{VNAT} is not the only de-alkalinization mechanism and implies that one of the three AgNHEs must be present there too. During molting there would be no amino acids in the lumen, no Na^+: amino acid symport and thus no NHE_{VNAT} so an AgNHE must be present to expel metabolic acids and de-alkalinize the lumen; indeed Gill and associates have identified one [48].

E. *Rectum.* Figure 7a is a dorsal view showing that Na^+/H^+ P-ATPase is not detectable in the dorsal anterior rectal cells (DAR cells) whereas it is present in the ventral view (Fig. 7B) [79]. In the posterior rectum before the luminal contents pass to the environment, we postulate that residual Na^+ is absorbed from the lumen by Na^+/K^+ P-ATPase and $KHCO_3$ is excreted to the environment while the scarce Na^+ is retained. Long ago, R.H. Stobbart showed that K^+ and HCO_3^- are excreted by the larvae of *A. aegypti* [80].

F. *Issues and resolutions.* I_{rev} changes but little with pH (Fig. 16). If the relevant ions in AgNHA1-over-expressing *Xenopus* oocytes are Na^+, K^+, and Cl^- in addition to H^+ then the I_{rev} will depend on their permeability's, equilibrium potentials, and concentrations. Since the concentration of the first three ions is in the millimolar range whereas that of H^+ between pH 6 and 8 is in the micromolar to nanomolar range, the effects of H^+ on I_{rev} would be only ~1/1,000th that of the other ions and would not be detectable. Moreover, voltages from H^+ V-ATPases and Na^+/K^+ P-ATPases would tend to stabilize the reversal potentials.

6 Conclusions

Proton and voltage gradients have dominated membrane bioenergetics since light energy was shown to be trapped as proton gradients to initiate photosynthesis. Proton electrochemical gradients generated by the mitochondrial electron transport system provide the energy that F-ATP synthase uses to make ATP. The same voltage gradient provides the energy that the electrophoretic EcNhA1 antiporter uses to drive $2H^+$ into cells and eject Na^+ that has leaked in from the alkaline environment. Paradoxically, hydrogen ions are available at the membrane surface because after ejection by the electron transport system they remain at the surface 1,000-times longer than they move to the bulk solution (reviews: Cherepanov et al. [20, 60, 61]). We postulate that a similar proton-rich layer exists at the surface of membranes to which the H^+ V-ATPase has ejected them. A rare exception to the proton rule is the Na^+ gradient that is generated by the Na^+/K^+ P-ATPase. The Na^+ gradient drives electroneutral Na^+/H^+ exchangers (NHEs) that translocate $1Na^+$ into the cells while ejecting $1H^+$ and thus removing metabolic acids. It has always seemed unlikely that metazoans would be the only taxa that rely solely on Na^+ gradients and do not use voltage gradients for secondary ion transport, especially since all metazoans that have been studied possess electrogenic H^+ V-ATPases and NHAs which may be electrophoretic. We have cloned and localized AgNHA1 and deduced that it uses the voltage from an H^+ V-ATPase to drive $2H^+$ into cells and $1Na^+$ or 1 K^+ out [66, 76]. Simultaneously, Xiang et al. [93] showed that HsNHA2 purchased from the *H. sapiens* genome project plays a role in Li^+/Na^+ exchange. Recently, Day et al. [23] cloned both DmNHA1 and DmNHA2 from *D. melanogaster* and implicated them in ion and pH regulation in MTs. Since all published genomes contain NHAs it is reasonable to ask if all of them, like EcNhA1 and tentatively AgNHA1, are electrically driven mediators of nH^+ influx and Na^+ or K^+ efflux. If so then the vast new field of electrical coupling in eukaryotic plasma membranes that was introduced by work on caterpillar midguts many years ago will have been opened to post-genomic exploitation.

Acknowledgments We thank Dr. Sergio Grinstein for the fluorescent image of AgNHA1-GFP expressing CHO cells, Dr. Mark R. Rheault for the data on hemolymph K^+ activities in *A. gambiae* larvae and Drs. Olga Vitavska and Helmut Wieczorek for the information on immunolabeling caterpillar membranes. We thank Dr. Grinstein, Dr. Helmut Wieczorek, Dr. Nathan Nelson, Dr. Walter Boron, Dr. Subrata Tripathi, and Dr. David Price for many helpful discussions and suggestions but absolve them of any responsibility for controversial aspects ofthis chapter. This work was supported in part by Research Grants AI-52436 and AI-30464 from NIH and by funds from the Whitney Laboratory, the Emerging Pathogens Institute and the Department of Epidemiology and Biostatistics at the University of Florida.

References

1. Ahearn, G. A., Franco, P. and Clay, L. P. (1990). Electrogenic 2 $Na^+/1$ H^+ Exchange in Crustaceans. *J Membrane Biol* **116**, 215–226.
2. Anderson, E. and Harvey, W. R. (1966). Active transport by the Cecropia midgut. II. Fine structure of the midgut epithelium. *J Cell Biol* **31**, 107–137.
3. Arkin, I. T., Xu, H., Jensen, M. O., Arbely, E., Bennett, E. R., Bowers, K. J., Chow, E., Dror, R. O., Eastwood, M. P., Flitman-Tene, R. et al. (2007). Mechanism of Na^+/H^+ antiporting. *Science* **317**, 799–803.
4. Azuma, M., Harvey, W. R. and Wieczorek, H. (1995). Stoichiometry of K^+/H^+ antiport helps to explain extracellular pH 11 in a model epithelium. *FEBS Lett* **361**, 153–156.
5. Berridge, M. J. (1966). Metabolic pathways of isolated malpighian tubules of the blowfly functioning in an artificial medium. *J Insect Physiol* **12**, 1523–1538.
6. Berridge, M. J., Gupta, B. L., Hall, T. A., Maddrell, S. H., Moreton, R. B. and Wall, B. J. (1977). Electron microprobe studies of electrolyte distribution in fluid transporting epithelia [proceedings]. *J Physiol* **266**, 32P–33P.
7. Beyenbach, K. W. (2001). Energizing epithelial transport with the vacuolar H($^+$)-ATPase. *News Physiol Sci* **16**, 145–151.
8. Boudko, D. Y., Kohn, A. B., Meleshkevitch, E. A., Dasher, M. K., Seron, T. J., Stevens, B. R. and Harvey, W. R. (2005a). Ancestry and progeny of nutrient amino acid transporters. *Proc Natl Acad Sci USA* **102**, 1360–1365.
9. Boudko, D. Y., Moroz, L. L., Harvey, W. R. and Linser, P. J. (2001a). Alkalinization by chloride/bicarbonate pathway in larval mosquito midgut. *Proc Natl Acad Sci U S A* **98**, 15354–15359.
10. Boudko, D. Y., Moroz, L. L., Linser, P. J., Trimarchi, J. R., Smith, P. J. and Harvey, W. R. (2001b). In situ analysis of pH gradients in mosquito larvae using non-invasive, self-referencing, pH-sensitive microelectrodes. *J Exp Biol* **204**, 691–699.

11. Boudko, D. Y., Stevens, B. R., Donly, B. C. and Harvey, W. R. (2005b). Nutrient Amino acid and Neurotransmitter transporters. In *Comprehensive Molecular Insect Science*, vol. 4 (ed. K. Iatrou, S. Gill, Lawrence I. Gilbert), pp. 255–309. Amsterdam: Elsevier.
12. Brett, C. L., Donowitz, M. and Rao, R. (2005). Evolutionary origins of eukaryotic sodium/proton exchangers. *Am J Physiol Cell Physiol* **288**, C223–239.
13. Brett, C. L., Donowitz, M. and Rao, R. (2006). Does the proteome encode organellar pH? *FEBS Lett* **580**, 717–719.
14. Brown, D. (1989). Vesicle recycling and cell-specific function in kidney epithelial cells. *Annu Rev Physiol* **51**, 771–784.
15. Brown, D. and Breton, S. (1996). Mitochondria-rich, proton-secreting epithelial cells. *J Exp Biol* **199**, 2345–2358.
16. Brown, D., Smith, P. J. S. and Breton, S. (1997). Role of V-ATPase-rich cells in acidification of the male reproductive tract. *J Exp Biol* **200**, 257–262.
17. Castagna, M., Shayakul, C., Trotti, D., Sacchi, V. F., Harvey, W. R. and Hediger, M. A. (1998). Cloning and characterization of a potassium-coupled amino acid transporter. *Proc Natl Acad Sci USA* **95**, 5395–5400.
18. Chao, A. C., Moffett, D. F. and Koch, A. (1991). Cytoplasmic pH and goblet cavity pH in the posterior midgut of the tobacco hornworm *Manduca sexta*. *J Exp Biol* **155**, 403–414.
19. Chatterjee, D., Chakraborty, M., Leit, M., Neff, L., Jamsakellokumpu, S., Fuchs, R., Bartkiewicz, M., Hernando, N. and Baron, R. (1992). The Osteoclast Proton Pump Differs in Its Pharmacology and Catalytic Subunits from Other Vacuolar H^+-ATPases. *J Exp Biol* **172**, 193–204.
20. Cherepanov, D. A., Junge, W. and Mulkidjanian, A. Y. (2004). Proton transfer dynamics at the membrane/water interface: dependence on the fixed and mobile pH buffers, on the size and form of membrane particles, and on the interfacial potential barrier. *Biophys J* **86**, 665–680.
21. Cidon, S. and Nelson, N. (1982). Properties of a Novel ATPase Enzyme in Chromaffin Granules. *J Bioenerg Biomembr* **14**, 499–512.
22. Cioffi, M. and Wolfersberger, M. G. (1983). Isolation of separate apical, lateral and basal plasma membrane from cells of an insect epithelium. A procedure based on tissue organization and ultrastructure. *Tissue Cell* **15**, 781–803.
23. Clements, A. N. (1992). The Biology of Mosquitoes. London: Chapman and Hall Press.
24. Day, J. P., Wan, S., Allan, A. K., Kean, L., Davies, S. A., Gray, J. V. and Dow, J. A. (2008). Identification of two partners from the bacterial Kef exchanger family for the apical plasma membrane V-ATPase of Metazoa. *J Cell Sci* **121**, 2612–2619.
25. Dean, R. B. (1941). Theories of electrolyte equilibrium in muscle. *Biological Symposia* **3**, 331.
26. Donini, A., Gaidhu, M. P., Strasberg, D. R. and O'Donnell, M. J. (2007). Changing salinity induces alterations in hemolymph ion concentrations and Na^+ and Cl^- transport kinetics of the anal papillae in the larval mosquito, *Aedes aegypti*. *J Exp Biol* **210**, 983–992.
27. Dow, J. A. (1984). Extremely high pH in biological systems: a model for carbonate transport. *Am J Physiol* **246**, R633–R636.
28. Dow, J. A., Gupta, B. L., Hall, T. A. and Harvey, W. R. (1984a). X-ray microanalysis of elements in frozen-hydrated sections of an electrogenic K^+ transport system: the posterior midgut of tobacco hornworm (*Manduca sexta)* in vivo and in vitro. *J Membr Biol* **77**, 223–241.
29. Dow, J. A. T. and Peacock, J. M. (1989). Microelectrode evidence for the electrical isolation of goblet cell cavities in *Manduca sexta* middle midgut. . *J Exper Biol* **143**, 101–114.
30. Ehrenfeld, J., Garcia-Romeu, F. and Harvey, B. J. (1985). Electrogenic active proton pump in *Rana esculenta* skin and its role in sodium ion transport. *J Physiol* **359**, 331–355.
31. Fernandez Moran, H., Oda, T., Blair, P. V. and Green, D. E. (1964). A macromolecular repeating unit of mitochondrial structure and function. correlated electron microscopic and biochemical studies of isolated mitochondria and submitochondrial particles of beef heart muscle. *J Cell Biol* **22**, 63–100.

32. Gluck, S., Kelly, S. and Al-Awqati, Q. (1982). The proton translocating ATPase responsible for urinary acidification. *J Biol Chem* **257**, 9230–9233.
33. Gruber, G., Radermacher, M., Ruiz, T., Godovac-Zimmermann, J., Canas, B., Kleine-Kohlbrecher, D., Huss, M., Harvey, W. R. and Wieczorek, H. (2000). Three-dimensional structure and subunit topology of the V(1) ATPase from M*anduca sexta* midgut. *Biochemistry* **39**, 8609–8616.
34. Gupta B. L., Berridge M. J. (1966). A coat of repeating subunits on the cytoplasmic surface of the plasma membrane in the rectal papillae of the blowfly, *Calliphora Erythrocephala* (Meig.), studied in situ by electron microscopy. *J Cell Biol 29,* 376–382.
35. Harvey, W. R. (1992). Physiology of V-ATPases. *J Exp Biol :* **172**, 1–17.
36. Harvey, W. R. (2009). Voltage coupling of primary H^+ V-ATPases to secondary Na^+- or K^+-dependent transporters. *J Exp Biol* **212**, 1620–1629.
37. Harvey, W. R., Boudko, D. Y., Rheault, M. R. and Okech, B. A. (2009). NHEVNAT: an H^+ V-ATPase electrically coupled to a Na^+:nutrient amino acid transporter (NAT) forms an Na^+/H^+ exchanger (NHE). *J Exp Biol* **212**, 347–357.
38. Harvey, W. R., Cioffi, M., Dow, J. A. and Wolfersberger, M. G. (1983). Potassium ion transport ATPase in insect epithelia. *J Exp Biol* **106**, 91–117.
39. Harvey, W. R., Cioffi, M. and Wolfersberger, M. G. (1981). Portasomes as coupling factors in active ion transport and oxidative phosphorylation. *Amer Zool* **21**, 775–791.
40. Harvey, B. J. and Ehrenfeld, J. (1986). Regulation of intracellular sodium and pH by the electrogenic H^+ pump in frog skin. *Pflugers Arch – Eur J Physiol* **406**, 362–366.
41. Harvey, W. R. and Nedergaard, S. (1964). Sodium-independent active transport of potassium in the isolated midgut of Cecropia silkworm. *Proc Natl Acad Sci USA* **51**, 757–765.
42. Hidalgo, C. (1982). Lipid-protein interactions and calcium transport in sarcoplasmic reticulum. *Ann N Y Acad Sci* **402**, 561–562.
43. Hunke, C., Chen, W. J., Schafer, H. J. and Gruber, G. (2007). Cloning, purification, and nucleotide-binding traits of the catalytic subunit A of the V1VO ATPase from *Aedes albopictus*. *Protein Expr Purif* **53**, 378–383.
44. Hunte, C., Screpanti, E., Venturi, M., Rimon, A., Padan, E. and Michel, H. (2005). Structure of a Na^+/H^+ antiporter and insights into mechanism of action and regulation by pH. *Nature* **435**, 1197–1202.
45. Ianowski, J. P. and O′Donnell, M. J. (2004). Basolateral ion transport mechanisms during fluid secretion by Drosophila Malpighian tubules: Na^+ recycling, Na^+:K^+:2Cl^- cotransport and Cl^- conductance. *J Exp Biol* **207**, 2599–2609.
46. Ianowski, J. P. and O′Donnell, M. J. (2006). Electrochemical gradients for Na^+, K^+, Cl^- and H^+ across the apical membrane in Malpighian (renal) tubule cells of *Rhodnius prolixus*. *J Exp Biol* **209**, 1964–1975.
47. Jefferies, K. C., Cipriano, D. J. and Forgac, M. (2008). Function, structure and regulation of the vacuolar (H^+)-ATPases. *Arch Biochem Biophys* **476**, 33–42.
48. Kang′ethe, W., Aimanova, K. G., Pullikuth, A. K. and Gill, S. S. (2007). NHE8 mediates amiloride-sensitive Na^+/H^+ exchange across mosquito Malpighian tubules and catalyzes Na^+ and K^+ transport in reconstituted proteoliposomes. *Am J Physiol Renal Physiol* **292**, F1501–F1512.
49. Kell, D. B. (1979). On the functional proton current pathway of electron transport phosphorylation. An electrodic view. *Biochim Biophys Acta* **549**, 55–99.
50. Kell, D. B. (1992). The protonmotive force as an intermediate in electron transport-linked phosphorylation: problems and prospects. *Curr Top Cell Regul* **33**, 279–289.
51. Koefoed-Johnsen, V. and Ussing, H. H. (1958). The nature of the frog skin potential. *Acta Physiol Scand* **42**, 298–308.
52. Krulwich, T. A. (1986). Bioenergetics of alkalophilic bacteria. *J Membrane Biol* **89**, 113–125.
53. Krulwich, T. A. and Guffanti, A. A. (1992). Proton-coupled bioenergetic processes in extremely alkaliphilic bacteria. *J Bioenerg Biomembrane* **24**(6), 587–599.

54. Küppers, J. and Thurm, U. (1979). Active ion transport by a sensory epithelium. I. Transepithelial short circuit current, potential difference, and their dependence on metabolism. *J Comp Phyisol* **134**, 131–136.
55. Lepier, A., Azuma, M., Harvey, W. R. and Wieczorek, H. (1994). K^+/H^+ antiport in the tobacco hornworm midgut: the $K(^+)$-transporting component of the K^+ pump. *J Exp Biol* **196**, 361–373.
56. Maddrell, S. H. P. (1981). The functional design of the insect excretory system. *J Exp Biol* **90**, 1–15.
57. Maddrell, S. H., Pilcher, D. E. and Gardiner, B. O. (1969). Stimulatory effect of 5-hydroxytryptamine (serotonin) on secretion by malpighian tubules of insects. *Nature* **222**, 784–785.
58. Meleshkevitch, E. A., Assis-Nascimento, P., Popova, L. B., Miller, M. M., Kohn, A. B., Phung, E. N., Mandal, A., Harvey, W. R. and Boudko, D. Y. (2006). Molecular characterization of the first aromatic nutrient transporter from the sodium neurotransmitter symporter family. *J Exp Biol* **209**, 3183–3198.
59. Meleshkevitch, E. A., Robinson, M., Popova, L. B., Miller, M. M., Harvey, W. R. and Boudko, D. Y. (2009). Cloning and functional expression of the first eukaryotic Na^+-tryptophan symporter, AgNAT6. *J. Exp. Biol*, 212, 1559–1567.
60. Mulkidjanian, A. Y., Cherepanov, D. A., Heberle, J. and Junge, W. (2005). Proton transfer dynamics at membrane/water interface and mechanism of biological energy conversion. *Biochemistry (Mosc)* **70**, 251–256.
61. Mulkidjanian, A. Y., Heberle, J. and Cherepanov, D. A. (2006). Protons @ interfaces: implications for biological energy conversion. *Biochim Biophys Acta* **1757**, 913–930.
62. Murata, T., Yamato, I., Kakinuma, Y., Shirouzu, M., Walker, J. E., Yokoyama, S. and Iwata, S. (2008). Ion binding and selectivity of the rotor ring of the Na^+-transporting V-ATPase. *Proc Natl Acad Sci U S A* **105**, 8607–8612.
63. Murata, T., Yamato, I., Kakinuma, Y. and Yokoyama, K. (2007). [Unveiled multifunctionality of V-ATPase and the molecular mechanism revealed by X-ray crystal structures]. *Tanpakushitsu Kakusan Koso* **52**, 335–341.
64. Nelson, N. and Harvey, W. R. (1999). Vacuolar and plasma membrane proton-adenosinetriphosphatases. *Physiol Rev* **79**, 361–385.
65. Nessler, S., Friedrich, O., Bakouh, N., Fink, R. H., Sanchez, C. P., Planelles, G. and Lanzer, M. (2004). Evidence for activation of endogenous transporters in *Xenopus laevis* oocytes expressing the Plasmodium falciparum chloroquine resistance transporter, PfCRT. *J Biol Chem* **279**, 39438–39446.
66. Okech, B. A., Boudko, D. Y., Linser, P. J. and Harvey, W. R. (2008). Cationic pathway of pH regulation in larvae of *Anopheles gambiae*. *J Exp Biol* **211**, 957–968.
67. Onken, H. and Moffett, D. F. (2009). Revisiting the cellular mechanisms of strong luminal alkalinization in the anterior midgut of larval mosquitoes. *J Exp Biol* **212**, 373–377.
68. Onken, H., Patel, M., Javoroncov, M., Izeirovski, S., Moffett, S. B. and Moffett, D. F. (2009). Strong alkalinization in the anterior midgut of larval yellow fever mosquitoes (Aedes aegypti): involvement of luminal $Na(^+)/K(^+)$-ATPase. *J Exp Zool Part A Ecol Genet Physiol* **311**, 155–161.
69. Orlowski, J. and Grinstein, S. (2004). Diversity of the mammalian sodium/proton exchanger SLC9 gene family. *Pflügers Arch* **447**, 549–565.
70. Padan, E., Bibi, E., Ito, M. and Krulwich, T. A. (2005). Alkaline pH homeostasis in bacteria: new insights. *Biochim Biophys Acta* **1717**, 67–88.
71. Piermarini, P. M., Weihrauch, D., Meyer, H., Huss, M. and Beyenbach, K. W. (2009). NHE8 is an intracellular cation/H^+ exchanger in renal tubules of the yellow-fever mosquito Aedes aegypti. *Am J Physiol Renal Physiol* **296**, F730–F750.
72. Pullikuth, A. K., Filippov, V. and Gill, S. S. (2003). Phylogeny and cloning of ion transporters in mosquitoes. *J Exp Biol* **206**, 3857–3868.

73. Radermacher, M., Ruiz, T., Harvey, W. R., Wieczorek, H. and Gruber, G. (1999). Molecular architecture of *Manduca sexta* midgut V1 ATPase visualized by electron microscopy. *FEBS Lett* **453**, 383–386.
74. Ramsay, J. A. (1953a). Active transport of potassium by the malpighian tubules of insects. *J Exp Biol* **30**, 358–369.
75. Ramsay, J. A. (1953b). Exchanges of sodium and potassium in mosquito larvae. *J Exp Biol* **30**, 79–89.
76. Rheault, M. R., Okech, B. A., Keen, S. B., Miller, M. M., Meleshkevitch, E. A., Linser, P. J., Boudko, D. Y. and Harvey, W. R. (2007). Molecular cloning, phylogeny and localization of AgNHA1: the first Na^+/H^+ antiporter (NHA) from a metazoan, *Anopheles gambiae*. *J Exp Biol* **210**, 3848–3861.
77. Saroussi, S. and Nelson, N. (2008). Vacuolar $H(^+)$-ATPase-an enzyme for all seasons. *Pflügers Arch.*
78. Skou, J. C. (1997). The identification of the sodium-potassium pump. *Nobel Lectures, Chemistry 1996–2000, Editor Ingmar Grenthe, World Scientific Publishing Co., Singapore, 2003.*
79. Smith, K. E., Vanekeris, L. A. and Linser, P. J. (2007). Cloning and characterization of AgCA9, a novel {alpha}-carbonic anhydrase from *Anopheles gambiae* Giles *sensu stricto* (Diptera: Culicidae) larvae. *J Exp Biol* **210**, 3919–3930.
80. Stobbart, R. H. (1971). Evidence for Na plus-H plus and Cl minus-HCO_3 minus exchanges during independent sodium and chloride uptake by the larva of the mosquito *Aedes aegypti* (L.). *J Exp Biol* **54**, 19–27.
81. Taglicht, D., Padan, E. and Schuldiner, S. (1993). Proton-sodium stoichiometry of NhaA, an electrogenic antiporter from Escherichia-Coli. *J Biol Chem* **268**(8), 5382–5387.
82. Terra, W. R. and Ferreira, C. (1994). Insect digestive enzymes: properties, compartmentalization and function. *Comp Biochem Physiol B Biochem Mol* Biol **109**, 1–62.
83. Thurm, U. and Wessel, G. (1979). Metabolism-dependent transepithelial potential differences at epidermal receptors of arthropods. I- Comparative data. *J Comp Physiol* **134**, 119–130.
84. Turbeck, B. O., Nedergaard, S. and Kruse, H. (1968). An anion-stimulated adenosine triphosphatase from the potassium-transporting midgut of the larva of *Hyalophora cecropia*. *Biochim Biophys Acta* **163**, 354–361.
85. Uchida, E., Ohsumi, Y. and Anraku, Y. (1985). Purification and properties of H^+-translocating Mg^{2+} -adenosine triphosphatase from vacuolar membranes of *Saccharomyces cerevisiae*. *J Biol Chem* **260**, 1090–1095.
86. Ussing, H. H. and Zerahn, K. (1951). Active transport of sodium as the source of electric current in the short-circuited isolated frog skin. *Acta Physiol Scand* **23**, 110–127.
87. Voss, M., Vitavska, O., Walz, B., Wieczorek, H. and Baumann, O. (2007). Stimulus-induced phosphorylation of V-ATPase by protein kinase A. *J Biol Chem.*
88. Weber, W. (1999). Ion currents of *Xenopus laevis* oocytes: state of the art. *Biochim Biophys Acta* **1421**, 213–233.
89. Wieczorek, H., Brown, D., Grinstein, S., Ehrenfeld, J. and Harvey, W. R. (1999). Animal plasma membrane energization by proton-motive V-ATPases. *Bioessays* **21**, 637–648.
90. Wieczorek, H., Putzenlechner, M., Zeiske, W. and Klein, U. (1991). A vacuolar-type proton pump energizes K^+/H^+ antiport in an animal plasma membrane. *J Biol Chem* **266**, 15340–15347.
91. Wieczorek, H., Weerth, S., Schindlbeck, M. and Klein, U. (1989). A vacuolar-type proton pump in a vesicle fraction enriched with potassium transporting plasma membranes from tobacco hornworm midgut. *J Biol Chem* **264**, 11143–11148.
92. Wieczorek, H., Wolfersberger, M. G., Cioffi, M. and Harvey, W. R. (1986). Cation-stimulated ATPase activity in purified plasma membranes from tobacco hornworm midgut. *Biochimica et Biophysica Acta* **857**, 271–281.

93. Xiang, M., Feng, M., Muend, S. and Rao, R. (2007). A human Na^+/H^+ antiporter sharing evolutionary origins with bacterial NhaA may be a candidate gene for essential hypertension. *Proc Natl Acad Sci U S A* **104**, 18677–18681.
94. Zhuang, Z., Linser, P. J. and Harvey, W. R. (1999). Antibody to $H(^+)$ V-ATPase subunit E colocalizes with portasomes in alkaline larval midgut of a freshwater mosquito (*Aedes aegypti*). *J Exp Biol* **202**, 2449–2460.

CFTR-Dependent Anion Transport in Airway Epithelia

J.W. Hanrahan

1 Introduction

Vertebrates use a variety of anion transport systems to drive transepithelial fluid secretion, each regulated by distinct second messenger systems. Tissues such as the pancreatic duct and colon respond to peptide hormones and transmitters that elevate intracellular levels of cAMP or cGMP (see Chap. 9 by J. F. White, this volume), whereas most exocrine glands have a dominant calcium-dependent system, which is activated by acetylcholine or other calcium mobilizing agonist. Some epithelia have both mechanisms, which may allow the integration of several physiological signals or provide more precise control by secretagogues that activate multiple signaling pathways [1, 2]. Epithelial secretion usually involves activating anion efflux through a rate-limiting channel or exchanger in the apical membrane. This chapter focuses on perhaps the most extensively studied secretory transporter, the cystic fibrosis transmembrane conductance regulator (CFTR). CFTR is a 1480 amino acid membrane glycoprotein which when mutated causes the autosomal recessive disease cystic fibrosis (CF) [3–5]. Many mutations in the CFTR gene lead to misfolding and retention of the mutant protein in the endoplasmic reticulum. Despite extensive studies of CFTR structure and function, its precise role in ion and fluid transport remain controversial, even in epithelia that are severely affected in cystic fibrosis [6]. Consequently it remains uncertain precisely how defective CFTR leads to the viscous secretions, chronic bacterial infections, airway inflammation, and respiratory failure that characterize CF.

CFTR has two membrane domains (TMD1 and TMD2), each comprised of six transmembrane segments. Nucleotide binding domains follow each TMD in the linear sequence (NBD1 and NBD2, respectively), placing CFTR in a large superfamily

---

J.W. Hanrahan (✉)
Department of Physiology, McGill University, Montréal, QC, Canada
e-mail: hanrahan@med.mcgill.ca

Work in the author's laboratory was supported by the Canadian Institutes of Health Research (CIHR), National Institutes of Health (NIDDK; DK54075-03), Canadian Cystic Fibrosis Foundation (CCFF) and the Cystic Fibrosis Foundation (USA).

G.A. Gerencser (ed.), *Epithelial Transport Physiology*,
DOI 10.1007/978-1-60327-229-2_7,

of "ATP binding cassette" (ABC) transport proteins that includes organisms from bacteria to humans. Members of the ABC superfamily transport a diverse range of substrates including peptides (e.g. STE6 [7], TAP1/TAP2 [8]), amino acids and sugars (e.g. HisP [9] and MalK [10], respectively), lipophilic drugs and xenobiotics (e.g. MDR [11] and MRP1 [12], respectively), and phospholipids (e.g. MDR2 [13]). CFTR is the only known ion channel in the superfamily and is also unique in having a central regulatory region called (rather loosely) the R domain. This chapter provides an overview of CFTR and an update on its physiological role.

2 CFTR

2.1 Evidence for Its Channel Activity

Heterologous expression of the *cftr* gene in mammalian and insect cells produced a cAMP-stimulated current [14] which was carried by ohmic anion-selective channels having relatively low (7–10 pS) unitary conductance [15, 16]. These and other single channel properties (slow gating, flickering at negative membrane potentials, weak sensitivity to external disulfonic stilbene inhibition) were identical to those of a nonrectifying, low-conductance anion channel which had been characterized previously in several epithelia [17–19], but was distinct from an outwardly rectifying anion channel that had been implicated in CF. On the basis of the correlation between CFTR protein expression and the appearance of channels and their function, it was proposed that *cftr* encodes the nonrectifying anion channel [15]. CFTR activation of an endogenous anion channel was ruled out when mutations in CFTR's transmembrane segments altered the selectivity of the macroscopic conductance [20]. The channel function of CFTR was formally proven when it was purified to homogeneity and reconstituted in planar lipid bilayers [21]. Early studies demonstrating CFTR to be the defective anion channel in cystic fibrosis have been reviewed in detail elsewhere [22–27].

2.2 Identification of Pore-Lining Residues

Localizing the pore is a goal when studying the structure and function of any ion channel, and amino acids that are likely to line the permeation pathway have been inferred by comparing the selectivity, conductance and blocker sensitivity of wild-type and mutated CFTR channels. A strong case for the involvement of the sixth transmembrane segment (TM6) has been based on finding that TM6 mutations affect the anion selectivity [20, 28], unitary conductance (30,31) [29], and multi-ion pore behavior [29] of CFTR and also its sensitivity to open channel blockers [30, 31]. A model has been proposed based on the accessibility of engineered cysteines to externally applied hydrophilic sulfhydryl methanethiosulfonate (MTS) reagents in which amino acids immediately following TM6 fold back into the membrane [32].

Although substituted-cysteine accessibility experiments have many caveats [33, 34], mutations at R358 in the proposed re-entrant loop do reduce anion:cation selectivity dramatically consistent with this model [35]. Reducing the net positive charge near the outer end of TM6 by mutagenesis (e.g. K335E, [36]) or by mutagenesis and addition of MTS reagents (R334C + MTSES, [37]) induces inward rectification of the current-voltage relationship consistent with electrostatic repulsion and lowering the concentration of permeant anions near the outer mouth of the pore. Introducing negative charge at this site also weakens binding and block by the lyotropic anion $Au(CN)^{2-}$ when applied externally [38]. The rates at which external MTS reagents react with cysteines inserted at 331 and 333 are slower in the open than in the closed state [39], however inferences regarding open vs. closed states must be interpreted with caution since flickery block by the pH buffer MOPS suggests that there are two conformationally distinct open states having different blocker sensitivities [40], and studies with sulfhydryl reagents indicate there may also be multiple closed conformations [41].

Mutations in TM5 and TM11 affect selectivity and channel gating suggesting these segments contribute to the pore [42–44], whereas similar studies of TM12 indicate that it does not contribute to the permeation pathway [45]. Although the CFTR pore is determined mainly by TMs 1–6 [46], some early observations remain unexplained. For example deletion of the amino terminus and TMs 1–4 reportedly had little effect on channel function [47], although several disease-associated mutations that strongly affect channel function (e.g. R117H) occur in this region [29]. TM1 mutations alter anion selectivity (K95 [20]) and rectification (G91 [34]), and cysteines inserted into this segment are accessible to extracellular hydrophilic MTS reagents [48], thus it is surprising that TMs 1–4 are not required for normal function and further studies of this region are needed to understand its contribution to the pore.

2.3 Selectivity of the Pore

When selectivity of the CFTR pore was studied using pancreatic duct [49] and thyroid epithelial cells [18], replacing external Na^+ with K^+, tetramethyl ammonium, or N-methyl-D-glucamine had little effect on the reversal potential (E_{rev}) whereas substituting extracellular Cl^- with gluconate or sulfate shifted E_{rev} by +26 mV [50, 17, 18]. Anion selectivity (i.e. $P_{Cl}/P_{cation} > 8$–14) was reported for channels in human colon T84 and the intestinal cell line CaCo2 [19, 51]. Halide permeability ratios (P_X/P_{Cl}) followed an (inverse) lyotropic series, with large, weakly hydrated ions having the highest permeability except iodide, which gave inconsistent results even within laboratories. Most single channel and macroscopic studies suggested $P_I/P_{Cl} < 1.0$ [49, 20, 52–54] , although I^- currents were not detected in cultured thyroid cells [18] and block by iodide has also been reported [55]. In one study, the P_I/P_{Cl} of single CFTR channels was high immediately after exposure to iodide and then declined [36].

Permeability ratios for polyatomic anions follow the sequence NO_3^- (1.73) > Cl^- (1.0) > HCO_3^- (0.25) >> gluconate$^-$ (0.03) when measured from the extracellular side with mixed solutions [49]. A similar sequence NO_3^- > Cl^- > HCO_3^- > formate$^-$ > acetate$^-$ was obtained under biionic conditions, although permeation by large kosmotropic (i.e. "water structuring") anions was higher from the cytoplasmic than extracellular side [28]. Thus, external pyruvate$^-$, propanoate$^-$, methane sulfonate$^-$, ethane sulfonate$^-$ and gluconate$^-$ were not measurably permeant (i.e. $P_X/P_{Cl} < 0.06$) when macroscopic current was measured using excised patches exposed to PKA and ATP, yet small currents carried by these anions were detectable from the cytoplasmic side indicating asymmetrical permeability to large monovalent ions [56]. Thiocyanate (SCN^-) is permeant in CFTR and plays an important physiological role in the lactoperoxidase/H_2O_2/SCN^- antimicrobial system of the airways [57–59]. In this host-defense mechanism, lactoperoxidase secreted onto the airway surface by submucosal glands catalyzes the oxidation of SCN^- by H_2O_2, which is generated at the apical membrane by the action of DUOX1, an NADPH oxidase. This oxidation yields $OSCN^-$ (hypothiocyanite), which is bactericidal and thus contributes to innate immunity.

The relationship between macroscopic permeability ratio and the ion radii suggests a minimum functional diameter for the pore of 5.3 Å from the outside [28] and ~12 Å from the inside [56, 28], suggesting that relatively large monovalent anions can permeate from the cytosolic side. Indeed, macroscopic currents carried by glutathione (GSH) efflux have been measured when excised patches were exposed to high concentrations of GSH on the cytosolic side [60]. CFTR channel blockers and mutations inhibited the GSH current, suggesting that CFTR mediates GSH flux, and this was later confirmed biochemically using purified CFTR protein [61]. Nevertheless, the CFTR-dependent GSH efflux from intact cells is low [62] and may depend on cytosolic Cl^- concentration rather than CFTR per se [63], thus it remains uncertain whether GSH permeation through CFTR contributes to cellular efflux under physiological conditions.

2.4 Mechanism of Anion Permeation

Several lines of evidence suggest that the CFTR pore can hold more than one ion at a time. The biionic permeability ratio P_I/P_{Cl} is concentration-dependent [55], a common feature of multi-ion pores [64, 65]. At least two distinct electrical distances have been calculated for voltage-dependent blocker sites, and the block by intracellular gluconate is relieved by raising external Cl^- concentration. When wild-type CFTR channels are bathed symmetrically in mixtures of Cl^- and SCN^-, single channel conductance *decreases* from 7 to 2 pS as the mole fraction of SCN^- (the more permeant anion in the mixture) is increased from 0 to 7%, but then increases again as the SCN^- mole fraction is increased further to 97% [29]. This "anomalous mole fraction effect" (AMFE) and the voltage-dependent block of Cl^- currents by cytoplasmic SCN^- are both abolished when R347 is mutated to aspartate, and the AMFE becomes pH-dependent when the residue is substituted with a titratable histidine, suggesting positive charge at this site is necessary for the AMFE and also for the binding of internal SCN^-. A four barrier, three site (4B3S) Eyring rate theory model reproduced the current-voltage relationships quantitatively under a wide range of conditions, i.e. single channel conductances, reversal potentials, block by intracellular gluconate, and AMFEs in mixtures of SCN^- and Cl^- [66]. Loss of the AMFE (as seen in the R347D mutant) could be simulated using the 4B3S model by reducing the depth of the well corresponding to intracellular SCN^- block. Whether the positive charge at R347 contributes directly to SCN^- binding or indirectly through changes in pore architecture is not known [67]. The residual conductance of R347H and R347P mutant channels may explain the relatively mild phenotype associated with these mutations in patients.

3 Regulation of the CFTR Channel by Phosphorylation

3.1 PKA

Phosphorylation of CFTR's regulatory (R) domain by PKA or PKG stimulates CFTR activity [68, 69] and may enhance its interactions with other proteins. In COS cells, forskolin increases in vivo phosphorylation on CFTR by 1.8-fold [68] to an apparent stoichiometry of ~5 moles PO_4/mole protein [70], although this is probably an underestimate since all intracellular ATP is not radiolabeled when cells are incubated with $[^{32}P]PO_4$. Sites that are conspicuously phosphorylated under in vivo conditions are $RRNS_{660}I$, $KRKNS_{700}I$, $RKVS_{795}L$, and $RRLS_{813}$ Q. Sites that do not appear to be phosphorylated in vivo cannot be dismissed however, as they can support channel activity (~50%) and are highly conserved between species despite sequence divergence in the R domain generally. Six PKA sites are phosphorylated in resting Xenopus oocytes [71], and phosphorylation of S_{737} and S_{768} downregulates channel activity [72, 73, 71]. Mutating all nine dibasic (R-R/K-X-S/T) and five monobasic (R/K-X-S/T) consensus sequences for PKA on the R domain along with the dibasic site $RKTS_{422}N$ near the beginning of NBD1 abolishes channel responses to PKA when the mutant is incorporated into planar bilayers (82). Whether all functionally important PKA sites are situated on the R domain is unclear, since "split" channels lacking the R domain (defined as amino acids 634–836 inclusive) can still respond to PKA stimulation in patch clamp experiments, albeit weakly [74]. The Type II isoform of cyclic GMP-dependent protein kinase (PKG Type II or cGKII) activates CFTR channels in intestinal epithelium and mediates the secretory diarrhea induced by heat stable enterotoxins [75]. Activation of CFTR channels by cGKII in excised patches is somewhat slower than by PKA, however cGKII incubation produces similar CFTR phosphopeptide maps suggesting that the same sites are phosphorylated by both kinases [76].

3.2 PKC

Protein kinase C (PKC) causes a small stimulation of CFTR channels but its main effects are to enhance both the rate and magnitude of activation by PKA in excised patches [69], maintaining CFTR in a PKA-responsive state [77]. Without PKC exposure excised channels become refractory to PKA stimulation after ~10 min, presumably because critical PKC sites have become dephosphorylated by a membrane-associated phosphatase (Fig. 1). Loss of PKA responsiveness can be reversed by adding PKC + ATP to the bath and prevented if PKC is continuously present. PKC belongs to a family of 11 related, lipid-dependent serine/threonine kinases having homologous catalytic domains. The isotypes are not specific in their ability to stimulate CFTR in vitro [78], nevertheless PKCξ has been identified as the one regulating CFTR in the airway submucosal gland cell line Calu3 [79]. Mutant channels lacking all nine PKC consensus sequences on the R domain have drastically reduced responses to PKA stimulation [80]. Most of this decrease in activation

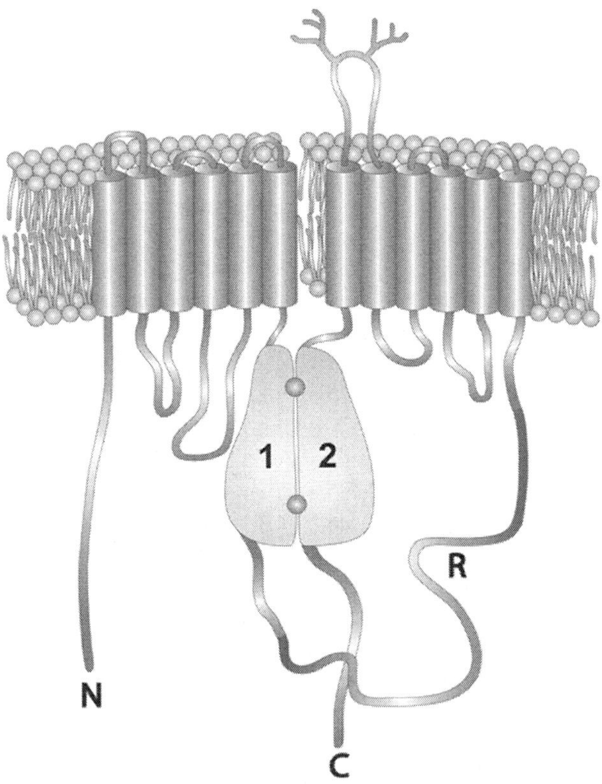

Fig. 1 Cartoon of CFTR showing the amino terminus (*N*), nucleotide binding domains *1* and *2* (*1*, *2*), the regulatory domain (*R*) and carboxy-terminus (*C*). Note the two membrane domains each containing six transmembrane segments and the asparagine-linked glycosylation of extracellular loop 4

is caused by loss of the T582, T604, and S686 near the N-terminal end of the R domain, whereas mutating PKC sites S641 and T682 to alanine enhanced responses to PKA and PKC, respectively, suggesting they are normally inhibitory when phosphorylated. How does the R domain regulate CFTR? Studies of domain-domain interactions when CFTR channels are assembled from three polypeptides indicate that phosphorylation does not cause release of the R domain as originally proposed. In fact, association of the R domain with other parts of CFTR is *enhanced* by phosphorylation when studied using channels that are assembled in situ from three polypeptides [81]. The R domain is pulled down more efficiently when prephosphorylated by protein kinase A, both in vivo, and when lysates expressing split channels lacking the R domain are incubated with a GST-R domain fusion protein. Similar results have been obtained for PKC phosphorylation of the R domain [82]. Thus, phosphorylation regulates CFTR by promoting association of the R domain with other domains rather than by causing its release from an inhibitory site.

3.3 Dephosphorylation

Phosphatases also control CFTR activity and cause channels to run down soon after they are excised from cAMP-stimulated cells [50, 69]. This spontaneous decline in open probability (P_o) has been attributed to phosphatase activity in the excised patch because channels could be reactivated by exposure to MgATP + PKA [69]. Rundown was not inhibited significantly by high concentrations of okadaic acid, and did not require Ca or calmodulin, indicating that it was not mediated by PP1, PP2A, or PP2B. Rundown was sensitive to the alkaline phosphatase inhibitors levamisole and bromotetramisole [83], but at concentrations (>200 μM) that we now know inhibit all four protein phosphatase types [84]. Reducing Mg^{2+} concentration slows rundown, consistent with the dependence of PP2C on millimolar concentrations of Mg^{2+} [85]. Calyculin A also had a small effect on CFTR, but in T_{84} monolayers the deactivation of cAMP-stimulated currents was insensitive to okadaic acid and calyculin A, suggesting most CFTR deactivation in epithelial cells is mediated by PP2C [86, 85]. Other cells may regulate CFTR differently however, for example in cardiac ventricular myocytes about 50% of CFTR deactivation is sensitive to the PP2A inhibitors microcystin and okadaic acid [87] suggesting an important role for PP2A in those cells. More recently two different PP2A subunits have been shown to interact with CFTR. An association between CFTR and the B'epsilon subunit of PP2A [88] was demonstrated using proteomics, and a yeast two hybrid screen revealed an interaction between the R domain of CFTR and the PP2A regulatory subunit PR65 [89].

4 CFTR and Its Neighbors

CFTR functions in a macromolecular complex with other proteins, which may explain why it has been reported to influence the epithelial sodium channel (ENaC) and many other transport proteins. Much effort is presently directed towards understanding these interactions since they may contribute to CF pathobiology.

4.1 Syntaxin 1A

Early studies identified interactions between the amino terminal tail (N-tail) of CFTR and syntaxin 1A [90, 91], which is part of the vesicle fusion apparatus at neuronal synapses [92]. Binding to the N-tail of CFTR inhibits its channel activity, apparently by disrupting interaction of the N-tail with the R domain. The N-tail/R domain interaction depends on a cluster of acidic amino acids between aa 46 and 63 of the N-tail which are predicted to lie on one side of an alpha helix. Interaction between the N-tail (or a recombinant N-tail-GST fusion protein) and the R domain increases open probability by prolonging the open state. The N-tail interacts preferentially with the amino-terminal region of the R domain between aa 595–740, however a more distal region of the R domain appears to stabilize the interaction. The N-tail also influences phosphorylation control of CFTR since mutating the acidic residues to alanines slows channel activation by forskolin and IBMX and accelerates the deactivation that occurs after forskolin and IBMX are washed out [93]. Another membrane protein, MUNC18, competes with CFTR for binding to syntaxin 1A and thereby relieves the inhibition of CFTR by syntaxin 1A [94]. A proximal region of the N-tail also interacts with the cytoskeletal protein filamin, and a disease-associated mutation (S13F) disrupts this interaction, causing the mutant CFTR to be retrieved more rapidly from the cell surface and degraded [95].

4.2 PDZ Binding Domain Proteins

Perhaps the best-known CFTR interacting proteins are those with PDZ (Post synaptic density 95, Discs large, ZO-1) domains such as NHERF1 (sodium-hydrogen exchange regulatory factor type1, also called EBP50 for "ezrin binding protein of 50 kiloDaltons"), which bind to the C terminal DTRL motif of CFTR and act as scaffolding/adaptor proteins. Mutating threonine or leucine in the DTRL motif of CFTR leads to partial redistribution of the protein from the apical to lateral plasma membrane in polarized epithelial cells, and abolishes binding to PDZ1 of EBP-50 in vitro [96, 97]. The structure of PDZ1 from EBP50 (solved at 1.7 Å resolution when bound to the C terminal five amino acids of CFTR) has revealed two hydrogen bonds and two salt bridges between the penultimate arginine and residues in PDZ1 [98]. However NHERF2, a related PDZ-domain protein, is also situated at the apical membrane of epithelia and can be immunoprecipitated with CFTR [99] whereas a PDZ domain protein called CAL (CFTR associated ligand) interacts with CFTR during its maturation in the trans-Golgi compartment [100]. A protein with four PDZ domains called CAP70 was identified using an affinity column with the C-tail of CFTR as bait [101]. Adding recombinant CAP70 to excised patches had a biphasic effect on P_O, which increased at low CAP70 concentrations and decreased at high concentrations suggesting that low concentrations of CAP70 enhance CFTR dimerization, and similar results have been obtained with fragments of EBP50 that contain two PDZ domains [102].

Interactions with PDZ domain proteins determine the lateral mobility, distribution, and recycling of CFTR at the plasma membrane. Most CFTR channels are normally immobile on the cell surface when studied on a time scale of minutes using fluorescence recovery after photobleaching, image correlation spectroscopy, or single particle tracking of fluorescently tagged channels. However, when PDZ binding is disrupted they become highly mobile, occasionally entering zones of transient confinement that may reflect other types of protein-protein interactions [103]. Overexpression of CFTR and an insufficient number of PDZ partners may explain the high mobility of CFTR reported under control conditions in previous studies [104].

4.3 Local Signaling

Huang et al. [105] found that CFTR channels could sometimes be activated by adding cpt-cAMP and ATP to patches that had been excised from Calu-3 airway epithelial cells, indicating that PKA holoenzyme must be associated with the membrane. The catalytic and type II (RII) regulatory subunits of PKA both co-immunoprecipitate with CFTR [99], and endogenous kinase activity in the precipitates can be abolished by Ht31, a peptide derived from the amino terminus of RII which specifically disrupts interaction with A-kinase anchoring proteins (AKAPs). The AKAP involved in this process remains to be established, although ezrin is one candidate since it is expressed at the apical membrane of Calu-3 and T84 cells and can be co-immunoprecipitated with CFTR. Ezrin also interacts with E3KARP and thus could provide a link between E3KARP-CFTR and PKA [99]. Protein complexes may also regulate the level of cAMP near CFTR through local regulation of cAMP export. The multidrug resistance protein 4 (MRP4) is a cAMP exporter which associates functionally and physically with CFTR through PDZK1, another PDZ domain protein, and suppresses its normal activation [106]. PKC is also anchored near CFTR by "Regulators of C kinase" or RACKS [107]. The importance of membrane targeting/tethering is well illustrated by comparison of CFTR activation by soluble and membrane-bound forms of cGMP-dependent protein kinase (Types I vs. II, respectively; [108]). Type I does not activate channels significantly whereas a chimera of Type I and Type II N-terminal anchoring domain does associate with the membrane and leads to activation of CFTR channels.

4.4 AMPK

In addition to PDZ domain proteins, CFTR can associate with $\alpha 1$ and $\alpha 2$ catalytic subunits of AMP-activated protein kinase (AMPK) [109], which are responsive to the AMP:ATP ratio and may serve as metabolic sensors. $\alpha 1$-AMPK binds to the

C-tail of CFTR between residues 1420 and 1457 and this interaction is disrupted by mutations at either of two protein trafficking motifs that are situated in this region. AMPK phosphorylates CFTR in vitro and inhibits CFTR currents by 35–50% when coexpressed in *Xenopus* oocytes. Whether this is due to direct phosphorylation of CFTR or an alteration in metabolism remains to be established. AMPK and CFTR are co-localized at the apical membrane of nasal epithelium.

4.5 Does CFTR Function as a Channel for Chloride, Bicarbonate, or Both?

The role of CFTR in HCO_3^- secretion has received much attention recently because HCO_3^- is the main ion transported by the small pancreatic ducts which are severely affected in cystic fibrosis and is also an important component of airway and intestinal secretions [110]. The human airway cell line Calu3, a widely used model for submucosal gland serous cells, actively secretes HCO_3^- rather than Cl^- [111, 112] yet the final bicarbonate concentration in secreted fluid is variable [113]. The relative contributions of HCO_3^- and Cl^- to fluid secretion have recently been quantified under open circuit conditions and the role of active HCO_3^- transport in driving fluid secretion and its dependence on CFTR has been confirmed (J. Liao, R. Robert, J. Shan, J. Hanrahan, unpublished observations). Patch clamp experiments have demonstrated that bicarbonate can permeate through the CFTR channel pore [49, 114, 28], nevertheless the significance of CFTR-mediated efflux during bicarbonate secretion is still hotly debated [115, 112, 116]. The method that is widely used to study anion exchange involves replacement of extracellular Cl^- with an impermeant anion and measurement of intracellular alkalinization when HCO_3^- enters on the exchanger, however it is difficult to exclude electrical coupling with this technique when there is a large parallel anion conductance to both ions and depolarization induced by Cl^- removal would also drive HCO_3^- uptake. At present, HCO_3^- efflux through apical CFTR channels can adequately explain secretion by Calu-3 cells and there is no need to invoke anion exchangers in the apical membrane although anion exchangers in the SLC26A family have been proposed in other tissues, notably the pancreatic duct [117].

5 Conclusions

Much progress has been made since the cloning of the CFTR gene, however, many important questions remain, including the location of the CFTR channel pore, mechanisms of regulation by phosphorylation and by protein-protein interactions, and the precise function of CFTR during bicarbonate secretion. Structural studies of CFTR, proteomic identification of its associated regulatory proteins, and more detailed physiological studies of bicarbonate secretion should help answer these questions.

References

1. Choi JY, Joo NS, Krouse ME, Wu JV, Robbins RC, Ianowski JP et al. Synergistic airway gland mucus secretion in response to vasoactive intestinal peptide and carbachol is lost in cystic fibrosis. J Clin Invest 2007; 117:3118–3127.
2. Ianowski JP, Choi JY, Wine JJ, Hanrahan JW. Mucus secretion by single tracheal submucosal glands from normal and CFTR knock-out mice. J Physiol 2007; 580:301–314.
3. Rommens JM, Iannuzzi MC, Kerem B-S, Drumm ML, Melmer G, Dean M et al. Identication of the cystic fibrosis gene: Chromosome walking and jumping. Science 1989; 245: 1059–1065.
4. Kerem B-S, Rommens JM, Buchanan JA, Markiewicz D, Cox TK, Chakravarti A et al. Identification of the cystic fibrosis gene: Genetic analysis. Science 1989; 245:1073–1080.
5. Riordan JR, Rommens JM, Kerem B-S, Alon N, Rozmahel R, Grzelczak Z et al. Identification of the cystic fibrosis gene: Cloning and characterization of complementary DNA. Science 1989; 245:1066–1073.
6. Quinton PM. Cystic fibrosis: A disease in electrolyte transport. FASEB J 1990; 4: 2709–2717.
7. Berkower C, Taglicht D, Michaelis S. Functional and physical interactions between partial molecules of STE6, a yeast ATP-binding cassette protein. J Biol Chem 1996; 271:22983–22989.
8. Russ G, Esquivel F, Yewdell JW, Cresswell P, Spies T, Bennink JR. Assembly, intracellular localization, and nucleotide binding properties of the human peptide transporters TAP1 and TAP2 expressed by recombinant vaccinia virus. J Biol Chem 1995; 270:21312–21318.
9. Hung L-W, Wang IX, Nikaido K, Liu PQ, Ames GFL, Kim S-H. Crystal structure of the ATP-binding subunit of an ABC transporter, the histidine permease of Salmonella typhimurium. Nature 1998; 396:703–707.
10. Diederichs K, Diez J, Greller G, Muller C, Breed J, Schnell C et al. Crystal structure of MalK, the ATPase subunit of the trehalose/maltose ABC transporter of the archeon Thermococcus litoralis. EMBO J 2000; 19:5951–5961.
11. Ambudkar SV, Dey S, Hrycyna CA, Ramachandra M, Pastan I, Gottesman MM. Biochemical, cellular, and pharmacological aspects of the multidrug transporter. Annu Rev Pharmacol Toxicol 1999; 39:361–398.
12. Cornwell MM, Tsuruo T, Gottesman MM, Pastan I. ATP-binding properties of P glycoprotein from multidrug-resistant KB cells. FASEB J 1987; 1:51–54.
13. Reutz S, Gros P. Phosphatidylcholine translocase: A physiological role for mdr2. Cell 1994; 77:1071–1081.
14. Anderson MP, Rich DP, Gregory RJ, Smith AE, Welsh MJ. Generation of cAMP-activated chloride currents by expression of CFTR. Science 1991; 251:679–682.
15. Kartner N, Hanrahan JW, Jensen TJ, Naismith AL, Sun S, Ackerley CA et al. Expression of the cystic fibrosis gene in non-epithelial invertebrate cells produces a regulated anion conductance. Cell 1991; 64:681–691.
16. Berger HA, Anderson MP, Gregory RJ, Thompson S, Howard PW, Maurer RA et al. Identification and regulation of the cystic fibrosis transmembrane conductance regulator-generated chloride channel. J Clin Invest 1991; 88:1422–1431.
17. Gray MA, Harris A, Coleman L, Greenwell JR, Argent BE. Two types of chloride channel on duct cells cultured from human fetal pancreas. Am J Physiol 1989; 257:C240–C251.
18. Champigny G, Verrier B, Gérard C, Mauchamp J, Lazdunski M. Small conductance chloride channels in the apical membrane of thyroid cells. FEBS Lett 1990; 259:263–268.
19. Tabcharani JA, Low W, Elie D, Hanrahan JW. Low-conductance chloride channel activated by cAMP in the epithelial cell line T84. FEBS Lett 1990; 270:157–164.
20. Anderson MP, Gregory RJ, Thompson S, Souza DW, Paul S, Mulligan RC et al. Demonstration that CFTR is a chloride channel by alteration of its anion selectivity. Science 1991; 253:202–205.

21. Bear CE, Li C, Kartner N, Bridges RJ, Jensen TJ, Ramjeesingh M et al. Purification and functional reconstitution of the cystic fibrosis transmembrane conductance regulator(CFTR). Cell 1992; 68:809–818.
22. Riordan JR, Chang X-B. CFTR, a channel with the structure of a transporter. Biochim Biophys Acta Bio-Energetics 1992; 1101:221–222.
23. Hanrahan JW, Tabcharani JA, Grygorczyk R. Patch clamp studies of apical membrane chloride channels. In: Dodge JA, Brock DJH, Widdicombe JH, editors. Cystic Fibrosis-Current Topics: Vol.1. London: John Wiley and Sons, Ltd., 1993: 93–137.
24. Gadsby DC, Nagel G, Hwang T-C. The CFTR chloride channel of mammalian heart. Annu Rev Physiol 1995; 57:387–416.
25. Hanrahan JW, Tabcharani JA, Becq F, Mathews CJ, Augustinas O, Jensen TJ et al. Function and dysfunction of the CFTR chloride channel. In: Dawson DC, Frizzell RA, editors. Ion channels and genetic diseases. New York: Rockefeller University Press, 1995: 125–137.
26. Quinton PM. Physiological basis of cystic fibrosis: A historical perspective. Physiol Rev 1999; 79:S3–S22.
27. Sheppard DN, Welsh MJ. Structure and function of the CFTR chloride channel. Physiol Rev 1999; 79:S23–S45.
28. Linsdell P, Tabcharani JA, Rommens JM, Hou Y-X, Chang X-B, Tsui L-C et al. Permeability of wild-type and mutant cystic fibrosis transmembrane conductance regulator chloride channels to polyatomic anions. J Gen Physiol 1997; 110:355–364.
29. Tabcharani JA, Rommens JM, Hou Y-X, Chang X-B, Tsui L-C, Riordan JR et al. Multi-ion pore behaviour in the CFTR chloride channel. Nature 1993; 366:79–82.
30. McDonough S, Davidson N, Lester HA, McCarty NA. Novel pore-lining residues in CFTR that govern permeation and open-channel block. Neuron 1994; 13:623–634.
31. Linsdell P, Hanrahan JW. Disulphonic stilbene block of cystic fibrosis transmembrane conductance regulator Cl- channels expressed in a mammalian cell line and its regulation by a critical pore residue. J Physiol 1996; 496:687–693.
32. Cheung M, Akabas MH. Locating the anion-selectivity filter of the cystic fibrosis transmembrane conductance regulator(CFTR) chloride channel. J Gen Physiol 1997; 109:289–299.
33. Yang N, George Jr. AL, Horn R. Molecular basis of charge movement in voltage-gated sodium channels. Neuron 1996; 16:113–122.
34. Mansoura MK, Smith SS, Choi AD, Richards NW, Strong TV, Drumm ML et al. Cystic fibrosis transmembrane conductance regulator(CFTR) anion binding as a probe of the pore. Biophys J 1998; 74:1320–1332.
35. Guinamard R, Akabas MH. Arg352 is a major determinant of charge selectivity in the the cystic fibrosis transmembrane conductance regulator chloride channel. Biochemistry 1999; 38:5528–5537.
36. Tabcharani JA, Linsdell P, Hanrahan JW. Halide permeation in wild-type and mutant cystic fibrosis transmembrane conductance regulator chloride channels. J Gen Physiol 1997; 110:341–354.
37. Smith SS, Liu X, Zhang Z-R, Sun F, Kriewall TE, McCarty NA et al. CFTR: Covalent and noncovalent modification suggests a role for fixed charges in anion conduction. J Gen Physiol 2001; 118:407–431.
38. Gong X, Linsdell P. Molecular determinants and role of an anion binding site in the external mouth of the CFTR chloride channel pore. J Physiol 2003; 549:387–397.
39. Beck EJ, Yang Y, Yaemsiri S, Raghuram V. Conformational changes in a pore-lining helix coupled to CFTR channel gating. J Biol Chem 2008; [Epub ahead of print] PMID: 18056267
40. Ishihara H, Welsh MJ. Block by MOPS reveals a conformational change in the CFTR pore produced by ATP hydrolysis. Am J Physiol Cell Physiol 1997; 273:C1278–C1289.
41. Fatehi M, Linsdell P. State-dependent access of anions to the cystic fibrosis transmembrane conductance regulator chloride channel pore. J Biol Chem 2007; [Epub ahead of print] PMID: 18167343.

42. Hou Y-X, Cui L, Riordan JR, Chang X-B. Allosteric interactions between the two non-equivalent nucleotide binding domains of multidrug resistance protein MRP1. J Biol Chem 2000; 275:20280–20287.
43. Zhang Z-R, McDonough SI, McCarty NA. Interaction between permeation and gating in a putative pore domain mutant in the cystic fibrosis transmembrane conductance regulator. Biophys J 2000; 79:298–313.
44. McCarty NA. Permeation through the CFTR chloride channel. J Exp Biol 2000; 203: 1947–1962.
45. Gupta J, Evagelidis A, Hanrahan JW, Linsdell P. Asymmetric structure of the cystic fibrosis transmembrane conductance regulator chloride channel pore suggested by mutagenesis of the twelfth transmembrane region. Biochemistry 2001; 40:6620–6627.
46. Ge N, Muise CN, Gong X, Linsdell P. Direct comparison of the functional roles played by different transmembrane regions in the cystic fibrosis transmembrane conductance regulator chloride channel pore. J Biol Chem 2004; 279:55283–55289.
47. Carroll TP, Morales MM, Fulmer SB, Allen SS, Flotte TR, Cutting GR et al. Alternate translation initiation codons can create functional forms of cystic fibrosis transmembrane conductance regulator. J Biol Chem 1995; 270:11941–11946.
48. Akabas MH, Kaufmann C, Cook TA, Archdeacon P. Amino acid residues lining the chloride channel of the cystic fibrosis transmembrane conductance regulator. J Biol Chem 1994; 269:14865–14868.
49. Gray MA, Pollard CE, Harris A, Coleman L, Greenwell JR, Argent BE. Anion selectivity and block of the small conductance chloride channel on pancreatic duct cells. Am J Physiol Cell Physiol 1990; 259:C752–C761.
50. Gray MA, Greenwell JR, Argent BE. Secretin-regulated chloride channel on the apical plasma membrane of pancreatic duct cells. J Membr Biol 1988; 105:131–142.
51. Bear CE, Reyes EF. cAMP-activated chloride conductance in the colonic cell line, Caco-2. Am J Physiol Cell Physiol 1992; 262:C251–C256.
52. Yang ICH, Cheng T-H, Wang F, Price EM, Hwang T-C. Modulation of CFTR chloride channels by calyculin A and genistein. Am J Physiol Cell Physiol 1997; 272:C142–C155.
53. Cliff WH, Frizzell RA. Separate Cl- conductances activated by cAMP and Ca2+ in Cl-secreting epithelial cells. Proc Natl Acad Sci USA 1990; 87:4956–4960.
54. Bell CL, Quinton PM. T84 cells: Anion selectivity demonstrates expression of Cl- conductance affected in cystic fibrosis. Am J Physiol Cell Physiol 1992; 262:C555–C562.
55. Tabcharani JA, Chang X-B, Riordan JR, Hanrahan JW. The cystic fibrosis transmembrane conductance regulator chloride channel: Iodide block and permeation. Biophys J 1992; 62: 1–4.
56. Linsdell P, Hanrahan JW. Adenosine triphosphate-dependent asymmetry of anion permeation in the cystic fibrosis transmembrane conductance regulator chloride channel. J Gen Physiol 1998; 111:601–614.
57. Gerson C, Sabater J, Scuri M, Torbati A, Coffey R, Abraham JW et al. The lactoperoxidase system functions in bacterial clearance of airways. Am J Respir Cell Mol Biol 2000; 22: 665–671.
58. Fragoso MA, Fernandez V, Forteza R, Randell RH, Salathe M, Conner GE. Transcellular thiocyanate transport by human airway epithelia. J Physiol 2004; 561:183–194.
59. Moskwa P, Lorentzen D, Excoffon KJA, Zabner J, McCray PB, Jr., Nauseef WM et al. A novel host defense system of airways is defective in cystic fibrosis. Am J Respir Crit Care Med 2007; 175:174–183.
60. Linsdell P, Hanrahan JW. Glutathione permeability of CFTR. Am J Physiol Cell Physiol 1998; 275:C323–C326.
61. Kogan I, Ramjeesingh M, Li C, Kidd JF, Wang Y, Leslie EM et al. CFTR directly mediates nucleotide-regulated glutathione flux. EMBO J 2003; 22:1981–1989.
62. Gao L, Kim KJ, Yankaskas JR, Forman HJ. Abnormal glutathione transport in cystic fibrosis airway epithelia. Am J Physiol Lung Cell Mol Physiol 1999; 277:L113–L118.

63. Gao L, Broughman JR, Iwamaoto T, Tomich JM, Venglarik CJ, Forman HJ. Correction of glutathione secretion in cystic fibrosis airway epithelia by a chloride channel forming peptide and chlorzoxazone. Pediatric Pulmonology suppl. 20, 248. 2000.
64. Eisenman G, Horn R. Ionic selectivity revisited: the role of kinetic and equilibrium processes in ion permeation through channels. J Membr Biol 1983; 76:197–225.
65. Hille B. Ionic channels of excitable membranes. 2 ed. Sunderland, Massachusetts: Sinauer Associates Incorporated, 1992.
66. Linsdell P, Tabcharani JA, Hanrahan JW. A multi-ion mechanism for ion permeation and block in the cystic fibrosis transmembrane conductance regulator chloride channel. J Gen Physiol 1997; 110:365–377.
67. Cotten JF, Welsh MJ. Cystic fibrosis-associated mutations at arginine 347 alter the pore architecture of CFTR. J Biol Chem 1999; 274:5429–5435.
68. Cheng SH, Rich DP, Marshall J, Gregory RJ, Welsh MJ, Smith AE. Phosphorylation of the R domain by cAMP-dependent protein kinase regulates the CFTR chloride channel. Cell 1991; 66:1027–1036.
69. Tabcharani JA, Chang X-B, Riordan JR, Hanrahan JW. Phosphorylation-regulated Cl- channel in CHO cells stably expressing the cystic fibrosis gene. Nature 1991; 352:628–631.
70. Picciotto MR, Cohn JA, Bertuzzi G, Greengard P, Nairn AC. Phosphorylation of the cystic fibrosis transmembrane conductance regulator. J Biol Chem 1992; 267:12742–12752.
71. Csanady L, Seto-Young D, Chan KW, Cenciarelli C, Angel BB, Qin J et al. Preferential phosphorylation of R-domain Serine 768 dampens activation of CFTR channels by PKA. J Gen Physiol 2005; 125:171–186.
72. Wilkinson DJ, Strong TV, Mansoura ME, Wood DL, Smith SS, Collins FS et al. CFTR activation: additive effects of stimulatory and inhibitory phosphorylation sites in the R domain. Am J Physiol Lung Cell Mol Physiol 1997; 273:L127–L133.
73. Chan KW, Smith SS, Csanady L, Nairn AC, Dawson DC, Gadsby DC. PKA- and Ca2+-dependent regulation of WT and S768A mutant CFTR Cl- channels in Xenopus oocytes. Biophys J 1998; 74:A396.
74. Chan KW, Csanády L, Seto-Young D, Nairn AC, Gadsby DC. Severed molecules functionally define the boundaries of the cystic fibrosis transmembrane conductance regulator′s NH2-terminal nucleotide binding domain. J Gen Physiol 2000; 116(2):163–180.
75. Pfeifer A, Aszódi A, Seidler U, Ruth P, Hofmann F, Fässler R. Intestinal secretory defects and dwarfism in mice lacking cGMP-dependent protein kinase II. Science 1996; 274: 2082–2086.
76. French PJ, Bijman J, Edixhoven M, Vaandrager AB, Scholte BJ, Lohmann SM et al. Isotype-specific activation of cystic fibrosis transmembrane conductance regulator-chloride channels by cGMP-dependent protein kinase II. J Biol Chem 1995; 270:26626–26631.
77. Jia Y, Mathews CJ, Hanrahan JW. Phosphorylation by protein kinase C is required for acute activation of cystic fibrosis transmembrane conductance regulator by protein kinase A. J Biol Chem 1997; 272:4978–4984.
78. Berger HA, Travis SM, Welsh MJ. Regulation of the cystic fibrosis transmembrane conductance regulator Cl- channel by specific protein kinases and phosphatases. J Biol Chem 1993; 268:2037–2047.
79. Liedtke CM, Cole TS. Antisense oligonucleotide to PKC-ϵ alters cAMP-dependent stimulation of CFTR in Calu-3 cells. Am J Physiol Cell Physiol 1998; 275:C1357–C1364.
80. Chappe V, Hinkson DAR, Zhu T, Chang X-B, Riordan JR, Hanrahan JW. Phosphorylation of protein kinase C sites in NBD1 and the R domain control CFTR channel activation by PKA. J Physiol 2003; 548:39–52.
81. Chappe V, Irvine T, Liao J, Evagelidis A, Hanrahan JW. Phosphorylation of CFTR by PKA promotes binding of the regulatory domain. EMBO J 2005; 24:2730–2740.
82. Seavilleklein G, Evagelidis A, Amer N, Chappe F, Hanrahan JW, Chappe V. PKC phosphorylation is necessary for PKA-dependent binding of the R domain with the rest of CFTR. Ped Pulmonol 2007; suppl. 30:204.

83. Becq F, Jensen TJ, Chang X-B, Savoia A, Rommens JM, Tsui L-C et al. Phosphatase inhibitors activate normal and defective CFTR chloride channels. Proc Natl Acad Sci USA 1994; 91:9160–9164.
84. Luo J, Zhu T, Evagelidis A, Pato M, Hanrahan JW. Role of protein phosphatases in the activation of CFTR (ABCC7) by genistein and bromotetramisole. Am J Physiol Cell Physiol 2000; 279:C108–C119.
85. Luo J, Pato MD, Riordan JR, Hanrahan JW. Differential regulation of single CFTR channels by PP2C, PP2A, and other phosphatases. Am J Physiol Cell Physiol 1998; 274: C1397–C1410.
86. Travis SM, Berger HA, Welsh MJ. Protein phosphatase 2C dephosphorylates and inactivates cystic fibrosis transmembrane conductance regulator. Proc Natl Acad Sci USA 1997; 94:11055–11060.
87. Gadsby DC, Hwang T-C, Horie M, Nagel G, Nairn AC. Cardiac chloride channels: Incremental regulation by phosphorylation/dephosphorylation. Ann NY Acad Sci 1993; 707:259–274.
88. Thelin WR, Kesimer M, Tarran R, Kreda SM, Grubb BR, Sheehan JK et al. The cystic fibrosis transmembrane conductance regulator is regulated by a direct interaction with the protein phosphatase 2A. J Biol Chem 2005; 280:41512–41520.
89. Vastiau A, Cao L, Jaspers M, Owsianik G, Janssens V, Cuppens H et al. Interaction of the protein phosphatase 2A with the regulatory domain of the cystic fibrosis transmembrane conductance regulator channel. FEBS Lett 2005; 579:3392–3396.
90. Naren AP, Nelson DJ, Xie WW, Jovov B, Pevsner J, Bennett MK et al. Regulation of CFTR chloride channels by syntaxin and Munc18 isoforms. Nature 1997; 390: 302–305.
91. Naren AP, Quick MW, Collawn JF, Nelson DJ, Kirk KL. Syntaxin 1A inhibits CFTR chloride channels by means of domain-specific protein-protein interactions. Proc Natl Acad Sci USA 1998; 95:10972–10977.
92. Naren AP, Di A, Cormet-Boyaka E, Boyaka PN, McGhee JR, Zhou W et al. Syntaxin 1A is expressed in airway epithelial cells, where it modulates CFTR Cl- currents. J Clin Invest 2000; 105:377–386.
93. Naren AP, Cormet-Boyaka E, Fu J, Villain M, Blalock JE, Quick MW et al. CFTR chloride channel regulation by an interdomain interaction. Science 1999; 286: 544–548.
94. Naren AP, Kirk KL. CFTR chloride channels: Binding partners and regulatory networks. News Physiol Sci 2000; 15:57–61.
95. Thelin WR, Chen Y, Gentzsch M, Kreda SM, Sallee JL, Scarlett CO et al. Direct interaction with filamins modulates the stability and plasma membrane expression of CFTR. J Clin Invest 2007; 117:364–374.
96. Wang S, Raab RW, Schatz PJ, Guggino WB, Li M. Peptide binding consensus of the NHE-RF-PDZ1 domain matches the C-terminal sequence of cystic fibrosis transmembrane conductance regulator(CFTR). FEBS Lett 1998; 427:103–108.
97. Short DB, Trotter KW, Reczek D, Kreda SM, Bretscher A, Boucher RC et al. An apical PDZ protein anchors the cystic fibrosis transmembrane conductance regulator to the cytoskeleton. J Biol Chem 1998; 273:19797–19801.
98. Karthikeyan S, Leung T, Ladias JAA. Structural basis of the Na^+/H^+ exchanger regulatory factor PDZ1 interaction with the carboxyl-terminal region of the cystic fibrosis transmembrane conductance regulator. J Biol Chem 2001; 276:19683–19686.
99. Sun F, Hug MJ, Bradbury NA, Frizzell RA. Protein kinase A associates with cystic fibrosis transmembrane conductance regulator via an interaction with ezrin. J Biol Chem 2000; 275:14360–14366.
100. Cheng J, Moyer BD, Milewski M, Hazama A, Mickle JE, Cutting GR et al. CFTR associates with the PDZ domain protein CAL at the trans-Golgi network. Pediatric Pulmonology suppl. 19, 168–169. 1999.

101. Wang S, Yue H, Derin RB, Guggino WB, Li M. Accessory protein facilitated CFTR-CFTR interaction, a molecular mechanism to potentiate the chloride channel activity. Cell 2000; 103:169–179.
102. Raghuram V, Mak D-OD, Foskett JK. Regulation of cystic fibrosis transmembrane conductance regulator single-channel gating by bivalent PDZ-domain-mediated interaction. Proc Natl Acad Sci USA 2001; 98:1300–1305.
103. Bates IR, Hebert B, Luo Y, Liao J, Bachir A, Kolin DL et al. Membrane lateral diffusion and capture of CFTR within transient confinement zones. Biophys J 2006; 91:1046–1058.
104. Haggie PM, Kim JK, Lukacs GL, Verkman AS. Tracking of quantum dot-labeled CFTR shows near immobilization by C-terminal PDZ interactions. Mol Biol Cell 2006; 17: 4937–4945.
105. Huang P, Trotter K, Boucher RC, Milgram SL, Stutts MJ. PKA holoenzyme is functionally coupled to CFTR by AKAPs. Am J Physiol Cell Physiol 2000; 278:C417–C422.
106. Li C, Krishnamurthy PC, Penmatsa H, Marrs KL, Wang XQ, Zaccolo M et al. Spatiotemporal coupling of cAMP transporter to CFTR chloride channel function in the gut epithelia. Cell 2007; 131:940–951.
107. Yarwood SJ, Steele MR, Scotland G, Houslay MD, Bolger GB. The RACK1 signaling scaffold protein selectively interacts with the cAMP-specific phosphodiesterase PDE4D5 isoform. J Biol Chem 1999; 274:14909–14917.
108. Vaandrager AB, Smolenski A, Tilly BC, Houtsmuller AB, Ehlert EM, Bot AGM et al. Membrane targeting of cGMP-dependent protein kinase is required for cystic fibrosis transmembrane conductance regulator Cl- channel activation. Proc Natl Acad Sci USA 1998; 95:1466–1471.
109. Hallows KR, Raghuram V, Kemp BE, Witters LA, Foskett JK. Inhibition of cystic fibrosis transmembrane conductance regulator by novel interaction with the metabolic sensor AMP-activated protein kinase. J Clin Invest 2000; 105:1711–1721.
110. Quinton PM. The neglected ion: HCO_3^-. Nature Medicine 2001; 7:292–293.
111. Lee MC, Penland CM, Widdicombe JH, Wine JJ. Evidence that Calu-3 human airway cells secrete bicarbonate. Am J Physiol Lung Cell Mol Physiol 1998; 274:L450–L453.
112. Devor DC, Singh AK, Lambert LC, DeLuca A, Frizzell RA, Bridges RJ. Bicarbonate and chloride secretion in Calu-3 human airway epithelial cells. J Gen Physiol 1999; 113: 743–760.
113. Irokawa T, Krouse ME, Joo NS, Wu JV, Wine JJ. A "virtual gland" method for quantifying epithelial fluid secretion. Am J Physiol Lung Cell Mol Physiol 2004; 287:L784–L793.
114. Poulsen JH, Fischer H, Illek B, Machen TE. Bicarbonate conductance and pH regulatory capability of cystic fibrosis transmembrane conductance regulator. Proc Natl Acad Sci USA 1994; 91:5340–5344.
115. Lee MG, Choi JY, Luo X, Strickland E, Thomas PJ, Muallem S.Cystic fibrosis transmembrane conductance regulator regulates luminal Cl-/HCO_3^- exchange in mouse submandibular and pancreatic ducts. J Biol Chem 1999; 274:14670–14677.
116. Ko SBH, Shcheynikov N, Choi JY, Luo X, Ishibashi K, Thomas PJ et al. A molecular mechanism for aberrant CFTR-dependent HCO_3^- transport in cystic fibrosis. EMBO J 2002; 21:5662–5672.
117. Ko SBH, Zeng W, Dorward MR, Luo X, Kim KH, Millen L et al. Gating of CFTR by the STAS domain of SLC26 transporters. Nature Cell Biol 2004; 6:343–350.

Sulfate and Phosphate Transporters in Mammalian Renal and Gastrointestinal Systems

Daniel Markovich

Abstract This chapter will summarize the most recent data available for sulfate and phosphate transport in mammalian renal and gastrointestinal systems. Dietary derived sulfate and phosphate are absorbed in the intestines and their circulating levels are controlled by renal tubular mechanisms. Such processes are facilitated by sulfate and phosphate transporters that exist in the epithelial cells of the kidneys and intestines. Sulfate transporters belong to two gene families, the Na^+-coupled sulfate transporters (SLC13) and the sulfate anion exchangers (SLC26). Phosphate transporters belong to three gene families, Type I (SLC17), Type II (SLC34), and Type III (SLC20) of which all members are Na^+-coupled phosphate transporters. Tissue distribution for these transporters is diverse, with some being restricted to either renal or intestinal tissues and with others showing more broad tissue distribution. Various modes of regulation affect the expression of these genes and the proteins they encode in their respective tissues. Lessons from knock-out mice reveal the physiological (and pathophysiological) roles these proteins play in the body and their contributions to sulfate and phosphate homeostasis.

Keywords Sulfate · Phosphate · Transporter · Intestine · Kidney · Epithelia

1 Introduction

Inorganic sulfate (S_i) and phosphate (P_i) are two molecules that play important roles in the body [1, 2]. Serum S_i and P_i are carefully maintained by renal tubular mechanisms [1, 2]. These two essential anions are involved in many crucial processes in the body (for reviews see [2–5]). Because of their hydrophilic nature, every cell in the body requires a mechanism by which S_i and P_i can cross their plasma membranes. Both anions are freely filtered in the glomerulus and under normal conditions, up

D. Markovich (✉)
School of Biomedical Sciences, The University of Queensland, St Lucia, QLD 4072, Australia
e-mail: d.markovich@uq.edu.au

G.A. Gerencser (ed.), *Epithelial Transport Physiology*,
DOI 10.1007/978-1-60327-229-2_8,

to 80% of filtered P_i and S_i is reabsorbed in the proximal tubule, the principal site of P_i and S_i reabsorption. Despite the similarity in molecular structures for the two tetra-oxyanions, PO_4^{3-} and SO_4^{2-}, they do not share the same transport pathways nor are the proteins involved in their cellular translocation related in structure or function. This review will be divided into two sections: mammalian (A) S_i transporters and (B) P_i transporters in renal and gastrointestinal systems. In the early 1990s, cDNA molecules encoding these transporters were isolated by expression cloning using *Xenopus* oocytes [6]. The structural isolation of these proteins has allowed molecular and cellular studies to be performed in order to elucidate the regulatory mechanisms which control their expression in vivo. This review will focus on mammalian renal and intestinal S_i and P_i transporters.

2 Mammalian Sulfate Transporters

Inorganic sulfate (S_i) is an essential anion for normal body function. S_i is required for cartilage formation, proteoglycan synthesis and cell matrix formation [3, 5]. S_i is also involved in many physiological and pharmacological processes, including activation and detoxification of many endogenous and exogenous compounds (including catecholamines, bile salts, and steroid/thyroid hormones) [3, 5]. There are numerous fatal disorders that arise from disturbances in S_i metabolism or S_i transport: Hunter's syndrome, Morquio's syndrome, Maroteaux–Lamy syndrome, metachromatic leukodystrophy, congenital chloride diarrhea, and diastrophic dysplasia [3, 5, 7–9]. Since sulfate is a hydrophilic anion, it requires specialized transport proteins to facilitate movement across the plasma membranes of cells. In vertebrates, there are two structurally and functionally distinct families of sulfate transporters: (1) Na^+-dependent sulfate transporters, which contain two members NaS1 (SLC13A1) and NaS2 (SLC13A4) belonging to the SLC13 gene family, and (2) Na^+-independent sulfate transporters, which contain eight members, Sat1 (SLC26A1), DTDST (SLC26A2), DRA (SLC26A3), PAT1/CFEX (SLC26A6), SLC26A7, SLC26A8, SLC26A9, SLC26A11 that belong to the SLC26 gene family (Table 1) [10–12]. This paper will now describe in detail the properties of the various known sulfate transporters.

2.1 SLC13A1 (NaS1)

Three mammalian SLC13A1 orthologues have been cloned so far: human NaS1, rat NaS1, and mouse NaS1 [10, 13–15]. Northern blotting and RT-PCR detected hNaS1 only in kidney; whereas rNaS1 mRNA was detected by Northern blotting in kidney and small intestine and mNaS1 was detected by Northern blotting and RT-PCR in kidney, ileum, duodenum/jejunum, and colon, with weaker signals in cecum, testis, adrenal, and adipose tissue [13, 15]. The NaS1 protein was localized to the apical or brush border membrane (BBM) of renal proximal tubular cells

Table 1 Mammalian sulfate transporters

Protein name	Gene name*	Major tissue(s) expression	Transport substrate(s)	References
NaS1	SLC13A1	Kidney, intestine	Na^+ $SO_4{}^{2-}$	[13–16]
NaS2	SLC13A4	Placenta, brain, heart, testis	Na^+ $SO_4{}^{2-}$	[46–50]
Sat1	SLC26A1	Liver, kidney, brain	$SO_4{}^{2-}$, Cl^-, oxalate	[52–55, 57]
DTDST	SLC26A2	Ubiquitous	$SO_4{}^{2-}$, Cl^-	[7, 59–60]
DRA/CLD	SLC26A3	Intestine, pancreas, prostate	$SO_4{}^{2-}$, Cl^-, $HCO_3{}^-$, OH^-	[61–65]
CFEX/PaT1	SLC26A6		$SO_4{}^{2-}$, Cl^-, $HCO_3{}^-$, OH^-, oxalate, formate	[66, 73–76, 78–81]
SLC26A7	SLC26A7	Kidney	$SO_4{}^{2-}$, Cl^-, oxalate	[73, 88]
Tat1	SLC26A8	Brain, sperm	$SO_4{}^{2-}$, Cl^-, oxalate	[88–90]
SLC26A9	SLC26A9	Lungs	$SO_4{}^{2-}$, Cl^-, oxalate	[88]
SLC26A10	SLC26A10	Pseudogene?	Unknown	[51,73]
SLC26A11	SLC26A11	Ubiquitous	$SO_4{}^{2-}$	[73, 87]

* Gene name given by the Human Genome Organization Nomenclature Committee (http://www.gene.ucl.ac.uk/nomenclature/).

[16, 17]. The human NaS1 gene consists of 15 exons (spanning over 83 kb) on human chromosome 7q31–7q32 [15].

hNaS1 encodes a protein of 595 amino acids with 13 putative transmembrane domains [15]. The mammalian NaS1 transporters are approximately 66.1 kDa in size, comprising of 13 TMDs predicted by TopPred2, with an intracellular NH_2 terminus and an extracellular COOH terminus (Fig. 1) [18]. hNaS1 and rNaS1 proteins contain 595 amino acids, whereas mNaS1 is one amino acid shorter, lacking a glutamate at residue 314 [19, 20]. The significance of this is unknown. NaS1 proteins have one putative extracellular N-glycosylation site, near the end of the C-terminus (Fig. 1). All SLC13 family proteins contain numerous predicted consensus sites for phosphorylation (protein kinase C, cAMP and cGMP, casein kinase II, tyrosine kinase sites) and N-myristoylation sites, for which the functional significance has yet to be determined. In addition, there is a 17 amino acid long consensus sequence motif (TSFAFLLPVANPPNAIV) called the "Sodium:sulfate symporter family signature" (PROSITE PS01271) situated at amino acids 523–539 of rNaS1, starting in TMD12 (Fig. 1), which is highly conserved among the SLC13A1 proteins (100% identity), SLC13A2 (71–82% identity), SLC13A3 (59–77%), SLC13A4 (77% identity), and SLC13A5 (77–82% identity). This motif, the significance of which is unknown, is well conserved in SLC13-like genes in lower eukaryotes and prokaryotes [2].

NaS1 encodes an electrogenic pH-insensitive high-affinity Na^+-dependent $SO_4{}^{2-}$ transporter (Na^+–$SO_4{}^{2-}$ cotransporter/symporter), with substrate preferences for the anions sulfate ($K_m = 93\,\mu M$), thiosulfate ($K_m = 84\,\mu M$), selenate

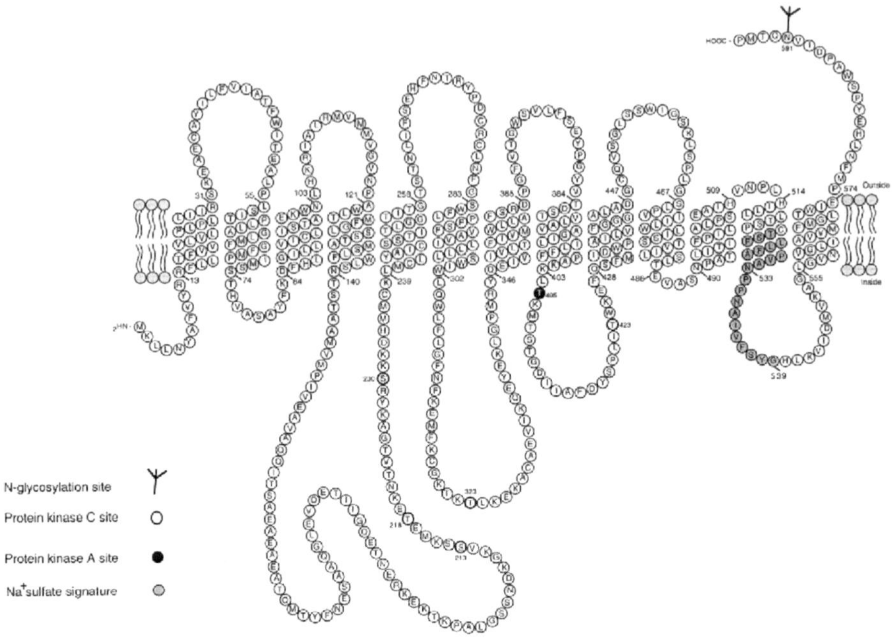

Fig. 1 Secondary structure model of the NaS1 protein (adapted from [2]). Topology is based on the prediction by the TopPred2 program. The N-terminus is intracellular and C-terminus is extracellular. One *N*-glycosylation site is in the extracellular C-terminus. Putative consensus sites are indicated

(K_m = 580 μM) and the cation Na^+ ($K_{m\,Na^+}$ = 16–24 mM) [21]. Significant *cis*-inhibition for NaS1-induced sulfate transport was observed with thiosulfate, selenate, tungstate, and molybdate (for all NaS1orthologs) and with succinate and citrate (for hNaS1 only) [15], suggesting possible subtle differences in substrates between the various NaS1 orthologs.

Although early transport studies in BBM vesicles and renal microperfusion studies have hypothesized that the BBM Na^+/S_i cotransporter, is responsible for regulating plasma S_i levels, up until recently, no metabolic and/or hormonal factors were identified as regulators of NaS_i-1 expression [22, 23]. Recent data from this laboratory (and others), has shown that several hormones and various dietary conditions regulate the expression of NaS_i-1 (and in some cases sat-1) protein in the kidney: (i) *Dietary sulfate intake*: Increased sulfate diet leads to reduced renal S_i reabsorption [23–25]. Chronically (1 week) adapted rats fed a high S_i diet, showed a 2-fold reduced BBM vesicular Na^+/S_i-uptake when compared to rats on a normal S_i diet. This change correlated closely with significant reductions in NaS_i-1 protein expression in renal proximal tubules [24]. (ii) *Dietary potassium intake*: Chronic potassium (K) depletion increases in proximal tubular citrate reabsorption leading to metabolic alkalosis, which has previously been shown to inhibit renal S_i reabsorption [26, 27]. NaS_i-1 protein levels were reduced by 4-fold, in rats fed on a

K-free diet, compared to control rats [17]. (iii) *Metabolic acidosis*: Rats subjected to chronic (10 days) metabolic acidosis (CMA) displayed a 2.5-fold decrease in BBM vesicular Na/S_i-cotransport, which correlated closely with a 2.7-fold decrease in BBM NaS_i-1 protein abundance in renal proximal tubules of CMA rats when compared to control rats [28]. (iv) *Glucocorticoids*: Dexamethasone (dex) treatment has been shown to reduce chicken renal BBM Na/S_i-cotransport without having an effect on BLM $SO_4{}^{2-}/HCO_3{}^{-}$ exchange [29]. Kinetic analysis of the dex effect on BBM Na/S_i-cotransport showed that the V_{max} was decreased by 2-fold (compared to controls), without affecting the K_m. Rats injected daily (for 10 days) with dex subcutaneously (0.1 mg/100 g body weight) showed a 2-fold reduction in BBM vesicular NaS_i-cotransport when compared to water-injected controls (unpublished data). This observation correlated with a 2-fold reduction in NaS_i-1 protein levels (on Western blots with BBM), in dex treated rats when compared to control rats (unpublished data). (v) *Thyroid hormone*: Serum sulfate levels are reduced in hypothyroid patients. Chemically induced hypothyroidism [using 6-propyl-2-thiouracil (PTU)] in rats led to significant decreases in serum sulfate concentrations, renal fractional reabsorption of sulfate and creatinine clearance [30]. Kidney BBM NaS_i-cotransport showed significant decreases in V_{max} values without changes in K_m in hypothyroid rats, with no significant differences in BLM sulfate/anion exchange transport V_{max} or K_m (sat-1 activity). NaS1 protein levels were significantly lower in the kidney cortex from hypothyroid rats, suggesting that PTU-induced hypothyroidism decreases Na/S_i-cotransport by downregulating NaS_i-1 expression [30]. (vi) *Vitamin D*: Vitamin D-deficient rats were shown to have lower plasma S_i levels than control animals, a 3-fold higher fractional S_i renal excretion, and a decrease in renal cortical BBM Na/S_i cotransport activity, associated with parallel decreases in both renal NaS1 protein and mRNA content [31]. Vitamin D supplementation resulted in a return to normal levels of plasma S_i, fractional S_i excretion and both renal NaS_i-1 mRNA and protein. All these effects were shown to be a direct effect of vitamin D, and not mediated by the effects of vitamin D on plasma calcium or PTH [31]. The identification of vitamin D as a modulator of S_i homeostasis may have important clinical implications: by regulating renal S_i transport and sulfatemia, vitamin D status would appear to be a major factor influencing the amount of S_i available for sulfation of proteoglycans. Vitamin D deficiency may thus result in a defect in the synthesis of these important matrix components, a mechanism that may contribute to producing some of the abnormalities observed in rickets and osteomalacia [31]. (vii) *Growth Hormone*: Growth hormone treatment of adult rats led to significant reductions in NaS_i-1 protein expression in rat kidneys [32]. (viii) *Nonsteroidal anti-inflammatory drugs (NSAIDs)*: NSAIDS, such as Ibuprofen (IBU), increase S_i renal clearance and decrease renal fractional reabsorption of S_i. The V_{max} for BBM vesicular BBM NaS_i-cotransport was lower (with no change in K_m) in IBU-treated rats when compared to control rats [33]. No differences were observed in BLM $SO_4{}^{2-}/HCO_3{}^{-}$ exchange in IBU rats compared to controls [33]. NaS1 protein levels in BBM were significantly lower in animals pretreated with IBU compared with that in control animals, indicating that IBU treatment alters Na/S_i-cotransport by downregulation of NaS_i-1 protein in renal BBMs [33].

(ix) *Post-natal expression*: Ontogeny of renal sulfate transporters NaS1 and sat-1 expression was measured during postnatal maturation [34]. Western blot analysis of renal membranes from rats showed no changes in NaS_i-1 protein expression in rats aged day 1 to day 77, whereas sat1 protein levels increased from day 1 to day 14, followed by a decrease in protein levels thereafter [34]. This would suggest, translational mechanisms may maintain constant numbers of NaS1 proteins on proximal tubular membranes during maturation, whereas sat-1 protein levels increase in the first two weeks followed by a decline into old age. (x) *Chronic Renal Failure (CRF)*: Hypersulfatemia occurs in CRF, by yet unknown mechanisms. NaS1 and sat-1 expression in adult male rats was assessed after 80% subtotal nephrectomy (Nx), a model of experimental CRF. NaS_i-1 protein expression was significantly increased in Nx rats compared to controls, whereas sat-1 protein decreased in Nx rats [35]. Although NaS1 protein was upregulated in CRF, renal S_i reabsorption was decreased, implying that other mechanisms may be involved in the renal handling of S_i. The increase in NaS1 and decrease in sat1 protein expression in renal tubules may result in intracellular S_i accumulation, thus leading to hypersulfatemia. The precise mechanisms which lead to changes in NaS_i-1 and sat-1 protein expression by CRF (and all the other conditions mentioned above), need to be resolved further at the cellular and molecular levels.

NaS1 mRNA and protein levels are downregulated in the renal cortex by high sulfate diet [24], hypothyroidism [30], vitamin D depletion [31], glucocorticoids [36], hypokalemia [17], metabolic acidosis [28], and NSAIDs [33], and upregulated by low sulfate diet [25], thyroid hormone [30], vitamin D supplementation [31], growth hormone [32], chronic renal failure [37], and during postnatal growth [38]. The mNaS1 gene (*Nas1*) has been shown to be significantly upregulated by vitamin D and thyroid hormone by binding to VDRE (at −525 bp) and T_3RE (at −436) on the *Nas1* promoter, respectively [39], and downregulated by glucocorticoids [14]. The hNaS1 gene (*NAS1*) is significantly upregulated by xenobiotic 3-methycholanthrene by binding to an XRE (at −2052 bp) on the *NAS1* promoter [40]. rNaS1 induced sulfate transport in *Xenopus* oocytes was completely inhibited by pharmacological activators of PKA (8br-cAMP) and PKC (di-octinoyl glycerol; DOG) [2, 41]. rNaS1-induced sulfate transport in *Xenopus* oocytes was inhibited by the heavy metals mercury, lead, chromium, and cadmium, suggesting that heavy metals inhibit renal brush border Na^+-sulfate cotransport via the NaS1 protein through various mechanisms and that this blockade may be responsible for sulfaturia following heavy metal intoxication [42].

The NaS1 (SLC13A1) proteins encode Na^+-$SO_4{}^{2-}$ cotransporters located on the apical/brush border membranes of epithelial cells lining the renal proximal tubule and small/large intestine [2]. Inorganic sulfate is absorbed in the lower small intestine (ileum) and up to 80% of filtered sulfate (through the glomerulus) is reabsorbed in the proximal tubule. Because of its cellular localization and distinct regulation under a variety of dietary and hormonal conditions (see below), NaS1 has been proposed to be a major regulator of serum sulfate concentrations at 0.3 mM in humans [2, 22]. To determine the functional role of NaS1 in sulfate homeostasis and the physiological consequences of its absence, we have recently generated a mouse lacking a functional NaS1 gene [43]. Nas1 null (*Nas1−/−*) mice exhibit increased urinary sulfate excretion and are therefore hyposulfatemic, with serum sulfate concentration being reduced by >75% in *Nas1−/−* mice when compared to wild-type *Nas1+/+* mice [43]. Using BBMV uptake studies, *Nas1−/−* mice were shown to have reduced renal and intestinal Na^+-sulfate cotransport. *Nas1−/−* mice have general growth retardation, with body weight being reduced by >20% when compared to *Nas1+/+* and *Nas1+/−* littermates at 2 weeks of age and remained so throughout adulthood [43]. *Nas1−/−* females had a lowered fertility, with a 60% reduction in litter size. Spontaneous clonic seizures were observed in *Nas1−/−* mice from 8 months of age. Behavioral studies of *Nas1−/−* mice demonstrated impaired memory and olfactory performance as well as decreased object-induced anxiety and decreased locomotor activity [44, 45]. The precise mechanisms of these features are currently being investigated. However, these data demonstrate that NaS1 is essential for maintaining sulfate homeostasis and its expression is necessary for a wide range of physiological functions.

2.2 *SLC13A4 (NaS2)*

Three SLC13A4 members have been isolated, human NaS2 (hNaS2) (previously known as SUT-1), rat NaS2 (rNaS2), and mouse NaS2 (mNaS2) [46–50]. hNaS2 mRNA is detected predominantly in human placenta, brain, testis, with weaker signals in heart, thymus, liver, and tonsillar epithelium [46, 47]. rNaS2 mRNA is found mainly in rat placenta, with lower levels in brain and liver, whereas mNaS2 mRNA is expressed predominantly in mouse placenta, brain, and lung, with weaker signals in the eye, thymus, heart, testis, and liver [48–50].

The hNaS2 gene (SLC13A4) contains 16 exons, spanning over 47 kb in length. Its 5'-flanking region contains CAAT- and GC-box motifs and a number of putative transcription factor binding sites, including GATA-1, AP-1, and AP-2 consensus sequences [46]. The rNaS2 gene slc13a4 maps to rat chromosome 4 and contains 17 exons, spanning over 46 kb in length [48]. This gene produces two alternatively spliced transcripts, of which the transcript lacking exon 2 is the most abundant form [48]. Its 5'-flanking region contains CAAT- and GC-box motifs and a number of putative transcription factor binding sites, including GATA-1, SP1, and AP-2 consensus sequences [48]. The mouse NaS2 cDNA spans 3,384 nucleotides and its open frame encodes a protein of 624 amino acids [50]. The mNaS2 gene Slc13a4 maps to mouse chromosome 6B1 and contains 16 exons, spanning over 40 kb in length [50]. Its 5'-flanking region contains CAAT- and GC-box motifs and a number of putative transcription factor binding sites, including GATA-1, MTF-1, STAT6, and HNF4 consensus sequences [50].

Expression of hNaS2 protein in *Xenopus* oocytes led to a Na^+-dependent transport of sulfate that was inhibited by thiosulfate, phosphate, molybdate, selenate, and tungstate, but not by oxalate, citrate, succinate, phenol red, and DIDS [46, 47]. Transport kinetics of hNaS2 determined a K_m for sulfate of 0.38 ± 0.10 mM, suggestive of a high affinity sulfate transporter. Na^+ kinetics determined a Hill coefficient of n = 1.6 ± 0.6, suggesting a Na:SO_4^{2-} stoichiometry of 2:1. Expression of rNaS2 protein in *Xenopus* oocytes led to a Na^+-dependent transport of sulfate that was inhibited by phosphate, thiosulfate, tungstate, selenate, oxalate, molybdate, chromate, and arsenate, but not by citrate, succinate, or DIDS [48, 49]. rNaS2 transport was found to be electrogenic, as evidenced from the inhibition of the uptake process by K^+-induced depolarization, with a K_m for sulfate of 0.15 or 1.26 mM [48, 49].

The hNaS2 protein encodes a Na^+-SO_4^{2-} cotransporter, expressed in high endothelial venules, where it was suggested to play an important role in sulfation, and in placenta, where it was proposed to facilitate sulfate transfer across the trophoblast from mother to foetus [47]. NaS2 was predicted to have 12 TMDs with both termini being extracellular, however no biochemical evidence exists to prove this model [47]. At present, no information is available on the regulation of NaS2 or on its physiological role in the body.

2.3 SLC26A1 (Sat1)

The first member of the SLC26 gene family, sat1 (sulfate anion transporter 1), was identified using a *Xenopus* oocyte expression cloning system from rat liver [38, 51]. More recently, mouse and human sat1 orthologues (named msat1 and hsat1, respectively) have been identified using a homology screening approach with the rat sat1 (rsat1) cDNA [52–54]. Northern blot analysis detected a single mRNA (3.8 kb) transcript for rsat1 very strongly in rat liver and kidney, with weaker signals in skeletal muscle and brain [54]. The abundance of rsat1 mRNA in the kidney was identical to its expression in the liver (the tissue from which it was cloned), prompting us to examine the role of sat1 in the kidney. msat1 mRNA expression was found to be strongest in the kidney and liver, with lower levels observed in cecum, calvaria, brain, heart, and skeletal muscle [53]. Two distinct transcripts were expressed in kidney and liver due to alternative utilization of the first intron, corresponding to an internal portion of the 5′-untranslated region. hsat1 mRNA expression was found most abundantly in the kidney and liver, with lower levels in the pancreas, testis, brain, small intestine, colon, and lung [52].

The msat1 and hsat1 proteins share 94 and 77% amino acid identities, respectively, with the rsat1 [2]. The coding region of the rat sat1 (rsat1) cDNA predicts a protein of 703 amino acids with a calculated molecular mass of 75.4 kDa. Hydrophobicity analysis of rsat1 protein suggests 12 putative transmembrane domains (TMDs) [2]. Interestingly, there are two consensus regions that are highly conserved between the sat1 proteins and other members of the SLC26 gene family. These are the "phosphopantetheine attachment site" (PROSITE PS00012) motif (QTQLSSVVSAAVVLLV) at residues 415–430 of rsat1 within the putative TMD9 and the "Sulfate transporter signature" (PROSITE PS01130) motif (PIYSLYTSFFANLIYFLMGTSR) at residues 98–119 within putative TMD3 of both rsat1 and msat1 [2]. Furthermore, there is a "sulfate transporter and anti-sigma antagonist" (STAS) domain found in the C-terminal cytoplasmic domain of the sat1 protein, which appears conserved in all members of the SLC26 family [51].

Rat kidney mRNA when injected into *Xenopus* oocytes induced a Na^+-independent sulfate transport activity that was inhibitable by DIDS, probenecid, and phenol red [55]. The degree of inhibition by these compounds was closely correlated with the inhibition pattern of the renal BLM sulfate anion exchanger, as well as with rsat1 cRNA-induced activity in *Xenopus* oocytes [54, 56]. Furthermore, using a hybrid depletion strategy, sat1 antisense oligonucleotides led to a complete abolition of the kidney mRNA-induced Na^+-independent sulfate transport activity (in *Xenopus* oocytes), confirming that rsat1 encodes the Na^+-independent sulfate transporter, whose function most closely correlates with the proximal tubular basolateral membrane (BLM) sulfate/bicarbonate anion exchanger [55]. This was confirmed by immunocytochemistry using rsat1 polyclonal antibodies showing sat1 protein localization restricted to the BLM of kidney proximal tubules [57]. The mouse sat1 gene (*Sat1*; Slc26a1) is approximately ~6 kb in length and consists of four exons, with an optional intron 1. The *Sat1* promoter is ~52% G + C rich and contains a number of well-characterized *cis*-acting elements, including sequences resembling responsive elements to thyroid hormone (T_3REs) and vitamin D (VDREs). *Sat1* promoter drove basal transcription in renal opossum kidney (OK) cells, and was stimulated by tri-iodothyronine (T_3), but not by 1,25-$(OH)_2$ vitamin D_3 (vitamin D). Site-directed mutagenesis of an imperfect T_3RE sequence at −454 bp led to a loss of *Sat1* promoter inducibility by T_3. This suggests that position −454 in the *Sat1* promoter contains a functional T_3RE that is responsible for transcriptional activation of *Sat1* by T_3. The human sat1 gene (*SAT1*; SLC26A1) is localized to human chromosome 4p16.3 and comprises of four exons stretching approximately 6 kb in length, with an alternative splice site formed from an optional exon (exon II) [52]. The *SAT1* promoter is ~60% G + C rich, contains a number of well-characterized *cis*-acting elements, but lacks any canonical TATA- or CAAT-boxes. The *SAT1* 5′ flanking region led to basal promoter activity in renal OK and LLC-PK1 cells. Using *SAT1* 5′ flanking region truncations, the first 135 bp was shown to be sufficient for basal promoter activity in both cells lines. Unlike in the mouse *Sat1* promoter, neither tri-iodothyronine (T_3), nor 1,25-$(OH)_2$ vitamin D_3 (vitamin D) could trans-stimulate *SAT1* promoter activity, despite the presence of sequences in its promoter resembling responsive elements to thyroid hormone (T_3REs) and vitamin D (VDREs). Mutation of the activator protein-1 (AP-1) site at position −52 in the *SAT1* promoter led to loss of transcriptional activity, suggesting the requirement of AP-1 for *SAT1* basal transcription. Unlike NaS1, very little information is available on the regulation of sat1 in vivo.

Functional characterization in *Xenopus* oocytes using radiotracer sulfate measurements, revealed rsat1 to encode a Na^+-independent sulfate transporter, with a high affinity ($K_m = 136\,\mu M$) for sulfate, that was strongly inhibited by the stilbene derivative DIDS (4,4′-diisothiocyanato-2,2′-disulfonate; $IC_{50\,DIDS} = 28\,\mu M$) and oxalate, but not by succinate or cholate [54]. These properties correlated closely with the functional activities of the sulfate/bicarbonate exchanger in liver canalicular membrane vesicles [58]. msat1 expression led to a ~20-fold induction in Na^+-independent sulfate transport, a ~6-fold induction in oxalate and chloride transport, but was unable to induce any formate, L-leucine, or succinate uptakes, when compared to water-injected (control) oocytes [53]. This suggests that msat1 transport was restricted to the anions SO_4^{2-}, oxalate, and chloride. As shown for rsat1, msat1-induced SO_4^{2-} transport occurred only in the presence of extracellular chloride, suggesting its requirement for chloride [53, 54]. Typical Michaelis–Menten saturation kinetics were observed for msat1-induced SO_4^{2-} uptake, with a calculated $K_m = 0.31 \pm 0.05$ mM and $V_{max} = 143.90 \pm 5.32$ pmol/oocyte/hour for SO_4^{2-} interaction [53]. These values are in close agreement with rsat1-induced SO_4^{2-} transport kinetics ($K_m = 0.14$ mM) [54]. msat1-induced SO_4^{2-} transport was significantly inhibited by molybdate, selenate, tungstate, DIDS, thiosulfate, phenol red, and probenecid, whereas citrate and glucose had no effect. These data suggest that msat1 encodes a functional SO_4^{2-}/chloride/oxalate anion exchanger, whose activity is blocked by anion-exchange inhibitors DIDS and phenol red, tetraoxyanions (selenate, molybdate, and tungstate) and thiosulfate. Similarly, using radiotracer measurements in *Xenopus* oocytes, hsat1 led to ~40-fold induction in Na^+-independent sulfate uptake, ~5-fold induction in chloride uptake, and ~6-fold induction in oxalate uptake, but was unable to induce formate uptake, when compared to the water-injected (control) oocytes [52]. This is in agreement with the transport activities for rsat1 and msat1, both of which transport sulfate, oxalate, and chloride, but not formate [53, 54]. Typical Michaelis–Menten saturation kinetics were observed for hsat1-induced sulfate uptake in *Xenopus* oocytes, with a calculated K_m for SO_4^{2-} of 0.19 ± 0.066 mM and a V_{max} of 52.35 ± 3.30 pmol/oocyte/hour [52]. Various compounds were tested as possible inhibitors of hsat1-induced sulfate uptake in *Xenopus* oocytes: the stilbene-derivative DIDS, phenol red, thiosulfate, molybdate, tungstate, and selenate all significantly inhibited hsat1-induced SO_4^{2-} transport, whereas citrate and glucose had no effect [52].

2.4 SLC26A2 (DTDST)

By positional cloning, a gene was identified on human chromosome 5q to be linked to an autosomal recessive osteochondrodysplasia called diastrophic dysplasia (DTD; OMIM 222600), with clinical features including dwarfism, spinal deformation, and abnormalities of the joints [7]. The human DTD gene (SLC26A2) is comprised of four exons and three introns spanning over 40 kb. DTD shares 48% and 33% amino acid identities with rsat-1 (Slc26a1) and DRA (Slc26a3), respectively [7]. Because of its high sequence similarity with rsat-1, the exact function of DTD was sought. Primary skin fibroblasts obtained from normal, carrier, and DTD patients (with mutations in the DTD gene) were cultured and assayed for sulfate uptake [7]. Cells from normal and carrier samples showed saturable sulfate transport, whereas cells from the DTD patients showed a greatly diminished sulfate uptake [7], suggesting that the normal (wild type) DTD gene encodes a functional sulfate transporter, which loses its transport activity when mutated (as in DTD patients). Thus the cDNA encoding the DTD gene became known as the diastrophic dysplasia sulfate transporter (DTDST) [7]. Translation of DTDST cDNA produced a protein of 739 amino acids, with a predicted molecular mass of 82 kDa, containing 12 putative TMDs [7]. DTDST mRNA expression was ubiquitous, detected by Northern blot analysis in all tissues of the body [7]. At least 30 different mutations have been identified in the human DTDST gene [11], which are responsible for a variety of recessively inherited chondrodysplasias, including diastrophic dysplasia (DTD), atelosteogenesis Type 2 (AO2) and a lethal condition known as achondrogenesis 1B (ACG1B) [7]. For a recent review on DTDST mutations and their contributions to these pathogenic states see [11].

Subsequently, the rat DTDST gene and cDNA (rdtdst) were cloned and its function was characterized in *Xenopus* oocytes [59]. rdtdst induced Na^+-independent sulfate transport, which was inhibited by chloride, unlike rsat-1 which required extracellular chloride for sulfate transport [59]. Furthermore, rdtdst-induced sulfate uptake was inhibited by thiosulfate, oxalate, and DIDS, with a *cis*-inhibition pattern identical to rsat-1 and DRA. No information is presently available on sulfate transport kinetics of DTDST. The mouse dtdst (mdtdst) cDNA (initially named *st-ob* for sulfate transporter in osteoblasts) was isolated from mouse fibroblastic cells C3H10T1/2 [60]. The mdtdst protein shares 80% amino acid identity with the human DTDST protein. Tissue distribution showed the mdtdst mRNA strongly expressed in the thymus, testis, calvaria, and the osteoblastic MC3T3-E1 cells, having a reduced expression in the undifferentiated C3H10T1/2 cells. The expression of mdtdst mRNA in C3H10T1/2 cells was shown to be increased by transforming growth factor-beta1 (TGF-beta1), retinoic acid, and dexamethasone, as well as BMP-2 [60]. BMP-2 also led to a 2-fold increase in sulfate uptake in C3H10T1/2 cells when compared to untreated (control) cells, suggesting that st-ob encodes a functional sulfate transporter, whose activity was upregulated by BMP-2. Since osteoblasts actively take up sulfate to synthesize proteoglycans (being major components of the extracellular matrix of bone and cartilage), the functional activity of mdtdst protein as a sulfate transporter may be important for osteoblastic differentiation.

2.5 SLC26A3 (DRA, CLD)

By subtractive hybridization using cDNA libraries from normal colon and adenocarcinoma tissues, a human cDNA which was downregulated in adenomas (DRA) was isolated [61]. Human DRA (hDRA) mRNA was found to be exclusively expressed in normal colon tissues, with its expression significantly decreased in colonic adenomas (polyps) and andenocarcinomas. hDRA mRNA was detected throughout the intestinal tract (duodenum, ileum, cecum, distal colon), but not in the esophagus or stomach [62]. The hDRA gene (SLC26A3) was mapped to human chromosome 7q22–q31.1 [63]. The predicted topology of this polypeptide was an 84.5 kDa protein, having either 10, 12, or 14 TMDs, with charged clusters of amino acids at its NH_2 and COOH termini, potential nuclear targeting motifs, an acidic transcriptional activation domain, and a homeobox domain [61]. These putative domains in the hDRA protein suggest it could interact with a transcription factor, which would be consistent with hDRA having a role in tissue-specific gene expression and/or as a candidate tumor suppressor gene. Using the baculovirus expression system, hDRA was expressed in insect Sf9 cells where it led to greater than 3-fold increase in sulfate uptake when compared to control cells [64]. In *Xenopus* oocytes, hDRA induced Na^+-independent sulfate, chloride and oxalate transport, which was DIDS-sensitive [9, 65, 66], suggesting that hDRA encodes a functional sulfate/chloride/oxalate anion exchanger. No information is presently available on the transport kinetics of DRA. The hDRA gene (SLC26A3) was found to be defective in patients with congenital chloride diarrhea (CLD; OMIM 214700) syndrome [9], a recessively inherited defect of intestinal chloride/bicarbonate exchange. CLD is a potentially fatal diarrhea characterized by a high chloride content [67, 68]. To date, over 30 different mutations have been identified in CLD patients, from various ethnic populations [68]. For a recent review on hDRA mutations and their contributions to a pathogenic states, see [11]. DRA gene was implicated as a positional and functional candidate for CLD [69, 70]. Due to its link with congenital chloride diarrhea, the hDRA gene became known as the CLD gene. The CLD gene spans ≈39 kb and is comprised of 21 exons [71]. Analysis of the putative CLD promoter region (570 bp) showed putative TATA and CCAAT boxes and multiple transcription factor binding sites for AP-1 and GATA-1 [71], however the functional significance of these sites is yet unknown. The mouse DRA (mDRA) cDNA orthologue was cloned using a combination of RT-PCR and RACE techniques [72]. When expressed in human embryonic kidney HEK293 cells, mDRA protein conferred Na^+-independent, electroneutral chloride/bicarbonate exchange activity. Northern blot analysis showed mDRA mRNA expressed at high levels in cecum and colon and at lower levels in small intestine [72].

2.6 SLC26A6 (CFEX, PAT1)

Human and mouse SLC26A6 were initially identified by searches of EST and genomic databases for novel homologues of previously known members of the SLC26 gene family [73–75]. SLC26A6 is relatively ubiquitously expressed in both epithelial and nonepithelial tissues with prominent expression in heart, placenta, liver, skeletal muscle, kidney, pancreas, stomach, and small intestine [73–76]. Immunolocalization studies demonstrated SLC26A6 expression on the apical membrane in several epithelia including proximal tubule [75, 77], pancreatic duct [73], and small intestine [76].

The SLC26A6 gene consists of 21 exons on human chromosome 3p21 or mouse chromosome 9 [74, 78]. Multiple splice isoforms of SLC26A6 have been identified in human and mouse due to use of alternative first exons, or alternative splice donor and acceptor sites [66, 73, 74, 78]. Differences in transport properties of isoforms functionally expressed successfully have not been detected, although there are conflicting results about whether the SLC26A6c and SLC26A6d isoforms are functional at all [66, 79]. Tissue-specific differences in expression among splice variants have been demonstrated [66].

The first functional expression studies were performed on mouse SLC26A6 and demonstrated Cl^--formate exchange activity [75]. Subsequent studies demonstrated that the transporter has broad specificity and can function in multiple DIDS-sensitive exchange modes involving sulfate, oxalate, Cl^-, formate, HCO_3^-, and OH^- as substrates [66, 76, 78–80]. Lactate and the anions of both short- and medium-chain fatty acids were also reported to be transported by SLC26A6 [81]. Cl^--oxalate exchange mediated by SLC26A6 is electrogenic, consistent with exchange of divalent oxalate for monovalent Cl^- [80]. Interestingly, Cl^--HCO_3^- exchange mediated by SLC26A6 is also electrogenic [78, 82], consistent with transport of more than a single HCO_3^- in exchange for Cl^-. Indeed, a careful comparison of Cl^- and HCO_3^- flux rates demonstrated that the HCO_3^-/Cl^- stoichiometry is 2:1 [83]. Comparison of the potency of various anions as inhibitors of SLC26A6 activity indicated that affinity for oxalate is appreciably greater than for Cl^-, HCO_3^-, sulfate, or formate [78, 80]. At equal substrate concentrations, transport of oxalate was twice as high as that of sulfate or formate, indicating that among these anions SLC26A6 has greatest activity as an oxalate transporter [80].

Given its ability to function as an anion exchanger with broad specificity for many substrates in addition to sulfate, a key issue is the physiological role(s) of SLC26A6 in the proximal tubule as well as in other tissues in which it is expressed. Studies in SLC26A6 null mice have provided insights into the physiological roles of SLC26A6 [84]. Comparison of wild-type and SLC26A6 null mice have indicated that SLC26A6 mediates all of the Cl^--oxalate exchange activity, a significant part of the sulfate-oxalate exchange activity, and a minor component of the Cl^--formate exchange activity in renal BBM vesicles [84]. Moreover, studies in SLC26A6 null mice demonstrated that SLC26A6 mediates a substantial portion of Cl^--OH^-/HCO_3^- exchange across the apical membrane of proximal tubule cells [85].

The possible role of SLC26A6 in the process of transtubular sulfate transport has not yet been assessed. The net direction of SLC26A6-mediated sulfate-anion exchange will depend on the gradients across the apical membrane of sulfate and any given exchanging anion. Given the presence in the same membrane of NaS1, which is theoretically capable of generating a cell to lumen gradient of sulfate larger than the corresponding gradient of any exchange partner, it is most likely that SLC26A6 would mediate backflux of sulfate from cell to lumen. Possible alterations in body sulfate homeostasis in SLC26A6 null mice have yet to be evaluated.

Recent studies have begun to define the role of SLC26A6-associated proteins in regulating transporter expression or function. The C-terminus of SLC26A6 has a predicted PDZ-binding motif [73, 75]. The C-terminus of SLC26A6 is capable of binding the PDZ domain containing proteins NHERF1, NHERF2, and PDZK1 (NHERF3) [66, 77, 86]. Interactions with these proteins were abolished when the predicted C-terminal PDZ binding motif of SLC26A6 was deleted [66, 77]. It should be noted that this PDZ binding motif is absent in the SLC26A6d splice variant [66]. The possible functional significance of interaction of CFEX with NHERF1 and NHERF2 in the kidney in vivo is not yet known. A dramatic loss of expression of SLC26A6 protein and functional activity was observed in the kidneys of PDZK1 null mice, indicating that this scaffolding protein is essential for the normal expression and function of SLC26A6 in the proximal tubule [77].

2.7 SLC26A7-A10

Using bioinformatics/genomic approaches, the last remaining five members of the SLC26 gene family (SLC26A7-A11) were identified by screening databases using sequence information of the first five SLC26 members [73, 87, 88]. Each gene encoding SLC26A7-A11 was found to reside on a different human chromosome, with a unique tissue distribution [51, 73]. Functional expression of the SLC26A7, A8, and A9 individually in *Xenopus* oocytes demonstrated significant radiotracer uptake of sulfate, chloride, oxalate, and formate, which was inhibited by DIDS [88]. Tat1 was found to be expressed in the testis [89], where its role was suggested to be in spermatogenesis [88]. Targeted disruption of the Tat1 gene SLC26A8 demonstrated that Tat1 null males were sterile due to complete lack of sperm motility and reduced sperm fertilization potential, suggesting that Tat1 is a critical component of the sperm annulus that is essential for proper sperm tail differentiation and motility [90]. SLC26A10 was found to be expressed in the brain [73], however, its function is yet unknown and it has been suggested to be an expressed pseudogene [51]. The SLC26A11 cDNA was isolated from human high endothelial venule endothelial cells (HEVEC) and when expressed in *Xenopus* oocytes led to Na^+-independent sulfate transport that was DIDS-sensitive [87]. No information is presently available on the sulfate transport kinetics or the precise physiological roles of the SLC26A7-A11 proteins.

3 Mammalian Phosphate Transporters

Inorganic phosphate (P_i) is an essential nutrient involved in many cellular processes, including glycolysis, gluconeogenesis, energy metabolism, and skeletal mineralization [91]. Three distinct families of phosphate transporters exist in vertebrates (Table 2), with all three being Na^+-coupled P_i transporters. (1) Type I Na-Pi cotransporters belong to the SLC17 gene family, which has four members NPT1, NPT3, NPT4, and SLC17A4 [92]. (2) Type II Na-Pi cotransporters belong to the SLC34 gene family, which has three members NaPi-IIa [93], NaPi-IIb [94], and NaPi-IIc [95]. (3) Type III Na-Pi cotransporters belong to the SLC20 gene family, which has two members Pit-1 and Pit-2 [96]. The following is a detailed description of the properties of the various known phosphate transporters.

Table 2 Mammalian phosphate transporters

Protein name	Gene name*	Tissue(s) expressed	Transport substrate(s)	References
NPT1 (NaPi-1)	SLC17A1	Kidney, liver	Na^+ $HPO_4{}^{2-}$, Cl^-, organic acids	[97–103]
NPT3	SLC17A2	Kidney, liver, pancreas, lung, brain, placenta, heart, muscle	Unknown	[92, 105]
NPT4	SLC17A3	Kidney, small intestine, liver, testis	Unknown	[92, 105]
SLC17A4	SLC17A4	Small intestine, colon, liver, pancreas	Unknown	[92, 105]
NaPi-IIa	SLC34A1	Kidney, osteoclasts, neurons	Na^+ $HPO_4{}^{2-}$	[93, 109–111]
NaPi-IIb	SLC34A2	Small intestine, liver, lung, testis	Na^+ $HPO_4{}^{2-}$	[94]
NaPi-IIc	SLC34A3	Kidney	Na^+ $HPO_4{}^{2-}$	[112]
Pit-1 (Glvr-1)	SLC20A1	Ubiquitous	Na^+ $HPO_4{}^{2-}$	[150]
Pit-2 (Glvr-2)	SLC20A2	Ubiquitous	Na^+ $HPO_4{}^{2-}$	[151–153]

* Gene name given by the Human Genome Organization Nomenclature Committee (http://www.gene.ucl.ac.uk/nomenclature/).

3.1 SLC17 (Type I Na^+-P_i Cotransporters)

The first Type I Na-Pi cotransporter, NaPi-1 (SLC17A1), was isolated by *Xenopus* oocyte expression cloning from a rabbit kidney cDNA library [97]. Rabbit NaPi-1 encodes a multi transmembrane domain protein of 465 amino acids which leads to Na^+-phosphate uptake in *Xenopus* oocytes with a K_m ~1 mM for P_i [97]. The human NaPi-1 orthologue NPT1 has similar transport characteristics with a slightly higher K_m of 0.3 mM for P_i [98]. Since the K_m for P_i transport by NaPi-1 was much lower than measured in native tissues (renal cortex), NaPi-1 was proposed not to encode the primary renal phosphate transporter. This was corroborated by the finding that NaPi-1 also transports organic anions, such as probenecid and penicillin, and has a higher K_m for organic anions (0.22 mM for benzylpenicillin) than for P_i [99]. Transport of organic anions by NaPi-1 and human NPT1 was found to be electrogenic, which was not influenced by pH [99–101]. The ionic coupling of these substrates has not been determined and the nature of the endogenous substrates remains unknown. Rabbit NaPi-1 mRNA is expressed on the BBM of the renal proximal tubule [102] and the sinusoidal membrane of hepatocytes [101, 103]. The other three NPT1-like proteins have been isolated through genomic screening: NPT3 (SLC17A2), NPT4 (SLC17A3), and a $Na^+/PO_4{}^{2-}$ cotransporter homologue (SLC17A4) [104, 105], with all four genes being localized to chromosome 6p21.3–p23 [104]. Northern blot analysis revealed high levels of NPT3 in the heart and muscle, with lower levels in brain, placenta, lung, liver, and kidney [92]. NPT4 was expressed highly in the liver and kidney, with lower levels in the small intestine and testis [92]. SLC17A4 was shown to be expressed in the liver, pancreas, small intestine, and colon [92]. Hepatocyte nuclear factor 1 alpha (HNF1 α) was shown to upregulate NPT1 and NPT4, however NPT3 and SCL17A4 were not tested [106]. It appears NPT1, 3, and 4 undergo alternative splicing, but the physiological significance of this has not been studied. No functional characterization of NPT3 (SLC17A2), NPT4 (SLC17A3), or SLC17A4 has been performed. Because of their lack of specificity for phosphate, the physiological roles of these SLC17 genes in the body remain to be elucidated.

3.2 SLC34 (Type II Na^+-P_i Cotransporters)

Na^+-dependent P_i transport was first demonstrated in renal and intestinal epithelia using membrane vesicle studies [107]. NaPi-IIa (SLC34A1) was the first member of this family to be isolated, from rat kidneys by expression cloning using *Xenopus laevis* oocytes [93]. NaPi-IIa protein was localized using immunohistochemistry to the apical BBM of renal proximal tubular cells [108] and was also detected in rat brain [109], osteoclasts [110], and osteoblast-like cells [111]. The second member NaPi-IIb (SLC34A2) was characterized on the BBM of enterocytes and was also detected in the lung, colon, testes, and liver [94]. The third member NaPi-IIc (SLC34A3) was detected on the BBM of the renal proximal tubule [112]. These three proteins share the highest sequence identities (over 80%) in the transmembrane-domains (TMDs), with the largest divergence found in the NH_2 and COOH termini.

The expression and regulation of SLC34 proteins in the kidneys and intestines have been studied extensively (for recent reviews see [1, 95, 113, 114]. The critical role of NaPi-IIa in P_i homeostasis was established by the hyperphosphaturia observed in the NaPi-IIa null mouse [115]. Furthermore, dysregulation of NaPi-IIa caused P_i deficiency disorders, including X-linked hypophosphatemia (XLH) and autosomal-dominant hypophosphatemic rickets (ADHR) [114]. However, no naturally occurring mutation in SLC34A1 has been linked to any human disease. In homozygous NaPi-IIa knockout mice, approximately 30% P_i reabsorption remains in the kidney, most likely due to NaPi-IIc, which was originally described as a growth-related P_i transporter [112]. Recent studies have described SLC34A3 mutations leading to hereditary hypophosphatemic rickets with hypercalciuria (HHRH) [116–118], suggesting that NaPi-IIc may play an important role in P_i homeostasis in humans. NaPi-IIb is responsible for transcellular P_i absorption in the small intestine and is regulated by dietary P_i [119] and metabolic acidosis [120]. In the liver, NaPi-IIb is involved in the reabsorption of P_i from primary hepatic bile [121]. In salivary glands, NaPi-IIb is involved in secreting P_i into saliva, where a high P_i content is important for remineralization of dental enamel [122].

The most studied member of this family in terms of regulation is NaPi-IIa. The BBM NaP_i-IIa is a highly regulated transporter both at the transcriptional and the post-transcriptional levels. Chronic (long term) administration of lipophilic hormones to animals have led to changes in Na/P_i-cotransport (NaP_i-IIa mRNA and protein levels) in isolated BBMs: thyroid hormones (T_3, T_4), 1,25-$(OH)_2$ vitamin D_3, and retinoic acid increase transport rate, whereas glucocorticoids and estradiol reduce transport activity [8, 123–127]. Other (peptide) hormones, such as insulin, insulin-like growth factor-1 (IGF-1) and growth hormone, lead to an increase in Na/P_i-cotransport rate [124, 127, 128]. Parathyroid hormone (PTH) and dietary P_i intake can modulate Type II NaP_i-cotransporter expression both acutely (within minutes to hours) and chronically (hours to days) by adaptive changes in renal BBM Na/P_i-cotransporter expression [127, 129, 130]. Dietary P_i intake regulates Type II Na/P_i-cotransporter in the BBMs of the proximal tubule [91, 127, 130, 131]. When the diet is changed from low P_i to high P_i, renal Na/P_i-cotransport activity is rapidly decreased [129]. This is characterized by a decrease in the apparent maximal transport rate (V_{max}) for P_i, with a concomitant decrease in the amount of NaP_i-IIa protein on the BBM [128]. Acutely, the effect of high P_i diet on transport is independent of de novo protein synthesis and is attributed to recycling of the endocytosed P_i into subapical compartments below the plasma membrane [132]. Dietary P_i overload then leads to lysosomal degradation of the transporter. Conversely, acute P_i deprivation leads to increases in NaP_i-IIa expression at the BBM, without increases in NaP_i-2 mRNA content, whereas chronic (long term) low P_i treatment leads to de novo mRNA and protein synthesis. A similar form of regulation of NaP_i-IIa is observed by PTH.

Transport kinetics of SLC34 proteins have been extensively studied in *X. laevis* oocytes (for a review, see [133, 134]). All three SLC34 proteins exhibit a strict dependence on external Na^+ as the driving substrate, with an apparent affinity for Na^+ of approx. 50 mM, a preference for divalent P_i ($HPO_4{}^{2-}$) as the driven substrate with an apparent affinity of $\geq$0.1 mM, and cotransport activity inhibitable by phosphonoformic acid (PFA) [135, 136]. NaPi-IIa and NaPi-IIb are electrogenic, and transport Pi with a 3:1 Na^+:$HPO_4{}^{2-}$ stoichiometry, whereas NaPi-IIc is electroneutral and operates with a 2:1 Na^+:$HPO_4{}^{2-}$ stoichiometry (for review, see [137]).

Structure–function studies of all three members of the Type II Na/Pi-cotransporters predict similar hydropathy profiles with eight transmembrane domains (TMDs) [133, 138]. Both N- and C-termini are located intracellularly and multiple N-glycosylation sites exist in a large extracellular loop between TMD3 and TMD4 (Fig. 2).[95, 137] The highest sequence identity (over 80%) between the three Type II Na/Pi-cotransporters is in the TMDs [133], with the least sequence similarity in the N- and C-termini as well as the large extracellular loops. Evidence suggests that the functional unit of NaPi-lla is the monomeric form of the protein [139].

Both NaPi-lla and NaPi-llc are located in the renal proximal tubule where they are responsible for P_i reabsorption and are regulated by a variety of hormones and metabolic factors (e.g., P_i diet or acidosis) [1, 112]. Regulation of proximal tubular P_i reabsorption has been described to take place by altering the abundance of NaPi-lla or NaPi-llc proteins in the BBM [95]. Many aspects of protein signaling have

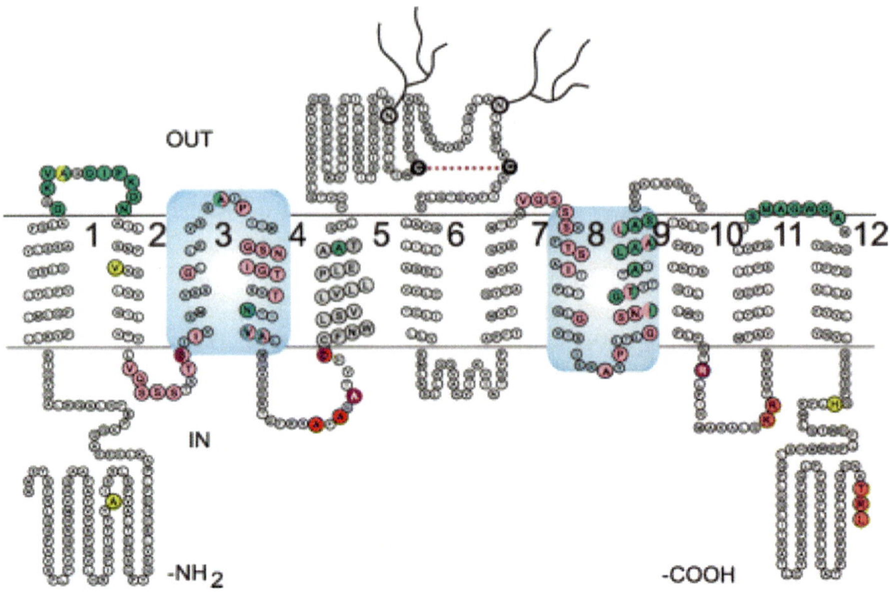

Fig. 2 Secondary structure model of the NaPi-IIa protein (adapted from [137]). Transmembrane domains were assigned according to TopPred2 program. The N- and C- termini are intracellular. Two *N*-glycosylation sites are in the large extracellular loop

been described for NaPi-IIa [1] but not for NaPi-IIb or NaPi-IIc, yet the precise mechanisms of the internalization of NaPi-lla and the mechanisms involved in the vesicular trafficking of NaPi-lla proteins have yet to be determined. In terms of regulation, both acute (2–4 h) and chronic (days to weeks) P_i depletion has been shown to increase the levels of NaPi-lla mRNA and proteins [129, 140, 141].

In renal proximal tubular cells, NaPi-lla interacts with the PDZ proteins NHERF-1 and PDZK1 [142], which are important for correct apical sorting and/or positioning of NaPi-lla, which has been shown by a NHERF-1 knock out model [143] and in OK cells [144].

NaPi-IIb has been shown to be responsible for transcellular P_i transport in the small intestine [94]. NaPi-llb has been shown to be upregulated by 1,25-$(OH)_2$ vitamin D_3 and low-P_i diet in the small intestine [95] by increasing its protein abundance via nontranscriptional mechanisms [145]. Furthermore, NaPi-llb has also been shown to be regulated in the small intestine by Epidermal Growth Factor [146], glucocorticoids [147], thyroid hormone [148], and stanniocalcin 1 [149].

3.3 SLC20 (Type III Na^+-P_i Cotransporters)

Type III Na^+/P_i cotransporters were identified initially as retroviral receptors Glvr-1 [150] and Ram-1 [151, 152]. Glvr-1 renders cells susceptible to infection by the gibbon ape leukemia virus and was found to have sequence identity with a phosphate permease in the filamentous fungus *Neurospora crassa* [150]. Subsequently, a related protein which renders cells susceptible to infection by the amphotropic murine retrovirus (Ram-1) was identified [151, 153]. Transport measurements showed that both Glvr-1 and Ram-1 encode electrogenic Na^+/P_i cotransporters, which were later renamed PiT-1 (SLC20A1) and PiT-2 (SLC20A2), respectively [152, 154, 155] and are present in all phyla. In animals and fungi, PiT-1- and PiT-2-induced P_i transport is Na^+ coupled whereas in prokaryotes and plants, it is coupled to H^+ transport [137]. The Type III Na^+/P_i cotransporters (PiT; SLC20) have a P_i transport preference for $H_2PO_4^-$ and therefore transport most efficiently in the acidic pH range and are not inhibited by PFA, whereas the Type II Na^+/P_i cotransporters (SLC34) which prefer HPO_4^{2-} transport most efficiently in the alkaline pH range and are inhibited by PFA [137].

Because of their ubiquitous tissue distribution, PiT proteins were suggested to have house keeping functions, but recently they were found to play an important role in providing P_i for the formation of mineralized bone [137]. PiT-1 mRNA was detected in osteoclasts and macrophages and the transfected protein was localizing to the BLM [110]. This subcellular localization was similar to that of NaPi-IIa, indicating an involvement of both NaPi-IIa and PiT-1 in P_i transport in bone-resorbing osteoclasts. In cultured osteoblast-like cells, P_i transport and PiT mRNA levels are regulated by P_i levels, epinephrine, platelet-derived growth factor (PDGF), insulin-like growth factor (IGF-1), and basic fibroblast growth factor (bFGF) [156–160] and in chondrogenic cells by P_i levels [161] and transforming growth factor-β (TGF-β) [162]. PiT proteins have been implicated in playing a role in normal calcification, as well as in pathological processes such as hyperphosphatemia-induced calcification of vascular tissue [163–165] and osteoarthritis [166]. PiT-1-mediated P_i transport was found to be regulated by P_i levels and PTH in human embryonic kidney cells (HEK-293) [35] and in rat parathyroid glands by plasma vitamin D and P_i levels [167]. Both PiT-1 and PiT-2 mRNAs were detected in mouse distal convoluted tubule (MDCT) cells suggesting that they may play a role in the distal segments of the kidney [168]. In the liver, PiT-1 and PiT-2 localize to the BLM in hepatocytes [121]. Since the identification of the retroviral receptors Glvr-1 and Ram-1 to encode are electrogenic Na^+/P_i cotransporters [152], very little is known of the transport kinetics for either PiT-1 or PiT-2 (for a recent review see [137]). Arsenate interacts with PiT-1 and PiT-2 by reducing P_i transport [152, 169]. Presently, there are no known inhibitors of PiT-mediated P_i transport. Both PiT-1 and PiT-2 proteins have a predicted topology of 12 TMDs, with the NH_2 and COOH termini both being extracellular, with PiT-2 having an *N*-glycosylated at Asp-81 [170].

4 Statement of Significance

Despite sharing chemical structural identity, the two tetra-oxyanions sulfate and phosphate, are substrates for structurally distinct membrane transporters that are differentially regulated in vivo and in various cell lines. The transporters responsible for transmembrane movement of these two anions have been mostly studied in cells originating from the mammalian renal and gastrointestinal systems. Much of the in vivo and in vitro regulation studies of these transport systems was performed at the cellular and system levels prior to the 1990s, when the structural identities of these proteins was yet unknown. During the 1990s, the molecular sequences of these proteins became apparent via the identification of genes encoding these transporters. Such structural information allowed the molecular characterization of both transport function and regulation of the respective transporters at both the protein and mRNA levels. This allowed previously hypothesized modes of regulation for these transport systems to be confirmed and also allowed new mechanisms for the molecular regulation of these transporters to be characterized. More recently, the generation of gene knock-out mice to these transporters, has permitted further characterization of their physiological and pathophysiological roles in vivo. The mouse null models and the use of other alternative species (such as Zebrafish, *Drosophila*, *Caenorhabditis elegans*, etc.) will allow future strategies to elucidate the regulation of these transporters in humans, in order to further define their physiological and pathophysiological roles in the body and their significance to whole body sulfate and phosphate homeostasis.

Acknowledgements Work in the author's laboratory is supported by the Australian Research Council and National Health and Medical Research Council.

References

1. Murer H, Hernando N, Forster I, Biber J. Proximal tubular phosphate reabsorption: molecular mechanisms. Physiological Reviews 2000;80:1373–409.
2. Markovich D. Physiological roles and regulation of mammalian sulfate transporters. Physiological Reviews 2001;81:1499–533.
3. Mulder GJ. Sulfate availability in vivo. In: Mulder GJ, ed. Sulfation of Drugs and Related Compounds. Boca Raton, FL: CRC; 1981:32–52.
4. Murer H. Transport of sulfate and phosphate in small intestine and renal proximal tubule: methods and basic properties. Comparative biochemistry and physiology A, Comparative Physiology 1988;90:749–55.
5. Tallgren LG. Inorganic sulphates in relation to the serum thyroxine level and in renal failure. Acta Medica Scandinavica Supplementum 1980;640:1–100.
6. Murer H, Markovich D, Biber J. Renal and small intestinal sodium-dependent symporters of phosphate and sulphate. The Journal of Experimental Biology 1994;196:167–81.
7. Hastbacka J, de la Chapelle A, Mahtani MM, et al. The diastrophic dysplasia gene encodes a novel sulfate transporter: positional cloning by fine-structure linkage disequilibrium mapping. Cell 1994;78:1073–87.
8. Taketani Y, Segawa H, Chikamori M, et al. Regulation of Type II Renal Na+-dependent Inorganic Phosphate Transporters by 1,25-Dihydroxyvitamin D3: Identification of a vitamin d-responsive element in the human NaPi-3 gene. Journal of Biological Chemistry 1998;273:14575–81.
9. Moseley RH, Hoglund P, Wu GD, et al. Downregulated in adenoma gene encodes a chloride transporter defective in congenital chloride diarrhea. American Journal of Physiology 1999;276:G185–92.

10. Markovich D, Murer H. The SLC13 gene family of sodium sulphate/carboxylate cotransporters. Pflugers Archiv: European Journal of Physiology 2004;447:594–602.
11. Dawson PA, Markovich D. Pathogenetics of the human SLC26 transporters. Current Medicinal Chemistr 2005;12:385–96.
12. Markovich D. Sulfate transport by SLC26 transporters. Novartis Foundation Symposium 2006;273:42–51; discussion -8, 261–4.
13. Markovich D, Forgo J, Stange G, Biber J, Murer H. Expression cloning of rat renal Na+/SO4(2-) cotransport. Proceedings of the National Academy of Sciences of the United States of America 1993;90:8073–7.
14. Beck L, Markovich D. The mouse Na(+)-sulfate cotransporter gene Nas1. Cloning, tissue distribution, gene structure, chromosomal assignment, and transcriptional regulation by vitamin D. Journal of Biological Chemistry 2000;275:11880–90.
15. Lee A, Beck L, Markovich D. The human renal sodium sulfate cotransporter (SLC13A1; hNaSi-1) cDNA and gene: organization, chromosomal localization, and functional characterization. Genomics 2000;70:354–63.
16. Lotscher M, Custer M, Quabius ES, Kaissling B, Murer H, Biber J. Immunolocalization of Na/SO4-cotransport (NaSi-1) in rat kidney. Pflugers Archiv European Journal of Physiology 1996;432:373–8.
17. Markovich D, Wang H, Puttaparthi K, et al. Chronic K depletion inhibits renal brush border membrane Na/sulfate cotransport. Kidney International 1999;55:244–51.
18. von Heijne G. Membrane protein structure prediction. Hydrophobicity analysis and the positive-inside rule. Journal of Molecular Biology 1992;225:487–94.
19. Pajor AM, Sun N, Bai L, Markovich D, Sule P. The substrate recognition domain in the Na+/dicarboxylate and Na+/sulfate cotransporters is located in the carboxy-terminal portion of the protein. Biochimica et Biophysica Acta 1998;1370:98–106.
20. Griffith DA, Pajor AM. Acidic residues involved in cation and substrate interactions in the Na+/dicarboxylate cotransporter, NaDC-1. Biochemistry 1999;38:7524–31.
21. Busch AE, Waldegger S, Herzer T, et al. Electrogenic cotransport of Na+ and sulfate in Xenopus oocytes expressing the cloned Na+SO4(2-) transport protein NaSi-1. Journal of Biological Chemistry 1994;269:12407–9.
22. Besseghir K, Roch-Ramel F. Renal excretion of drugs and other xenobiotics. Renal Physiology 1987;10:221–41.
23. Murer H, Manganel M, Roch-Ramel F. Tubular transport of monocarboxylates, Krebs cycle intermediates and inorganic sulphate. In: Winhager E, ed. Handbook of Physiology. Oxford: Oxford University Press; 1992:2165–88.
24. Markovich D, Murer H, Biber J, Sakhaee K, Pak C, Levi M. Dietary sulfate regulates the expression of the renal brush border Na/Si cotransporter NaSi-1. Journal of the American Society of Nephrology : JASN 1998;9:1568–73.
25. Sagawa K, DuBois DC, Almon RR, Murer H, Morris ME. Cellular mechanisms of renal adaptation of sodium dependent sulfate cotransport to altered dietary sulfate in rats. The Journal of Pharmacology and Experimental Therapeutics 1998;287:1056–62.
26. Levi M, McDonald LA, Preisig PA, Alpern RJ. Chronic K depletion stimulates rat renal brush-border membrane Na-citrate cotransporter. The American Journal of Physiology 1991;261:F767–73.
27. Frick A, Durasin I. Regulation of the renal transport of inorganic sulfate: Effects of metabolic changes in arterial blood pH. Pflugers Archiv : European Journal of Physiology 1986;407:541–6.
28. Puttaparthi K, Markovich D, Halaihel N, et al. Metabolic acidosis regulates rat renal Na-Si cotransport activity. The American Journal of Physiology 1999;276:C1398–404.
29. Renfro J, Clark N, Metts R, Lynch M. Glucocorticoid inhibition of Na/sulfate transport in chick renal brush border membranes. American Journal of Physiology 1989;256:1176–83.
30. Sagawa K, Murer H, Morris ME. Effect of experimentally induced hypothyroidism on sulfate renal transport in rats. The American Journal of Physiology 1999;276:F164–71.

31. Fernandes I, Hampson G, Cahours X, et al. Abnormal sulfate metabolism in vitamin D-deficient rats. The Journal of Clinical Investigation 1997;100:2196–203.
32. Sagawa K, Han B, DuBois DC, Murer H, Almon RR, Morris ME. Age- and growth hormone-induced alterations in renal sulfate transport. The Journal of Pharmacology and Experimental Therapeutics 1999;290:1182–7.
33. Sagawa K, Benincosa LJ, Murer H, Morris ME. Ibuprofen-induced changes in sulfate renal transport. The Journal of Pharmacology and Experimental Therapeutics 1998;287:1092–7.
34. Markovich D, Fogelis TS. Ontogeny of renal sulfate transporters: postnatal mRNA and protein expression. Pediatric Nephrology (Berlin, Germany) 1999; 13:806–11.
35. Fernandes I, Beliveau R, Friedlander G, Silve C. NaPO(4) cotransport Type III (PiT1) expression in human embryonic kidney cells and regulation by PTH. The American Journal of Physiology 1999;277:F543–51.
36. Lee HJ, Sagawa K, Shi W, Murer H, Morris ME. Hormonal regulation of sodium/sulfate co-transport in renal epithelial cells. Proceedings of the Society for Experimental Biology and Medicine Society for Experimental Biology and Medicine (New York, NY) 2000;225: 49–57.
37. Fernandes I, Laouari D, Tutt P, Hampson G, Friedlander G, Silve C. Sulfate homeostasis, NaSi-1 cotransporter, and SAT-1 exchanger expression in chronic renal failure in rats. Kidney International 2001;59:210–21.
38. Markovich D, Werner A, Murer H. Expression cloning with *Xenopus* oocytes. In: Hildebrandt F, Igarashi, P, ed. Techniques in Molecular Medicine. Heidelberg: Springer Verlag; 1999:310–8.
39. Dawson PA, Markovich D. Regulation of the mouse Nas1 promoter by vitamin D and thyroid hormone. Pflugers Arch 2002;444:353–9.
40. Lee A, Markovich D. Characterization of the human renal Na(+)-sulphate cotransporter gene (NAS1) promoter. Pflugers Arch 2004;448:490–9.
41. Markovich D. Molecular regulation and membrane trafficking of mammalian renal phosphate and sulphate transporters. European Journal of Cell Biology 2000;79:531–8.
42. Markovich D, Knight D. Renal Na-Si cotransporter NaSi-1 is inhibited by heavy metals. The American Journal of Physiology 1998;274:F283–9.
43. Dawson PA, Beck L, Markovich D. Hyposulfatemia, growth retardation, reduced fertility and seizures in mice lacking a functional NaS_i-1 gene. Proceedings of the National Academy Sciences USA 2003;100:13704–9.
44. Dawson PA, Steane SE, Markovich D. Impaired memory and olfactory performance in NaSi-1 sulphate transporter deficient mice. Behavioural Brain Research 2005;159:15–20, in press.
45. Dawson PA, Steane SE, Markovich D. Behavioural abnormalities of the hyposulphataemic Nas1 knock-out mouse. Behavioural Brain Research 2004;154:457–63.
46. Markovich D, Regeer RR, Kunzelmann K, Dawson PA. Functional characterization and genomic organization of the human Na(+)-sulfate cotransporter hNaS2 gene (SLC13A4). Biochemical and Biophysical Research Communications 2005;326:729–34.
47. Girard JP, Baekkevold ES, Feliu J, Brandtzaeg P, Amalric F. Molecular cloning and functional analysis of SUT-1, a sulfate transporter from human high endothelial venules. Proceedings of the National Academy of Sciences of the United States of America 1999;96:12772–7.
48. Dawson PA, Pirlo KJ, Steane SE, et al. The rat Na+-sulfate cotransporter rNaS2: functional characterization, tissue distribution, and gene (slc13a4) structure. Pflugers Arch 2005;450:262–8.
49. Miyauchi S, Srinivas SR, Fei YJ, et al. Functional characteristics of NaS2, a placenta-specific Na+-coupled transporter for sulfate and oxyanions of the micronutrients selenium and chromium. Placenta 2006;27:550–9.
50. Dawson PA, Pirlo KJ, Steane SE, Kunzelmann K, Chien YJ, Markovich D. Molecular cloning and characterization of the mouse Na+ sulfate cotransporter gene (Slc13a4): Structure and expression. Genes & Genetic Systems 2006;81:265–72.

51. Mount DB, Romero MF. The SLC26 gene family of multifunctional anion exchangers. Pflugers Arch 2004;447:710–21.
52. Regeer RR, Lee A, Markovich D. Characterization of the human sulfate anion transporter (hsat-1) protein and gene (SAT1; SLC26A1). DNA and Cell Biology 2003;22:107–17.
53. Lee A, Beck L, Markovich D. The mouse sulfate anion transporter gene Sat1 (Slc26a1): cloning, tissue distribution, gene structure, functional characterization, and transcriptional regulation thyroid hormone. DNA and Cell Biology 2003;22:19–31.
54. Bissig M, Hagenbuch B, Stieger B, Koller T, Meier PJ. Functional expression cloning of the canalicular sulfate transport system of rat hepatocytes. The Journal of Biological Chemistry 1994;269:3017–21.
55. Markovich D, Bissig M, Sorribas V, Hagenbuch B, Meier PJ, Murer H. Expression of rat renal sulfate transport systems in Xenopus laevis oocytes. Functional characterization and molecular identification. The Journal of Biological Chemistry 1994;269:3022–6.
56. David C, Ullrich KJ. Substrate specificity of the luminal Na+-dependent sulphate transport system in the proximal renal tubule as compared to the contraluminal sulphate exchange system. Pflugers Archiv 1992;421:455–65.
57. Karniski LP, Lotscher M, Fucentese M, Hilfiker H, Biber J, Murer H. Immunolocalization of sat-1 sulfate/oxalate/bicarbonate anion exchanger in the rat kidney. The American Journal Of Physiology 1998;275:F79–87.
58. Meier P, Valantinas J, Hugentobler G, Rahm I. Bicarbonate sulfate exchange in canalicular rat liver plasma membrane vesicles. The American Journal of Physiology 1987;253:461–8.
59. Satoh H, Susaki M, Shukunami C, Iyama K, Negoro T, Hiraki Y. Functional analysis of diastrophic dysplasia sulfate transporter. Its involvement in growth regulation of chondrocytes mediated by sulfated proteoglycans. The Journal of Biological Chemistry 1998;273:12307–15.
60. Kobayashi T, Sugimoto T, Saijoh K, Fukase M, Chihara K. Cloning of mouse diastrophic dysplasia sulfate transporter gene induced during osteoblast differentiation by bone morphogenetic protein-2. Gene 1997;198:341–9.
61. Schweinfest C, Henderson K, Suster S, Kondoh N, Papas T. Identification of a colon mucosa gene that is down-regulated in colon adenomas and adenocarcinomas. Proceedings of the National Academy of Sciences of the United States of America 1993;90:4166–70.
62. Byeon MK, Westerman MA, Maroulakou IG, et al. The down-regulated in adenoma (DRA) gene encodes an intestine-specific membrane glycoprotein. Oncogene 1996;12:387–96.
63. Taguchi T, Testa JR, Papas TS, Schweinfest C. Localization of a candidate colon tumor-suppressor gene (DRA) to 7q22–q31.1 by fluorescence in situ hybridization. Genomics 1994;20:146–7.
64. Byeon MK, Frankel A, Papas TS, Henderson KW, Schweinfest CW. Human DRA functions as a sulfate transporter in Sf9 insect cells. Protein Expression and Purification 1998;12: 67–74.
65. Silberg DG, Wang W, Moseley RH, Traber PT. The down regulated in adenoma (dra) gene encodes an intestine-specific membrane sulfate transport protein. Journal of Biological Chemistry 1995;270:11897–902.
66. Lohi H, Lamprecht G, Markovich D, et al. Isoforms of SLC26A6 mediate anion transport and have functional PDZ interaction domains. American Journal of Physiology Cell Physiology 2003;284:C769–C79.
67. Holmberg C. Congenital chloride diarrhoea. Clin Gastroenterol 1986;15:583–602.
68. Kere J, Lohi H, Hoglund P. Genetic Disorders of Membrane Transport III. Congenital chloride diarrhea. American Journal of Physiology 1999;276:G7–G13.
69. Hoglund P, Sistonen P, Norio R, et al. Fine mapping of the congenital chloride diarrhea gene by linkage disequilibrium. American Journal of Human Genetics 1995;57:95–102.
70. Hoglund P, Haila S, Scherer SW, et al. Positional candidate genes for congenital chloride diarrhea suggested by high resolution physical mapping in chromosome region 7q31. Genome Research 1996;6:202–10.

71. Haila S, Hoglund P, Scherer SW, et al. Genomic structure of the human congenital chloride diarrhea (CLD) gene. Gene 1998;214:87–93.
72. Melvin JE, Park K, Richardson L, Schultheis PJ, Shull GE. Mouse down-regulated in adenoma (DRA) is an intestinal Cl(–)/HCO(3)(–) exchanger and is up-regulated in colon of mice lacking the NHE3 Na(+)/H(+) exchanger. Journal of Biological Chemistry 1999;274: 22855–61.
73. Lohi H, Kujala M, Kerkela E, Saarialho-Kere U, Kestila M, Kere J. Mapping of five new putative anion transporter genes in human and characterization of SLC26A6, A candidate gene for pancreatic anion exchanger. Genomics 2000;70:102–12.
74. Waldegger S, Moschen I, Ramirez A, et al. Cloning and characterization of SLC26A6, a novel member of the solute carrier 26 gene family. Genomics 2001;72:43–50.
75. Knauf F, Yang CL, Thomson RB, Mentone SA, Giebisch G, Aronson PS. Identification of a chloride-formate exchanger expressed on the brush border membrane of renal proximal tubule cells. Proceedings of the National Academy of Sciences of the United States of America 2001;98:9425–30.
76. Wang Z, Petrovic S, Mann E, Soleimani M. Identification of an apical Cl(–)/HCO3(–) exchanger in the small intestine. American Journal of Physiology Gastrointestinal and Liver Physiology 2002;282:G573–9.
77. Thomson RB, Wang T, Thomson BR, et al. Role of PDZK1 in membrane expression of renal brush border ion exchangers. Proceedings of the National Academy of Sciences of the United States of America 2005;102:13331–6.
78. Xie Q, Welch R, Mercado A, Romero MF, Mount DB. Molecular characterization of the murine Slc26a6 anion exchanger: functional comparison with Slc26a1. American Journal of Physiology Renal Physiology 2002;283:F826–38.
79. Chernova MN, Jiang L, Friedman DJ, et al. Functional comparison of mouse slc26a6 anion exchanger with human SLC26A6 polypeptide variants: differences in anion selectivity, regulation, and electrogenicity. The Journal of Biological Chemistry 2005;280: 8564–80.
80. Jiang Z, Grichtchenko, II, Boron WF, Aronson PS. Specificity of Anion Exchange Mediated by Mouse Slc26a6. The Journal of Biological Chemistry 2002;277:33963–7.
81. Nozawa T, Sugiura S, Hashino Y, Tsuji A, Tamai I. Role of anion exchange transporter PAT1 (SLC26A6) in intestinal absorption of organic anions. Journal of Drug Targeting 2004;12:97–104.
82. Ko SB, Shcheynikov N, Choi JY, et al. A molecular mechanism for aberrant CFTR-dependent HCO(3)(–) transport in cystic fibrosis. EMBO Journal 2002;21:5662–72.
83. Shcheynikov N, Wang Y, Park M, et al. Coupling Modes and Stoichiometry of Cl-/HCO3-Exchange by slc26a3 and slc26a6. The Journal of General Physiology 2006;127: 511–24.
84. Jiang Z, Asplin JR, Evan AP, et al. Calcium oxalate urolithiasis in mice lacking anion transporter Slc26a6. Nature Genetics 2006;38:474–8.
85. Wang Z, Wang T, Petrovic S, et al. Renal and intestinal transport defects in Slc26a6-null mice. American Journal of Physiology Cell Physiology 2005;288:C957–65.
86. Gisler SM, Pribanic S, Bacic D, et al. PDZK1: I. a major scaffolder in brush borders of proximal tubular cells. Kidney International 2003;64:1733–45.
87. Vincourt JB, Jullien D, Amalric F, Girard JP. Molecular and functional characterization of SLC26A11, a sodium-independent sulfate transporter from high endothelial venules. FASEB Journal 2003;17:890–2.
88. Lohi H, Kujala M, Mäkelä S, et al. Functional characterization of three novel tissue-specific anion exchangers: SLC26A7, A8 and A9. Journal of Biological Chemistry 2002;277: 14246–54.
89. Toure A, Morin L, Pineau C, Becq F, Dorseuil O, Gacon G. Tat1, a novel sulfate transporter specifically expressed in human male germ cells and potentially linked to rhogtpase signaling. Journal of Biological Chemistry 2001;276:20309–15.

90. Toure A, Lhuillier P, Gossen JA, et al. The testis anion transporter 1 (Slc26a8) is required for sperm terminal differentiation and male fertility in the mouse. Human Molecular Genetics 2007;16:1783–93.
91. Murer H. Homer Smith Award. Cellular mechanisms in proximal tubular Pi reabsorption: some answers and more questions. Journal of the American Society of Nephrology : JASN 1992;2:1649–65.
92. Reimer RJ, Edwards RH. Organic anion transport is the primary function of the SLC17/Type I phosphate transporter family. Pflugers Arch 2004;447:629–35.
93. Magagnin S, Werner A, Markovich D, et al. Expression cloning of human and rat renal cortex Na/Pi cotransport. Proceedings of the National Academy of Sciences of the United States of America 1993;90:5979–83.
94. Hilfiker H, Hattenhauer O, Traebert M, Forster I, Murer H, Biber J. Characterization of a murine Type II sodium-phosphate cotransporter expressed in mammalian small intestine. Proceedings of the National Academy of Sciences of the United States of America 1998;95:14564–9.
95. Murer H, Forster I, Biber J. The sodium phosphate cotransporter family SLC34. Pflugers Arch 2004;447:763–7.
96. Collins JF, Bai L, Ghishan FK. The SLC20 family of proteins: dual functions as sodium-phosphate cotransporters and viral receptors. Pflugers Arch 2004 447: 647–52.
97. Werner A, Moore ML, Mantei N, Biber J, Semenza G, Murer H. Cloning and expression of cDNA for a Na/Pi cotransport system of kidney cortex. Proceedings of the National Academy of Sciences of the United States of America 1991;88:9608–12.
98. Miyamoto K, Tatsumi S, Yamamoto H, et al. Chromosome assignments of genes for human Na(+)-dependent phosphate co-transporters NaPi-3 and NPT-1. Tokushima Journal of Experimental Medicine 1995;42:5–9.
99. Busch AE, Schuster A, Waldegger S, et al. Expression of a renal Type I sodium/phosphate transporter (NaPi-1) induces a conductance in Xenopus oocytes permeable for organic and inorganic anions. Proceedings of the National Academy of Sciences of the United States of America 1996;93:5347–51.
100. Uchino H, Tamai I, Yamashita K, et al. p-aminohippuric acid transport at renal apical membrane mediated by human inorganic phosphate transporter NPT1. Biochemical and Biophysical Research Communications 2000;270:254–9.
101. Yabuuchi H, Tamai I, Morita K, et al. Hepatic sinusoidal membrane transport of anionic drugs mediated by anion transporter Npt1. The Journal of Pharmacology and Experimental Therapeutics 1998;286:1391–6.
102. Biber J, Custer M, Werner A, Kaissling B, Murer H. Localization of NaPi-1, a Na/Pi cotransporter, in rabbit kidney proximal tubules. II. Localization by immunohistochemistry. Pflugers Archiv European Journal of Physiology 1993;424:210–5.
103. Custer M, Meier F, Schlatter E, et al. Localization of NaPi-1, a Na-Pi cotransporter, in rabbit kidney proximal tubules. I. mRNA localization by reverse transcription/polymerase chain reaction. Pflugers Archiv European Journal of Physiology 1993;424:203–9.
104. Ruddy DA, Kronmal GS, Lee VK, et al. A 1.1-Mb transcript map of the hereditary hemochromatosis locus. Genome Research 1997;7:441–56.
105. Shibui A, Tsunoda T, Seki N, Suzuki Y, Sugane K, Sugano S. Isolation and chromosomal mapping of a novel human gene showing homology to Na+/PO4 cotransporter. Journal of Human Genetics 1999;44:190–2.
106. Cheret C, Doyen A, Yaniv M, Pontoglio M. Hepatocyte nuclear factor 1 alpha controls renal expression of the Npt1-Npt4 anionic transporter locus. Journal of Molecular Biology 2002;322:929–41.
107. Murer H, Evers J, Kinne R. Polarity of proximal tubular epithelial cells in relation to transepithelial transport. Current Problems in Clinical Biochemistry 1976;6: 173–89.

108. Custer M, Lötscher M, Biber J, Murer H, Kaissling B. Expression of Na-Pi cotransport in rat kidney: localization by RT-PCR and immunohistochemistry. The American Journal of Physiology 1994;266:F767–F74.
109. Mulroney SE, Woda CB, Halaihel N, et al. Central control of renal sodium-phosphate (NaPi-2) transporters. American Journal of Physiology Renal Physiology 2004;286: F647–52.
110. Khadeer MA, Tang Z, Tenenhouse HS, et al. Na+-dependent phosphate transporters in the murine osteoclast: cellular distribution and protein interactions. American Journal of Physiology Cell Physiology 2003;284:C1633–44.
111. Lundquist P, Murer H, Biber J. Type II Na+-Pi cotransporters in osteoblast mineral formation: regulation by inorganic phosphate. Cellular Physiology and biochemistry 2007;19: 43–56.
112. Segawa H, Kaneko I, Takahashi A, et al. Growth-related renal Type II Na/Pi cotransporter. Journal of Biological Chemistry 2002;277:19665–72.
113. Murer H, Hernando N, Forster I, Biber J. Regulation of Na/Pi transporter in the proximal tubule. Annual Review of Physiology 2003;65:531–42.
114. Tenenhouse HS. Regulation of phosphorus homeostasis by the Type IIa na/phosphate cotransporter. Annual Review of Nutrition 2005;25:197–214.
115. Beck L, Karaplis AC, Amizuka N, Hewson AS, Ozawa H, Tenenhouse HS. Targeted inactivation of Npt2 in mice leads to severe renal phosphate wasting, hypercalciuria, and skeletal abnormalities. Proceedings of the National Academy of Sciences of the United States of America 1998;95:5372–7.
116. Bergwitz C, Roslin N, Tieder M, et al. SLC34A3 mutations in patients with hereditary hypophosphatemic rickets with hypercalciuria predict a key role for the sodium-phosphate cotransporter NaPi-IIc in maintaining phosphate homeostasis. American Journal of Human Genetics 2006;78:179–92.
117. Ichikawa S, Sorenson A, Imel E, Friedman N, Gertner J, Econs M. Intronic deletions in the SLC34A3 gene cause hereditary hypophosphatemic rickets with hypercalciuria. Journal of Clinical Endocrinology and Metabolism 2006;91:4022–7.
118. Lorenz-Depiereux B, Benet-Pages A, Eckstein G, et al. Hereditary hypophosphatemic rickets with hypercalciuria is caused by mutations in the sodium-phosphate cotransporter gene SLC34A3. American Journal of Human Genetics 2006;78:193–201.
119. Radanovic T WC, Murer H, Biber J. Regulation of intestinal phosphate transport. I. Segmental expression and adaptation to low-P(i) diet of the Type IIb Na(+)-P(i) cotransporter in mouse small intestine. American Journal of Physiology Gastrointestinal and Liver Physiology 2005;288:G496–500. .
120. Stauber A, Radanovic T, Stange G, Murer H, Wagner C, Biber J. Regulation of intestinal phosphate transport. II. Metabolic acidosis stimulates Na(+)-dependent phosphate absorption and expression of the Na(+)-P(i) cotransporter NaPi-IIb in small intestine. American Journal of Physiology Gastrointestinal and Liver Physiology 2005;288: G501–6.
121. Frei P, Gao B, Hagenbuch B, et al. Identification and localization of sodium-phosphate cotransporters in hepatocytes and cholangiocytes of rat liver. American Journal of Physiology Gastrointestinal and Liver Physiology 2005;288:G771–8.
122. Homann V, Rosin-Steiner S, Stratmann T, Arnold WH, Gaengler P, Kinne RK. Sodium-phosphate cotransporter in human salivary glands: molecular evidence for the involvement of NPT2b in acinar phosphate secretion and ductal phosphate reabsorption. Archives of Oral Biology 2005;50:759–68.
123. Alcalde AI, Sarasa M, Raldua D, et al. Role of thyroid hormone in regulation of renal phosphate transport in young and aged rats. Endocrinology 1999;140:1544–51.
124. Beers KW, Thompson MA, Chini EN, Dousa TP. beta-Estradiol inhibits Na+-P(i) cotransport across renal brush border membranes from ovarectomized rats. Biochemical and Biophysical Research Communications 1996;221:442–5.

125. Brandis M, Harmeyer J, Kaune R, Mohrmann M, Murer H, Zimolo Z. Phosphate transport in brush-border membranes from control and rachitic pig kidney and small intestine. The Journal of Physiology 1987;384:479–90.
126. Loffing J, Lotscher M, Kaissling B, et al. Renal Na/H exchanger NHE-3 and Na-PO4 cotransporter NaPi-2 protein expression in glucocorticoid excess and deficient states. Journal of the American Society of Nephrology 1998;9:1560–7.*
127. Murer H, Biber J. A molecular view of proximal tubular inorganic phosphate (Pi) reabsorption and of its regulation. Pflugers Archiv : European Journal of Physiology 1997;433:379–89.
128. Caverzasio J, Bonjour JP. Growth factors and renal regulation of phosphate transport. Pediatric Nephrology (Berlin, Germany) 1993;7:802–6.
129. Markovich D, Verri T, Sorribas V, Forgo J, Biber J, Murer H. Regulation of opossum kidney (OK) cell Na/Pi cotransport by Pi deprivation involves mRNA stability. Pflugers Archiv European Journal of Physiology 1995;430:459–63.
130. Murer H, Forster I, Hernando N, Lambert G, Traebert M, Biber J. Posttranscriptional regulation of the proximal tubule NaPi-II transporter in response to PTH and dietary P(i). The American Journal of Physiology 1999;277:F676–84.
131. Murer H, Biber J. Renal sodium-phosphate cotransport. Current Opinion in Nephrology and Hypertension 1994;3:504–10.
132. Lotscher M, Wilson P, Nguyen S, et al. New aspects of adaptation of rat renal Na-Pi cotransporter to alterations in dietary phosphate. Kidney International 1996;49: 1012–8.
133. Forster IC, Kohler K, Biber J, Murer H. Forging the link between structure and function of electrogenic cotransporters: the renal Type IIa Na+/Pi cotransporter as a case study. Progress in Biophysics and Molecular Biology 2002;80:69–108.
134. Forster IC, Hernando N, Biber J, Murer H. Proximal tubular handling of phosphate: A molecular perspective. Kidney International 2006;70:1548–59.
135. Busch AE, Wagner CA, Schuster A, et al. Properties of electrogenic Pi transport by a human renal brush border Na+/Pi transporter. Journal of the American Society of Nephrology JASN 1995;6:1547–51.
136. Xu H, Inouye M, Missey T, Collins JF, Ghishan FK. Functional characterization of the human intestinal NaPi-IIb cotransporter in hamster fibroblasts and Xenopus oocytes. Biochimica et Biophysica Acta 2002;1567:97–105.
137. Virkki LV, Biber J, Murer H, Forster IC. Phosphate transporters: a tale of two solute carrier families. American Journal of Physiology Renal Physiology 2007;293:F643–54.
138. Lambert G, Traebert M, Hernando N, Biber J, Murer H. Studies on the topology of the renal Type II NaPi-cotransporter. Pflugers Archiv : European Journal of Physiology 1999;437:972–8.
139. Kohler K, Forster IC, Lambert G, Biber J, Murer H. The functional unit of the renal Type IIa Na+/Pi cotransporter is a monomer. Journal of Biological Chemistry 2000;275: 26113–20.
140. Levi M, Kempson SA, Lotscher M, Biber J, Murer H. Molecular regulation of renal phosphate transport. The Journal of Membrane Biology 1996;154:1–9.
141. Moz Y, Silver J, Naveh-Many T. Characterization of cis-acting element in renal NaPi-2 cotransporter mRNA that determines mRNA stability. American Journal of Physiology Renal Physiology 2003;284:F663–70.
142. Biber J, Gisler SM, Hernando N, Wagner CA, Murer H. PDZ interactions and proximal tubular phosphate reabsorption. American Journal of Physiology Renal Physiology 2004;287:F871–5.
143. Shenolikar S, Voltz JW, Minkoff CM, Wade JB, Weinman EJ. Targeted disruption of the mouse NHERF-1 gene promotes internalization of proximal tubule sodium-phosphate cotransporter Type IIa and renal phosphate wasting. Proceedings of the National Academy of Sciences of the United States of America 2002;99:11470–5.

144. Hernando N, Deliot N, Gisler SM, et al. PDZ-domain interactions and apical expression of Type IIa Na/P(i) cotransporters. Proceedings of the National Academy of Sciences of the United States of America 2002;99:11957–62.
145. Katai K, Miyamoto K, Kishida S, et al. Regulation of intestinal Na+-dependent phosphate co-transporters by a low-phosphate diet and 1,25-dihydroxyvitamin D3. The Biochemical Journal 1999; 343 Pt 3:705–12.
146. Xu H, Collins JF, Bai L, Kiela PR, Ghishan FK. Regulation of the human sodium-phosphate cotransporter NaP(i)-IIb gene promoter by epidermal growth factor. American Journal of Physiology Cell Physiology 2001;280:C628–36.
147. Arima K, Hines ER, Kiela PR, Drees JB, Collins JF, Ghishan FK. Glucocorticoid regulation and glycosylation of mouse intestinal Type IIb Na-P(i) cotransporter during ontogeny. American Journal of Physiology Gastrointestinal and Liver Physiology 2002;283:G426–34.
148. Cross HS, Debiec H, Peterlik M. Mechanism and regulation of intestinal phosphate absorption. Mineral and Electrolyte Metabolism 1990;16:115–24.
149. Madsen KL, Tavernini MM, Yachimec C, et al. Stanniocalcin: a novel protein regulating calcium and phosphate transport across mammalian intestine. The American Journal of Physiology 1998;274:G96–102.
150. Johann SV, Gibbons JJ, O'Hara B. GLVR1, a receptor for gibbon ape leukemia virus, is homologous to a phosphate permease of Neurospora crassa and is expressed at high levels in the brain and thymus. Journal of Virology 1992;66:1635–40.
151. Miller DG, Edwards RH, Miller AD. Cloning of the cellular receptor for amphotropic murine retroviruses reveals homology to that for gibbon ape leukemia virus. Proceedings of the National Academy of Sciences of the United States of America 1994;91:78–82.
152. Kavanaugh MP, Miller DG, Zhang W, et al. Cell-surface receptors for gibbon ape leukemia virus and amphotropic murine retrovirus are inducible sodium-dependent phosphate symporters. Proceedings of the National Academy of Sciences of the United States of America 1994;91:7071–5.
153. van Zeijl M, Johann SV, Closs E, et al. A human amphotropic retrovirus receptor is a second member of the gibbon ape leukemia virus receptor family. Proceedings of the National Academy of Sciences of the United States of America 1994;91:1168–72.
154. Kavanaugh MP, Kabat D. Identification and characterization of a widely expressed phosphate transporter/retrovirus receptor family. Kidney International 1996;49:959–63.
155. Olah Z, Lehel C, Anderson WB, Eiden MV, Wilson CA. The cellular receptor for gibbon ape leukemia virus is a novel high affinity sodium-dependent phosphate transporter. The Journal of Biological Chemistry 1994;269:25426–31.
156. Chien ML, Foster JL, Douglas JL, Garcia JV. The amphotropic murine leukemia virus receptor gene encodes a 71-kilodalton protein that is induced by phosphate depletion. Journal of Virology 1997;71:4564–70.
157. Suzuki A, Palmer G, Bonjour JP, Caverzasio J. Stimulation of sodium-dependent inorganic phosphate transport by activation of Gi/o-protein-coupled receptors by epinephrine in MC3T3-E1 osteoblast-like cells. Bone 2001;28:589–94.
158. Suzuki A, Palmer G, Bonjour JP, Caverzasio J. Stimulation of sodium-dependent phosphate transport and signaling mechanisms induced by basic fibroblast growth factor in MC3T3-E1 osteoblast-like cells. Journal of Bone and Mineral Research : The Official Journal of the American Society for Bone and Mineral Research 2000;15:95–102.
159. Zhen X, Bonjour JP, Caverzasio J. Platelet-derived growth factor stimulates sodium-dependent Pi transport in osteoblastic cells via phospholipase Cgamma and phosphatidylinositol 3′ -kinase. Journal of Bone and Mineral Research : The Official Journal of the American Society for Bone and Mineral Research 1997;12:36–44.
160. Zoidis E, Ghirlanda-Keller C, Gosteli-Peter M, Zapf J, Schmid C. Regulation of phosphate (Pi) transport and NaPi-III transporter (Pit-1) mRNA in rat osteoblasts. The Journal of Endocrinology 2004;181:531–40.

161. Wang D, Canaff L, Davidson D, et al. Alterations in the sensing and transport of phosphate and calcium by differentiating chondrocytes. The Journal of Biological Chemistry 2001;276:33995–4005.
162. Palmer G, Guicheux J, Bonjour JP, Caverzasio J. Transforming growth factor-beta stimulates inorganic phosphate transport and expression of the Type III phosphate transporter Glvr-1 in chondrogenic ATDC5 cells. Endocrinology 2000;141:2236–43.
163. Jono S, McKee MD, Murry CE, et al. Phosphate regulation of vascular smooth muscle cell calcification. Circulation Research 2000;87:E10–7.
164. Li X, Yang HY, Giachelli CM. Role of the sodium-dependent phosphate cotransporter, Pit-1, in vascular smooth muscle cell calcification. Circulation Research 2006;98:905–12.
165. Mizobuchi M, Ogata H, Hatamura I, et al. Up-regulation of Cbfa1 and Pit-1 in calcified artery of uraemic rats with severe hyperphosphataemia and secondary hyperparathyroidism. Nephrology, dialysis, transplantation : official publication of the European Dialysis and Transplant Association – European Renal Association 2006;21:911–6.
166. Cecil DL, Rose DM, Terkeltaub R, Liu-Bryan R. Role of interleukin-8 in PiT-1 expression and CXCR1-mediated inorganic phosphate uptake in chondrocytes. Arthritis and Rheumatism 2005;52:144–54.
167. Tatsumi S, Segawa H, Morita K, et al. Molecular cloning and hormonal regulation of PiT-1, a sodium-dependent phosphate cotransporter from rat parathyroid glands. Endocrinology 1998;139:1692–9.
168. Tenenhouse HS, Gauthier C, Martel J, Hesek FA, Coutermarsh BA, Friedman PA. Na+ -phosphate cotransport in mouse distal convoluted tubule cells: evidence for Glvr-1 and Ram-1 gene expression. Journal of Bone and Mineral Research : the Official Journal of the American Society for Bone and Mineral Research 1998;13:590–7.
169. Bai L, Collins JF, Ghishan FK. Cloning and characterization of a Type III Na-dependent phosphate cotransporter from mouse intestine. American Journal of Physiology Cell Physiology 2000;279:C1135–43.
170. Salaun C, Marechal V, Heard JM. Transport-deficient Pit2 phosphate transporters still modify cell surface oligomers structure in response to inorganic phosphate. Journal of Molecular Biology 2004;340:39–47.

Role of H^+-K^+ ATPase, Na^+-K^+-$2Cl^-$ and Na^+-Cl^--HCO_3^- Transporters in Vertebrate Small Intestine

John F. White

Abstract Small intestinal function is multifaceted, characterized as it is by several functional states including those of the absorptive and postabsorptive states. In the postabsorptive state urodele small intestinal brush border $Na^+K^+2Cl^-$ cotransport operates in parallel with Cl^--HCO_3^- exchange and K^+-H^+ ATPase antiport and in series with basolateral membrane Na^+-$nHCO_3^-$-anion exchange to absorb Na^+, K^+, and Cl^- ions and regulate luminal fluid pH. In the absorptive state nonmetabolizable cotransported solutes such as galactose and valine in the lumen inhibit $Na^+K^+2Cl^-$ cotransport and Na^+-$nHCO_3^-$-anion exchange to block transcellular Cl^- absorption while simultaneously stimulating Na^+ absorption. When under the influence of catecholamines K^+-H^+ ATPase activity is supplanted by enhanced Na^+-H^+ and Cl^--HCO_3^- exchange. In contrast, agents which elevate intracellular cyclic AMP inhibit $Na^+K^+2Cl^-$ cotransport, Na^+-H^+ and Cl^--HCO_3^- exchange and reduce K^+-H^+ ATPase antiport while promoting HCO_3^- ion secretion via ion channels in the brush border membrane. Progress has been made in distinguishing the respective roles of mature villus enterocytes and immature intervillus (i.e., crypt region) enterocytes in these different transepithelial ion transport processes.

Keywords Small intestine · Cl^- absorption · Intestinal H^+ secretion · HCO_3^- secretion · $Na^+K^+2Cl^-$ cotransport · Intestinal K^+-H^+ ATPase · $Na^{+-}Cl^-$-HCO_3^- exchange · Na^+-$nHCO_3^-$-anion exchange · Na-sugar cotransport · Na-amino acid cotransport · K^+ conductance · Cl^- conductance · Villus enterocytes · Intervillus enterocytes · Disulfonic stilbene · Carbonic anhydrase · Metabolic stress

1 Introduction

This chapter provides a description of key ion transport processes operating in the enterocytes lining the mucosa of vertebrate small intestine. Special emphasis will

J.F. White (✉)
Department of Physiology, School of Medicine, Emory University, Atlanta, GA, 30322, USA
e-mail: jfwhite@physio.emory.edu

G.A. Gerencser (ed.), *Epithelial Transport Physiology*,
DOI 10.1007/978-1-60327-229-2_9,

be placed on those transporters present in urodeles (the order of caudates or tailed amphibian that includes newts and salamanders) because two carnivorous salamanders, the congo eel *Amphiuma means tridactylum* (a legless salamander) and the axolotl *Ambystoma mexicanum* have been found to be especially useful animal models for studies of mucosal physiology.

The vertebrate small intestinal mucosa aids in the digestion of a complex diet of plant and animal matter and oversees the absorption from the lumen into the blood of a multitude of different nutrients. It supports these functions by maintaining a luminal environment optimized for digestion of the food and absorption of the nutrients. Absorptive and secretory ion transport processes have evolved that maintain the pH, ion concentrations, and osmolarity of the luminal fluid within narrow limits. Furthermore, because digestion and absorption are episodic, wherein periods of feeding are followed by periods when no nutrients are absorbed, the mucosa must modify its function, assuming a postabsorptive state of indefinite duration that limits energy utilization while maintaining the barrier function which keeps blood electrolytes and nutrients from leaking across the intestinal wall into the lumen. Figure 1 depicts the major states through which the intestinal mucosa cycles and for which specific and appropriate ion transporters must operate. Emphasis is placed on the transitions from the postabsorptive state to either (1) the absorptive state, or (2) the state in which catecholamines influence salt absorption, or (3) the state in which cyclic AMP drives ion secretion. Progress has been made in characterizing the different ion transport processes operating in these distinct functional states and in defining the functional differences between the immature crypt (intervillus) cells and the mature (villus) cells lining the small intestinal villi.

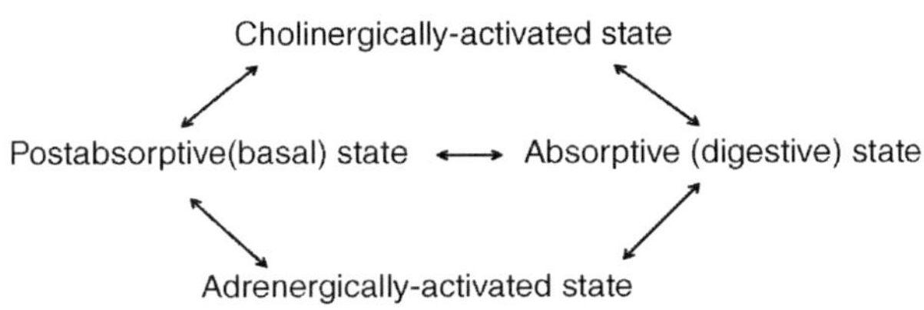

Fig. 1 The small intestine transitions through several major states of activity, each characterized by a specific set of ion transport processes

Our understanding of epithelial transport physiology has advanced as new techniques have been developed. In this chapter three technical advances are described, the villus tissue chamber, the villus sheet preparation and isolation and primary culture of intestinal cell nests. These were possible because of the unique architecture of the urodele intestinal mucosa. The technical advances allowed characterization of the developmental changes in ion and solute transport as immature enterocytes emerged from cell nests, inserted in the intervillus epithelium and advanced up the villus. Better characterization of the ion transport properties of the basolateral membrane, on the "dark side" of the epithelial cell, also became possible.

2 Ion Transport Processes in the Basal (Postabsorptive) State: Apical $Na^+K^+2Cl^-$ Cotransport and Basolateral Na^+-Cl^- -HCO_3^- Exchange

2.1 Electrogenic Cl^- Absorption Requires Bicarbonate and Occurs in Villus Absorptive Cells

Chloride absorption by urodele small intestine is electrogenic when the luminal fluid is devoid of metabolizable substrate. Isolated duodenal segments of salamander small intestine, stripped of their serosal muscle layers and clamped into Ussing chambers generate a transepithelial electrical potential difference (V_{ms}), serosal (s) side electrically negative to the mucosal (m) side. When voltage-clamped to zero V_{ms} the requisite short-circuit current (I_{sc}) is consistent with net anion absorption when the intestine only has endogenous substrates to metabolize and the bathing medium is buffered with HCO_3^- (25 mM) and gassed with 95% O_2, 5% CO_2 to achieve a pH of 7.40 [1]. Replacement of bath HCO_3^-/CO_2 with another buffer or addition to the medium of acetazolamide, an inhibitor of the enzyme carbonic anhydrase (CA), eliminated the serosa-negative V_{ms} and the I_{sc}. CA catalyzes the hydration of CO_2 to form H^+ and HCO_3^-. A serosa-negative V_{ms} associated with HCO_3^--dependent Cl^- absorption has also been reported for flounder small intestine [2, 3]. Measurement of radiolabeled bidirectional Cl^- fluxes (using ^{36}Cl) through paired urodele intestinal segments revealed a net Cl^- absorptive flux ($J^{Cl}{}_{net}$) in the presence but not in the absence of HCO_3^- in the medium; acetazolamide eliminated $J^{Cl}{}_{net}$. $J^{Cl}{}_{net}$ exceeded the I_{sc}. As described below, the difference is accounted for by a flow of HCO_3^- ions in the secretory (s→m) direction [4]. Hence, $I_{sc} = J^{Cl}{}_{net} - J^{HCO3-}{}_{net}$, where $J^{HCO3-}{}_{net}$ is the net secretory HCO_3^- flux. When tissues were initially bathed bilaterally in a Cl^--free (SO_4^{2-}-based) medium, addition of increasing amounts of Cl^- ions (added as choline chloride) to the mucosal medium produced stepwise increases in I_{sc} which approached a maximum at 15 mM Cl^-, exhibited a *Km* of 5.4 mM, and was blocked when the disulfonic stilbene SITS, an inhibitor of anion exchangers, was present in the serosal medium [5]. Other anions evoked an equal or smaller current than Cl^-. Thus, the selectivity sequence for the anion-evoked current was $Cl^- = Br^- > SCN^- > NO_3^- > F^- = I^-$.

Special Ussing-style tissue chambers (villus chambers) were employed which allowed physical isolation of 7.4 mm^2 of either the villus or intervillus mucosa and measurement of the transepithelial voltages each generated [6]. Mucosal segments stripped of serosal smooth muscle layers were stretched mildly and oriented over a 1 mm × 7 mm slit to clamp a tissue segment consisting predominantly of villus or intervillus mucosa. The serosa-negative V_{ms} was discovered to arise from the villus mucosa, which is lined by fully mature enterocytes; the intervillus mucosa generated no voltage. This and other differences in transport behavior between villus and intervillus mucosa, delineated in several studies to be described, are listed in Table 1.

Table 1 Transepithelial transport properties of mature villus mucosa and less mature intervillus mucosa in urodele intestine are compared. Each ion transport effector produced a signature change in electrical behavior (ΔI_{sc}) which was first characterized extensively in whole mucosa. Paired villus chambers were then employed to distinguish whether one or both regions of the mucosa contributed to the response [6]. SGLT1, sodium glucose cotransporter 1; SAAT, sodium amino acid cotransporter; NHE, Na^+-H^+ exchanger. † Basal, electrogenic H^+ secretion is equivalent to absorption of HCO_3^- ions

ΔI_{sc}, transporter (references)	Intervillus epithelium	Villus epithelium
Na-glucose cotransport, SGLT 1 [6, 14, 24]	–	+
Na-valine cotransport, SAAT [6, 24]	+	+
Basal, electrogenic Cl^- absorption, $Na^+K^+2Cl^-$ [6]	–	+
cAMP-stimulated HCO_3^- secretion [6]	+	+
Basal, electrogenic H^+ secretion†, K^+-H^+ ATPase [14]	–	+
NE-stimulated H^+ secretion, NHE [24]	+	+
Bidirectional symmetric HCO_3^- diffusion potentials [6]	+	–

2.2 Cl^- Influx is Saturable, Requires Na^+ and K^+ Ions and is Inhibited by Furosemide

^{36}Cl influx into isolated segments of urodele intestine was measured in order to characterize the luminal membrane pathways of Cl^- uptake [7]. ^{36}Cl influx from the mucosal medium into the cells ($J^{Cl}{}_{mc}$) in ueq/g tissue water in 5 min was linear with time and reached a maximum as luminal bath [Cl^-] was increased. The calculated K_m of 5.3 mM is in excellent agreement with the measured apparent K_m for the Cl^--evoked current, noted above, of 5.4 mM. Although SITS added to the serosal fluid blocked the Cl^--evoked current it did not reduce $J^{Cl}{}_{mc}$, indicating that $J^{Cl}{}_{mc}$ does not reflect transcellular Cl^- movement (through the cell into the serosal medium). Monovalent anions blocked $J^{Cl}{}_{mc}$ in the order $I^- = SCN^- = NO_3^- > Br^- > F^-$. Substitution of mucosal medium Na^+ with choline or N-methyl glucamine reduced $J^{Cl}{}_{mc}$ by 60–70%. Removal of medium K^+ reduced $J^{Cl}{}_{mc}$ 50%. After medium Na^+ and K^+ were replaced re-exposure to Na^+ added alone did not stimulate Cl^- influx whereas reexposure to $Na^+ + K^+$ stimulated Cl^- influx. $J^{Cl}{}_{mc}$ was not sensitive to the diuretic amiloride (which can block Na^+ channels and Na^+/H^+ exchange) or to SITS in the mucosal medium. However, $J^{Cl}{}_{mc}$ was reduced 34% by furosemide in the luminal medium. Furosemide is a loop diuretic which inhibits Na^+-K^+-$2Cl^-$ cotransport [8]. It was concluded that Cl^- ions enter the cell by electroneutral transport, most likely Na-K-2Cl cotransport, as illustrated in Fig. 2 [7]. A Na^+- and K^+-dependent Cl^- transporter has been reported in several regions of the renal tubule and macula densa of the kidney [9, 10] and designated either NK2C [9] or BSC1 [10]. Na-K-2Cl cotransport has also been reported for flounder small intestine [11].

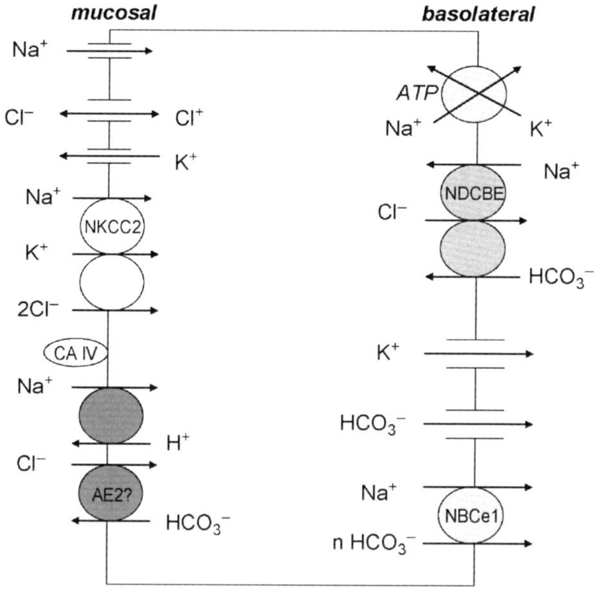

Fig. 2 Proposed cell model of ion transport processes in the mucosal (brush border) and basolateral (serosal) membranes of mature villus enterocytes in urodele small intestine (duodenum) responsible for electrogenic Cl^- absorption linked to secretion of HCO_3^- ions. About one-half of mucosal Cl^- influx occurs via a furosemide-sensitive electroneutral Na^+-K^+-$2Cl^-$ cotransporter (presumably the NK2C isoform); the other half occurs via a stilbene-insensitive Cl^--HCO_3^- exchange mediated by an anion exchanger (AE) isoform not yet established [30]. Serosal Cl^- exit is envisioned as largely via a stilbene-inhibitable Na^+-Cl^--HCO_3^- exchange, most likely the NDCBe transport isoform [30]. The stoichiometry of the Na-Cl-HCO_3 exchanger is not established. Carbonic anhydrase type IV (CAIV) is presumed to be associated with the brush border membrane and to catalyze the formation of CO_2 and H_2O; CO_2 enters the cell

2.3 Cl^- Efflux into the Serosal Fluid Requires Na^+

In a related study, bilateral addition of Na^+ (as Na gluconate) to Na^+-free mucosal and serosal media stimulated the I_{sc} and $J^{Cl}{}_{net}$ [5]. Addition of Na^+ only to the serosal bath markedly increased the mucosal to serosal (m→s) $^{36}Cl^-$ flux. Another disulfonic stilbene, DIDS, when added to the serosal medium blocked the Na-dependent $^{36}Cl^-$ flux. These results are consistent with a model in which Cl^- ions exit the cell at the serosal membrane by a Na^+-dependent anion exchange process. This process is characterized further below (Sect. 2.6).

2.4 Intestinal Cells Alkalinize the Luminal Fluid and Acidify the Serosal Fluid

Segments of urodele duodenum, bathed bilaterally in normal (Cl^--containing) media, alkalinize an unbuffered mucosal bath as they acidify the serosal fluid [4].

When the serosal fluid contains HCO_3^- and is gassed with 95% O_2 and 5% CO_2 then the unbuffered mucosal bath becomes alkalinized. A secretory HCO_3^- flux ($J^{HCO3}{}_{sm}$) is measurable by acid titration of the unbuffered mucosal fluid. It is assumed that simultaneously an equal flux of protons occurs from cell to serosal fluid. Acetazolamide (0.1 mM) inhibited $J^{HCO3}{}_{sm}$ on average by 43% in several tissues [4].

2.5 $J^{HCO3}{}_{sm}$ is Influenced by the Serosal Bath Na^+ Concentration

$J^{HCO3}{}_{sm}$ was increased when the NaCl concentration of the serosal medium was increased in increments in replacement for choline chloride. The K_m for stimulation of $J^{HCO3}{}_{sm}$ by serosal bath Na was 3–4 mM. In contrast, addition of Na to the mucosal medium was without effect on $J^{HCO3}{}_{sm}$ [12].

2.6 Cl^- Ions Exit into the Serosal Medium via a Saturable Path, Not a Conductive One

The serosal membrane potential (V_s) of villus enterocytes was measured directly by employing a novel dissection procedure which transforms a three-dimensional villus into a flat two-dimensional sheet using a microsurgical procedure that maintains the integrity of the epithelium [13]. The "villus sheet preparation" was then mounted in a tissue chamber, serosal surface upward, thereby allowing direct access of microelectrodes to the serosal membrane of enterocytes normally residing at the tip of the villus in the intact epithelium. Stable recordings of serosal membrane potential (V_s) of many minutes duration were possible. V_s was depolarized by elevating serosal bath [K^+] or lowering bath [HCO_3^-] but was largely unaltered by changes in serosal bath [Cl^-]. The absence of a significant conductive pathway for Cl^- ions suggests that they exit the cell by a carrier-mediated route.

Measurements with Cl^--sensitive microelectrodes provided evidence that the serosal transport of Cl^- ions is by a saturable pathway. Intracellular Cl^- activity ($a_{Cl}{}^i$) was measured using double-barreled Cl^- sensitive microelectrodes. One barrel sensed the cytoplasmic Cl^- activity and the other barrel sensed the membrane potential (V_m). The stilbene-sensitive portion of the absorptive Cl^- flux, which reflects the carrier-mediated Cl^- exit into the serosal bath, approached a maximum as $a_{Cl}{}^i$ increased [14]. The calculated K_m of the basolateral transporter for cytoplasmic Cl^- was determined to be 4–6 mM. It was concluded that Cl^- ions absorbed across the brush border membrane subsequently exit the enterocyte into the serosal fluid by a pathway dependent upon Na^+ and HCO_3^-. It is envisioned that an electrogenic carrier utilizes the transmembrane Na gradient to simultaneously energize Cl^- exit and Na^+ and HCO_3^- entry (or H^+ export). One possible stoichiometry is 1:1:1 Na^+-Cl^--HCO_3^- exchange (NDCBe) as illustrated in Fig. 2. The basolateral exchange carrier may also regulate luminal membrane Cl^- uptake by influencing cytoplasmic pH or HCO_3^- concentration.

2.7 Electrogenic Cl^- Transport Depends Upon Apical Membrane K^+ Conductance and Basolateral Membrane Anion Export

Even though Cl^- transport is electrogenic the evidence above indicates that luminal Cl^- uptake occurs via two parallel electroneutral processes, Na^+-K^+-$2Cl^-$ cotransport and Cl^--HCO_3^- exchange. The dilemma is resolved if K^+ ions moving through channels in the luminal membrane carry positive current out of the cell and supply K^+ ions at the outer face of the brush border membrane to support Na^+-K^+-$2Cl^-$ cotransport. Support for the existence of luminal K^+ channels is seen in the depolarization of the mucosal membrane by 30.3 mV when luminal bath [K^+] was elevated from 2.5 to 25 mM [13]. At the basolateral membrane an inward positive current may be carried by the net export of an anion with each cycle of the Na-Cl-HCO_3^- exchange carrier. Moreover, using double-barreled pH-sensitive microelectrodes, the luminal membrane was observed to hyperpolarize and the cytoplasm alkalinize when medium HCO_3^-/CO_2 was replaced with HEPES at constant pH [15]. This latter finding suggests that cytoplasmic pH, influenced by HCO_3^- entry across the basolateral membrane, couples the two membranes functionally by regulating luminal K^+ conductance and hence the availability of K^+ ions in the luminal fluid for Na-K-2Cl cotransport.

3 Electroneutral Cl^--HCO_3^- Exchange at the Brush Border

3.1 Luminal Alkalinization is Linked to Cl^- Absorption

The results described above indicate that a fraction of Cl^- absorption is linked to HCO_3^- secretion. The titration technique was employed to investigate the linkage further. $J^{HCO3}{}_{sm}$ was stimulated in a stepwise manner when the Cl^- concentration of the luminal medium was increased in increments in replacement for sulfate [12]. As little as 1 mM Cl^- was sufficient to stimulate $J^{HCO3}{}_{sm}$. The calculated K_m was 1–2 mM, a value considerably lower than that for electrogenic Cl^- absorption (K_m=5.4 mM), as noted above. Addition of Cl^- ions to the serosal medium (in equimolar replacement for SO_4^{2-} ions) was largely without effect on $J^{HCO3}{}_{sm}$. In contrast with these findings in urodele intestine, in rabbit ileum HCO_3^- secretion was not altered by bath Cl^- replacement [16].

3.2 Cl^- Absorption Exceeds HCO_3^- Secretion

A detailed examination of the relation between the unidirectional absorptive flux of ^{36}Cl and the secretory flux of HCO_3^- in individual isolated intestinal segments revealed that if luminal membrane Cl^- influx is linked to HCO_3^- efflux then the anion exchange ratio cannot be 1 for 1 [17]. The correlation coefficient between $J_{Cl}{}^{net}$ and $J^{HCO3}{}_{sm}$ was 0.85 and highly significant. The slope relating $J^{Cl}{}_{net}$ and $J^{HCO3}{}_{sm}$ was 1.8, and was also highly significant, indicating nearly twice as many millimoles of Cl^- absorbed per millimole of HCO_3^- secreted. This suggested the presence of a 1 for 1 Cl^--HCO_3^- exchange process in parallel with a second, HCO_3^--independent Cl^- uptake process (Na^+-K^+-$2Cl^-$ cotransport) discussed above. In support, whether SITS, DIDS, furosemide, or piretanide was added to the serosal medium, to inhibit a basolateral Cl^- exit pathway, significant inhibition of the Cl^- absorptive (ms) flux and $J^{HCO3}{}_{sm}$ resulted; the change in $J^{Cl}{}_{net}$ exceeded the change in $J^{HCO3}{}_{sm}$ in every series. SITS added to the mucosal medium had no effect on Cl^- absorption indicating that neither mucosal membrane Cl^- influx process is stilbene-sensitive. Lastly, $J^{HCO3}{}_{sm}$ was not altered when Na^+ in the luminal medium was replaced with another cation [12].

3.3 Intracellular pH is Influenced by Cl^- Ions in the Luminal Bath

Further evidence for luminal membrane Cl^-/HCO_3^- exchange was obtained using double-barreled pH sensitive microelectrodes to measure intracellular pH (pH_i) and serosal membrane potential [18]. When tissues were initially bathed in Cl^--free (SO_4^{2-}-based) medium then exposed on their mucosal surface to Cl^- ions at 86 mM, in order to activate the putative Cl^--HCO_3^- exchanger, pHi declined 0.34 +/– 0.08 units. The acidification of the cytoplasm is consistent with a decline in intracellular [HCO_3^-]. The serosal membrane potential (V_s) fell 29.6 mV. The stilbene, DNDS, in the luminal medium did not block acidification; DNDS in the serosal medium or furosemide in the luminal medium reduced electrogenic Cl^- absorption but did not alter cytoplasmic acidification. Additional measurements indicated that a threshold pHi of 7.1 exists for the exchanger; at higher pH_i net HCO_3^- loss from the cell occurred via the brush border anion exchanger.

4 K^+-H^+ ATPase Oversees Jejunal Electrogenic K^+-Dependent H^+ Secretion

There is considerable published evidence that during the postabsorptive period protons are actively secreted by the mature cells that line the villi of urodele jejunum by a mechanism which is demonstrably electrogenic but linked to the active uptake of K^+ ions. This transporter is functionally equivalent to that identified in the gastric mucosa [19], colon [20], and kidney [21] and acidifies the luminal fluid while actively accumulating K^+ ions into the enterocyte. The evidence follows.

4.1 Proton Secretion is Active and Occurs in Mature Villus Absorptive Cells

When the mid-region of the small intestine (the jejunum) of the urodele is bathed in a Cl^--free, HCO_3^--buffered medium (pH 7.4) in which sulfate is the major anion it generates a transepithelial voltage (V_{ms}) [22]. Under short-circuit conditions the I_{sc} was consistent with a secretory flux of cations such as protons. The I_{sc} was not accounted for by a net secretory Na flux when unidirectional fluxes were measured using ^{22}Na. Moreover, the I_{sc} was inhibited by acetazolamide. Measurements employing the pH-stat technique revealed a flux of alkali from the HCO_3^--buffered mucosal fluid (pH = 7.4) to the unbuffered serosal medium. The rate of serosal alkalinization was quantified as the rate of addition of H_2SO_4 required to maintain serosal bath pH constant and was expressed as an absorptive HCO_3^- flux ($J^{HCO3}{}_{ms}$). $J^{HCO3}{}_{ms}$ was equivalent to the I_{sc}. It was inferred that the serosal alkalinizing flux ($J^{HCO3}{}_{ms}$) was associated with an equal and opposite flux of protons into the mucosal fluid, i.e. a proton secretory flux (J^H), hence, that the measured flux, $J^{HCO3}{}_{ms}$, equaled the proton flux, i.e., $J^{HCO3}{}_{ms} = J^H$. Exposure to dinitrophenol (1 mM) or to anoxia conditions (by gassing the media with 100% N_2) each greatly inhibited the H^+ flux. The H^+ flux increased as a function of luminal bath [HCO_3^-] and maximized at a bath [HCO_3^-] of 35 mM at constant CO_2. The villus chambers were employed to establish that electrogenic proton transport was localized exclusively in the mature absorptive cells lining the villi of the jejunal mucosa [23]. This is noted in Table 1.

4.2 J^H is Dependent Upon Luminal Bath K^+ and Influences Cell K^+ Content

J^H was significantly reduced when K^+ was replaced in the luminal bathing medium; replacement of bath Na^+ did not influence J^H [24]. Readdition of KCl to a K^+-free mucosal medium stimulated J^H in a saturable manner while addition of choline Cl had no effect. J^H was maximal when medium [K^+] = 5 mM. The calculated K_m for luminal bath K^+ to stimulate J^H equaled 2.2 mM. Addition of KCl to the serosal medium had little effect on J^H. Hence, H^+ secretion was linked to luminal bath K^+. The results are consistent with the presence of an active, K^+-dependent H^+ secretory flux in jejunal absorptive cells.

To further investigate the role of K^+ ions in H^+ secretion intracellular K activity ($a_K{}^i$) was measured using double-barreled microelectrodes in which one barrel was sensitive to a_K. When HCO_3^--buffered medium bathed both apical and serosal tissue surfaces $a_K{}^i$ equaled 58.5 mM. The value is consistent with active K^+ accumulation above electrochemical equilibrium by the cell. When ouabain, the inhibitor of Na^+-K^+ ATPase, was added to the serosal medium $a_K{}^i$ declined over 2 h to 16 mM, a value consistent with passive K^+ distribution at the prevailing membrane potential. In contrast, when bicarbonate was removed from the serosal fluid, thereby inducing a H^+ secretory flux, $a_K{}^i$ was elevated to 83.6 mM. Upon subsequent exposure to serosal ouabain $a_K{}^i$ declined but remained above electrochemical equilibrium at 33 mM after 4 h. Furthermore, tissues bathed in Na^+-free medium containing ouabain retained cell K^+ ($a_K{}^i$ = 46 mM), even after 4 h, as long as the HCO_3^- gradient was maintained. Hence, a K^+ uptake process is operating at the luminal membrane which does not involve Na^+ but is dependent upon HCO_3^-/CO_2.Unilateral replacement of K^+ in either the mucosal or serosal medium reduced $a_K{}^i$. Acetazolamide reduced $a_K{}^i$ over 2 h in Na-free medium and concomitantly reduced the H^+ current. It was concluded that K^+ ions are actively accumulated across both luminal and basolateral membranes of the enterocytes and that the luminal uptake process is linked to H^+ secretion [24]. K^+-dependent intestinal H^+ secretion has also been reported for rat jejunum [25].

4.3 J^H is Blocked by Inhibitors of Gastric H^+-K^+ ATPase

Further evidence of intestinal H^+-K^+ ATPase activity was seen in the effects of the gastric H^+-K^+ ATPase inhibitors omeprazole and SCH28080 on intestinal H^+ secretion and K^+ accumulation [26, 27]. Thus, over 1 h omeprazole (136 μg/ml of luminal bathing medium) lowered I_{sc} and J^H by 40% and reduced $a_K{}^i$ from 74 to 59 mM. $a_K{}^i$ reverted to control values beginning about 30 min after exposure. The treated tissues subsequently exhibited a normal increase in I_{sc} when exposed to galactose to initiate Na-sugar cotransport indicating some degree of selectivity in the effect of omeprazole in H^+ transport. Similarly, the potent gastric H^+-K^+ ATPase inhibitor SCH28080 (2-methyl-(8-phenylmethoxy)imidazo(1,2a)pyridine-3-acetonitrile) inhibited J^H and I_{sc} completely over 90 min when added to the mucosal medium to a final concentration of 20 μM [28]. The calculated IC_{50} for the inhibitor was 6.5 μM. The electrical response to glucose added subsequently to the mucosal bath was equal to that in tissues not exposed to SCH28080.

4.4 pH-Sensitive K^+ Channels Reside in the Brush Border Membrane

A conductive pathway that is selective for K^+ ions resides in the brush border membrane and plays a role in active proton secretion [13, 15]. Using the villus sheet preparation membrane potentials were measured with microelectrodes by impaling jejunal villus cells across their serosal membrane. The mucosal membrane potential (V_m) was calculated as the difference between the transepithelial potential (V_{ms}) and the measured serosal membrane potential (V_s), i.e., $V_m = V_{ms} - V_s$. As noted above (Sect. 2.7), 10-fold elevation of mucosal bath [K^+] reduced the mucosal membrane potential (V_m) by 30.3 mV [13]. When jejunal mucosae were bathed in Cl^--free (SO_4^{2-}-based) media and exposed to 25 mM HCO_3^- and 5% CO_2 (medium pH = 7.4) V_m was –77.2 mV [15]. When the fluid was changed to one buffered at pH 7.4 using HEPES many, but not all, cells became hyperpolarized, indicative of enhanced K^+ conductance. The average membrane hyperpolarization was -11.8 ± 1.6 mV. Intracellular pH, measured with double-barreled pH-sensitive microelectrodes, averaged 7.16 ± 0.05 in HCO_3^-/CO_2-buffered medium. When the buffer was changed to HEPES at the same pH a rapid alkalinization of the cytoplasm occurred. Hence, the hyperpolarization is likely due to enhanced K^+ conductance when the cytoplasm is alkalinized. Cytoplasmic alkalinization would be expected "behind" the activated proton pump; enhanced K^+ efflux through the K^+ conductive pathway into the luminal medium would hyperpolarize the membrane and serve to provide K^+ ions in the luminal fluid to support K^+-H^+ antiport.

4.5 J^H Depends Upon HCO_3^- Ions Exiting the Serosal Membrane

H^+ secretion by salamander enterocytes is driven by active, K^+-dependent H^+ export coupled to an ion-exchange process in the basolateral membrane that promotes HCO_3^- ion exit into the serosal fluid linked to the uptake of a Na^+ ion and an anion, as illustrated in Fig. 3. The exchange process is in parallel with a conductive pathway for HCO_3^- ions. Using the villus sheet preparation the serosal membrane potential (V_s) was measured directly using microelectrodes [29]. In Cl^--free SO_4^{2-}-based medium buffered with HCO_3^-/CO_2 V_s equaled −86.5 mV. Lowering the serosal bath [HCO_3^-] from 25 to 2.5 mM depolarized the membrane by 6.7 mV; elevation of the [HCO_3^-] to 50 mM hyperpolarized the membrane by 7.5 mV. The calculated transference number for HCO_3^- was 0.14. The value of V_s correlated strongly ($r = 0.987$) with the log of the serosal bath [HCO_3^-]. The disulfonic stilbene DNDS reduced by one-half the effect of lowering serosal bath [HCO_3^-] on V_s. The effect is attributable to inhibition of the carrier-mediated pathway of HCO_3^- exit. Replacement of all but 1 mM serosal bath Na^+ with N-methylglucamine produced a partially reversible depolarization of the serosal membrane of 35.8 mV that was not accompanied by a change in fractional membrane resistance (hence did

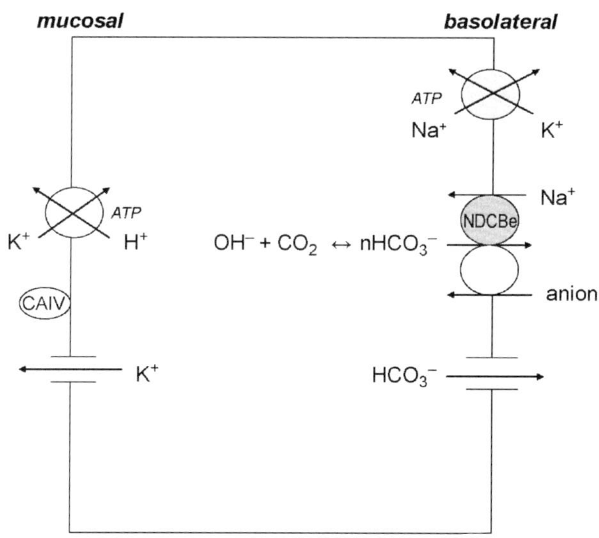

Fig. 3 Cell model of H^+-K^+ ATPase activity underlying H^+ secretion in mature villus enterocytes in the jejunum of urodele small intestine. CAIV is carbonic anhydrase type IV. Outward current carried by K^+ through a brush border channel is matched by inward current carried by the outward flow of HCO_3^- ions through a conductive channel and through a symporter (NDCBe) linked to Na^+ and an anion

not apparently alter membrane conductance) and was nearly completely blocked by 1 mM DNDS. A 10-fold elevation of serosal bath [K^+] reduced the serosal membrane potential (V_s) by 44.5 mV. The combined fractional conductance of K^+ ($t_K = 0.81$) and HCO_3^- ($t_{HCO3} = 0.14$) accounted for 95% of the serosal membrane conductance under Cl^--free conditions. The results are consistent with the involvement of a conductive (ion channel) pathway and a parallel nonconductive exchange pathway that couples the exit of one or more HCO_3^- ions to the entry of a Na^+ ion and an anion at the serosal membrane, i.e., Na^+-nHCO_3^--anion symport, as illustrated in Fig. 3. The family of HCO_3^- transporters is described in recent reviews [30, 31].

4.6 The Basolateral Exchanger Prefers Certain Anions

The Na^+-nHCO_3^--anion symporter prefers the anions Cl^-, Br^-, I^-, or HSO_4^- over gluconate or benzene sulfonate [32]. Thus, the rate of alkalinization of the serosal fluid was unchanged upon replacement of bath Cl^- with Br^-, I^-, or SO_4^{2-} but was reduced by replacement with gluconate or benzene sulfonate. Curiously, DIDS only partially inhibited the Na^+-nHCO_3^--anion symporter and the inhibition was reversed upon washout of DIDS. In comparison, in the same study the inhibition by DIDS of electrogenic Cl^- absorption via inhibition of the NDCBe exchanger (Fig. 2) was complete and was not reversed upon washout of DIDS.

4.7 Carbonic Anhydrase Plays a Key Role in Proton Secretion

The CA inhibitor acetazolamide inhibits the proton secretory flux (J^H) of the basal state at a low concentration (0.1 mM). A more slowly diffusing analog, benzolamide, inhibited J^H more rapidly when added to the mucosal medium than when added to the serosal medium, raising the possibility that a subtype of CA is associated with the brush border membrane. Acetazolamide reduced the intracellular potassium activity ($a_K{}^i$) from 66 to 42 mM and inhibited the ability of K^+ ions added to the mucosal fluid to stimulate J^H [24]. Lastly, acetazolamide reduced the magnitude of the depolarization of the basolateral membrane induced by quickly lowering the serosal bath [HCO_3^-] from 50 to 0.5 mM [29]. In summary, the CA activity may be associated with the brush border membrane as illustrated in Fig. 2 and be of the CA IV subtype [33]. A direct interaction between CA and the Na/HCO_3 cotransporter has been postulated [34].

4.8 Electrogenic H^+ Secretion Derives from Linked Mucosal and Serosal Conductive Processes

An electroneutral 1 for 1 cotransport of a proton and a K^+ ion could generate a short-circuit current whereby $I_{sc} = J^H$ if, in the steady state, K^+ ions exiting across the brush border membrane through K^+-selective channels carry an outward current which is matched at the serosal membrane by the inward current carried by the conductive exit of HCO_3^- ions through anion channels and through Na^+- $nHCO_3^-$-anion symporters as illustrated in Fig. 3. In this view K^+-H^+ ATPase generates the transmembrane K^+ and HCO_3^- gradients which are required for current generation.

4.9 The H^+ Secretory Process is Inhibited by Methacholine and Cyclic AMP

When added to the serosal medium the cholinergic agonist, methacholine, reduced K^+-dependent proton secretion by more than 50% after a delay averaging 8 min. Methacholine was ineffective when added to the mucosal bath. The $EC_{50} = 3.7 \times 10^{-7}$ M for the inhibitory effect. Preexposure of the tissue to the muscarinic receptor antagonist atropine blocked the effect [27]. Furthermore, the beta adrenergic antagonist propanolol had no effect on J^H indicating that the intestinal K^+-H^+ ATPase is not under the influence of local ß-adrenergic agonists [35]. In contrast, H^+ ion secretion by gastric parietal cells is stimulated by acetylcholine [36]. In an earlier study it was observed that H^+ secretion was rapidly reduced 34% upon exposure of isolated segments of urodele duodenum to 1 mM theophylline [37]. Exogenous cyclic AMP and dibutyryl cAMP produced similar effects.

4.10 Summary

The H^+-K^+ ATPase transporter operates in the basal state. Substantial evidence is provided above that in the basal (postabsorptive) state urodele jejunal villus cells secrete H^+ ions via a brush border ion transporter that utilizes ATP to accumulate luminal K^+ ions into the cytoplasm and acidify the luminal fluid while alkalinizing the serosal fluid. A H^+-K^+ ATPase transporter is most likely responsible [38]. Because Na^+-H^+ exchange does not operate in the basal state, which is characterized by limited supplies of endogenous energy, the H^+-K^+ ATPase transporter can acidify the luminal fluid and accumulate cell K^+ without obligating the cell to perform transepithelial Na^+ absorption. Active H^+ secretion linked to K^+ uptake is reduced by elevating cyclic AMP or by exposure to methacholine. As discussed below, catecholamines also reduce H^+-K^+ ATPase activity. In addition to promoting HCO_3^- absorption the intestinal H^+-K^+ ATPase may function to absorb K^+ ions and acidify an unstirred layer adjacent to the brush border to facilitate H^+-dependent oligopeptide absorption [39]. An acid microclimate has been observed to reside at the brush border surface of rat jejunal enterocytes [40].

5 Catecholamines Stimulate Jejunal Na^+ and Cl^- Absorption

5.1 *Basal, Active K^+-H^+ Transport is Supplanted by Amiloride-Sensitive Na^+-H^+ Exchange*

When jejunal segments of urodele small intestine are bathed in Cl^--free and substrate-free medium luminal acidification occurs due to an active K^+-H^+ antiport (K^+-H^+ ATPase) in the brush border membrane, as described above (Sect. 4). The process occurs in the mature villus cells (Table 1). There is no net Na^+ transport occurring under this condition [22]. When norepinephrine (NE) is added to a Cl^--free serosal medium bathing jejunal segments engaged in active H^+ secretion by this mechanism an increase in I_{sc} and luminal acidification occurs which is no longer dependent upon luminal K^+ but is instead dependent upon Na^+ in the luminal bath [41]. Net Na^+ absorption also develops. As noted in Table 2, responsiveness to NE resides in both villus and intervillus mucosal regions. Very low concentrations of NE ($EC_{50} = 3.7 \times 10^{-7}$ M) produce the response. Na^+-dependent luminal acidification requires HCO_3^- ions and CO_2 in the luminal medium. The transport was not inhibited by omeprazole but was inhibited by amiloride (1 mM), an inhibitor of Na^+-H^+ exchange. Under the Cl^--free conditions in which it was studied Na^+-dependent H^+ transport was electrogenic overall, rendering the serosal bath electrically negative to the mucosal bath. When the tissue was voltage-clamped the I_{sc} was equal to the proton flux. In Table 2 the characteristics of the proton secretory flux in the basal state (basal J^H) are compared with those when the mucosa is under the influence of NE

Table 2 Two mechanisms of H^+ secretion in urodele small intestine are compared. The H^+ secretory flux (J^H) in the basal state, i.e., basal J^H, is due to K^+-H^+ ATPase. Norepinephrine (NE) induces Na^+-H^+ exchange. In both conditions $I_{sc} = J^H$. J^H values were measured while tissues were bathed on mucosal and serosal surfaces with Cl^--free media. ↓ reflects inhibition

Characteristic (reference)	Basal J^H	NE-stimulated J^H
Locus in intestinal mucosa [14, 24]	Villus	Villus and intervillus
Bath ion dependence [12, 17, 21, 24]	Luminal K^+	Luminal Na^+
Effect of acetazolamide [12, 24]	↓	↓
Effect of methacholine [21]	↓	No Δ
Effect of omeprazole [18, 24]	↓	No Δ
Effect of amiloride [24]	No Δ	↓
Effect of yohimbine [25]	No Δ	↓
Effect of propanolol [25]	No Δ	No Δ

5.2 *NE Stimulates Cl^--HCO_3^- Exchange*

As described above, when urodele intestinal segments were bathed in a substrate-free medium containing Cl^- as the major anion there was net Cl^- absorption accompanied by a residual flux attributable to a smaller net secretory HCO_3^- flux [1]. This result was consistent with measurements establishing Cl^-/HCO_3^- exchange by urodele small intestine [17]. The flux values reflect the presence of apical electrogenic Na^+-K^+-$2Cl^-$ cotransport operating in parallel with apical Cl^- absorption linked to HCO_3^- secretion via 1 for 1 Cl^--HCO_3^- exchange (Sect. 2.2). The two Cl^- transport processes are of about equal magnitude [7]. When tissues were exposed to NE, stimulation of the absorptive Cl^- flux and net Cl^- absorption were observed without alteration of the residual flux or the short-circuit current [35]. It was concluded that in parallel with elevated electroneutral Na^+/H^+ exchange NE also stimulates electroneutral Cl^-/HCO_3^- exchange without altering the parallel process of Na^+-K^+-$2Cl^-$ symport.

5.3 NE Operates Through Alpha$_2$ Adrenergic Receptors

Using several alpha and beta adrenergic agonists and antagonists it was demonstrated that NE operates through the α_2 subtype of adrenergic receptors residing on the basolateral membrane of enterocytes [35]. The α adrenergic antagonist yohimbine blocked NE-stimulated acid secretion but had virtually no effect on basal (spontaneous) acid secretion driven by H^+-K^+ ATPase (Table 1). The cytoplasmic effector activated by NE binding to α_2 adrenergic receptors is unidentified. The phosphodiesterase inhibitor theophylline elevates cytoplasmic cAMP and inhibits net Cl^- absorption [42] and ^{36}Cl influx into the intestinal mucosa by 60% [7]. Hence, it is reasonable to assume that the α_2 adrenergic receptor does not act to elevate cAMP.

5.4 NE Produces Changes in Enterocyte Ultrastructure

Norepinephrine transforms the ultrastructure of urodele enterocytes in the cellular region immediately below the apical plasma membrane. NE produced a significant increase in the density of small vesicular and tubular profiles in the cytoplasm immediately below the brush border of the enterocytes [43]. The use of horseradish peroxidase revealed that the apical tubulovesicular compartments were physically contiguous with the luminal extracellular space. This observation points to the direct involvement of the brush border membrane in proton secretion; the tubules and vesicles are elements of a dynamic pool of transporting membranes under the control of catecholamines. The morphological alterations in urodele enterocytes upon stimulation of proton transport is analogous to those changes reported in gastric parietal cells [44, 45], turtle urinary bladder [46, 47], and rat distal colon [48] when these tissues were stimulated to increase their proton transport.

6 Ion Transport Processes in Urodele Intestine Under the Influence of cAMP

6.1 Does Elevated Cyclic AMP Activate Net Cl^- Secretion?

Evidence is considerable that elevation of cyclic AMP inhibits salt and water absorption by in vitro intestinal segments isolated from a variety of vertebrates. Furthermore, bidirectional ion flux measurements under short-circuit conditions indicate that secretion of Cl^- ions is simultaneously elevated [49]. The response is thought to underlie intestinal function in secretory diarrheas such as cholera and in diseases such as cystic fibrosis. However, Lucas has shown that the physical meaning of a unidirectional ion flux measured under short-circuit conditions is more complicated than previously assumed.

6.2 Opposing Fluxes Interact with Each Other

Lucas argued that the interpretation of the results is flawed in previous studies which purported to demonstrate that enterotoxins stimulate net Cl^- secretion in in vitro segments of vertebrate intestine [49]. These studies employed the technique of short-circuit current while measuring unidirectional ^{36}Cl fluxes. Analytical solutions to the differential equations describing fluxes between compartments to demonstrate that "a measured "unidirectional" flux contains information about both absorptive and secretory processes, regardless of which flux is measured." As a result, a condition which decreases the absorptive (m→s) flux of an ion will cause an apparent increase in the secretory (s→m) flux. In essence, because an isotope that is transported from m→s can return to the intermediate compartment by transport processes operating in the reverse direction, then a change in one unidirectional flux cannot occur without affecting the reverse flux. Hence, the observed increase in the s→m Cl^- flux in isolated intestinal preparations reported after exposure to different agents, in numerous studies catalogued by Lucas, may actually be a consequence solely of an agent-induced reduction in the m→s flux of Cl^- ions. In most of the 17 studies examined, the increase in the s→m Cl flux after exposure to an agent was paralleled by a decrease in the absorptive (m→s) flux. Lucas concluded that a measured increase in the secretory flux of Cl^- after exposure to a secretagogue is not unequivocal evidence for direct stimulation of a chloride secretory process.

6.3 Cyclic AMP Does Not Alter the s→m $^{36}Cl^-$ Flux in the Urodele Intestine

The implications of the Lucas hypothesis with respect to measurements in urodele intestine are seen in two studies. In the first, ouabain increased the s→m and reduced the m→s Cl^- flux while eliminating net Cl^- absorption [1]. According to the Lucas hypothesis, because the m→s flux declined, the observed increase in the opposite (s→m) flux may only be apparent. In another study which related specifically to the question, theophylline reduced the m→s Cl^- flux and eliminated net Cl^- absorption in duodenal segments but did not change the s→m Cl^- flux. Instead, the residual flux, which is a measure of HCO_3^- secretion in the conditions employed, was increased [42]. In confirmation, it was demonstrated using the pH-stat technique that theophylline stimulated an electrogenic secretory flux of HCO_3^- when tissues were bathed in Cl^--free medium [50, 51]. Furthermore, acetazolamide inhibited the ongoing HCO_3^- secretory flux. A recent study has provided evidence that dibutyryl cAMP induces electrogenic HCO_3^- secretion in normal human duodenum but not in duodenum of cystic fibrosis patients. It was concluded that HCO_3^- secretion normally occurs through cystic fibrosis transmembrane regulator (CFTR) channels which are conductive to Cl^- and HCO_3^- ions [52]. Studies delineating the role that CFTR plays in intestinal fluid secretion have been recently reviewed [53]. In summary, these data in urodele enterocytes provide support for an elevated secretory flux of HCO_3^- but not Cl^- upon elevation of cAMP. Nevertheless, the evidence for the existence of Cl^- channels in the brush border of mammalian small intestinal cells is extensive. A recent review provides a description of several different Cl^-, Na^+, and K^+ channels in gastrointestinal tissues [54].

6.4 Cyclic AMP Reduces J^H

Theophylline, the inhibitor of cell phosphodiesterase, reduced J^H 34%. When bathed in normal medium the residual flux (J^R), previously shown to equate to the net Cl^- absorptive flux, also fell to zero indicating that the Cl^--absorptive flux was eliminated by theophylline [37]. Exogenous 1 mM cyclic AMP (cAMP) produced similar effects. Whether this response reflected a primary effect at the luminal membrane (H^+-K^+ ATPase antiport) or at the serosal membrane (Na^+–$nHCO_3^-$-anion exchange) or both is not clear.

6.5 Cyclic AMP Activates a Basolateral Membrane Channel Conductive to Cl^-

Outwardly rectifying single channels displaying a reversal potential consistent with Cl-selective channels were observed in cell-attached patches on the serosal membrane of enterocytes in isolated segments of axolotl small intestine exposed to theophylline [55]. The channels were not evident in tissues not exposed to theophylline. The villus sheet preparation was employed [13]. The serosal surface was exposed for 1 h to collagenase. The basal lamina and adherent collagen were removed by manual dissection to expose the serosal membrane of villus enterocytes. Single channel slope conductance was 24 pS from –80 to 20 mV and 131 pS from 20 to 140 mV. When the patch was depolarized, open probability increased from 0.07 ($V_p = -80\,mV$) to 0.7 ($V_p = 140\,mV$). Because Cl^- ions are normally actively transported from the cytoplasm to the serosal bath a theophylline-activated Cl^- channel would allow a backflux of Cl^- ions into the cytoplasm and thereby reduce net Cl^- movement from cytoplasm to serosal fluid. This response could explain the reduction in net Cl^- absorption caused by theophylline (Sect. 6.3). In summary, elevated cAMP inhibits intestinal Cl^- absorption and H^+ secretion and stimulates HCO_3^- secretion (Fig. 4).

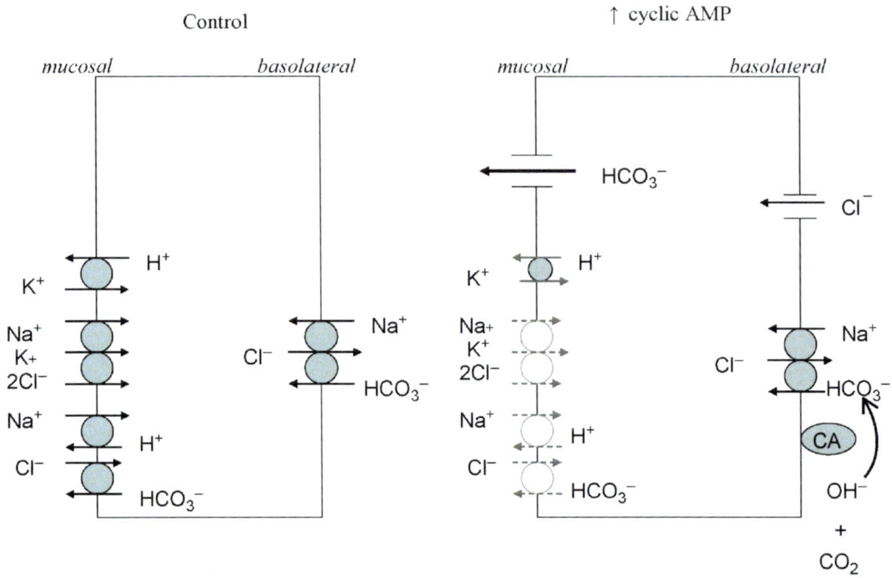

Fig. 4 Elevated cAMP inhibits several brush border transporters of urodele enterocytes either partially (luminal K^+-H^+ ATPase) or completely (*right, faded open circles*), while stimulating net HCO_3^- secretion and inhibiting net Cl^- absorption. Carbonic anhydrase (CA) provides about one-third of transported HCO_3^-. For simplicity, basolateral Na^+-K^+ATPase, operative under both conditions, was not illustrated

7 Differences in Transcellular Transport Between Villus and Intervillus Regions of Urodele Small Intestinal Mucosa

7.1 Intervillus Transepithelial Ion and Solute Transport Differs from that in the Villus Epithelium

Lacking crypts of Lieberkuhn, the germinative foci within mammalian small intestinal mucosa, the urodele intestine derives new cells from multicellular "nests," clusters of germinative cells within the lamina propria of the intervillus region.

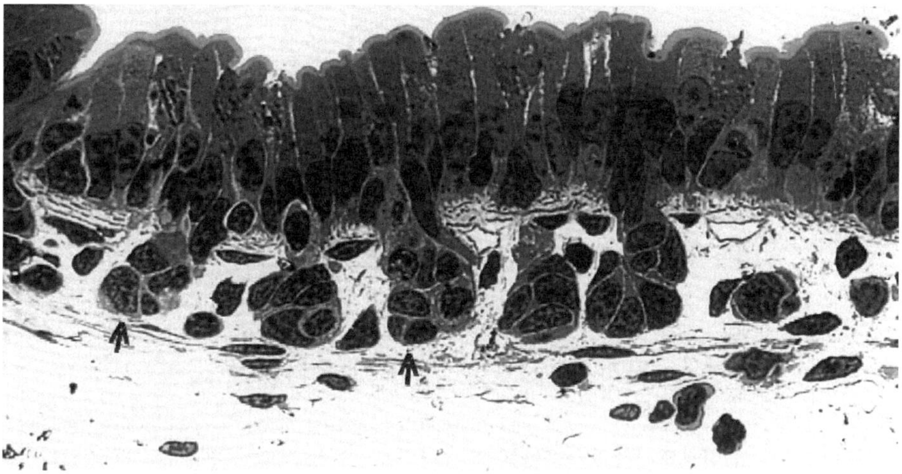

Fig. 5 Histological section of *Amphiuma jejunum* at the base of a fold [43]. *Arrows* point to multicellular nests in the subepithelial space. Some nests adhere to the overlying epithelium. The epithelium consists of columnar epithelial cells. The brush border, comprised of microvilli, is evident at the apical end of the cells (*top end* of micrograph). Magnification × 240. This figure used with permission of the journal

In histological sections it is seen that some of these nests adhere to the overlying epithelium, as illustrated in Fig. 5. The nest cells insert into the epithelium, migrate up the villus and reside there before eventually being desquamated as the result of programmed cell death [56]. These migratory events, from intervillus to villus mucosa, are associated with very significant changes in the capabilities of the enterocytes to effect transcellular ion and solute transport.

As noted above (Sect. 2.7), the villus chamber was employed to isolate the villus and intervillus epithelia and quantify the relative contribution of the two regions to the electrical responses measured in the whole, intact intestinal mucosa. In Table 1 it is seen that the immature intervillus enterocytes possess some but not all of the transport capabilities of the fully mature villus enterocytes. For example, the intervillus enterocytes can cotransport Na and amino acids. However, they lack the ability to cotransport Na and glucose. Whole cell currents measured in cell-attached patches on monolayers grown from cell nests confirm and extend these observations.

7.2 Cell Nests Can Be Isolated, Grown into Monolayers in Tissue Culture

Scrapes of the small and large intestinal mucosa were separately incubated with collagenase for 30 min, filtered through mesh, allowed to settle, and washed five or more times with fresh culture medium before plating them onto collagen-coated tissue culture dishes under sterile conditions [52]. A phase-contrast image of isolated,

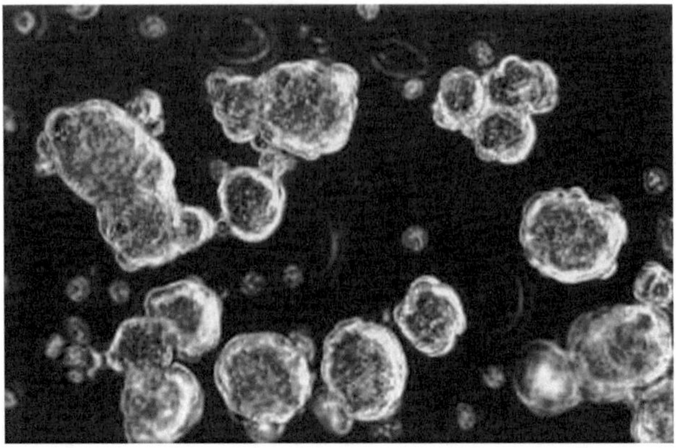

Fig. 6 Phase-contrast image of intact round or slightly oval nests of cells isolated from Amphiuma small intestine by collagenase digestion [57]. Goblet cells, possessing prominent mucinogen granules, comprise about 3% of the population. Magnification × 160. This figure used with permission of the journal

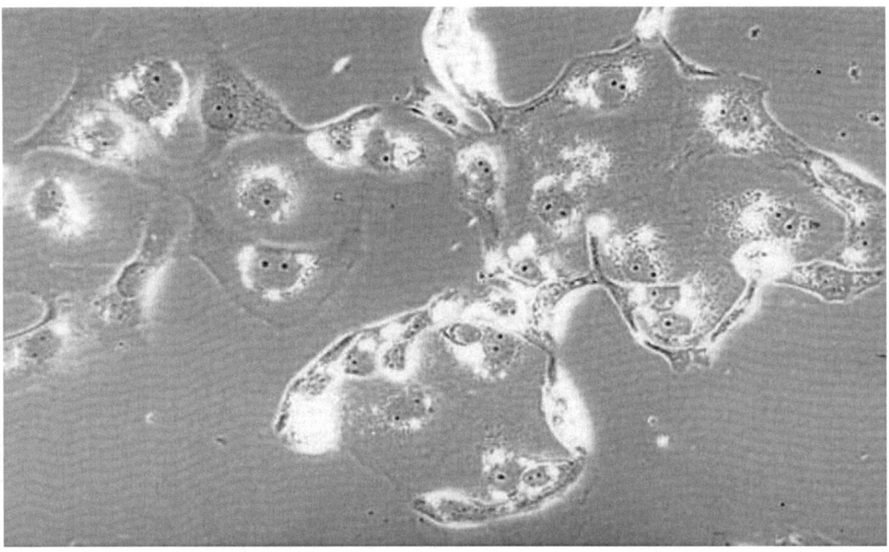

Fig. 7 A typical small intestinal cell monolayer one week after plating the nests in tissue culture medium [57]. Tight junctional complexes link adjacent epithelial cells. Electron microscopy of the monolayer cells revealed short microvilli sparsely distributed over their surface. Magnification × 1,000. This figure used with permission of the journal

intact small intestinal nests is seen in Fig. 6. Electron micrographs revealed absorptive, mucus and endocrine cells. A typical small intestinal cell monolayer, one week after plating the nests in tissue culture medium, is seen in bright-field in Fig. 7.

7.3 Whole Cell Currents in Monolayers Establish Na^+-Amino Acid Cotransport

Monolayers were bathed in a buffered medium which included 104.7 meq/L of Na^+ [57]. Patch electrodes were filled with a buffered medium containing 105 meq/L of K^+. As seen in Fig. 8 there was a substantial current of about 20 picoamps (pA) before exposure to valine. This is likely due to Na^+ currents through brush border Na^+ channels [58]. Equimolar replacement of bath mannitol with L-valine caused an abrupt inward current of about 30 pA (I_{aa}) which was maintained until valine was washed away. D-valine produced no effect, in keeping with the known preference of the amino acid transporters for L- and not D- isomers of the amino acid. L-alanine and L-phenylalanine, two amino acids which share affinity for the Na-amino acid cotransporter, produced I_{aa} of similar magnitude as L-valine. Increasing the bath valine concentration from 0 to 50 mM caused an increase in I_{aa} which approached a maximum; the slope of the relationship between the change in I_{aa} and the reciprocal of the change in valine concentration indicated a K_m of 90 mM and a V_{max} of 100 pA. This is indicative of a low affinity, high capacity carrier for valine. I_{aa} was zero when bath Na^+ was replaced with the cation $Tris^+$ as expected if sodium ions are required to carry the current. Also, colonic cell monolayers grown from cell nests isolated from colonic tissue did not exhibit inward currents when exposed to valine. The colon does not engage in active, transcellular absorption of amino acids from the lumen. Neither glucose nor galactose produced changes in whole cell current in small intestinal monolayers. This indicates that Na-glucose cotransporters

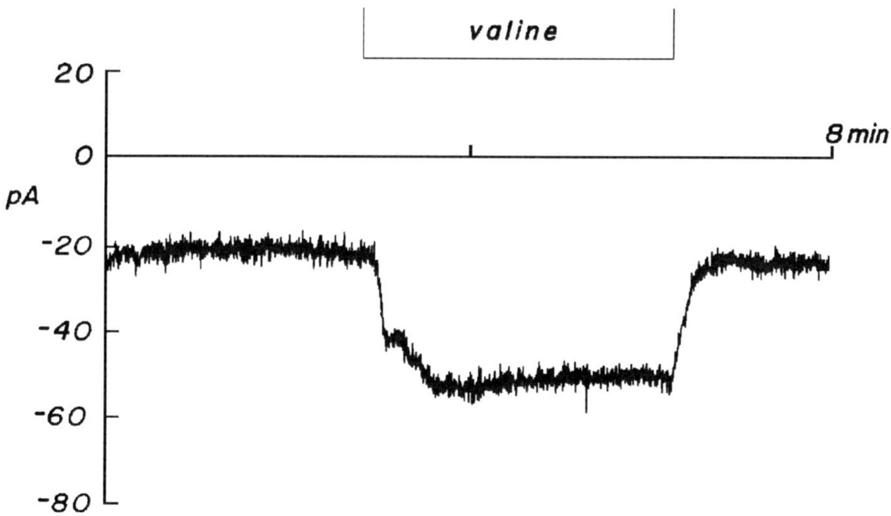

Fig. 8 Replacement of bath mannitol with L-valine caused an abrupt inward current (in picoamps, pA) in monolayer cells which was maintained until valine was washed away [57]. D-valine produced no effect on the whole cell current. This figure used with permission of the journal

are not present in these cells, a finding which is entirely consistent with earlier work employing the villus chamber, in which it was demonstrated that a short-circuit current consistent with Na^+ absorption was generated by the intervillus epithelium when exposed to amino acids such as valine while glucose had no effect on the I_{sc} (Table 1). Glucose did stimulate such a current in villus epithelia, indicating that Na-glucose cotransport only occurs in the most mature, differentiated enterocytes [6]. This result suggests that when grown into monolayers the nest cells retain the transport properties of intervillus cells and, under the conditions of their growth in vitro, do not differentiate fully enough to express Na-glucose cotransporters. In conclusion, low affinity, high capacity Na-amino acid cotransporters reside in small intestinal cell monolayers grown from cell nests. These cells are not able to cotransport Na^+ and glucose. These results are consistent with the conclusions derived from the data in Table 1, namely, that the intervillus epithelium is capable of Na^+-amino acid but not Na^+-glucose cotransport. Hence, these cells behave phenotypically as intervillus enterocytes.

The technique of isolating cell nests and preparing monolayers from them should be very helpful in further characterization of the ion and solute transport capabilities of the intervillus enterocytes. For example, previous studies indicate that they should exhibit sensitivity to catecholamines and to secretion-inducing agents such as cholera toxin (see Table 1). Perhaps more importantly, the cell culture techniques could be exploited to investigate potential growth factors and growth conditions which might stimulate differentiation of the intervillus enterocytes to express the phenotype of the fully mature villus enterocyte as a way to understand how differentiation is regulated. Lastly, the stem cells, which presumably generate the cell nests, must be isolated and the factors regulating their division and their ability to generate cell nests should be defined.

7.4 Summary: No Support for a Secretory Response Limited to the Intervillus Epithelium

Evidence from two experimental approaches indicate that the immature intervillus cells express Na-amino acid cotransport and Na-H exchange, and are responsive to α_2-adrenergic agonists and agents which elevate intracellular cAMP. In contrast, the fully mature villus enterocytes possess the additional capabilities of Na-glucose cotransport, Na-K-2Cl symport and H^+-K^+ antiport. The tight junctions between the intervillus enterocytes are relatively leaky to HCO_3^- ions while the villus tight junctions are less leaky [6]. This observation is consistent with a theoretical study which concluded that the mammalian intestinal villus tight junctions have higher electrical resistance than those in the crypts [59]. Given the role of exogenous glucose to provide the energy for electrogenic Cl^- absorption, described below (Sect. 8), the expression of Na-glucose cotransporters in the mature villus cells would provide them with a metabolic energy form (glucose) capable of powering the additional active Na^+, K^+, H^+, and Cl^- transport which occurs with the expression of Na-K-2Cl cotransport and K^+-H^+ antiport.

In contrast with the differential response to amino acids, enterocytes at all stages of development possess the ability to respond to adrenergic and cholinergic agonists, which powerfully influence intestinal salt and water transport, and respond to agonists which initiate secretory diarrheas. In this way a large epithelial response can be mustered by the intestine.

This measure of the relative contribution of the mature villus epithelium and the less mature intervillus epithelium to transepithelial ion and solute absorption and secretion affords important insights into the developmental changes occurring in urodele enterocytes as they migrate up the villus. They contribute to our understanding of the comparative transport capabilities of the two regions with respect to a wide variety of ions and solutes. The overall model contrasts sharply with that proposed for rabbit intestine, which assigns the crypt epithelium as the predominant site of cAMP-induced anion and water secretion [39]. In the urodele mucosa both regions are capable of cyclic AMP-stimulated HCO_3^- secretion.

8 Net Cl^- Absorption is Reduced when Metabolic Energy is Limited by the Need to Absorb Na^+

8.1 Certain Solutes Inhibit Electrogenic Cl^- Absorption

One of the most remarkable transitions that the small intestine undergoes is that which occurs when food is introduced into the intestinal lumen. Prior to this, during the basal (postabsorptive) state, access of the small intestine to metabolites is limited to its own endogenous metabolic reserves and metabolites available in the blood. The urodele intestine actively transports Cl^- ions and generates a serosa-negative transepithelial voltage which supports the passive, paracellular absorption of Na^+ ions from the lumen to the blood and simultaneously deters the back leak of Na^+ ions from blood to lumen. As noted in Fig. 9, left and right, when the isolated intestinal segments are voltage clamped a negative short-circuit current is evident and is due to electrogenic Cl^- absorption. It is also evident in the same figure that addition of the sugars galactose or 3-o-methylglucose or the amino acid valine to the mucosal medium at 20 mM (in substitution for 20 mM mannitol) was followed by a reversal in the sign of the I_{sc} consistent with disappearance of net electrogenic Cl^- absorption and simultaneous induction of net Na^+ absorption due to cotransport into the enterocyte of Na^+ with the sugar or amino acid. This interpretation was validated by measuring bidirectional ^{22}Na and ^{36}Cl fluxes [60]. Simultaneously, as seen in Fig. 10, intracellular Na^+ activity ($a_{Na}{}^{i}$), measured using Na^+-sensitive microelectrodes, was elevated upon exposure to galactose or valine, a result which was substantiated in a subsequent study [58]. Also, active accumulation of Cl^- ions above electrochemical equilibrium was eliminated. In contrast, addition to the luminal fluid of the metabolizable sugar glucose, which is also cotransported with Na^+ across the brush border membrane [60], does not inhibit net Cl^- absorption (Fig. 9, left) or the ability of the luminal cell membrane to actively accumulate Cl^- ions into

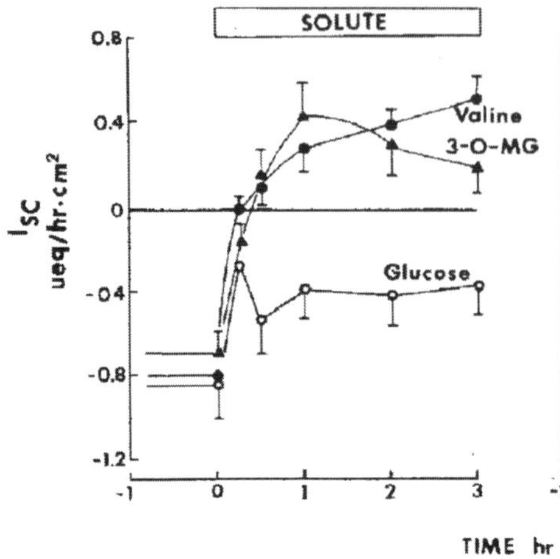

Fig. 9 Effect of transported solutes on electrogenic Cl^- absorption [60]. The negative short-circuit current (I_{sc}) preceding time = 0 reflects electrogenic Cl^- absorption. The I_{sc} becomes positive upon exposure to valine or 3-o-methylglucose at time = 0 (*left*) or to galactose (*right*). Addition of glucose to the mucosal bath reduced but did not abolish the negative I_{sc} (*left*) and, when present in the serosal bath, prevented the reversal of the I_{sc} caused by galactose (*right*). All solutes were employed at 20 mM. The figure used with permission of the journal

the cytoplasm of the enterocytes. Furthermore, $a_{Na}{}^{i}$ is not reduced (Fig. 10). The I_{sc} became less negative, a fact which is explained by activation of glucose-stimulated electrogenic Na^+ absorption in parallel with on-going electrogenic Cl^- absorption. The results of unidirectional labeled ^{22}Na and ^{36} Cl flux measurements confirmed that net Cl^- absorption was inhibited by galactose and valine but not by glucose. As seen in Fig. 9, right, galactose was not able to reverse the I_{sc} (i.e., inhibit electrogenic Cl^- absorption) when glucose (20 mM) was present in the serosal fluid. When the serosal bath glucose concentration was elevated in steps after galactose inhibited the I_{sc} the negative I_{sc} was partially restored as seen in Fig. 11. The glucose effect maximized when serosal bath glucose concentration was 15 mM; the calculated K_m for the response was 6.2 mM. Of several potential metabolic substrates tested (glucose, fructose, mannose, Na butyrate, Na glutamate) only glucose restored the Cl^- absorptive current.

8.2 In Metabolic Stress Na^+ Transport Has Priority Over Cl^- Transport

It is proposed that galactose and valine are not metabolized and therefore do not provide energy for basolateral Na transport via Na^+-K^+ ATPase. They elevate $a_{Na}{}^{i}$ because apical Na-solute cotransport is not matched by Na^+ export at the basolateral membrane. The elevated $a_{Na}{}^{i}$ reduces the driving force for Na^+ entry across both apical and basolateral cell membranes. This will reduce luminal Cl^- entry

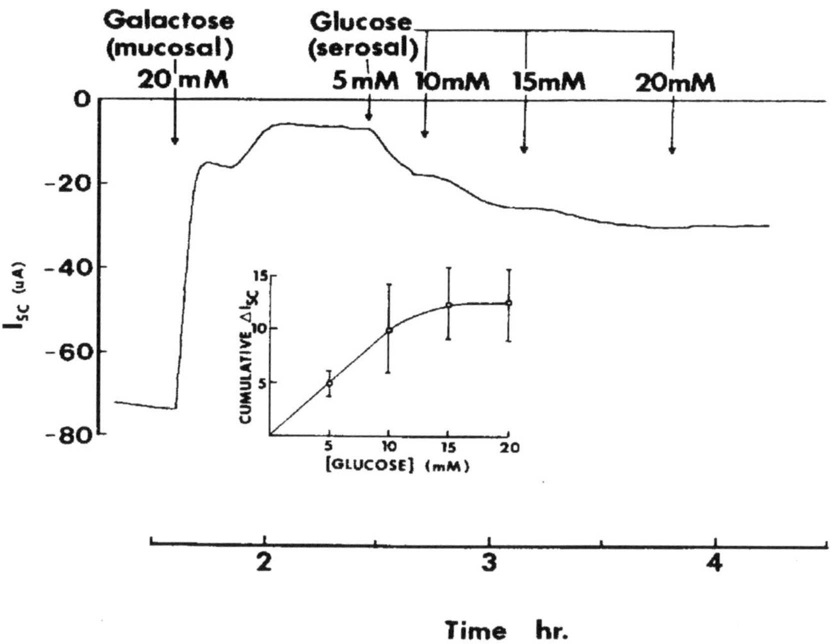

Fig. 10 Effect of transported solutes on intracellular Na$^+$ activity ($a_{Na}{}^{i}$) and mucosal membrane potential (γ_m). $a_{Na}{}^{i}$ is low before solute addition (*open bars*). $a_{Na}{}^{i}$ is significantly elevated after addition to the mucosal bath of 20 mM galactose or valine (*filled bars*) whereas glucose did not significantly elevate $a_{Na}{}^{i}$ [60]. The figure used with permission of the journal

Fig. 11 Typical electrical response of isolated intestine to sequential additions of glucose to the serosal medium after initial exposure to galactose added to the mucosal medium [60]. The negative-going I_{sc} in microamps (μA) reflects stimulation of electrogenic Cl$^-$ absorption by glucose. *Inset*: average change in I_{sc} of five tissues to cumulative additions of glucose to the serosal medium. The figure used with permission of the journal

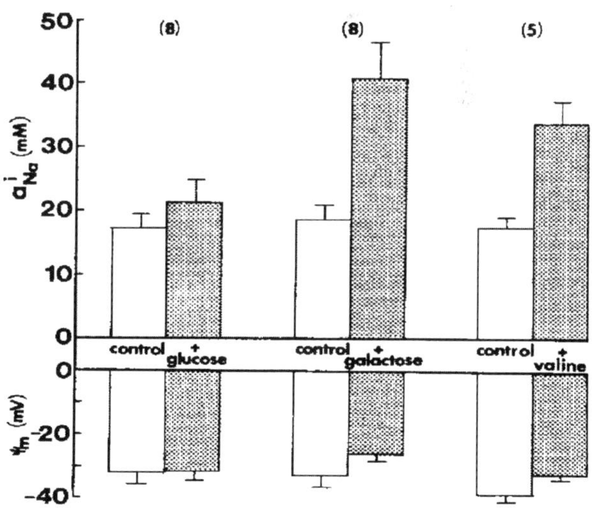

via Na-K-2Cl cotransport. Apical Na^+-H^+ exchange would be similarly reduced and this would in turn inhibit apical Cl^--HCO_3^- exchange because of the reduced production of HCO_3^- ions. Basolateral Na^+-Cl^--HCO_3^- exchange would also be inhibited because of the reduced transmembrane Na^+ gradient. Hence, the energetics of Cl^- transport across both cell membranes becomes unfavorable and Cl^- accumulation above electrochemical equilibrium and transcellular electrogenic Cl^- absorption is reduced. In contrast, glucose is metabolizable and thereby provides an energy source for basolateral membrane Na^+ transport to keep $a_{Na}{}^{i}$ low, maintain the transmembrane Na gradients, and allow apical and basolateral Na-dependent transporters to operate normally and oversee electrogenic Cl^- absorption. Hence, intestinal Cl^- transport is sacrificed when solutes in the intestinal lumen must be transported into the blood under conditions of limited availability of energy.

In summary, endogenous cellular metabolic reserves of in vitro urodele intestine are inadequate to maintain Cl^- absorption at normal levels when coupled Na-solute transport is activated in parallel. Net electrogenic Na^+ absorption has priority over net electrogenic Cl^- absorption in this circumstance. Glucose is a critical energy source for the enterocyte whether it originates from the blood or the diet. Recent studies have offered evidence that tissues deprived of metabolites develop a state of "metabolic stress" which may induce a response of the tissue that involves a metabolism-sensing kinase.

8.3 The Intestinal Response to Metabolic Stress Response May Involve AMP-Activated Protein Kinase (AMPK)

AMPK is a metabolic-sensing kinase which is activated in metabolic stress in a wide variety of tissues and is particularly sensitive to the ratio of cytoplasmic AMP/ATP [61]. A reduction in cell ATP relative to AMP activates AMPK which turns on ATP-generating metabolic pathways. It increases expression of GLUT1 (Na-glucose cotransport), decreases expression of CFTR Cl^- channels and epithelial Na^+ channels, and inhibits $Na^+K^+2Cl^-$ cotransport activity. The resulting cellular response would serve to produce many of the changes in transport observed when urodele intestine is exposed to galactose, for example. Because galactose induces Na-galactose cotransport, the cell is obligated to stimulate basolateral, electrogenic ATP-dependent Na^+ cotransport. It is too soon to relate these findings to the observed ability of certain sugars and amino acids to inhibit urodele Cl^- transport but the potential role of AMPK and other signaling pathways in the response requires investigation.

9 Significance

The studies described in this Chapter had multiple aims, foremost of which was the identification and characterization of the membrane transporters operating in the normal vertebrate small intestine to move ions and solutes across the intestinal mucosa in the absorptive and secretory directions. Bilateral fluxes of salts and water are necessary for absorption of ingested foods and minerals from intestinal lumen to the blood and for secretion of salts and water from the blood into the intestinal lumen; the latter flux maintains the fluidity, pH, and osmolality of the intestinal contents. The contributions to transepithelial ion and solute transport of the mature absorptive cells that line the upper villus and the less mature cells residing on the lower villus and intervillus surfaces are distinctly different. These studies describe the changes in the expression of membrane transporters as enterocytes mature during their migration toward the villus tip. For example, very early in their development the cells at the base of the villi display the ability to cotransport Na^+ and amino acids but only acquire the capabilities of Na^+-glucose cotransport and electrogenic Cl^- transport as they reach full maturity at the villus tip. Identification of the specific membrane transporters in the luminal (brush border) membrane and the basolateral membrane of the enterocyte are essential for obtaining insight into the manner in which cells control net transcellular absorptive and secretory fluxes. Electrogenic Cl^- absorption occurring during the lumen-empty, postabsorptive state is blocked when gut function shifts to the absorptive state in response to loading the luminal fluid with glucose or amino acids which are cotransported with Na^+. Depending on its ability to be metabolized for energy, a luminal organic solute will either promote or inhibit electrogenic Cl^- absorption while stimulating Na^+ absorption. Electrogenic absorption of Cl^- ions is energy dependent and can be supported by glucose provided in the diet or in the blood. Intracellular signaling pathways such as AMP-protein kinase (AMPK), activated in metabolic stress, may influence gut transport by altering distinct membrane transporters in one or both of the enterocyte cell membranes. These studies are pertinent to the goal of understanding how dietary organic solutes promote intestinal salt and water absorption. In addition, understanding the response of the intestinal mucosa to neuromodulators such as norepinephrine is essential for defining the control of the autonomic nervous system over gut function. Lastly, the mechanisms by which HCO_3^- secretion by the intestinal mucosa is enhanced are relevant to understanding the cellular response in clinical conditions characterized by diarrhea and how the duodenal mucosa protects itself from luminal hydrochloric acid secreted by the stomach.

References

1. White JF. Bicarbonate-dependent chloride absorption in small intestine: Ion fluxes and intracellular chloride activities. J Membrane Biol 1980; 53: 95–107.
2. Field M, Karnaky KJ Jr, Smith PL, Bolton JE, Kinter WB. Ion transport across the isolated mucosa of the winter flounder, Pseudopleuronectes americanus: I. Functional and structural properties of cellular and paracellular pathways for Na and Cl. J Membrane Biol 1978; 41:265–293.
3. Huang, KC, Chen TST. Ion transport across intestinal mucosa of winter flounder Pseudopleuronectes americanus. Am J Physiol 1971; 220:1734–1738.
4. Imon MA, White JF. Intestinal HCO_3^- secretion measured by pH stat in vitro: relationship to metabolism and transport of Na^+ and $C1^-$. J Physiol (London) 1981; 314:429–443.

5. White JF, Ellingsen D, Burnup K. Electrogenic Cl^- absorption by Amphiuma small intestine: dependence on serosal Na from tracer and Cl microelectrode studies. J Membrane Biol 1984; 78: 223–233.
6. Gunter-Smith PJ, White JF. Contribution of villus and intervillus epithelium to transepithelial potential difference and response to theophylline and sugar. Biochim Biophys Acta 1979; 557: 425–435.
7. White JF. Characteristics of chloride ion influx in Amphiuma small intestine. Am J Physiol 1989; 256 (Gastrointest Liver Physiol. 19):G166–G177, 1989.
8. Palfrey HC, Feit PN, Greengard P. cAMP-stimulated cation transport in avian erythrocytes: inhibition by loop diuretics. Am J Physiol 1980; 238 (Cell Physiol 7): C139–148.
9. Haas M, Forbush B III. The Na-K-Cl cotransporters. J. Bioenerg Biomembr 1998; 30:161–172.
10. Gamba, G. Electroneutral chloride-coupled co-transporters. Curr Opin Nephrol Hypertens 2000; 9:535–540.
11. Musch MW, Orellana SA, Kimberg LS, Field M, Halm DR, Krazny EJ, Frizzell RA. Na^+-K^+-Cl^- cotransport in the intestine of a marine teleost. Nature (London) 1982; 300:351–353
12. White JF, Imon MA. Intestinal HCO_3^- secretion: stimulation by mucosal Cl^- and serosal Na^+. J Membrane Biol 1982; 68:207–214.
13. White JF, Ellingsen D. Basolateral impalement of intestinal villus cells. Electrophysiology of chloride transport. Am J Physiol 1989; 256 (Cell Physiol 25):C1022–C1032.
14. White JF. Modes of Cl transport across mucosal and serosal membranes of urodele intestinal cells. J Membrane Biol 1986; 92: 75–89.
15. White JF. pH-sensitive K^+ conductance and HCO_3^- conductance in intestinal epithelial cells. FASEB J 1989; 3:A860.
16. Sheerin HE, Field M. Ileal HCO_3^- secretion: relationship to Na^+ and Cl^- transport and effect of theophylline. Am J Physiol 1975; 228:1065–1074.
17. White JF. Evidence against luminal 1 for 1 $C1^-$-HCO_3^- exchange. Am J Physiol 1986; 251:G230–G236.
18. White, JF. Threshold intracellular pH for apical Cl^-/HCO_3^- exchange in enterocytes. FASEB J 1991; 5:A691.
19. Shull GE, Lingrel JB. Molecular cloning of the rat stomach H^+-K^+ ATPase. J Biol Chem 1986; 261:16788–16791.
20. Asano S, Hoshina S, Nakaie Y, Watanabe T, Sato M, Suzuki Y, Takeguchi N. Functional expression of putative H^+,K^+-ATPase from guinea pig distal colon. Am J Physiol 1998; 275: C669–C674.
21. Wingo CS, Smolka AJ. Function and structure of H^+-K^+ATPase in the kidney. Am J Physiol 1995; 269:F1–F16.
22. White JF, Imon MA. HCO_3^- absorption by in vitro Amphibian small intestine. Am J Physiol 1981; 241: G389–396.
23. White JF. Intestinal electrogenic HCO_3^- absorption localized to villus epithelium. Biochim Biophys Acta 1982; 687:343–345.
24. Imon MA, White JF. Association between HCO_3^- absorption and K^+ uptake by Amphiuma jejunum: relations among HCO_3^- absorption, luminal K^+ and intracellular K^+ activity. Am J Physiol 1984; 246:G732–G744.
25. Lucas ML. The association between acidification and electrogenic events in the rat proximal jejunum. J Physiol (London) 1976; 257:645–662.
26. White JF. Omeprazole inhibits H^+ secretion by Amphiuma jejunum. Am J Physiol 1985; 248: G256–G259.
27. White JF, Britanisky R. Adrenergic agents stimulate and cholinergic agents inhibit H^+ secretion by amphibian jejunum. Am J Physiol 1986; 251 (Gastrointest Liver Physiol 14):G405–G412.
28. Hinton CF, White JF. Inhibition of jejunal H^+ secretion in Amphiuma by SCH 28080. FASEB J 1988; 2:A734.

29. White JF. Conductive pathways for HCO_3^- in the basolateral membrane of salamander intestinal cells. Am J Physiol 1989; 257(Cell Physiol 26):C252–C260.
30. Romero MF, Fulton CM, Boron WF. The SCL4 family of HCO_3^- transporters. Pflugers Arch – Eur J Physiol 2004; 447:495–509.
31. Romero MF. Molecular pathophysiology of SLC4 bicarbonate transporters. Curr. Opin Nephrol Hypertens 2005; 14:495–501.
32. White JF, Imon MA. A role for basolateral anion exchange in active jejunal absorption of HCO_3^-. Am J Physiol 1983; 244: G397–G405.
33. Sly WS, Hu PY. Human carbonic anhydrases and carbonic anhydrase deficiencies. Ann Rev Biochem 1995; 64:375–401.
34. Lu J, Daly CM, Parker MD, Gill HS, Piermarini PM, Pelletier MF, Boron WF. Effect of human carbonic anhydrase II on the activity of the human electrogenic Na/HCO_3 cotransporter NBC31-A in Xenopus oocytes. J Biol Chem 2006; 281:19241–19250.
35. Hinton CA, White JF. Alpha$_2$-receptors mediate norepinephrine stimulated acid secretion in Amphiuma jejunum. Am J Physiol 1988; 255 (Gastrointest Liver Physiol 18):G640–G646.
36. Schubert ML. Gastric secretion. Curr Opin Gastroenterol 2005; 21:636–643.
37. Imon MA, White JF. The effect of theophylline on intestinal HCO_3^- transport measured by pH stat in Amphiuma. J Physiol (London) 1981; 321: 343–354.
38. Dubose TD Jr, Gitomer J, Codina J. H^+-K^+ ATPase. Curr Opinion Nephrol Hypertens 1999; 8:597–602.
39. Field M. Intestinal ion transport and the pathophysiology of diarrhea. J Clin Invest 2003; 111:931–943.
40. Lucas K, Schneider W, Halerich F, Blair J. Direct measurement by pH-microelectrode of the pH microclimate in rat proximal jejunum. Proc R Soc London B 1975; 192:39–48.
41. White JF, Hinton CF. Norepinephrine induces Na^+-H^+ and Cl^--HCO_3^- exchange in Amphiuma intestine: locus and response to amiloride. Am J Physiol 1988; 255 (Gastrointest Liver Physiol 18): G18–G26.
42. White JF. Chloride transport and intracellular chloride activity in the presence of theophylline in Amphiuma small intestine. J Physiol (London) 1981; 321:331–341.
43. Yu Q-C, White JF. Link between cell apical morphology and H^+ secretion in salamander small intestine. Am J Physiol 1991, 261(Gastrointest Liver Physiol 24): G50–G56.
44. Forte TM, Machen TE, Forte JG. Ultrastructure and physiological changes in piglet oxyntic cells during histamine stimulation and metabolic inhibition. Gastroenterology 1975; 69:1208–1222.
45. Helander HF, Hirschowitz BI. Quantitative ultrastructural studies on inhibited and on partly stimulated gastric parietal cells. Gastroenterology 1974; 67:447–452.
46. Husted RF, Mueller AL, Kessel RG, Steinmetz PR. Surface characteristics of carbonic anhydrase-rich cells in turtle urinary bladder. Kidney Int 1981; 19:491–502.
47. Stetson DL, Steinmetz PR. Role of membrane fusion in CO_2 stimulation of proton secretion by turtle bladder. Am J Physiol 1983; 245(Cell Physiol):C113–C120.
48. McLaughlin ML, Ucci AA, Perrone RD. Aldosterone induces luminal membrane vesicles and H^+ secretion in rat distal colon. Clin Res 1985; 33:324A.
49. Lucas ML. Amendments to the theory underlying Ussing chamber data of chloride secretion after bacterial enterotoxin exposure. J Theoretical Biol 2005; 234:21–37.
50. Gunter PJ, White JF. Evidence for electrogenic bicarbonate transport in Amphiuma small intestine. Biochim Biophys Acta 1978; 507:549–551.
51. Gunter-Smith PJ, White JF. Response of Amphiuma small intestine to theophylline: effect on bicarbonate transport. Am J Physiol 1979; 236:E775–783.
52. Pratha VS, Hogan DL, Martenson BA, Bernard J, Zhou R, Isenberg JI. Identification of transport abnormalities in duodenal mucosa and duodenal enterocytes from patients with cystic fibrosis. Gastroenterology 2000; 118:1051–1060.
53. Thiagarajah JR, Verkman AS. CFTR pharmacology and its role in intestinal fluid secretion. Current Opinion in Pharmacology 2003; 3:594–599.

54. Feranchak AP. Ion channels in digestive health and disease. J Ped Gastroenterol Nutr 2003; 37(3):230–241.
55. White JF. Chloride channels in epithelial cells of intestine. In: Gerencser G, ed. Symposium: Electrogenic Cl^- Transporters in Biological Membranes. Berlin Heidelberg: Springer-Verlag. Adv Comp Env Physiol 1994:221–237.
56. Scheving LA, Jin WH, Chong KM, Gardner W, Cope FO. Dying enterocytes downregulate signaling pathways converging on Ras: rescue by protease inhibition. Am J Physiol 1998; 274: C1363–372.
57. White JF. Primary cultures of salamander intestinal epithelial cells from mucosal nests: whole cell Na^+ currents induced by valine. Am J Physiol 1994;267 (Gastrointest. Liver Physiol. 30): G59–G66.
58. White JF, Ellingsen D, Mayer S. Na microelectrode study of the pathways of Na entry into Amphiuma intestinal absorptive cells. Am J Physiol 1987; 252 (Cell Physiol 21):C505–514.
59. Marcial MA, Carlson SL, Madara JL. Partitioning of paracellular conductance along the ileal crypt-villus axis: a hypothesis based on structural analysis with detailed consideration of tight junction structure-function relationships. J Membr Biol 1984; 80:59–70.
60. White JF, Burnup K, Ellingsen D. Effect of sugars and amino acids on amphibian intestinal Cl^- transport and intracellular Na^+, K^+, and Cl^- activity. Am J Physiol 1986; 250:G109–G117.
61. Hallows KR. Emerging role of AMP-activated protein kinase in coupling membrane transport to cellular metabolism. Curr Opin Nephrol Hypertens 2005; 14:464–471.

The H^+- and H^+, K^+-ATPases of the Collecting Duct

Brian D. Cain, Michelle L. Gumz, Deborah L. Zies, and Amanda K. Welch

Abstract The kidney functions to maintain ion balance via the activity of various channels, exchangers and adenosine triphosphate (ATP)-driven pumps. In the collecting duct, a H^+-ATPase and two distinct H^+,K^+-ATPases have been localized on the apical membrane in the acid-secreting type A intercalating cells. All three of these pumps contribute to acid–base balance, and the H^+,K^+-ATPases may also participate in potassium conservation. The expression levels and activities of the ATP-driven H^+ transporters may vary depending on physiological conditions in the organism. The goal of this chapter is to consider the molecular properties of the plasma membrane H^+-ATPase and the H^+,K^+-ATPases of the collecting duct. The H^+-ATPase is a member of the V-ATPase family. These enzymes are very large multi-subunit enzyme complexes that appear to have a common overall structure and a rotary mechanism similar to the mitochondrial F_1F_0-ATP synthases. In contrast, the H^+,K^+-ATPases are P-Type ATPases and specifically members of the P-type IIC family of cation translocating enzymes. The P-type IIC ATPases consist of an α and β subunit pair with the α subunit housing both the catalytic site and the ion channels through the membrane.

Keywords ATPase · V-ATPase · F-ATPase · H-ATPase · H,K-ATPase · Ion pumps

1 Introduction

Regulation of both electrolyte and acid–base balance are two of the most important functions of the kidney. In healthy individuals, the bloodstream is maintained within a narrow range around pH 7.4 with potassium (K^+) levels between 3.5 and 5.5 mEq. In order to achieve homeostasis, the kidney employs a variety of channels,

B.D. Cain (✉)
Department of Biochemistry and Molecular Biology, University of Florida, 32610-0245, Gainesville, FL, USA
e-mail: bcain@ufl.edu

G.A. Gerencser (ed.), *Epithelial Transport Physiology*,
DOI 10.1007/978-1-60327-229-2_10,

ion exchangers and adenosine triphosphate (ATP)-driven proton (H^+) pumps. H^+ translocating ATPases fall into three general classes. These are the mitochondrial F-type F_1F_0 ATP synthase, the V-type H^+-ATPase, and the P-type H^+,K^+-ATPase. In the collecting duct of the kidney, a V-type H^+-ATPase and two distinct P-type H^+,K^+-ATPases have been localized on the apical membrane in the acid-secreting type A intercalating cells. All three of these pumps contribute to acid–base balance, and the H^+,K^+-ATPases also play an important role in K^+ conservation (Fig. 1). Tsuruoka and Schwatrz [1] suggested that at least 65% of acid secretion in the outer medullary collecting duct was the product of H^+-ATPase activity and the rest was attributable to the H^+,K^+-ATPases, while others estimate the contribution of the H^+,K^+-ATPases to be somewhat higher. The relative levels of activity should not be considered fixed in absolute terms, in part, because the estimates were made using differing approaches and assumptions. More importantly, the expression levels and activities of the ATP-driven H^+ transporters are probably highly dynamic depending on physiological conditions in the organism, such as electrolyte balance and hormonal status [2]. For example, when K^+ is restricted, H^+,K^+-ATPase activity is stimulated to promote recovery of K^+ from the lumen. The primary goal of this chapter is to consider the molecular properties of the plasma membrane H^+-ATPase and the H^+,K^+-ATPases in the collecting duct.

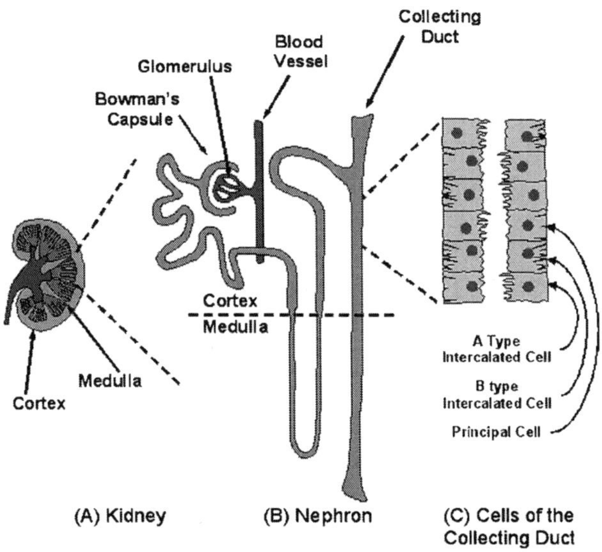

Fig. 1 Location of H^+- and H^+,K^+-ATPases in intercalated cells. **A**, cross section of the mammalian kidney. **B**, a single nephron. **C**, cells of the collecting duct

Mammalian H^+-ATPases are members of the V-type family of ATP-driven H^+ pumps [3–5]. The V-family name was coined in reference to the original discovery of this family of H^+ pumps in the yeast vacuole. The V-type ATPases are an ancient enzyme family, and the eukaryotic enzymes are evolutionarily related to archaea A-ATPases from the microbial world. The V-ATPases are very large multiple subunit enzyme complexes that share a common overall structure and a rotary mechanism with the F-ATPases. The V-type H^+-ATPases are best known for acidification of eukaryotic cell intracellular membrane compartments, such as the Golgi, endosomes, and lysosomes. In mammals, H^+-ATPases are also found in the plasma membranes of a few specialized epithelia [6]. The best examples are osteoclasts, the male reproductive tract, the inner ear, and the intercalated cells of the collecting duct.

H^+,K^+-ATPases are members of the P-type family of membrane bound, cation translocating enzymes [7]. The P-family name refers to direct phosphorylation of the large α subunit common to all family members during the catalytic cycle. The apical membrane H^+,K^+-ATPases of collecting duct intercalated cells use the energy of ATP hydrolysis to drive absorption of K^+ ions from the lumen of the nephron in exchange for H^+ against the concentration gradient. The H^+,K^+ ATPases are grouped with the Na^+,K^+-ATPases into the P-type IIC ATPase sub-class [7], or more simply the X^+,K^+-ATPases. The distinguishing characteristic of the P-type IIC ATPases is an α and β subunit pair forming the functional unit.

2 P-Type H^+,K^+-ATPases

2.1 Detection of H^+,K^+-ATPase

The earliest evidence for H^+,K^+-ATPase activity in the collecting duct came from microperfusion studies of outer medullary collecting ducts from rabbits fed low K^+ diets [8]. H^+ secretion and K^+ flux were determined and shown to be sensitive to the H^+,K^+-ATPase specific inhibitor omeprazole. Over the past 12 years, numerous physiology, molecular biology, and biochemistry studies have established that collecting duct H^+,K^+-ATPase activity is the sum of at least two distinctive H^+,K^+-ATPases. The two pumps were first identified in gastrointestinal tissues and are distinguished by their catalytic subunits, the "gastric" HKα1 and "colonic" HKα2 [9–10]. These proteins are the products of separate genes and the pumps containing them have clearly different biochemical properties and regulatory mechanisms. For example, sensitivity to either Sch28080 or relatively high concentrations of ouabain have been widely used by physiologists to distinguish HKα1 H^+,K^+-ATPase from HKα2 H^+,K^+-ATPase activity, respectively. Our recent review provides a more thorough treatment of the pharmacological inhibitors for the H^+,K^+-ATPases [11].

Although expression levels were much lower in renal tissues than in the gastrointestinal tract, mRNAs encoding both the "gastric" HKα1 and "colonic" HKα2 were readily detectable in the kidney. HKα1 subunit mRNA was found in the connecting tubule and in intercalated cells of the rat cortical and medullary collecting ducts by polymerase chain reaction (PCR) and in situ hybridization [12–13]. The HKβ subunit mRNA was also observed by in situ hybridization [14], and the HKβ promoter directed expression of a β-galactosidase reporter gene in the collecting duct of transgenic mice [15]. In situ hybridization and PCR were also used to detect HKα2 mRNA in the rat cortical collecting duct [16–17]. Moreover, the original cDNAs for the HKα2 subunits from human [18] and rabbit [19] were cloned using kidney-derived mRNA for the templates.

Immunohistochemistry has been used to localize H^+,K^+-ATPase proteins in collecting ducts and other segments [20]. Antibodies raised against the human HKα1 (ATP4a) and HKα2 (Atp12a also referred to as ATP1AL1) proteins were employed to probe human cortical and medullary collecting duct tissue. Strong staining was seen in the apical membranes of intercalated cells in both cortical and medullary collecting ducts using the anti-HKα1 antibody. Light staining was observed in the principal cells. The anti-HKα2 antibody yielded light staining of the intercalated cells with occasional labeling of the principal cells. Verlander et al. [21] achieved generally similar results in the rabbit using an antibody raised against the HKα2c (see below) subunit specific antibody. Intense staining was observed for the apical membranes of both A and B type intercalated cells of the collecting duct. A weaker signal was also seen on apical membranes of the principal cells. These two reports suggested that H^+,K^+-ATPase activity was not necessarily restricted to the acid-secreting type A intercalated cell, but also present in low abundance in all cell types in the collecting duct.

2.2 Subunits of the H^+,K^+-ATPases

Mammalian H^+,K^+-ATPases are composed of two subunits, α and β in a 1:1 stoichiometry (Fig. 2). The active form of the enzymes appear to be heterotetramers of two $\dot{\alpha}$ and two $\dot{\beta}$ subunits [22, 23].

In the mammalian kidneys, three HKα subunits have been identified: the "gastric" HKα1, the "colonic" HKα2a, and in some species a variant of the colonic subunit called HKα2c. The HKα2 subunits share 64% primary sequence identity with HKα1 [24]. They are all approximately 115 kDa and span the membrane with ten transmembrane α-helices (M1–M10). The membrane domain houses the ion transport channels. Approximately one-third of the mass of the α subunit is in a large cytoplasmic loop between M4 and M5 that contains both the nucleotide binding domain and the phosphorylation site. The HKα2c subunit is essentially identical to the well-characterized "colonic" HKα2a but with an extended N-terminal domain [25]. In newborn rats, HKα2 mRNA and protein are expressed in abundance [26]. Western blot analysis indicated that the protein product was slightly larger than HKα2a subunit and recognized by an HKα2c-specific antibody. The rabbit HKα2c variant is 61 amino acids longer than the HKα2a subunit. In rabbits, the HKα2c subunit mRNA results from the use of a downstream transcription start site for the HKα2 gene promoter and an alternative splicing event [27]. Although several potential phosphorylation sites exist in the N-terminal extension, it is not known whether the HKα2c H^+,K^+-ATPase has activity or regulation distinct from the HKα2a enzyme. There are probably as yet unknown factors involved in assembly, intracellular trafficking,

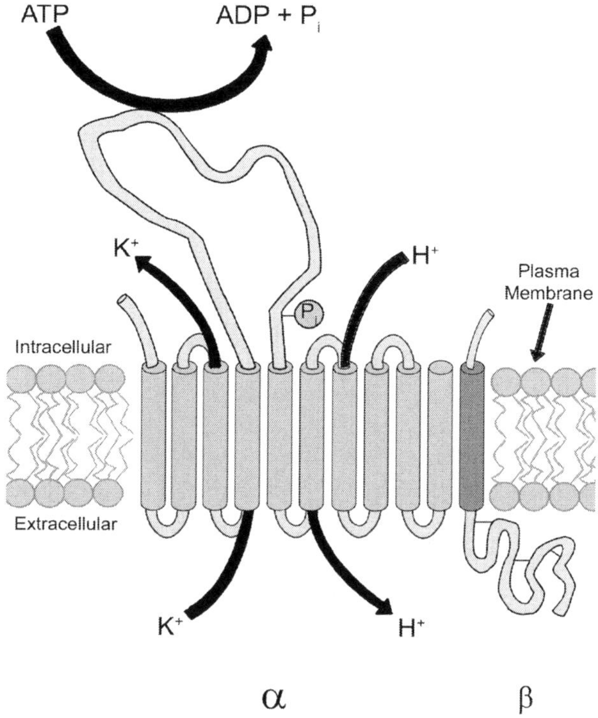

Fig. 2 Subunit structure of H^+,K^+-ATPase. The plasma membrane topology of the HKα and HKβ subunits is shown. The HKβ subunit likely interacts with HKα transmembrane helices M7, M9, and M10 [47]. The HKα subunit houses the catalytic site and the ion translocation mechanism

activation, and eventual removal of H^+,K^+-ATPases from the plasma membrane. One of the recently identified factors is protein kinase A-mediated phosphorylation of a conserved serine in the C-terminal segment of HKα2 [28]. This post-translational modification is involved with maturation of the subunit in the assembly of the enzyme complex. At the other end of the functional life of an HKα2 H^+,K^+-ATPase, the pump interacts with tetraspanin protein CD63 that participates in recycling the pump from the plasma membrane [29].

The β subunits of H^+,K^+-ATPases are catalytically inert intrinsic membrane proteins. The deduced proteinaceous components of β subunits are approximately 30 kDa in size and span the membrane only once. Each has an extensive extracellular domain with three disulfide bridges and varying numbers of glycosylation sites. Glycosylation and the disulfide bridges are important for the β subunit to participate in assembly of an active enzyme complex [30–31]. The β subunit is responsible for the structural and functional maturation of the α subunit and trafficking of the enzyme from the endoplasmic reticulum to the plasma membrane [32]. It is clear that the HKβ subunit associates with the HKα1 subunit to form the gastric H^+,K^+-ATPase in the stomach [23]. There is every reason to believe that the same pairing exists in the collecting duct. However, the evidence is not so clear for the HKα2 H^+,K^+-ATPases.

There are four candidate β subunits that may pair with HKα2 subunits. These are the gastric H^+,K^+-ATPase HKβ subunit and the Na^+,K^+-ATPase subunits NKβ1,

NKβ2, and NKβ3. Examination of the human genome data base reveals no other apparent X^+,K^+-ATPase β-like genes. All four candidate β subunit proteins were co-immunoprecipitated with the HKα2c subunit when expressed in COS-1 cells [33]. Published works from several independent research groups employing a variety of different expression systems indicated that HKα2 was able to form functional, stable complexes with HKβ, NKβ1, and NaKβ3 [34–36]. The best case has been made for a HKα2/NKβ1 H^+,K^+-ATPase. In a convincing study, Pestov et al. [37] used several approaches to demonstrate NKβ1 protein at the apical membrane of the coagulating gland of the rat anterior prostate. The Na^+,K^+-ATPase was only present at the basolateral membrane, so association with HKα2 best explains the presence of NKβ1 at the apical membrane. Immunohistochemistry provided direct evidence for apical colocalization of the HKα2 and NKβ1 subunits. Li et al. [38] demonstrated that HKα2 and NKβ1 both localized to the apical plasma membrane of epithelial cells in rat distal colon. A series of experiments were performed in HEK-293 cells in order to assess plasma membrane delivery. A green fluorescent protein-HKα2 fusion protein was paired with either NKβ1 or NKβ3 to follow delivery of the pump to the plasma membrane. H^+,K^+-ATPase activity was also measured by Rb^{86} uptake. The results clearly showed that NKβ1 was the more efficient partner for HKα2.

2.3 Mechanism of ATP-Driven Ion Translocation

H^+,K^+-ATPases are nonelectrogenic, transporting H^+ and K^+ in opposite directions in a 2:2 ratio. There is also sufficient evidence to suggest that the HKα2 H^+,K^+-ATPase can act as a Na^+,K^+-ATPase [39–41]. During the catalytic cycle (Fig. 3), a H^+,K^+-ATPase in the E1-H^+ conformation binds Mg^{+2}-ATP in the nucleotide binding pocket [7]. Transfer of a phosphate from ATP to a conserved aspartic acid generates a conformational change from the E1 to the E2 state. Conversion to the E2 state blocks intracellular access to the ion binding sites, and then opens extracellular access allowing release of H^+ into the collecting duct. Following release of H^+, extracellular K^+ binds. Hydrolysis releases inorganic phosphate from the phosphorylation site. The E2-K^+ bound form of the ATPase undergoes the conformational shift back to the E1 state as K^+ is released inside the cell. Release of K^+ into the cytoplasm allows binding of intracellular H^+ completing the catalytic cycle.

Probably the most important steps toward understanding structure–function relationships in the P-type ATPases during the current decade were contributed by the Toyoshima laboratory. Both a 2.6 Å resolution crystal structure of the sarcoplasmic reticulum Ca^{+2}-ATPase in the E1 conformation and a 3.1 Å crystal structure of the Ca^{+2}-ATPase in the E2 state were determined [42–43]. The structures proved conclusively that there were ten transmembrane helices and four of them contribute to coordination sites for the cations during translocation (M4–6, M8). Three structural domains were defined within the cytoplasmic region of the enzyme designated P (phosphorylation), N (nucleotide), and A (actuator). The N domain is located in the large cytoplasmic loop between M4 and M5 and it contains the nucleotide-binding pocket. The P domain houses the phosphorylated aspartic acid and consists of two segments of the large loop separated by the N domain. The A

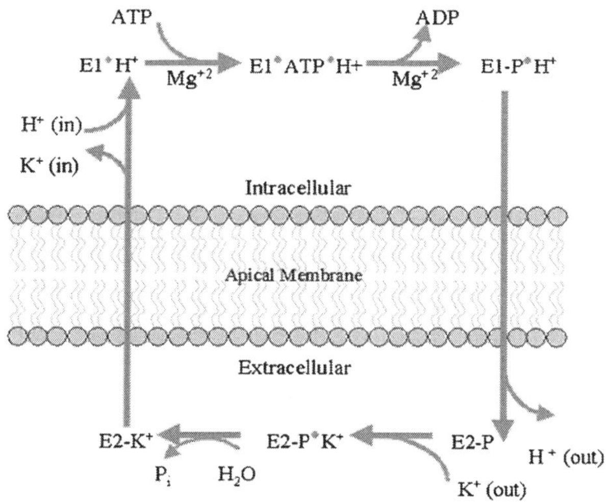

Fig. 3 Mechanism of ion translocation by H^+,K^+-ATPase. H^+,K^+-ATPase cycles through two major conformational states, referred to as E1 and E2, during ATP-driven ion translocation. High-resolution structures of the Ca^{+2}-ATPase in both the E1 and the E2 conformations have been determined [42–43]

domain is composed of the cytosolic N-terminus of the enzyme and the loop located between M2 and M3.

The differences between the E1 and E2 structures are striking and lend considerable insight into the mechanism. Dramatic rearrangements of the cytoplasmic and transmembrane domains of Ca^{+2}-ATPase take place as the enzyme transitions from the E1 to the E2 state. The E2 conformation is far more compact. Of the four transmembrane helices that coordinate Ca^{+2} ions only M8 does not move during catalysis. M4 moves downward, M5 bends toward M4 and an unwound portion of the M6 helix rotates by nearly 90°. M4 and M5 move in concert with motions in the cytoplasmic P domain. The domain movements result in N and P domain shifts of 50 and 23 Å, respectively. In the E2 structure, the A domain appears to stabilize the conformation by protein–protein interactions with both N and P domains. The mechanistic importance of these radical conformational changes becomes clear when ion binding and phosphorylation are considered together. For example, in the cytoplasmic region the enzyme can bind ATP in the absence of Ca^{+2} [44]. However, futile ATP hydrolysis is prevented because Ca^{+2} binding is required prior to the phosphorylation step. As the enzyme transitions from the E1-Ca^{+2} state to the E2 state the ion binding sites disappear due to the rearrangement of the membrane transmembrane helices. One of the more unusual features of the Ca^{+2}-ATPase structures is that neither the E1 nor E2 conformation revealed a water-filled channel needed for ion movement from the coordination sites to the membrane surfaces [45]. Very recently, Pedersen et al. [46] reported the structure of a P-type III plasma membrane H^+-ATPase from *Arabidopsis thaliana*. This 3.5 Å structure has the expected water-filled cavity leading to the extracellular surface of the membrane.

Morth et al. [47] published a 3.5-Å resolution structure for the pig renal Na^+,K^+-ATPase. The obvious major difference between the Na^+,K^+-ATPase structure and the higher resolution Ca^{+2}-ATPase structures is the presence of a NKβ1 subunit.

The NKβ1 subunit is associated with M7 and M10 of the NKα subunit in the intact enzyme complex.

To date, no high-resolution structure of either an HKα1 or an HKα2 H^+,K^+-ATPase has been reported. However, the availability of the Ca^{+2}-ATPase structures has provided the means to consider H^+,K^+-ATPase structure through homology modeling. We developed the first HKα2 H^+,K^+-ATPase model using the coordinates of the Ca^{+2}-ATPase in the E1 conformation as template [48]. The HKα2 H^+,K^+-ATPase model retained the overall shape of the Ca^{+2} ATPase. All three cytoplasmic structural domains and the ten transmembrane spans were evident. Importantly, the distance measured between the phosphorylation site aspartic acid (D351) and a conserved lysine (K515) in the nucleotide-binding site was very similar to the distance measured in the Ca^{+2}-ATPase E1 structure [42]. This shared feature suggests that an H^+,K^+-ATPase probably undergoes the same dramatic domain rearrangements observed in the Ca^{+2}-ATPase as the enzyme undergoes the conformational change from the E1 to the E2 state. HKα1 H^+,K^+-ATPase models in both the E1 and E2 conformations were used together with biochemical experiments to evaluate the docking of K^+-competitive inhibitors in the pump [49] and the role of a salt bridge required for K^+ binding [50].

The most comprehensive modeling studies on the HKα1 subunit came from the Sachs laboratory and were based on the E1-Ca^{+2}, the E2-thapsigargin structure and the E2-magnesium fluoride structures [51–52]. Munson et al. used a sophisticated combination of energy minimization and molecular dynamics to generate homology models of the HKα1 H^+,K^+-ATPase, and then correlated the models with the large body of biochemical data on the stomach acid pump. In the first paper, Munson et al. [51] modeled three hydronium ions (H_3O^+) into the E1 HKα1 model and compared it to the E2-K^+ model. The modeling suggested that one of the H_3O^+interacts with two acidic residues in M6 (D824, E820) and another in M5 (E795) in the E1 conformation. The H_3O^+ion is displaced by insertion of a lysine (M5 K791) into this region in the E2 model. Movement of K791 allows K^+occupation of the former H_3O^+ site, dephosphorylation of the pump, and release of K^+ to the cytosol. This mechanism was summarized in a review on the HKα1 H^+,K^+-ATPase as a drug target [53]. Specific binding sites for the covalent proton pump inhibitors (i.e. pantoprazole) and for the K^+ competitive inhibitors (i.e. SCH28080) were also suggested. In the second model, Munson et al. [52] refined the binding properties of the inhibitors to consider the ion translocation pathway. A hydrated vestibule formed by ion-binding helices M4-6 and M8 allowed for entry of the K^+-competitive inhibitors. Docking of the inhibitor would prevent access of K^+ to the ion-binding site. The K^+ entry portal was predicted to occur near the extracellular loop between M5 and M6, and the K^+ exit pathway was modeled between the intracellular surfaces of M1 and M2.

2.4 Patients and Knockout Mice

The physiological roles of the two H^+,K^+-ATPase subtypes in the kidney are apparently quite different. HKα1 H^+,K^+-ATPase appears to be the subtype that contributes to normal homeostasis while the HKα2 enzyme responds to changes in physiological conditions, such as K^+ depletion. HKα1 was readily detectable in the collecting duct of animals fed a normal diet, but HKα2 was present at only very low levels [11, 26]. However, expression of HKα2 increased in response to low K^+, low Na^+, and changes in acid–base balance. No human kidney disorder can be directly attributed to a specific defect in H^+,K^+-ATPase. However, a case study of a novel form of distal renal tubular acidosis was suggestive of a possible deficiency in HKα2 H^+,K^+-ATPase [54]. This single patient, a two-year-old child, suffered from severe hypokalemia, metabolic acidosis, and hypomagnesaemia.

In the absence of a literature on human pathology, a series of knockout mice have been constructed to investigate the functional roles of the HKα1 H^+,K^+-ATPase and HKα2 H^+,K^+-ATPase. In HKα1 knockout mice the Type I Sch-28080-sensitive H^+,K^+-ATPase activity is lost from the kidney [55]. Predictably the major effects of the HKα1 knockout mice appeared in the stomach [56]. The animals survived but they had achlorhydria, hypergastrinemia, and metaplasia of the gastric epithelium. No overt renal phenotype was observed in the HKα1 null mice on a normal diet. However, a subsequent publication reported a possible compensating H^+,K^+-ATPase activity [57]. Measurement of K^+-dependent proton secretion in the cortical collecting duct of HKα1 knockout animals revealed an apparently SCH28080- and ouabain-insensitive acid–base transporter. Perusal of the human genome for a third Type IIC P-type ATPase α subunit gene did not reveal a likely candidate gene. Therefore, despite apparent ouabain-insensitivity of the compensating activity, it seems most reasonable that HKα2 H^+,K^+-ATPase is involved. One might imagine differential properties for the pump based on alternative β subunit selection or post-translational modification.

In HKα2 knockout animals the Type III ouabain-sensitive H^+,K^+-ATPase activity is missing [55]. K^+ retention has been studied in HKα2 null mice under control and low dietary K^+ conditions [58]. Surprisingly, no significant differences in K^+balance were observed between wild type and the HKα2-deficient mice fed a normal diet. However, when fed a K^+-free diet for 18 days, the HKα2 knockout mice lost twice as much body weight and had lower plasma and muscle K^+ levels than the wild-type mice. Although the K^+ wasting was primarily in the form of fecal K^+, these studies demonstrated that the HKα2 H^+,K^+-ATPase has a pivotal role in maintaining K^+ balance. Interestingly, the kidney of the knockout mouse was still capable of reducing K^+ loss by 100-fold suggesting the presence of a compensating mechanism. It should be noted that HKα1 H^+,K^+-ATPase is expressed in the kidney but not in the colon where the K^+ wasting phenotype was apparent. Pestov et al. [59] investigated the acidification of anterior prostate fluid in the HKα2 null mice. The pH of prostate fluid in the HKα2 knockout animals was 6.96 compared to 6.38 in wild-type mice. This loss of acidification strongly suggested that the HKα2 H^+,K^+-ATPase functions as a proton pump for acidification of luminal prostate fluids. Like the HKα1 H^+,K^+-ATPase animals, the HKα2 H^+,K^+-ATPase mice lacked a readily apparent renal phenotype.

Lynch et al. [60] addressed this apparent conundrum by studying HKα1, HKα2, and double HKα1/HKα2 knockout mice. Rates of pH recovery and acid extrusion in response to an acute acid load were determined for the knockout animals and normal control mice. Acid extrusion was measurably slower in both type A and type B intercalated cells in the collecting ducts of the single knockout mice than in normal controls. As expected, a much more dramatic result was obtained with the HKα1/HKα2 double-knockout animals. Clearly both HKα1 and HKα2 H^+,K^+-ATPases contributed individually to normal acid–base balance. Moreover, the results suggest that the presence of one pump can partially compensate for the loss of the other in the collecting duct, and this may account for the absence of a renal phenotype in the HKα1 and HKα2 knockout mice.

3 V-Type H^+-ATPase

3.1 Detection of H^+-ATPase

Although V-type H^+-ATPases pump H^+at the expense of ATP hydrolysis, the only common feature between an H^+-ATPase and an H^+,K^+-ATPase is a nucleotide-binding site. The two pumps share no other similarities in terms of structure or mechanism (Fig. 4). Indeed the H^+-ATPases bear an unmistakable resemblance to the F-type F_1F_0-ATP synthases [5]. By convention, subunits in the V_1sector are designated in capital letters (i.e. subunit A), and V_0 subunits are written in the lower case (i.e. subunit a). Mammalian H^+-ATPases contain at least 13 different subunits including V_1 A-H subunits and V_0 a–e plus c″ subunits [3, 6]. Mammals have multiple genes encoding several isoforms for a number of these subunits. For each subunit, there is a ubiquitously expressed gene encoding a protein found in all cells. In addition, there are two genes in mammals for subunits B (i.e. B1 and B2), C, E, d, and e, three G subunit genes, and four genes encoding the a subunit [6, 61].

V-type H^+-ATPases are found in all mammalian cells where they primarily function to acidify intracellular compartments. In addition to their internal H^+-ATPases, specialized epithelial cells, including the intercalated cells of the collecting duct, have plasma membrane H^+ pumps. The presence of an H^+-ATPase in the renal epithelium has been recognized for nearly two decades [62], and a substantial body of literature has accumulated describing tissue-specific expression and cellular localization of H^+-ATPase subunits. It should be noted that the intercalated cells of the collecting duct possess the highest levels of H^+-ATPase expression found in the kidney [63].

The existence of multiple sites of H^+-ATPase action in intercalated cells limits the usefulness of mRNA detection. Methods used to detect mRNAs such as Northern blots, in situ hybridization and PCR provide information on what H^+-ATPase proteins are present, but not where in the cell the subunits function. For example, Smith et al. [64] and Sun-Wada et al. [65] identified multiple isoforms of subunits C, d, and G in human and mouse, respectively. PCR experiments [64] and Northern blot analysis [65] established that subunit G3 was expressed exclusively in kidney,

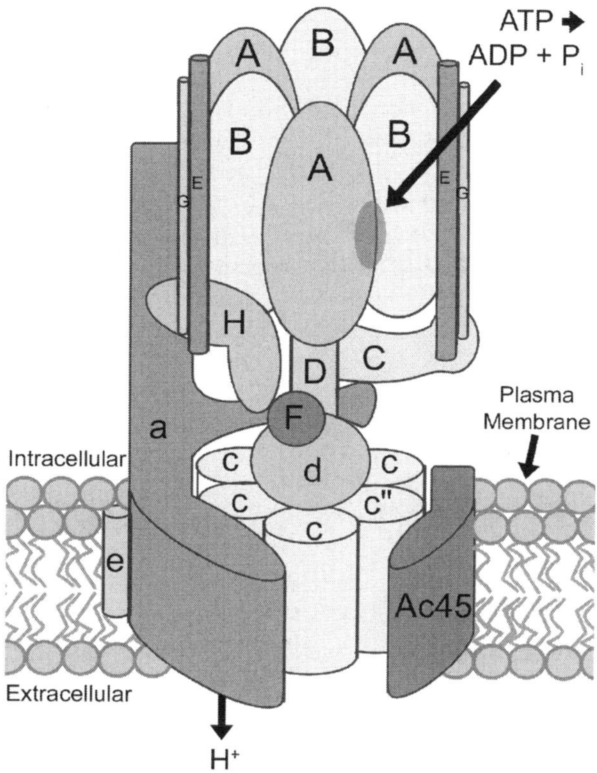

Fig. 4 Subunit structure of H^+-ATPase. The subunit organization of the plasma membrane V_1V_0-ATPase is shown. The Ac45 protein has been characterized in osteoclast H^+-ATPase, but it has not been shown to be present in the kidney. The V_1V_0rotor is composed of subunits DFdc_5c''. The catalytic site is at the interface between A and B subunits, and the a and c_5c'' directly participate in H^+ translocation

while d2 and C2 were expressed in kidney and in a few other tissues. Moreover, C2 transcripts undergo tissue-specific alternative splicing that yields lung- and kidney-specific isoforms called C2-a and C2-b, respectively [66]. In contrast, mRNAs for the ubiquitously expressed isoforms C1, d1, and G1 were found in all tissue samples tested. H^+-ATPase proteins known to be expressed in a tissue-specific manner within the kidney are the B1, C2, G3, a2, a4,d2 and e2 subunits [6, 61].

As might be expected, the tissue-specific expression of these subunits results from transcriptional regulation. This was demonstrated for subunit B1 by construction of a transgenic mouse with an enhanced green fluorescent protein (EGFP) reporter gene under control of the H^+-ATPase B1 gene promoter [67]. Reporter gene expression was found in kidney, lung, and epididymis of transgenic animals mirroring localization studies of the B1 subunit. Reporter gene expression in the collecting duct was restricted only to type A and type B intercalated cells and was not observed in principal cells.

Immunohistochemistry has been the method of choice for defining the specific subunits in the plasma membrane H^+-ATPases [6, 63]. For intercalated cells, the most extensively studied localization has been performed on the subunit B isoforms. Subunit B2 is a constituent of the internal membrane H^+-ATPases, while the B1 subunit is the principal plasma membrane isoform. Interestingly, subunit B1 has been observed at the apical membrane of type A cells and the basolateral membranes of type B cells suggesting complex cell type-specific trafficking mechanisms. Intracellular localization of H^+-ATPases is governed by interactions with the cytoskeleton [6, 68]. A major regulatory action for plasma membrane H^+-ATPase involves regulated cycling pumps between the plasma membrane and sub-apical membrane vesicles [69]. Staining for subunit B1 in type A intercalated cells appeared relatively indistinct under normal conditions. However, vivid apical membrane B1 staining was observed in response to metabolic acidosis produced by chronic acetazolamide reflecting mobilization of H^+-ATPase to the plasma membrane. Mobilization was not necessarily restricted to B1 H^+-ATPases, because B2 was also detected in response to the drug [70]. Other subunits specifically associated with plasma membrane H^+-ATPase include C2-b, a4, d2, and probably e2 [61, 66, 71–74].

3.2 Subunits of the H^+-ATPases

A mammalian V-type H^+-ATPase is a massive 900-kDa enzyme complex containing 13 different subunits in a complex stoichiometry. At the present time, the subunit stoichiometry of an H^+-ATPase appears to be $A_3B_3CDE_2G_2H$ in V_1 and $ac_5c''de$ in V_0 (Fig. 4). A search of mammalian genomes suggest that higher organisms lack the c' subunit found in yeast H^+-ATPases (Ge Hong Sun-Wada, personal communication). It is not clear whether expression of a protein called Ac45 recently identified in the osteoclast enzyme has been adequately investigated in the kidney [75]. In general, it appears that the 13 ubiquitously expressed subunits assemble into H^+-ATPases that are destined to function in the intracellular compartments. The tissue-specific subunits replace ubiquitously expressed subunit isoforms in the plasma membrane H^+-ATPases associated with the specialized functions of acid-pumping cells like osteoclasts and intercalated cells. Therefore, it can be inferred that the most common type A intercalated cell apical membrane H^+-ATPase might incorporate the B1, C2b, G3, a4, d2, and e2 subunits. The immunohistochemical evidence provided the experimental support for this idea, and recently, coordinate regulation during murine kidney development has been noted for subunits B1, G3, C2b, and a4 [76–77].

The functional roles of many of the individual H^+-ATPase subunits has been defined primarily in studies in *Saccharomyces cerevisiae*(reviewed in [4–5]). In many cases expression of a mammalian subunit will functionally complement a yeast strain deficient in the homologous subunit. This was most recently demonstrated for a newly cloned human e2 subunit cDNA [61]. Genetic and biochemical approaches have established that the enzyme has three catalytic nucleotide-binding sites located at the interfaces between AB subunit pairs (Fig. 4). Although both subunits provide determinants for ATP binding, most of these residues are contributed by the A subunit. Subunits C and H are involved in association of V_1 with V_0. V_0 H^+ translocation is mediated by subunit a in association with a ring consisting of the c_5c'' subunits. The V_1 and V_0 sectors are linked by a central rotary stalk and two peripheral stalks [4]. The peripheral stalks consist of V_1 subunits E and G in association with H, C and/or the extramembranous domain of V_0 subunit A. A direct G3-a4 protein–protein interaction has been confirmed in human kidney H^+-ATPase [78]. The rotary stalk has V_1 D and F proteins along with V_0 subunit d.

3.3 *Mechanism of ATP-Driven Ion Translocation*

Investigation of the rotary mechanism of H^+-ATPase has closely followed work on F_1F_0-ATP synthase. Researchers interested in both enzyme systems have been hampered by the lack of a high-resolution structure for either the V_0 or F_0 sectors, respectively. Although the atomic details await the structures, a common overall mechanism has emerged for ATP-driven proton pumping. In this respect, it is more convenient to consider the two enzymes in terms of rotor and stator elements rather than V_1 and V_0, or F_1 and F_0. The rotor subunits of eukaryotic F_1F_0 are $\gamma\delta\epsilon c_{10}$, and for V_1V_0 $DFdc_5c''$ (Fig. 4).

The origins of the rotary mechanism date to Paul Boyer's proposal of the binding change mechanism for F_1-ATPase in the late 1970s [79]. The mechanism predicted that all three catalytic sites exist in different conformational states at any moment in time, and that all three cycled in succession through each of the three states as a consequence of catalysis. Physical confirmation of the first prediction was obtained by the Walker laboratory's solution of a 2.8 Å structure for bovine F_1 [80]. The structure clearly showed three distinct catalytic site conformations. Satisfying the second prediction demanded evidence of physical rotation within the enzyme.

The most striking experiment was observation of single molecule rotation reported by Noji et al. [81]. They immobilized isolated bacterial F_1 $\alpha_3\beta_3\gamma$ on a cover slip and attached a fluorescent actin filament to the rotor subunit γ. Unidirectional ATP-driven rotation of the actin filament was observed using a fluorescence microscope demonstrating rotation of the rotor stalk subunit. Technically more sophisticated experiments subdivided each 120° step into an 80° step associated with ATP binding at one catalytic site and a 40° step related to phosphate dissociation from a different site [82–83]. Rotation was reproduced using a prokaryotic V_1-ATPase-like sector from *Thermus thermophilus*[84]. The A_3B_3DF complex was fixed to the cover slip, and synthetic beads attached to either the D or F rotor subunits to follow ATP-dependent movement. The experiment has not been performed in a eukaryotic V_1because subunit H bridges the rotor D subunit and the peripheral stalk in isolated V_1 inhibiting ATP hydrolysis by preventing rotation [85]. Therefore, the next step was to look at intact enzyme complexes. This was accomplished in yeast V_1V_0 by immobilizing the rotor c-subunit ring on the cover slip and attaching an actin filament to the stator peripheral stalk subunit G [86]. In this experimental design, the stator elements would spin around an immobile rotor. As before, ATP-dependent unidirectional rotation was observed. Together these experiments and many others confirmed the major predictions of the binding change mechanism for both F- and V-type ATPases.

Therefore, the prevailing model for ATP-driven H^+ translocation through an intercalated cell apical membrane H^+-ATPase is based on the binding change mechanism. ATP hydrolysis provides the energy for a conformational shift within the catalytic sites in the $V_1A_3B_3$ hexamer. The conformational change yields approximately 40 pN-nm of torque used to drive a 120° rotation of subunit D, and this rotational motion is transmitted through the rotor stalk subunits to the V_0 c_5c'' ring. Rotation of the ring makes a conserved carboxylic acid residue in each c (or c'') subunit available for protonation from the cytoplasmic side of the membrane, probably via a water-filled half-channel in the a subunit. Protonated c subunits cycle around the ring with subsequent ATP hydrolysis events. After a nearly complete circuit, the proton is released into a second a subunit half-channel leading to the lumen of the nephron. An essential arginine located in the membrane domain of the a subunit facilitates protonation and/or deprotonation of the c subunits by altering the pKa of each passing carboxylic acid residue in succession. The peripheral stalks act as the major structural features of the stator by holding the $V_1A_3B_3$ catalytic sites and the V_0 a subunit H^+ channels in place against the movement of the rotor.

3.4 Patients and Knockout Mice

Although an acid- and calcium-sensitive growth phenotype is readily apparent, yeast strains defective in the V-type ATPase genes are viable and can be maintained under laboratory conditions [4]. However, the V-type H^+-ATPases of the intracellular membrane systems have important housekeeping functions in every mammalian cell. The essential nature of the enzyme was demonstrated by an attempt to construct a targeted disruption of the V_0 c subunit gene in mice [87]. The homozygous $c^{-/-}$ knockout was lethal, and the embryos were detectable for only 3.5 days after fertilization.

The genes expressed in a tissue-specific manner were more interesting. Two of the genes associated with human recessive distal renal tubular acidosis (dRTA) encode kidney-expressed subunits of the H^+-ATPase [88–90]. Children displaying high urinary pH and metabolic acidosis were screened and linkage analysis employed to locate candidate genes. The H^+-ATPase genes were identified as the subunit B1 gene (ATP6V1B1) and the a4 gene (ATP6V0A4). In both genes, many of the mutations resulted in major disruptions of the coding sequence by introduction of stop codons or frameshifts. In other cases, missense mutations affected codons for highly conserved amino acids. For the B1-deficient patients dRTA was consanguineous with deafness, and this makes sense because subunit B1 is also expressed in the inner ear [91]. However, $B1^{-/-}$ knockout mice had normal urinary acidification and hearing [92]. Only challenging the animals with an acid load resulted in metabolic acidosis. The reason for the $B1^{-/-}$ knockout phenotype was, in part, attributed to a net dietary alkaline load from normal lab rodent chow. However, there was also a molecular mechanism capable of partially suppressing the disease phenotype. The B2 H^+-ATPase was present in the type A cell medullary collecting duct apical membrane [93]. Patients with subunit a4-deficient dRTA had a different clinical presentation. Although subunit a4 is expressed in the ear, deafness was not initially associated with ATP6V0A4 dRTA patients [89]. However, late onset hearing loss has been noted in some kindreds [90]. Apparently the compensating mechanisms for a4 deficiency in the ear are more efficient than those in the kidney.

4 Summary

The past decade has seen an enormous improvement in our understanding of the physiology, regulation, and biochemistry of the V-type H^+-ATPases and the P-type H^+,K^+-ATPases. Representatives of both classes of enzymes are expressed in the intercalated cells of the collecting duct where they contribute to renal functions, such as acid–base balance and K^+conservation. Despite the superficial similarity in their activities, H^+-ATPases and H^+,K^+-ATPases are completely different in terms of molecular architecture and enzymatic mechanism. The H^+-ATPase is a very large, multisubunit enzyme complex that pumps H^+ using a rotary mechanism. In contrast, the H^+,K^+-ATPase is a relatively simple two subunit pump that carries out ion translocation as the result of a conformational shift within a single subunit.

References

1. Tsuruoka S, Schwartz GJ. Metabolic acidosis stimulates H^+ secretion in the rabbit outer medullary collecting duct (inner stripe) of the kidney. J. Clin. Invest. 1997;99:1420–31.
2. Wingo CS, Cain BD. The renal H-K-ATPase: physiological significance and role in potassium homeostasis. Ann. Rev. Physiol. 1993;55:323–47.
3. Forgac M. Vacuolar ATPases: rotary proton pumps in physiology and pathophysiology. Nature Rev. 2007;8:917–29.
4. Kane PM. The when, where, and how of organelle acidification by the yeast vacuolar H^+-ATPase. Microbiol. Mol. Biol. Rev. 2006;70:177–91.
5. Drory O, Nelson N. The emerging structure of vacuolar ATPases. Physiology 2007;21:317–25.
6. Breton S, Brown D. New insights into the regulation of V-ATPase-dependent proton secretion. Am. J. Physiol. 2007;292:F1–10.
7. Kuhlbrandt, W. Biology, structure and mechanism of P-type ATPases. Nature Rev. 2004;5:282–95.
8. Wingo CS. Active proton secretion and potassium absorption in the rabbit outer medullary collecting duct. Functional evidence for proton-potassium-activated adenosine triphosphatase. J. Clin. Invest. 1989;84:361–5.
9. Sachs G, Shin JM, Vagin O, Lambrecht N, Yakubov I, Munson K. The gastric ATPase as a drug target: past, present, and future. J. Clin. Gastroenterol. 2007;41:S226–42.
10. Codina J., Dubose TD. Molecular regulation and physiology of the H^+,K^+-ATPase in the kidney. Semin. Nephrol. 2006;26:345–51.
11. Zies DL, Gumz ML, Wingo CS, Cain BD. The renal H^+,K^+-ATPases as therapeutic targets. Expert Opin. Therap. Targets. 2007;11:881–90.
12. Ahn KY, Kone BC. Expression and cellular localization of mRNA encoding the "gastric" isoform of H^+-K^+-ATPase alpha subunit in rat kidney. Am. J. Physiol. 1995;268:F99–109.
13. Ahn KY, Turner PB, Madsen KM, Kone BC. Effects of chronic hypokalemia on renal expression of the "gastric" H^+-K^+-ATPase alpha subunit gene. Am. J. Physiol. 1996;270:F557–66.

14. Campbell-Thompson ML, Verlander JW, Curran KA et al. In situ hybridization of the H-K-ATPase β subunit mRNA in rat and rabbit. Am. J. Physiol. 1995;269:F345–54.
15. Callaghan JM, Tan S, Khan MA, et al. Renal expression of the gene encoding the gastric H^+,K^+-ATPase β-subunit. Am. J. Physiol. 1995;268:F363–74.
16. Ahn KY, Park KY, Kim KK, Kone BC. Chronic hypokalemia enhances expression of the H^+,K^+-ATPase alpha 2-subunit gene in renal medulla. Am. J. Physiol. 1996;271:F314–21.
17. Marsy S, Elalouf JM, Doucet A. Quantitative RT-PCR analysis of mRNAs encoding a colonic putative H-K-ATPase alpha subunit along the rat nephron: effect of K^+ depletion. Phlugers Arch. 1996;432:494–500.
18. Grishin AV, Sverdlov VE, Kostina MB, Modyanov NN. Cloning and characterization of the entire cDNA encoded by ATP1AL1 – a member of the human Na,K/H,K-ATPase gene family. FEBS Lett. 1994; 349:144–50.
19. Campbell WG, Weiner ID, Wingo CS, Cain, BD. H-K-ATPase in the RCCT-28A rabbit cortical collecting duct cell line. Am. J. Physiol. 1999;276:F237–45.
20. Kraut JA, Helander KG, Helander HF, Iroezi ND, Marcus EA, Sachs G. Detection and localization of H^+,K^+-ATPase isoforms in human kidney. Am. J. Physiol. 2001;281:F763–8.
21. Verlander JW, Moudy RM, Campbell WG, Cain BD, Wingo CS. Immunohistochemical localization of H-K-ATPase α_{2c}-subunit in rabbit kidney. Am. J. Physiol. 2001;281:F357–65.
22. Lutsenko S, Kaplan JH. Organization of P-type ATPases: significance of structural diversity. Biochemistry 1995;34:15607–13.
23. Shin JM, Grundler G, Senn-Bilfinger J, Simon WA, Sachs G. Functional consequences of the oligomeric form of the membrane-bound gastric H,K-ATPase. Biochemistry 2005;44: 16321–2.
24. Caviston TL, Campbell WG, Wingo CS, Cain BD. Molecular identification of the renal H^+,K^+-ATPases. Semin. Nephrol. 1999;19:431–7.
25. Campbell WG, Weiner ID, Wingo CS, Cain BD. H-K-ATPase in the RCCT-28A rabbit cortical collecting duct cell line. Am. J. Physiol. 1999;276:F237–45.
26. Codina, J, Dubose TD. Molecular regulation and physiology of the H^+,K^+-ATPase in kidney. Semin. Nephrol. 2006;26:345–51.
27. Zies DL, Gumz ML, Wingo CS, Cain BD. Characterization of the rabbit *HKα2* gene promoter. Biochim. Biophys. Acta. 2006;1759:443–50.
28. Codina, J, Liu J, Bleyer AJ, Penn RB, Dubose TD. Phosphorylation of S^{955} at the protein kinase A consensus promotes maturation of the α subunit of the colonic H^+,K^+-ATPase. J. Am. Soc. Nephrol. 2006;17:1833–40.
29. Codina J, Li J, Dubose TD. CD63 interacts with the carboxy terminus of the colonic H^+,K^+-ATPase to increase plasma membrane localization and $^{86}Rb^+$ uptake. Am. J. Physiol. 2005;288:C1279–86.
30. Asano S, Kawada K, Kimura T, Grishin AV, Caplan MJ, Takeguchi N. The roles of carbohydrate chains of the beta-subunit on the functional expression of H^+,K^+-ATPase. J. Biol. Chem. 2000;275:8324–30.
31. Kimura T, Tabuchi Y, Taguchi N, Asano S. Mutational study on the roles of disulfide bonds in the beta subunit of gastric H^+,K^+-ATPase. J. Biol. Chem. 2002;277:20671–7.
32. Geering K. The functional role of beta subunits in oligomeric P-type ATPases. J. Bioenerg. Biomembr. 2001;33:425–38.
33. Gumz ML. The colonic (HKα2) H^+,K^+ ATPase and the effects of aldosterone in the kidney. Doctoral dissertation, University of Florida.
34. Grishin AV, Bevensee MO, Modyanov NN, Rajendran V, Boron WF, Caplan MJ. Functional expression of the cDNA encoded by the human ATP1AL1 gene. Am. J Physiol. 1996;271:F539–51.
35. Asano S, Hoshina S, Nakaie Y et al. Functional expression of the putative H^+,K^+-ATPase. Am. J. Physiol. 1998;275:C669–74.
36. Sangan P, Thevananther S, Sangan S, Rajendran VM, Binder HJ. Colonic H-K-ATPase α- and β-subunits express ouabain-insensitive H-K-ATPase. Am. J. Physiol. 2000;278:C182–9.

37. Pestov NB, Korneenko AN, Zhao et al. Identification of the β-subunit for non-gastric H-K-ATPase in rat anterior prostate. Am. J. Physiol. 2004;286:C1229–37.
38. Li J, Codina J, Petroske E, Werle MJ, Willingham MC, Dubose TD. The effect of beta-subunit assembly on function and localization of the colonic H^+,K^+-ATPase alpha-subunit. Kidney Int. 2004;66:1068–75.
39. Grishin AV, Caplan MJ. ATP1AL1, a member of the non-gastric H,K-ATPase family, functions as a sodium pump. J. Biol. Chem. 1998;273:27772–8.
40. Codina J, Pressley TA, Dubose TD. The colonic H^+,K^+-ATPase functions as a Na^+-dependent $K^+(NH_4^+)$-ATPase in apical membranes from rat distal colon. J. Biol. Chem. 1999;274:19693–8.
41. Rajendran VM, Sangan P, Geibel J, Binder HJ. Ouabain-sensitive H,K-ATPase functions as a Na,K-ATPase in apical membranes of rat distal colon. J. Biol. Chem. 2000;275: 13035–40.
42. Toyoshima C, Nakasako M, Nomura H, Ogawa H. Crystal structure of the calcium pump of sarcoplasmic reticulum at 2.6 Å resolution. Nature. 2000;405:647–55.
43. Toyoshima C, Nomura H. Structural changes in the calcium pump accompanying the dissociation of calcium. Nature. 2002;418:605–11.
44. Toyoshima C, Nomura H, Sugita Y. Structural basis of ion pumping by Ca^{+2}-ATPase in the sarcoplasmic reticulum membrane. FEBS Lett. 2003;555:106–10.
45. Gouaux E., MacKinnon R. Principles of selective ion transport in channels and pumps. Science 2005;310:11461–5.
46. Pedersen BP, Buch-Pedersen MJ, Morth JP, Palmgren MG, Nissen P. Crystal structure of the plasma membrane proton pump. Nature 2007;450:1111–4.
47. Morth, JP, Pedersen BP, Toustrup-Jensen MS et al. Crystal structure of the sodium-potassium pump. Nature 2007; 450:1043–9.
48. Gumz ML, Duda D, McKenna R, Wingo CS, Cain BD. Molecular modeling of the rabbit colonic (HKα2a) H^+,K^+-ATPase. J. Molec. Model. 2003;9:283–9.
49. Asano, S, Yoshida A, Yashiro H et al. The cavity structure for docking the K^+-competitive inhibitors in the gastric proton pump. J. Biol. Chem. 2003;279:13968–75.
50. Koenderink JB, Swarts HG, Willems PH, Krieger E, De Pont JJ. A conformation-specific interhelical salt bridge in the K^+ binding site of gastric H,K-ATPase. J. Biol. Chem. 2004;279:16417–24.
51. Munson K, Garcia R, Sachs G. Inhibitor and ion binding sites on the gastric H,K-ATPase. Biochemistry. 2005;44:5267–84.
52. Munson K, Law RJ, Sachs G. Analysis of the gastric H,K-ATPase for ion pathways and inhibitor binding sites. Biochemistry 2007;46:5398–417.
53. Shin JM, Sachs G. Gastric H-K-ATPase as a drug target. Dig. Dis. Sci. 2006;51:823–33.
54. Simpson AM, Schwartz GJ. Distal renal tubular acidosis with severe hypokalemia, probably caused by colonic H^+-K^+-ATPase deficiency. Arch. Dis. Child. 2001;84: 504–7.
55. Dherbecourt O, Cheval L, Bloch-Faure M, Meneton P, Doucet A. A molecular identification of Sch28080-sensitive K-ATPse activities in mouse kidney. Phlugers Arch. 2006;451:769–75.
56. Spicer Z, Miller ML, Andringa A et al. Stomachs of mice lacking the gastric H,K-ATPase alpha subunit have achlorhydria, abnormal parietal cells and ciliated metaplasia. J. Biol. Chem. 2000;275:21555–65.
57. Petrovic S, Spicer Z, Greeley T, Shull GE, Soleimani M. Novel Schering and ouabain-insensitive potassium-dependent proton secretion in the mouse cortical collecting duct. Am. J. Physiol. 2002;282:F133–43.
58. Meneton P, Shultheis PJ, Greeb J et al. Increased sensitivity to K^+ deprivation in colonic H,K-ATPase-deficient mice. J. Clin. Invest. 1998;101:536–42.
59. Pestov NB, Korneenko TV, Shakhparonov MI, Shull GE, Modyanov NN. Loss of acidification of anterior prostate fluids in Atp12a-null mice indicates that nongastric H-K-ATPse functions as proton pump in vivo. Am. J. Physiol. 2006;291:C366–74.

60. Lynch, IJ, Rudin A, Xia SL et al. Impaired acid secretion in cortical collecting duct intercalated cells from H,K-ATPase-deficient mice: role of HK isoforms. Am. J. Physiol. 2008; epub ahead of print.
61. Blake-Palmer KG, Su Y, Smith AN, Karet FE. Molecular cloning and characterization of a novel form of the human vacuolar H^+-ATPase e-subunit: an essential proton pump component. Gene 2007;393:94–100.
62. Brown D, Hirsch S, Gluck S. An H^+-ATPase in opposite plasma membrane domains in kidney epithelial cell subpopulations. Nature. 1988;331:622–4.
63. Wagner CA, Finberg KE, Breton S, Marshansky V, Brown D, Geigel JP. Renal vacuolar H^+-ATPase. Physiol. Rev. 2004;84:1263–314.
64. Smith AN, Borthwick KJ, Karet FE. Molecular cloning and characterization of novel tissue-specific isoforms of the human vacuolar H^+-ATPase C, G and d-subunits, and their evaluation in autosomal recessive distal renal tubular acidosis. Gene 2004;297:169–77.
65. Sun-Wada GH, Yoshimizu T, Imai-Senga Y, Wada Y, Futai M. Diversity of mouse proton-translocating ATPase: presence of multiple isoforms of the C, d and G subunits. Gene 2003;302:147–53.
66. Sun-Wada GH, Murata Y, Namba M, Yamamoto A, Wada Y, Futai M. Mouse proton pump ATPase C subunit isoforms (C2-a and C2-b) specifically expressed in kidney and lung. J. Biol. Chem. 2003;278:44843–51.
67. Miller RL, Zhang P, Smith M et al. V-ATPase B1-subunit promoter drives expression of EGFP in intercalated cells of kidney, clear cells of epididymis and airway cells of lung in transgenic mice. Am. J. Physiol. 2005;288:C1134–44.
68. Holliday LS, Bubb MR, Jiang J, Hurst IR, Zuo J. Interactions between vacuolar H^+-ATPases and microfilaments in osteoclasts. J. Bioenerg. Biomembr. 2005;37:419–23.
69. Paunescu TG, Da Silva N, Russo LM. Association of soluble adenylyl cyclase with the V-ATPase in renal epithelial cells. Am. J. Physiol. 2008;294:F130–8.
70. Paunescu TG, Da Silva N, Marshansky V, McKee M, Breton S, Brown D. Expression of the 56-kDa B2 subunit isoform of the vacuolar H^+-ATPase in proton-secreting cells of the kidney and epididymis. Am. J. Physiol. 2004;287:C149–62.
71. Oka T, Murata Y, Namba M et al. a4, a unique kidney-specific isoform of mouse vacuolar H^+-ATPase subunit a. J. Biol. Chem. 2001;276:40050–4.
72. Stehberger, Schulz N, Finberg KE et al. Localization and regulation of the ATP6V0A4 vacuolar H^+-ATPase subunit defective in an inherited form of distal renal tubular acidosis. J. Am. Soc. Nephrol. 2003;14:3027–38.
73. Smith AN, Jouret F, Bord S et al. Vacuolar H^+-ATPase d2 subunit: molecular characterization, developmental regulation, and localization to specialized proton pumps in kidney and bone. J. Am. Soc. Nephrol. 2005;16:1245–56.
74. Schulz N, Dave MH, Stehberger PA, Chau T, Wagner CA. Differential localization of vacuolar H^+-ATPases containing a1, a2, a3, or a a4 (ATP6V0A1-4) subunit isoforms along the nephron. Cell. Physiol. Biochem. 2007;20:109–20.
75. Feng H, Cheng T, Pavlos NJ et al. Cytoplasmic terminus of a vacuolar proton pump accessory subunit Ac45 is required for proper interaction with V_0 domain subunits and efficient osteoclastic bone resorption. J. Biol. Chem. 2008; epub ahead of print.
76. Jouret F, Auzanneau C, Debaix H et al. Ubiquitous and kidney-specific subunits of vacuolar H^+-ATPase are differentially expressed during nephrogenesis. J. Am. Soc. Nephol. 2005;16:3235–46.
77. Kawasaki-Nishi S, Yamaguchi A, Forgac M, Nishi T. Tissue specific expression of the splice variants of the mouse vacuolar proton-translocating ATPase a4 subunit. Biochem. Biophys. Res. Comm. 2007;364:1032–6.
78. Norgett EE, Borthwick KJ, Al-Lamki RS, Su Y, Smith AN, Karet FE. V_1 and V_0 domains of the human H^+-ATPase are linked by an interaction between the G and a subunits. J. Biol. Chem. 2007;282:14421–7.
79. Boyer P. A research journey with ATP synthase. J. Biol. Chem. 2002;277:39045–61.

80. Abrahams JP, Lutter R, Leslie AG, Walker JE. Structure at 2.8 Å resolution of F_1-ATPase from bovine heart mitochondria. Nature 1994;370:621–8.
81. Noji H, Yasuda R, Yoshida M, Kinosita K. Direct observation of the rotation of F_1-ATPase. Nature 1997;386:299–302.
82. Hirono-Hara Y,Ishizuka K, Kinsoita K, Yoshida M, Noji H. Activation of pausing F_1 motor by external force. Proc. Natl. Acad. Sci. 2005;102:4288–93.
83. Adachi K, Oiwa K, Nishizaka T et al. Coupling of rotation and catalysis in F_1-ATPase revealed by single-molecule imaging and manipulation. Cell 2007;130:309–21.
84. Imamura H, Nakano M, Noji H et al. Evidence for rotation of V_1-ATPase. Proc. Natl. Acad. Sci. 2003;100:2312–5.
85. Jefferies KC, Forgac M. Subunit H of the V-ATPase inhibits ATP hydrolysis by free V_1 domain by interaction with the rotary subunit F. J. Biol. Chem. 2008; epub ahead of print.
86. Hirata T, Iwamoto-Kihara, Sun-Wada GH, Okajima T, Wada Y, Futai M. Subunit rotation of vacuolar-type proton pumping ATPase. J. Biol. Chem. 2003;278:23714–9.
87. Inoue H, Noumi T, Nagata M, Murakami H, Kanazawa H. Targeted disruption of the gene encoding the proteolipid subunit of mouse vacuolar H^+-ATPase leads to early embryonic lethality. Biochim. Biophys. Acta. 1999;1413:130–8.
88. Karet FE, Finberg KE, Nelson RD et al. Mutations in the gene encoding B1 subunit of H^+-ATPase cause renal tubular acidosis with sensorineural deafness. Nat. Genet. 1999;21:84–90.
89. Karet FE, Finberg KE, Nelson RD et al. Localization of a gene for autosomal recessive distal renal tubular acidosis with normal hearing (rdRTA2) to 7q33–34. Am. J. Hum. Genet. 1999;65:1656–65.
90. Stover EH, Borthwick KJ, Bavalia C et al. Novel ATP6V1B1 and ATP6V0A4 mutations in autosomal recessive distal renal tubular acidosis with new evidence of hearing loss. J. Med. Genet. 2003;39:796–803.
91. Lang F, Vallon V, Knipper M, Wangemann P. Functional significance of channels and transporters expressed in the inner ear and kidney. Am. J. Physiol. 2007;293:C1187–208.
92. Finberg, KE Wagner CA, Bailey MA et al. The B1 subunit of the H^+-ATPase is required for maximal urinary acidification. Proc. Natl. Acad. Sci. 2005;102:13616–21.
93. Paunescu TG, Russo LM, Da Silva N. et al. Compensatory membrane expression of the V-ATPase B2 subunit isoform in medullary intercalated cells of B1-deficient mice. Am. J. Physiol. 2007;293:F1915–26.

Acid/Base Regulation in Renal Epithelia by H,K-ATPases

I. Jeanette Lynch and Charles S. Wingo

Abstract The renal H,K-ATPases are integral membrane proteins, located in the renal collecting duct (CD) and other segments of the nephron, that participate in the regulation of proton secretion and cation reabsorption. Together with the apical Na/H exchangers and the H-ATPases, the H,K-ATPases, function as proton secretory mechanisms that are responsible for reclamation of filtered HCO_3, regeneration of HCO_3 consumed in the oxidation of the products of metabolism, and the generation of new HCO_3 in the kidney. This chapter examines the role of H,IC-ATPases in acidbase and volume homeostasis. We discuss the H,K-ATPase holoenzyme structure followed by a review of studies regarding renal H,K-ATPase localization, function, and regulation. Finally, a recent report using a genetic approach to quantify the contribution of two a-subunit isoforms of the H,K-ATPase to acid secretion has confirmed years of pharmacological studies from many laboratories and demonstrates that both isoforms of this enzyme are normally active in the CD.

Keywords Sodium · Potassium · P-TYPE ATPase · Transport · Secretion · Reabsorption · ATPases

1 Introduction

Serum potassium concentration $[K]_s$ is critically important for normal function of most cells, notably cardiac, renal, muscular, and neuronal cells. The kidney adapt to large changes in potassium (K) intake and maintains $[K]_s$ within a narrow concentration range (3.5–5.0 mEq/L). Normal kidney and lung function are critical in maintaining arterial blood pH within a narrow range (~7.35–7.45 pH units), and both organs respond to integrative stimuli to maintain blood pH within this range.

C.S. Wingo (✉)
North Florida/South Georgia Veterans Health System and College of Medicine, University of Florida, Gainesville, FL 32608, USA
e-mail: charles.wingo@va.gov

G.A. Gerencser (ed.), *Epithelial Transport Physiology*,
DOI 10.1007/978-1-60327-229-2_11,

The renal regulation of blood pH involves reabsorption of filtered HCO_3, primarily in the proximal tubule, but also in the distal nephron and collecting duct (CD).

The renal H,K-ATPases are integral membrane proteins of the CD and other segments (see below) that participate in the regulation of proton secretion and cation reabsorption. There is substantial agreement from physiological and genetic disruption studies that the CD is critical in the final regulation of electrolyte excretion. Figure 1 illustrates the nephron and CD system.

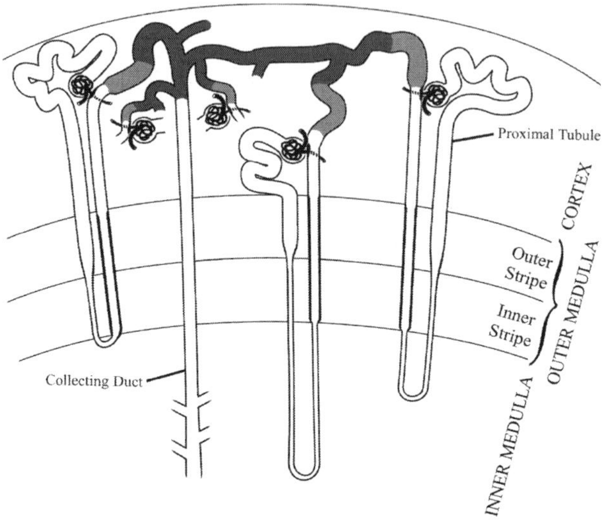

Fig. 1 Review of the CD. Distal nephron segments [distal convoluted tubule (DCT, *light stippling*) and connecting segments (CNT, *dark stippling*)] merge into CDs that start in the cortex and extend through the medulla to the papillary ducts. The medulla is divided into inner and outer regions. A CD from the cortex, outer medulla, or inner medulla is referred to as a CCD, OMCD, or IMCD, respectively. The outer medulla is divided into inner and outer stripe regions relative to the position of neighboring proximal tubules (thick descending in the outer stripe and thin descending in the inner stripe)

Together with the apical Na/H exchangers and the H-ATPases, these three categories of proton secretory mechanisms are responsible for reclamation of filtered HCO_3, regeneration of HCO_3 consumed in the oxidation of the products of metabolism, and the generation of new HCO_3 in the kidney. This review examines the role of H,K-ATPases in acid/base and volume homeostasis. The vacuolar or H-ATPases also contribute to proton secretion both in the proximal tubule and in the CD. The relative role of H,K-ATPase and H-ATPase will be considered below.

The H,K-ATPases translocate protons to the extracellular space in exchange for the uptake of cations with the greatest affinity for K. The H,K-ATPases play an important role in K absorption especially under conditions of hypokalemia. Hypokalemia, which is common in metabolic alkalosis in rats and humans, increases ammoniagenesis, distal acidification, and is associated with reduced serum aldosterone, a potent blood pressure regulator.

We shall briefly discuss the H,K-ATPase holoenzyme structure followed by a review of studies regarding renal H,K-ATPase localization, function, and regulation. The regulation of H,K-ATPase involves many factors including hypokalemia, acidosis, alkalosis, renal ammonia concentration, ambient pCO_2, and hormones. In addition, the global disruption of sodium/hydrogen exchange (NHE-3), ischemic-reperfusion injury, and drug-induced HCO_3 wasting upregulate H,K-ATPase mRNA or activity. However, dietary sodium (Na) restriction also activates H,K-ATPase, and it has been scientifically established that the gastric H,K-ATPase can accept Na on the K binding site [1]. Therefore, assuming that H,K-ATPase functions solely in proton and K transport may fail to appreciate the full contribution of this enzyme to renal function. Lastly, a recent report using a genetic approach to quantify the contribution of two α-subunit isoforms of the H,K-ATPase to acid secretion has confirmed years of pharmacological studies from many laboratories and demonstrates that both isoforms of this enzyme are normally active in the CD.

2 The H,K-ATPase Genes

The H,K-ATPases belong to the P-type cation-transporting ATPase superfamily [2]. This superfamily can be classified into several families including the Ca-ATPases, Na,K-ATPases, and H,K-ATPases. Like the Ca-ATPase, the H,K-ATPases have four major protein domains, ten membrane spanning helices, and phosphorylation and nucleotide-binding sites. Unlike the Ca-ATPase, both the H,K-ATPase and Na,K-ATPases require regulatory beta (HKβ) subunits in addition to the catalytic alpha (α) subunits for enzymatic function. Data from two groups suggest that in its native state, the H,K-ATPase forms a dimer $(\alpha\beta)_2$ [3, 4]

There are at least two known genes that are transcriptionally competent and express α subunits in the kidney, $HK\alpha_1$ and $HK\alpha_2$ [5]. Table 1 summarizes the nomenclature used to describe H,K-ATPase genes and gene products. $HK\alpha_1$, also referred to as the gastric isoform, is encoded by the Atp4a gene. Similarly, $HK\alpha_2$, historically called the colonic or nongastric isoform, is encoded by the Atp12a gene and is expressed in many tissues. Isoforms of both α subunits have been cloned from

Table 1 Genes and gene products of the H,K-ATPase holoenzyme complex

Gene	Protein (species)	Proposed enzyme (minimal complex)
ATP4a	$HK\alpha_1$	$HK\alpha_1$-HKβ-H,K-ATPase
ATP12a, ATP1AL1	$HK\alpha_{2a}$ (rat, rabbit)	$HK\alpha_2$-$NaK\beta_1$-H,K-ATPase
	$HK\alpha_{2b}$ (rat)	
	$HK\alpha_{2c}$ (rabbit)	
	$HK\alpha_4$ (human)	
ATP4b	HKβ	$HK\alpha_1$-HKβ-H,K-ATPase
ATP1B1	$NaK\beta_1$	$HK\alpha_2$-$NaK\beta_1$-H,K-ATPase

mammalian species including human, rat, mouse, dog, rabbit, pig, cow, and monkey. $HK\alpha_4$ is the human α subunit product of the ATP12A gene and is the nongastric ortholog of the rodent $HK\alpha_2$. ATP12A is also referred to as ATP1AL1.

Splice variants for $HK\alpha_2$ have been reported in rat, $HK\alpha_{2a}$ [6] and $HK\alpha_{2b}$ [7], and rabbit, $HK\alpha_{2a}$ and $HK\alpha_{2c}$ [8]. Structural models of both α isoforms have been published [9–11]. Figure 2 illustrates the proposed model of $HK\alpha_2$ based on an alignment with the Ca-ATPase, a similar ATPase for which a crystal structure has been reported. H,K-ATPase α subunits are phosphorylated at their N- and C- terminal domains by protein kinases that may modulate the enzymatic activity [12] and/or regulatory subunit binding [13]. The gastric β subunit is a 294 amino acid glycoprotein with a core molecular weight of 33.6 kDa, a single putative transmembrane domain and N-linked potential glycosylation sites. $HK\beta$ is required for membrane trafficking, protection from degradation, and ion-binding properties [5, 14]. The renal $HK\alpha_1$ associates with the gastric-type β subunit ($HK\beta$ encoded by the Atp4b gene) which is essential for acid-secretory activity [15] and K-ATPase activity (see below).

Genetic disruption of this $HK\beta$ abolishes K-ATPase activity in gastric membrane vesicles, but not in colonic membrane vesicles, which indicates both the importance of the $HK\beta$ subunit in gastric H,K-ATPase activity and that this subunit is not the endogenous partner for the $HK\alpha_2$ subunit of the colonic H,K-ATPase (Shao, et al. submitted for publication). Indeed, considerable evidence supports the role of the β_1 subunit of the Na,K-ATPase (encoded by the ATP1β1 gene) as the physiological β-subunit for $HK\alpha_2$ [5].

$HK\alpha_1$ and $HK\alpha_2$ are structurally similar, yet differences at the K-binding site may explain differing sensitivities to K-competitive inhibitors [10]. $HK\alpha_1$ H,K-ATPase is sensitive to micromolar concentrations of imidazopyridine, 2-methyl-8-(phenylmethoxy)imidazol[1,2a]-pyridine-3-acetonitrile (Sch-28080) [16], whereas the pharmacological profile of $HK\alpha_2$ H,K-ATPase is not completely understood. Ouabain, a classic inhibitor of the Na, K-ATPase, has been used to assess the physiological contribution of $HK\alpha_2$ [5].

The literature remains divided regarding the specificity of the $HK\alpha_2$ H,K-ATPase to ouabain. For instance, when mice lacking functional $HK\alpha_1$ protein were maintained on a K-deplete diet, which upregulates the $HK\alpha_2$ isoform, both Sch-28080-sensitive and ouabain-sensitive K-ATPase activity was observed (Type III) [17]. Type III activity did not exist in $HK\alpha_2$-deficient mice suggesting that the $HK\alpha_2$ H,K-ATPase may be inhibited by both compounds at least under K-deplete conditions. The authors interpreted these data to suggest that enzyme complexes containing $HK\alpha_2$ are inhibited by Sch-28080 and ouabain during K-deplete conditions (see below). However, the $HK\alpha_2$ H,K-ATPase is insensitive to ouabain in isolated colonic mucosal vesicle preparations (Shao, et al. submitted for publication). There are also reports of a novel renal H,K-ATPase in normokalemic mice which is supported by evidence of a K-dependent proton secretory mechanism in the cortical CD (CCD) intercalated cells (ICs) that is insensitive to both Sch-28080 and ouabain [18]. Therefore, assessing the contribution of each isoform to ion transport using the traditional pharmacological approach may be less than ideal.

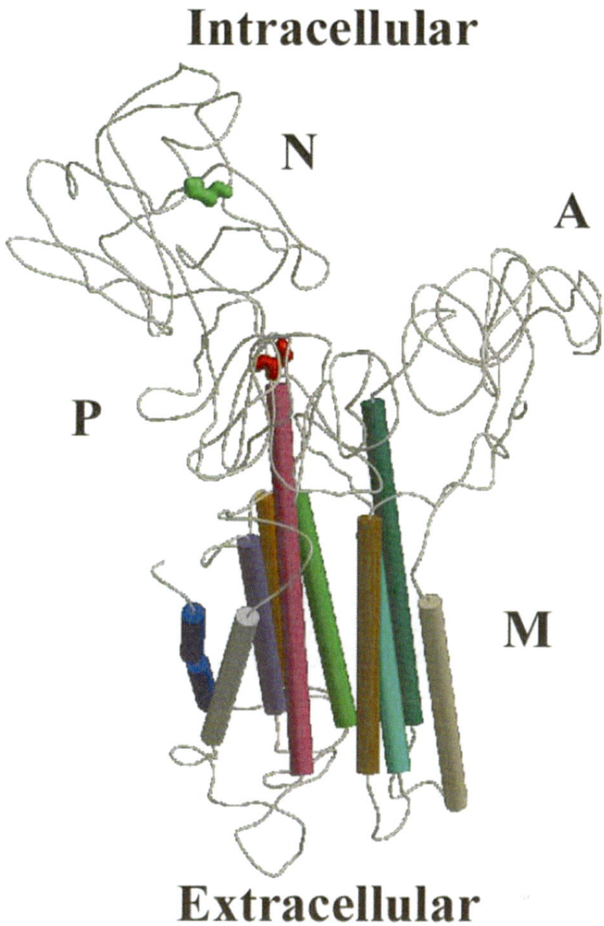

Fig. 2 Proposed architecture of the HKa2a subunit. A molecular model of the HKa2a subunit was built based on the coordinates from the 2.6-A Ca+2 ATPase structure (Toyoshima et al. Nature 2000). The model contains four readily visible domains: The P domain, containing the phosphorylation site (amino acid D351 is shown in 3D), the N domain, containing the nucleotide-binding site (K515 lies at the bottom of the nucleotide-binding pocket and is shown in 3D), the A domain, containing the N-terminus that acts as an actuator in the Ca+2 ATPase, and the M domain which contains ten transmembrane helices. The b subunit of the HK ATPase is thought to interact with the C-terminal region, near transmembrane helices 7, 8, and 10 [9]

2.1 HKα $_1$ H,K-ATPase

Physiological experiments in multiple animal models collectively indicate that the H,K-ATPase enzyme functions to reabsorb cations and secrete acid in the CD. Functional studies have provided convincing evidence for the existence of Sch-28080-sensitivity in the cortical CD (CCD) [19–23], the OMCD [24–28], and the IMCD [25, 29–32]. Specifically in the ICs from CCD, there is physiological [18,

33–40], immunohistochemical [41, 42], and molecular [43] data that collectively support the role of an active HKα_1 H,K-ATPase.

2.2 *HKα_2 H,K-ATPase*

Few physiological studies support a role for the highly regulated HKα_2 H,K-ATPase under normal conditions; however, immunohistochemistry demonstrates HKα_2 to be present and a recent report confirms that the ATP12a gene is functionally active in the normal mouse kidney (discussed below). HKα_2 is upregulated by K depletion and is ouabain-sensitive at least in intact cells at relatively high concentrations [32, 44]. Furthermore, Type III K-ATPase is partially Sch-28080 and ouabain sensitive and is reported to be mediated by HKα_2 H,K-ATPase [17]. The pharmacological profile of the HKα_2 H,K-ATPase is not well defined, and ineffective concentrations of inhibitor may have been used in some studies where HKα_2 H,K-ATPase was not observed under normal conditions. For example, in the rabbit CCD under K-restricted conditions that upregulates HKα_2, only luminal Sch-28080-sensitive K reabsorption was measurable and luminal ouabain (0.1 mM) had no effect [45]. There was also no effect of 0.1-mM luminal ouabain to inhibit H,K-ATPase mediated pH_i recovery in rabbit B-type ICs [37]. Specifically, the concentration used in these experiments may be less than the effective concentration needed to inhibit the HKα_2 H,K-ATPase [5], or there are species differences in the sensitivity of this pump.

There is strong evidence for the upregulation of this isoform at both the mRNA and protein level under low K conditions [46–51]. Functional studies have provided convincing evidence for the induction of HKα_2 H,K-ATPase activity in the cortical CD (CCD) under low Na dietary conditions [39], and in the OMCD [46] and tIMCD under dietary K restriction [31, 32]. The presence of HKα_2 in the ICs of CCD is supported by physiological [33, 39, 40], immunohistochemical [52], and molecular [8] data. In addition, metabolic alkalosis has been reported to upregulate HKα_2 H,K-ATPase [53].

3 Localization

3.1 *K-Dependent ATPase Activities*

K-stimulated ATPase activities have been reported in the CD of rabbit and rat and are summarized in Table 2. [54]. In brief, K-dependent ATPase activity was classified as either Type I or Type III. Type I was Sch-28080-sensitive and ouabain-insensitive and was present in the CCD and OMCD of rat maintained on a normal K diet [55]. Type III activity, defined as partially Sch-28080- and ouabain-sensitive, replaced Type I activity in rats fed a low K diet. A recent study reported the existence of Type I and Type III activities in mouse and further characterized both activities in

Table 2 The K-ATPase activities of the kidney and their response to hypokalemia. K-activated ATPase (K-ATPase) activity has been identified in isolated microdissected nephron segments, but the molecules responsible for the K-ATPase activity remain to be determined. All of the activities are sensitive to Sch-28080, which suggest that they arise from H,K-ATPase subunits. Three distinct enzymatic subtypes of K-ATPase activity have been described [8], which exhibit different characteristics as shown below

	Type I K-ATPase	Type II K-ATPase	Type III ATPase
IC_{50} Sch-28080	0.25 μM	1.7 μM	0.85 μM
IC_{50} ouabain	Insensitive	6.4 μM	20 μM
Cation activation	K	K	K, Na
Effect K depletion	Decreases	Decreases	Increases
Localization	Collecting duct	Proximal tubule, TALH	Collecting duct

mice genetically deficient in $HK\alpha_1$ or $HK\alpha_2$ H,K-ATPase [17]. Dherbecourt et al. reported that Type I activity was no longer detectable in the $HK\alpha_1$-knockout mice; whereas the Type III activity was preserved. These data suggest that Type I activity requires the expression of the $HK\alpha_1$ subunit. In contrast, Type III activity was no longer detectable in the $HK\alpha_2$-knockout mice; whereas, the Type I activity was preserved, which suggests that Type III activity requires the expression of the $HK\alpha_2$ subunit.

Another pharmacologically distinct K-ATPase activity (Type II) was discovered in the proximal tubule (PT) and thick ascending limb (TAL) of rat [56]. This unique activity was Sch-28080- and ouabain-sensitive and was markedly reduced during K depletion [55]. Type II activity was similar to Type III in pharmacology, yet different because (1) it had lower sensitivity to ouabain and Sch-28080, (2) was almost completely absent during K-depletion, and (3) was not stimulated by Na. Beltowski and Wójcicka confirmed the existence of Type II activity in the cortex and medulla by an independent method; however, it was not possible to conclude in which nephron segment(s) the activity existed [57]. The presence of Type II activity has not been described in mice genetically deficient in $HK\alpha_1$ or $HK\alpha_2$ H,K-ATPase.

3.2 *H,K-ATPase Isoforms at the Level of Message (mRNA) and Protein*

Studies verify that H,K-ATPase mRNA and protein are present in the kidney of many mammalian species including humans. Several experiments identify the presence of this enzyme in the luminal region of the connecting tubule (CNT) and/or CD [41, 49, 52, 58, 59]. Functional studies have also demonstrated its apical location where it could facilitate urinary acidification and K reabsorption. This concept is well supported by years of physiological experimentation in the isolated, microperfused CD under various conditions. Some of these important functional studies are summarized later.

Using in situ hybridization, all cells of the rat CD were reported to express HKα_1 mRNA under normal and low K dietary conditions [43, 60]. During hypokalemia, there was hypertrophy of the OMCD$_{is}$ and two-fold higher HKα_1 mRNA in cells of the cortex. This is of interest because the majority of regulation of H,K-ATPases at the mRNA and protein level appears to involve HKα_2, and most evidence indicates that the HKα_2 isoform(s) is more highly regulated by hypokalemia than HKα_1. Mice lacking functional HKα_2 have increased sensitivity to K deprivation suggesting that HKα_2 plays a critical role in the K conservation [61]. It is well established that the HKα_2 mRNA is upregulated due to K depletion [46, 49–51, 62]. The intrarenal distribution of the HKα_2 mRNA was assessed in normal diet rabbit [63] and was selectively expressed in the CNT, CCD, and OMCD with the highest expression in CNT cells. Fejes-Toth et al. later confirmed the presence of the HKα_2 protein in the CNT as well (see below) [59]. Using Northern blot analysis and in situ hybridization, the relative expression of HKα_2 was compared in normal and K-deprived rat kidney [47]. Rats on a normal diet had low amounts of the HKα_2 mRNA in the cortical and medullary TAL, distal convoluted tubule, CNT, and the entire CD. However, when rats were placed on a low K diet, there was a five-fold increase in the HKα_2 mRNA in the outer and inner medulla compared to control. The in situ hybridization signal was increased in the ICs of the OMCD$_{is}$ and the proximal IMCD. These results are the opposite to those observed for the HKα_1 subunit and suggest that axial heterogeneity is an important element of regulation of different H,K-ATPase α subunit isoforms. There is evidence that the protein as well as the mRNA of HKα_2 is increased due to dietary K restriction. Using antibodies to HKα_1 (HK 12.18) and HKα_2 (AS 31.7), microsomes from whole rat kidney were enriched, the protein was isolated and then immunoblotted. Low K diet increased the abundance of HKα_2, but not HKα_1 protein [48].

3.3 Immunohistochemistry

Immunohistochemistry confirms the presence of HKα_1 and HKα_2 H,K-ATPase protein in the CD. Using mouse monoclonal antibodies (HK 9.15 and HK 4.18) against the hog HKα_1 H,K-ATPase, rat and rabbit ICs of the CCD and OMCD from animals on a normal diet had diffuse cytoplasmic staining [42]. HKα_1 protein in ICs was confirmed in a later study in rat, also on a normal K diet, using a rabbit polyclonal antibody to the N-terminus of the hog gastric H,K-ATPase α subunit (HKαN2) [41]. Interestingly, this study reported a similar localization pattern previously reported for the H-ATPase, apical staining in A-type ICs and basolateral staining in the B-type ICs. A rabbit polyclonal antibody to HKα_{2a}, designated M-1, was produced and localized to the apical membrane of rat colon [58]. The M-1 antibody (anti-HKα_2) was tested in rat kidney under dietary Na or K depletion [49]. Dietary K depletion, but not Na depletion, increased the protein level in the outer medulla. Importantly, immunohistochemistry localized the protein to the apical membrane of the OMCD PCs (not ICs) and confirmed that the protein was upregulated during dietary K depletion. Three monoclonal antibodies against three different epitopes

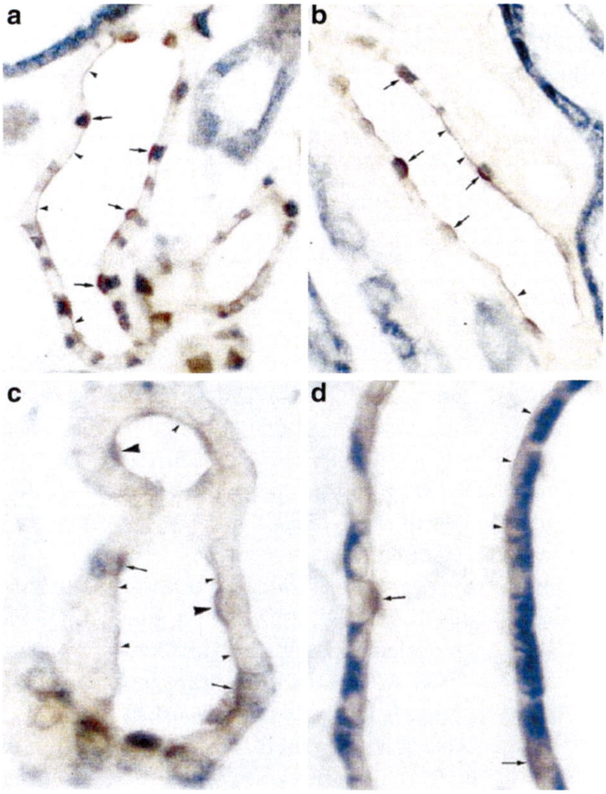

Fig. 3 Immunohistochemical localization of H,K-ATPase α_{2c}-subunit in rabbit kidney. **a** and **b**: Light micrographs of rabbit cortex (**a**) and outer medulla (**b**) labeled for both $HK\alpha_{2c}$ and carbonic anhydrase II (CA II). Intercalated cells are identified by cytoplasmic CA II immunoreactivity (*blue label*). Apical $HK\alpha_{2c}$ immunoreactivity (*brown label*) is present in both CA II-positive (*arrows*) and CA II-negative cells (*arrowheads*), but the intensity and width of the apical $HK\alpha_{2c}$ label are greater in the intercalated cells. **c**: Light micrograph of rabbit connecting segment labeled for both $HK\alpha_{2c}$ and Cl^-/HCO_3^- exchanger (AE1) protein. Type A intercalated cells are identified by AE1-immunoreactivity (*blue label*). Apical $HK\alpha_{2c}$ immunoreactivity (*brown label*) is present in both AE1-positive (*arrows*) and AE1-negative cells. Although the majority of AE1-negative cells have only a thin band of $HK\alpha_{2c}$ immunolabel (*small arrowheads*), occasional AE1-negative cells exhibit broad, intense apical $HK\alpha_{2c}$ immunostaining (*large arrowheads*). **d**: Light micrograph of rabbit connecting segment labeled for both $HK\alpha_{2c}$ and $NaK\alpha_1$. Connecting segment cells are identified by extensive basolateral $NaK\alpha_1$ immunoreactivity (*blue label*). Apical $HK\alpha_{2c}$ immunoreactivity (*brown label*) is present in both $NaK\alpha_1$-positive (*arrowheads*) and $NaK\alpha_1$-negative cells (*arrows*). However, the intensity and width of the apical $HK\alpha_{2c}$ label are greater in the $NaK\alpha_1$-negative cells, which represent intercalated cells [113]

of the rabbit $HK\alpha_2$ subunit that reacted with the apical membrane region of rabbit colon were tested in the kidney of rabbits on a normal K diet using laser scanning confocal microscopy [59]. All three antibodies reacted to the apical membrane region of CNT cells. In another study that is shown in Fig. 3 also under normal dietary conditions, intense apical immunolocalization in A- and B-type ICs and thin apical labeling of CNT cells and PCs of the rabbit CNT, CCD, OMCD was reported for the rabbit variant HKα_{2c} [52].

In summary, the mRNA and protein of both H,K-ATPase subtypes are present along the CD, and the upregulation of the HKα_2 H,K-ATPase by hypokalemia in the renal medulla is well supported. Apical localization of the enzyme in immunohistochemical studies supports the concept of the H,K-ATPase as an important K reabsorptive and acid-secretory mechanism.

3.4 Expression in Human Kidney

In 1991, Modyanov and co-workers reported that ATP1AL1 (the human ortholog of murine ATP12a) was functionally active and expressed in human brain and kidney [64]. It is also present and functional in the microvillous plasma membrane of the transporting epithelia of the human placenta [65]. cDNA from ATP1AL1 was isolated and sequenced from human skin and kidney [66] and later cloned from human skin and colon [67]. A qualitative estimation of the relative expression levels was performed by RT-PCR in a number of tissues from mouse, rat, rabbit, dog, and human [68]. Significant expression levels were observed in the skin, kidney, and distal colon of all species except for human colon. Interestingly, the abundance of ATP1AL1 mRNA was reported to be greater in the kidney than in the distal colon of humans. Different H,K-ATPase isoforms have been detected and localized to the human kidney [69]. Specifically, ICs of human CD (cortical and medullary regions) had cytoplasmic staining with antibodies against HKα_1 (monoclonal 1218), and PCs were also stained. An antibody to HKα_4, the human ortholog of HKα_2, stained ICs of the CD (cortical and medullary regions) and occasionally PCs. Codina et al., 2006 recently evaluated the abundance of this gene and gene product in human kidney (SU-PO001). Using in situ hybridization and immunolocalization, they confirmed the presence of ATP1AL1 mRNA in human kidney. Intriguingly, immunolocalization was at both the apical and basolateral membranes of the CD.

4 Regulation

4.1 Acid/Base Regulation of the H,K-ATPase

4.1.1 K and HCO_3 Reabsorption Along the CD

$^{86}Rb^+$ is similar in size and charge to K^+ and has been used in many physiological experiments to mimic K. Tracer efflux of this isotope in renal microperfusion experiments revealed the unidirectional K absorptive flux in the distal tubule and CD. Measuring the total CO_2 (tCO_2) in luminal perfusate (collected fluid from microperfused nephron segments) allows direct quantitation of proton secretion. With increased rates of net luminal acid secretion, tCO_2 decreases as HCO_3 is titrated in the perfusate. Thus, stoichiometric net HCO_3 reabsorption can be equated with proton secretion. An important distinction to consider is the difference between measuring tCO_2 flux which is a measure of luminal proton secretion during steady state pH_i and measuring pH_i recovery due to an acute acid load (discussed in the next section). These conditions are very different and the contribution of different proton pumps to pH homeostasis during each condition may differ significantly. Moreover pH_i recovery experiments in the absence of pharmacological inhibitors does not distinguish between apical and basolateral proton secretion. Such considerations may clarify the relative contribution of H-ATPase and H,K-ATPase activities on net proton secretion as measured by these two different methods. For example, when Na-H exchange is inhibited, nanomolar concentrations of bafilomycin A_1 dramatically reduce pH_i recovery from an acid load in many cell types suggesting an important role for the vacuolar H-ATPase for the acute regulation of pH_i. In contrast, based on pharmacological inhibitors, measurements of tCO_2 flux have consistently demonstrated a greater contribution of H,K-ATPase to net proton secretion than H-ATPase in several [25–27] (for a detailed review, see [70]) but not all laboratories [71]. Nevertheless, all of these cited studies show a significant (>35%) contribution of H,K-ATPase to net luminal acidification.

4.2 CCD

Steady-state J_{tCO2} flux experiments clearly demonstrate that an apical H,K-ATPase plays an important role in K reabsorption and acid secretion in all segments of the CD. In the rabbit CCD under K-restricted conditions, luminal Sch-28080-sensitive K reabsorption was measured [45]. K reabsorption was reduced by 39% by Sch-28080. Additionally, Zhou et al. reported that basolateral ouabain, which inhibits the Na,K-ATPase, increased K reabsorption and this stimulation was Sch-28080-sensitive (i.e. an H,K-ATPase activity). This suggests that up-regulation of H,K-ATPase activity may compensate for decreasing intracellular K due to inhibition of the Na,K-ATPase. In a separate study in the CCD of rabbits adapted to a K-restricted diet, acute in vitro acidosis (10% CO_2) stimulated Sch-28080-sensitive K reabsorption [19].

4.3 OMCD

The OMCD is a major site for H,K-ATPase mediated H secretion under K-replete [24, 72, 73] and K-deplete dietary conditions [28, 50, 74, 75]. Similar to the CCD, luminal Sch-28080 inhibited K reabsorption by 41% in the inner stripe of the OMCD ($OMCD_{is}$) of rabbits adapted to a K-restricted diet [24]. In addition, the stimulation of K reabsorption by 10% CO_2 was reduced by 44% in the presence of luminal Sch-28080. Under normal K and low K conditions, basolateral Sch-28080 significantly decreased HCO_3 flux by 76 and 62% respectively, suggesting that acid secretion or bicarbonate reabsorption is mediated by H,K-ATPase under both dietary conditions. To resolve questions concerning the specificity and membrane permeability of Sch-28080, experiments in the $OMCD_{is}$ of K-replete rabbits were performed investigating the effect of luminal A80915a, structurally different to Sch-28080, and the less membrane permeable H,K-ATPase inhibitor, H224/25 [26]. Both of these inhibitors reduced HCO_3 reabsorption; importantly, H224/25 only inhibited HCO_3 reabsorption when applied to the lumen. This is direct physiological evidence that H,K-ATPase is functioning at the apical membrane under the conditions of those studies.

4.4 IMCD

Finally, in the most distal segment of the CD there is H,K-ATPase-mediated acid secretion. Wall et al. concluded that the H,K-ATPase mediates acid secretion in microperfused terminal IMCD (tIMCD) of rats with chronic metabolic acidosis [30]. Specifically, steady-state acid secretion (measured as J_{tCO2}) was determined to be luminal K-dependent and Sch-28080 sensitive. Furthermore, K uptake in tIMCD cells in primary culture was also Sch-28080 sensitive further supporting the case for H,K-ATPase in that segment. Also in the tIMCD, dietary K deficiency was reported to induce a Sch-28080-sensitive HCO_3 reabsorption mechanism that was not observed under normal diet conditions [32]. Additionally, high concentrations of luminal ouabain (5 mM) reduced this HCO_3 reabsorption in the presence or absence of Sch-28080 indicating independent acid secretion mechanisms.

4.4.1 Acute and Chronic Acid Loading

A method for measuring intracellular pH (pH_i) in CD cells using the pH-sensitive fluorescent dye 2′,7′-bis-(2-carboxyethyl)-5-(and-6)-carboxyfluorescein (BCECF), was described by Weiner and Hamm [76]. This method has been used to evaluate proton secretion as the pH_i recovery from an acute acid load using in vitro microperfused or split-opened CD segments and cultured cells. Renal epithelial cells are acid-loaded with addition and removal of NH_3/NH_4, and the subsequent rate of proton translocation is measured. HCO_3-free solutions are used to exclude the contribution of HCO_3 exchange to pH_i regulation. To assess Na-independent pH_i recovery, Na/H-exchange activity is inhibited by the application of exchanger inhibitors or the removal of basolateral Na. H,K-ATPase activity has been defined as Na-independent and K-dependent pH recovery.

The CD exhibits a significant Sch-28080-sensitive and K-dependent mechanism in response to acute acid loading [8, 28, 29, 34, 36, 37, 77, 78]. pH_i recovery in ICs, but not PCs, of rabbit CCD was determined to be K-dependent and Sch-28080-sensitive [36] suggesting H,K-ATPase mediated pH recovery in these cells. Chronic metabolic acidosis stimulates a Sch-28080-sensitive acid secretion mechanism in both rat and rabbits [30, 35, 38]. Silver et al. compared the K-dependent pH_i recovery rates in CCD ICs from rabbits on a normal diet and those given NH_4Cl supplemented drinking water to induce chronic metabolic acidosis [35]. The K-dependent pH_i rates were three times higher in ICs from the animals with chronic metabolic acidosis compared to control animals and were Sch-28080-sensitive. In a separate study, Silver et al. confirmed that pH_i recovery from acute acid loading in rabbit CCD ICs was K-dependent and Sch-28080 sensitive and significantly increased in animals with chronic metabolic acidosis [38]. However, no upregulation of the $HK\alpha_1$ or $HK\alpha_2$ mRNA was seen in rat kidney from chronic metabolic acidosis treated animals [50], suggesting that activity was modulated by either redistribution of existing pumps or changes in the kinetic activity of the pump. This is consistent with the report of Wall et al. that H,K-ATPase mediates acid secretion in the tIMCD of rats with chronic metabolic acidosis [30].

The contribution of H,K-ATPase to pH recovery in A-type and B-type ICs has been established in rabbit and mouse [18, 34, 37, 40]. Sch-28080 or luminal K removal completely inhibited H-ATPase-independent (NEM-insensitive) and Na-independent pH recovery in the B-type IC of rabbit CCD [37]. In the rabbit A-type IC, a Sch-28080-sensitive or luminal K-dependent pH recovery mechanism was less apparent [34]. More recently, H,K-ATPase activity has been confirmed as EIPA- and bafilomycin A_1-insensitive pH_i recovery from acute acid loading in both A-type and B-type ICs from two separate mouse strains [18, 79].

Kawahara et al. investigated the functional activity of H,K-ATPase in individual cells of the OMCD under normal and K deletion [28]. H,K-ATPase activity was Sch-28080-sensitive, K-dependent, present on the luminal membrane of ICs and OMCDi cells, and increased more than two-fold by dietary K depletion.

4.4.2 Ammonia and CO_2/pH

Ammonia increases HCO_3 reabsorption (acid secretion) in the rat CCD [80] and IMCD [81]. The term ammonia will be used to refer to the total NH_3/NH_4 species; the proportion of these two species is pH-dependent. The effects of ammonia on acid/base balance were investigated in the rabbit CCD [33, 82, 83]. Ammonia stimulates a bafilomycin A_1-insensitive, luminal Sch-28080- and ouabain-sensitive acid secretion mechanism suggesting that both isoforms of the H,K-ATPase are involved in mediating this response. Further studies demonstrate that ammonia stimulates the CCD H,K-ATPase through an intracellular calcium-dependent, microtubule-dependent, and an apical v-SNARE-dependent process signifying membrane vesicle insertion into the apical plasma membrane [83]. These observations are consistent with the effects of MAPTAM, calmodulin, W-7, and colchicine on the stimulation of K reabsorption by acute acidosis (10% CO_2) in the CCD of K-deplete rabbits [19]. Combined, these findings suggest that ammonia, acute acidosis, or increased pCO_2 stimulate a calcium-dependent mechanism of proton extrusion via an H,K-ATPase. The activation of H,K-ATPase by ammonia and by pCO_2/pH would be physiologically adaptive since both are associated with adaptation to an acid load or the induction of intracellular acidosis. Of note, several K channels are activated by increasing pCO_2, which is consistent with the role of K-recycling as a mechanism to activate H,K-ATPase activity [84]. Whether ammonia also activates these channels and whether there is a common signal (decreased pH_i) or multiple signals for H,K-ATPase activation is poorly understood and requires systematical examination.

4.4.3 Alkalosis

Under normal conditions, HKα_1 and HKα_2 were both detectable in rabbit CCD cells, and the HKα_2 mRNA was reported to be expressed 4.5-times higher under conditions of alkalosis compared with acidosis [53]. In another study, HCO_3 secretion was induced in the CCD of normal rats and was discovered to be sensitive to Sch-28080 when applied to the basolateral side [25]. In this study, Gifford et al. reported that acid secretion in the OMCD and IMCD of normal rats was sensitive to Sch-28080 but only when applied to the lumen, and basolateral Sch-28080 did not inhibit acid secretion in these segments. The authors concluded that the effect of Sch-28080 to block HCO_3 secretion in the CCD should be acting predominantly basolaterally as opposed to apically in the OMCD and IMCD, which correlates with the rat HKα_1 protein localization pattern reported by Bastani et al. [41]. H,K-ATPase localization and level of expression, likely regulated by dietary K or some factor regulated by dietary K, may be cell-type as well as species specific.

4.4.4 Ischemic-Reperfusion Injury and NHE-3 Deficiency

H,K-ATPase have been implicated as a compensatory mechanism to other renal perturbations. HKα_2 H,K-ATPase has been demonstrated to be upregulated by ischemic-reperfusion injury or acetazolamide treatment. Rats subjected to ischemic-reperfusion injury substantially up-regulate cortical HKα_2 H,K-ATPase, and not HKα_1, perhaps to compensate for the significant ($\sim$75%) down-regulation of the cortical and medullary NHE-3 [85]. NHE-3 is an apical Na/H-exchanger that facilitates over half of the acid secretion that occurs in the proximal tubule. Rats treated with acetazolamide which causes renal HCO_3 wasting also upregulate cortical HKα_2. Of particular interest, given the findings discussed below regarding the specificity of the cation absorptive cycle of the H,K-ATPases, NHE-3-deficient mice have profoundly reduced acid secretion in the proximal tubule with only slightly lower than normal serum HCO_3 levels [86]. In addition, the NHE-3 deficient mice have five-fold higher serum aldosterone, higher renin mRNA, increased ENaC activity and massively induced HKα_2 mRNA in the colon. Upregulation of H,K-ATPase in the colon may also compensate for loss of H secretion by NHE-3 as well as compensate for the loss of K during Na reabsorption. In an attempt to resolve the compensatory acid secretion mechanism limiting the develop of substantial metabolic acidosis in the NHE-3-deficient mice, Nakamura et al. studied net HCO_3 absorption in the OMCD and concluded that a HKα_1 H,K-ATPase is involved in the adaptive changes in NHE-3-deficient mice [87]. Specifically, there was a greater Sch-28080-sensitive acid secretion and upregulation of HKα_1 mRNA in the OMCD from NHE-3 knockout mice compared to the control. Nakamura et al. reported that renal HKα_2 H,K-ATPase does not play a compensatory role in the NHE-3 knockout since the HKα_2 mRNA was undetectable in kidney, and there was no effect of 1.0-mM luminal ouabain. However, the axial heterogeneity of each CD segment and the small mass of CD in the whole kidney or cortical sections limits the interpretation of these data.

4.5 Hormonal Regulation and Intracellular Signaling

Acid and base excretion by the kidney is hormonally regulated and involves the renin-angiotensin II-aldosterone system, endothelin, and thyroid hormones [88]. Aldosterone is a potent blood pressure regulator that activates Na reabsorption. Although chronic administration of this potent mineralocorticoid stimulates K secretion in the cortical CD (CCD), it does not acutely stimulate K excretion within the physiological range [89]. Mineralocorticoids are known to stimulate H secretion in the CCD [90], OMCD [91–93], and IMCD [94]. Deoxycorticosterone acetate (DOCA) administration profoundly increased Na-independent pH_i recovery in majority (PCs) and minority cells (A-type ICs) of the $OMCD_{is}$ of rabbits on a normal diet [93]. H,K-ATPase is known to function in acid secretion in the OMCD; however, Kuwahara et al. reported that H,K-ATPase did not appear to play a role in the stimulation of acid secretion by DOCA in the ICs of the rabbit outer OMCD [92]. The OMCD PCs, known to express H,K-ATPase and respond to DOCA treatment, were not investigated in this study. Additionally, no upregulation of $HK\alpha_2$ mRNA or protein was seen in the OM of rats on low Na diet with elevated aldosterone levels [49]. As opposed to the absence of upregulated $HK\alpha_2$ mRNA in the medulla of rats with elevated aldosterone levels, Sangan et al. reported that low Na increased $HK\alpha_2$ mRNA in the cortex that correlated with a slight increase in diffuse cytoplasmic fluorescence seen under the control conditions. A low Na diet (12–14 days) has been shown to upregulate Sch-28080-sensitive H,K-ATPase mediated pH recovery in rat CCD ICs that is also sensitive to ouabain [39]. Considering these studies, if $HK\alpha_2$ H,K-ATPase mediates aldosterone stimulated acid secretion, the action may take place in the rat CCD and not the medulla.

Na/H exchange and H-ATPase, but not H,K-ATPase, activity were reported to be responsible for increased H secretion due to endothelin-mediated increased aldosterone activity in rats [95–97]. This study utilized micropuncture of accessible distal tubules and did not characterize the role of H,K-ATPase-mediated acid secretion in clearly identified nephron segments known to express this important acid secretory mechanism (i.e. CDD, OMCD, and IMCD). The plausibility of an H,K-ATPase mediating aldosterone and endothelin stimulated H secretion warrants further investigation.

Hormones may be involved in the cascade of events needed to upregulate the $HK\alpha_2$ H,K-ATPase during K depletion since hypophysectomized rats exhibited an attenuated response [51]. Neither NHE isoforms nor the $HK\alpha_1$ isoform was upregulated with K depletion in this study. Hormonal regulation of H,K-ATPase activities has been studied in CD [98–100]. Calcitonin, 1-deamino-8-D-arginine vasopressin (dD-AVP), and isoproterenol have been demonstrated to stimulate H,K-ATPase activity. In the CCD, adenylate cyclase (AC) was stimulated by vasopressin, calcitonin, and isoproterenol in PCs, A-type ICs, and B-type ICs, respectively. In the OMCD, AC was stimulated by vasopressin in PCs and by glucagon in A-type ICs. Using these cell-specific regulatory hormones, it was determined that cyclic AMP (cAMP) stimulates Type I activity in A- and B-type ICs of the CCD and Type III activity in PCs of the CCD and OMCD and ICs of the OMCD [98].

Furthermore, the signaling pathway involved in the stimulation of H,K-ATPase by calcitonin that occurred in the A-type ICs involved a cAMP/EpacI/Rap-1/Raf-B/ERK cascade and was independent of protein kinase A (PKA) [100]. However, the stimulation of H,K-ATPase by isoproteronol that occurred in the B-type ICs was PKA dependent and involved the phosphorylation of β-adrenergic receptors and the switch from G_s to G_i coupling leading to the tyrosine kinase/Ras/Raf-1 cascade and ERK1/2 activation [99]. An independent study concluded that the cAMP-PKA pathway mediated the stimulation of ouabain-sensitive H,K-ATPase in the cortex and medulla [101]. Finally, the stimulation of protein kinase C has been reported to mediate the internalization of the human $HK\alpha_2$ H,K-ATPase [102].

4.6 Na Transport

There appears to be a significant interaction of Na and K for reabsorption by an apparently common mechanism at least for the $HK\alpha_1$. Moreover, studies under different physiological conditions support a significant interaction of Na and K with a Sch-28080-sensitive absorptive mechanism. Studies in the CCD concluded that K and Na compete for the same reabsorptive pathway [22]. Possible competitive binding between Na and K on the H,K-ATPase are supported by the observation that two structurally different H,K-ATPase inhibitors, Sch-28080 and A80915a, can inhibit Na reabsorption in the rabbit CCD under conditions of dietary Na depletion [103]. This observation suggests that the $HK\alpha_1$ isoform can mediate Na absorption since Sch-28080 is the classic gastric-type inhibitor. Dherbecourt et al. reported K-ATPase activity (Type III) in $HK\alpha_1$-knockout mice that is sensitive to both Sch-28080 and ouabain [17]. Direct evidence distinguished Type III activity from Type I activity by several characteristics, but of particular note, Type III activity can be stimulated by Na which suggests direct binding of Na on this enzyme [55]. The studies of Swarts et al. support a role for Na (in place of K) in the catalytic reaction kinetics of gastric membrane vesicles (Fig. 4). If the $HK\alpha_1$ H,K-ATPase can transport Na under physiological conditions, then the disruption of the apical NHE-3 would lead to enhanced Na luminal delivery and plausibly an upregulation of $HK\alpha_1$ as observed by Nakamura et al.

Dietary K depletion has been shown to produce a volume- and Na-dependent form of hypertension despite reduction of plasma aldosterone concentration. As noted previously evidence suggests that the H,K-ATPase could participate in neutral salt uptake. Experimental evidence supports the functional coupling of an H,K-ATPase and an apical Cl/HCO_3 exchanger in the CCD, possibly in the B-type IC [103]. Under conditions of K restriction it is plausible that either enzyme or both enzymes may be involved in sodium retention.

There is some physiological evidence that one or more H,K-ATPase isoforms are stimulated by dietary Na depletion which would be consistent with a role for H,K-ATPase in Na reabsorption. The ICs from the split open CCD of rats on a low Na diet with elevated serum aldosterone levels had two-fold higher K-dependent pH_i recovery from an acid pulse than normal Na diet rats [39]. The control K-dependent

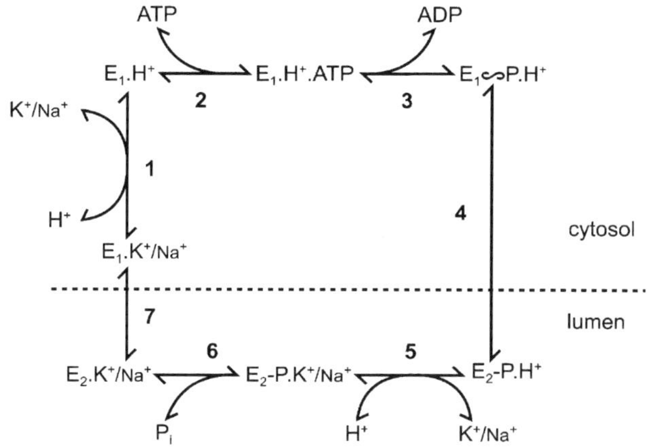

Fig. 4 A modified Post–Albers scheme for $HK\alpha_1$ containing "H,K-ATPase." The proposed normal reaction cycle of the $HK\alpha_1$ H,K-ATPase (modified from Swarts et al. [114]) uses the Post–Albers scheme originally derived from studies of Na,K-ATPase. The smaller font for Na^+ reflect the fact that Na^+ can be transported albeit at lesser efficiency than K^+ by the enzyme complex. The enzyme cycle normally turns clockwise

pH recovery was blocked by Sch-28080, but abolishment in the ICs from the low Na animals required the application of both Sch-28080 and ouabain. These results suggested that NaCl depletion induces an additional H,K-ATPase isoform. However, in separate studies, dietary Na-depletion did not upregulate the kidney $HK\alpha_2$ mRNA [49, 104] or protein [104] which would suggest that yet a third isoform (or a heterodimer enzyme complex) is induced under these conditions (or that the increase in pump activity occurred as a post-translational event). Such studies are compatible with the conclusions of Petrovic et al. that suggest that the kidney of $HK\alpha_1$ knockout mice possess a novel H,K-ATPase that was neither $HK\alpha_1$ or $HK\alpha_2$ [18].

4.7 Dietary K – Functional Modeling and the Role of K Channels

Functional studies lead to the development of an H,K-ATPase model governed by dietary K status. This model suggests that the H,K-ATPase functions in coordination with either apical or basolateral K channels depending on the availability of K. Under normal dietary K-conditions, the H,K-ATPase functions in proton secretion and K recycling in coordination with apical K channels. Figure 5 models the major transporters involved in K and acid/base transport in the three cell types of the CD. Data from K-tracer, HCO_3 flux, and pH_i recovery experiments clearly demonstrate that an apical H,K-ATPase functions to reabsorb K and secrete protons in all segments of the CD [19, 20, 24, 26, 32, 45]. Additional in vitro microperfusion studies measuring H,K-ATPase-mediated acid secretion in the CCD in response to increased ambient pCO_2 revealed that the enzyme's activity was dependent on the application of barium, a K-channel blocker [20, 21]. The side of the blockade

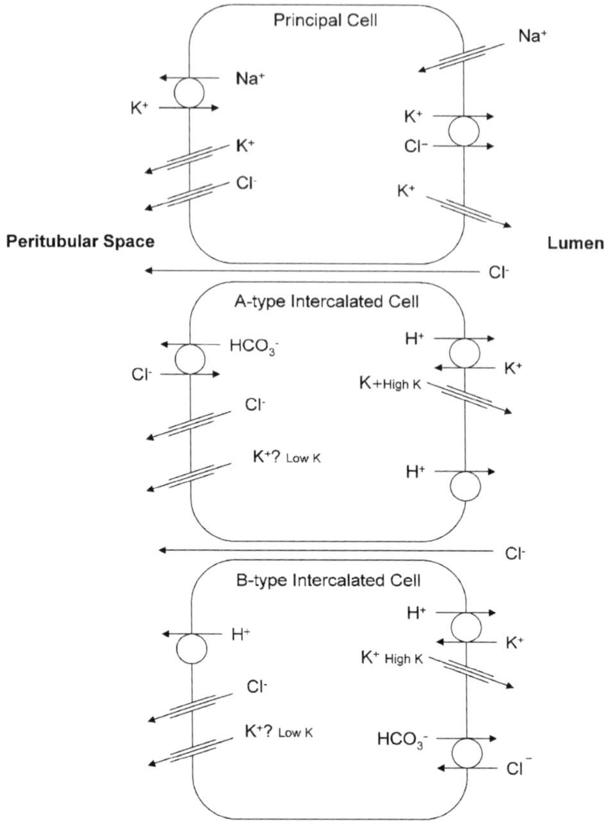

Fig. 5 The major transporters involved in K and acid/base transport in principle cells, A-type intercalated cells, and B-type intercalated cells of the CD. Apical recycling of K is important for acid secretion during K replete conditions. During K depletion, K reabsorption and acid secretion are dependent on basolateral K channels

depended on the dietary K intake of the rabbit. Under K-replete conditions, barium blocked acid secretion in the CCD when applied to the lumen. In contrast, under K-deplete conditions, peritubular application of barium blocked acid secretion. Also, under low-K conditions, K reabsorption was inhibited by peritubular barium. From this data, a model was proposed coordinating the H,K-ATPase with apical and basolateral K channels suggesting apical membrane recycling of K during K repletion and K reabsorption via basolateral K channels during K depletion.

There are multiple K channel classes at the apical membrane of the CD [105]; therefore, it might be expected that more than one single channel type would partner with the H,K-ATPase. Interestingly, in stomach where H,K-ATPase is essential for acid secretion, the inhibition of the K channel KCNQ1 or disruption of its expression inhibits acid secretion [106–108]. KCNQ1 is present in the kidney including the CD [109]. KCNQ1 mRNA was expressed intensely and basolaterally in the DCT and CNT. Diffuse cellular labeling was apparent in the ICT and CCD with some cells exhibiting discrete apical labeling. These cells were primarily PCs and B-type ICs and a few A-type ICs. In the medulla, there was diffuse staining of ICs and basolateral staining in PCs. Future investigations will assess the effect of dietary K on KCNQ1 localization and whether selective inhibitors of KCNQ1 affect H,K-ATPase activity.

5 Newer Observations and Future Direction

Mice lacking functional expression of HKα_1, HKα_2, or both HKα_1 and HKα_2 (double knockout, HK$\alpha_{1,2}$) provide a unique opportunity to evaluate the effects of specific HKα gene product deficiency at the cellular level in the absence of pharmacological inhibitors. The deletion of HKα_1 causes significant morphological changes in the gastric epithelia and impairs acid secretion [110]. The deletion of HKα_2 causes colonic K wasting that is exacerbated under dietary K restriction [61].

The effects of these single or the combined disruption of HKα_1 and HKα_2 on acid secretion in the ICs of the CCD has been recently published [79]. As mentioned previously, assessing the contribution of each isoform to acid extrusion using the traditional pharmacological approach may be less than ideal. This study used a genetic approach to determine the contribution of HKα_1 and HKα_2 to CCD IC proton transport in contrast to a pharmacological approach. Rates of pH$_i$ recovery and acid secretion were investigated in response to acute acid-loading in A-type and B-type ICs of the CCD of wild-type mice and in mice with genetic disruption of HKα_1, HKα_2, or both HKα_1 and HKα_2. IC types were defined based on the pH$_i$ change in response to basolateral chloride removal. The results confirm a significant and separate contribution of both HKα_1 and HKα_2 to acid secretion in both A-type and B-type ICs in mice fed a normal diet. The deletion of each individual α subunit resulted in an impairment in the ability of ICs of the CCD to secrete protons. In mice lacking both α subunits, there was an additional significant reduction in pH recovery in the A-type IC. These data provide direct evidence that H,K-ATPases participate in H transport under normal dietary conditions and unambiguously demonstrate that mice fed a normal diet clearly possess two separate mechanisms of H extrusion, one which can be attributed to HKα_1 and a second, functionally distinct mechanism that can be attributed to HKα_2. Future studies will be needed to examine the physiological contribution of each H pump under different physiological stimuli.

The CD is capable of K secretion or reabsorption, and it is well established that the renal HKα_2 H,K-ATPase is preferentially upregulated during dietary K restriction (previously discussed). Interestingly, HKα_2 is not upregulated by K depletion in the colon [49, 61]. Yet, Meneton et al. report that increased sensitivity to K deprivation in the HKα_2 knockout mice was due to fecal K wasting. They do not rule out that the renal HKα_2 may contribute to K conservation and note that HKα_2 knockout mice had consistently higher mean urinary K excretion values, but the differences were small. It should be emphasized that both isoforms are active in the kidney, and such data suggests that one isoform may compensate for the absence of the other. No significant effect of renal HKα_1 or HKα_2 knockout on whole animal K-homeostasis has been observed, but in each case these animals had another intact HKα isoform.

Initial investigations suggest that no overt renal phenotype exist in either individual knock out animal [61, 104, 110]. Enzymes with similar function could compensate for the deletion of a given HKα subunit including other H,K-ATPase α subunits. H-ATPase has been reported to be upregulated in the HKα_1 knock out mice [111]. Similarly, HKα_1 has been reported to compensate for the deletion of NHE [87]. To date, there are no reports that HKα_2 is upregulated with the deletion of HKα_1; however, it has been reported that another K-dependent proton secretory mechanism exists in the HKα_1 knockout mouse [18]. Clearly there is a need for additional investigations in these H,K-ATPase α subunit-deficient mice especially in mice deficient in both HKα_1 and HKα_2. Arterial blood gas and electrolyte analysis of urine, feces, and blood under various conditions of physiological stress (ex. low dietary K or Na and chronic metabolic acidosis) are needed to fully understand the role of this enzyme in renal function. Clearly, defects in compensatory acid-extrusion mechanisms require a complete understanding of the H,K-ATPase's role in renal function.

Few, if any, studies have examined the effects of maneuvers that change either HKα_1 or HKα_2 on the regulation of HKβ. In the gastric parietal cells this protein is essential for gastric acid secretion [15], K-ATPase activity (Shao, unpublished observations), and localization to the plasma membrane [111]. Hence, one of the unresolved issues is which protein is rate limiting to the production of an intact holoenzyme complex. Given the complexity of the kidney and the multiple other renal transporters it is clear that substantial work needs to be done to define the role of either HKα_1 or HKα_2 in the kidney and the organism.

In summary, at least two, and probably more H,K-ATPase isoforms exist in the kidney. The initial rationale and current prevailing model proposes a role for both pumps in K reabsorption depending on dietary K intake and luminal proton secretion under both normal and K-restricted conditions. The magnitude of net tCO_2 absorption (net steady state bicarbonate absorption) that is sensitive to specific H,K-ATPase inhibitors is substantial. However, the evidence reviewed herein directly supports a role for HKα_1 H,K-ATPase in CD Na reabsorption and indirectly for HKα_2 H,K-ATPase. If substantial Na transport occurs in vivo via these pumps, the previous perplexing data may be more clearly understood. Equally important, such observations may have profound impact on our understanding of acid/base, volume, and systemic blood pressure regulation. Whereas knockout studies have unambiguously identified both pumps in the kidney in animals on a normal diet, compensatory mechanisms or nonrenal actions of H,K-ATPase make it important to establish how these flux studies translate to the intact organism. Logically, selective disruption of the expression of these pumps strictly in the CD is needed to establish whether global knockout studies do not have additional effects outside the kidney. If CD specific knockout exhibit a different phenotype then it can be inferred that these pumps have additional functions outside their established and classical sites of action. The case report by Simpson and Schwartz of severe hypokalemia and distal renal tubular acidosis may provide a clinical correlation for future studies [112].

References

1. Swarts HGP, Klaassen CHW, Schuurmans Stekhvoen FMAH, De Pont JJHHM. Sodium acts as a potassium analog on gastric H-K-ATPase. J Biol Chem 1995; 270:7890–7895.
2. Palmgren MG, Axelsen KB. Evolution of P-type ATPases. Biochim Biophys Acta 1998; 1365(1–2):37–45.
3. Shin JM, Sachs G. Dimerization of the gastric H^+,K^+-ATPase. J Biol Chem 1996; 271(4):1904–1908.
4. Ivanov AV, Modyanov NN, Askari A. Role of the self-association of beta subunits in the oligomeric structure of Na+/K+-ATPase. Biochem J 2002; 364(Pt 1):293–299.
5. Codina J, DuBose TD, Jr. Molecular regulation and physiology of the H+,K+ -ATPases in kidney. Semin Nephrol 2006; 26(5):345–351.
6. Crowson MS, Shull GE. Isolation and characterization of a cDNA encoding the putative distal colon H^+,K^+-ATPase. Similarity of deduced amino acid sequence to gastric H^+,K^+-ATPase and Na^+,K^+-ATPase and mRNA expression in distal colon, kidney, and uterus. J Biol Chem 1992; 267:13740–13748.
7. Kone BC, Higham SC. A novel N-terminal splice variant of the rat H^+-K^+-ATPase $\alpha 2$ subunit. J Biol Chem 1998; 273(5):2543–2552.
8. Campbell WG, Weiner ID, Wingo CS, Cain BD. H-K-ATPase in the RCCT-28A rabbit cortical collecting duct cell line. Am J Physiol 1999; 276(2 Pt 2):F237–F245.
9. Gumz ML, Duda D, McKenna R, Wingo CS, Cain BD. Molecular modeling of the rabbit colonic (HKalpha2a) H(+), K(+) ATPase. J Mol Model (Online) 2003; 9(5): 283–289.
10. Asano S, Morii M, Takeguchi N. Molecular and cellular regulation of the gastric proton pump. Biol Pharm Bull 2004; 27(1):1–12.
11. Asano S, Yoshida A, Yashiro H, Kobayashi Y, Morisato A, Ogawa H et al. The cavity structure for docking the K(+)-competitive inhibitors in the gastric proton pump. J Biol Chem 2004; 279(14):13968–13975.
12. Togawa K, Ishiguro T, Kaya S, Shimada A, Imagawa T, Taniguchi K. Reversible phosphorylation of both Tyr7 and Tyr10 in the alpha-chain of pig stomach H^+, K^+ ATPase by a membrane-bound kinase and a phosphatase. J Biol Chem 1995; 270(26): 15475–15478.
13. Feschenko MS, Wetzel RK, Sweadner KJ. Phosphorylation of Na,K-ATPase by protein kinases. Sites, susceptibility, and consequences. Ann NY Acad Sci 1997; 834: 479–488.
14. Courtois-Coutry N, Roush D, Rajendran V, McCarthy JB, Geibel J, Kashgarian M et al. A tyrosine-based signal targets H/K-ATPase to a regulated compartment and is required for the cessation of gastric acid secretion. Cell 1997; 90(3):501–510.
15. Scarff KL, Judd LM, Toh BH, Gleeson PA, Van Driel IR. Gastric H^+,K^+-Adenosine Triphosphatase beta Subunit Is Required for Normal Function, Development, and Membrane Structure of Mouse Parietal Cells. Gastroenterology 1999; 117(3):605–618.
16. Wallmark B, Briving C, Fryklund J, Munson K, Jackson R, Mendlein J et al. Inhibition of gastric H^+,K^+-ATPase and acid secretion by SCH 28080, a substituted pyridyl (1,2a)imidazole. J Biol Chem 1987; 262:2077–2084.
17. Dherbecourt O, Cheval L, Bloch-Faure M, Meneton P, Doucet A. Molecular identification of Sch28080-sensitive K-ATPase activities in the mouse kidney. Pflugers Arch 2006; 451(6):769–775.
18. Petrovic S, Spicer Z, Greeley T, Shull GE, Soleimani M. Novel Schering and ouabain-insensitive potassium-dependent proton secretion in the mouse cortical collecting duct. Am J Physiol Renal Physiol 2002; 282(1):F133–F143.
19. Zhou X, Nakamura S, Xia SL, Wingo CS. Increased CO_2 stimulates K/Rb reabsorption mediated by H-K-ATPase in CCD of potassium-restricted rabbit. Am J Physiol Renal Physiol 2001; 281(2):F366–F373.

20. Zhou X, Lynch IJ, Xia SL, Wingo CS. Activation of H^+-K^+-ATPase by CO_2 requires a basolateral Ba^{2+}-sensitive pathway during K restriction. Am J Physiol 2000; 279(1):F153–F160.
21. Zhou X, Wingo CS. Stimulation of total CO_2 flux by 10% CO_2 in rabbit CCD: role of an apical Sch-28080- and Ba-sensitive mechanism. Am J Physiol 1994; 267(Renal 36):F114–F120.
22. Zhou X, Wingo CS. Mechanisms of rubidium permeation by rabbit cortical collecting duct during potassium restriction. Am J Physiol 1992; 263:F1134–F1141.
23. Zhou X, Wingo CS. Mechanisms for enhancement of Rb efflux by 10% CO_2 in cortical collecting duct (CCD). Clin Res 40(2), 179A. 1992. Ref Type: Abstract.
24. Wingo CS, Armitage FE. Rubidium absorption and proton secretion by rabbit outer medullary collecting duct via H-K-ATPase. Am J Physiol 1992; 263:F849–F857.
25. Gifford JD, Rome L, Galla JH. H^+-K^+-ATPase activity in rat collecting duct segments. Am J Physiol 1992; 262:F692–F695.
26. Armitage FE, Wingo CS. Luminal acidification in the K-Replete $OMCD_i$: Contributions of H-K-ATPase and bafilomycin-A_1-sensitive H-ATPase. Am J Physiol 1994; 267: F450–F458.
27. Armitage FE, Wingo CS. Luminal acidification in the potassium-replete $OMCD_i$: Inhibition of bicarbonate absorption by K removal and luminal Ba. Am J Physiol 1995; 269(Renal 38):F116–F124.
28. Kuwahara M, Fu WJ, Marumo F. Functional activity of H-K-ATPase in individual cells of OMCD: localization and effect of K^+ depletion. Am J Physiol 1996; 270:F116–F122.
29. Ono S, Guntupalli J, DuBose TD, Jr. Role of H^+-K^+-ATPase in pH_i regulation in inner medullary collecting duct cells in culture. Am J Physiol 1996; 270(Renal 39): F852–F861.
30. Wall SM, Truong AV, DuBose TD, Jr. H^+-K^+-ATPase mediates net acid secretion in rat terminal inner medullary collecting duct. Am J Physiol 1996; 271(Renal 40):F1037–F1044.
31. Nakamura S, Amlal H, Galla JH, Soleimani M. Colonic H-K-ATPase is induced and mediates increased HCO3- reabsorption in inner medullary collecting duct in potassium depletion. Kidney Int 1998; 54(4):1233–1239.
32. Wall SM, Mehta P, DuBose TD, Jr. Dietary K+ restriction upregulates total and Sch-28080-sensitive bicarbonate absorption in rat tIMCD. Am J Physiol 1998; 275(4 Pt 2):F543–F549.
33. Frank AE, Weiner ID. Effects of ammonia on acid-base transport by the B-type intercalated cell 233. J Am Soc Nephrol 2001; 12(8):1607–1614.
34. Milton AE, Weiner ID. Intracellular pH regulation in the rabbit cortical collecting duct A-type intercalated cell. Am J Physiol 1997; 273(Renal 42):F340–F347.
35. Silver RB, Mennitt PA, Satlin LM. Stimulation of apical H-K-ATPase in intercalated cells of cortical collecting duct with chronic metabolic acidosis. Am J Physiol 1996; 270(3 Pt 2):F539–47.
36. Silver RB, Frindt G. Functional identification of H-K-ATPase in intercalated cells of cortical collecting tubule. Am J Physiol 1993; 264(Renal 33):F259–F266.
37. Weiner ID, Milton AE. H^+-K^+-ATPase in rabbit cortical collecting duct B-type intercalated cell. Am J Physiol 1996; 270(Renal 39):F518–F530.
38. Silver RB, Frindt G, Mennitt P, Satlin LM. Characterization and regulation of H-K-ATPase in intercalated cells of rabbit cortical collecting duct. J Exp Zool 1997; 279(5):443–455.
39. Silver RB, Choe H, Frindt G. Low NaCl diet increases H-K-ATPase in intercalated cells from rat cortical collecting duct. Am J Physiol 1998; 44(1):F94–F102.
40. Lynch IJ, Rudin A, Xia SL, Stow LR, Shull GE, Weiner ID et al. Impaired acid secretion in cortical collecting duct intercalated cells from H-K-ATPase-deficient mice: role of HK{alpha} isoforms. Am J Physiol Renal Physiol 2008; 294(3):F621–F627.
41. Bastani B. Co-localization of H-ATPase and H,K-ATPase immunoreactivity in the rat kidney. J Am Soc Nephrol 1995; 5:1476–1482.
42. Wingo CS, Madsen KM, Smolka A, Tisher CC. H-K-ATPase immunoreactivity in cortical and outer medullary collecting duct. Kidney Int 1990; 38:985–990.

43. Ahn KY, Kone BC. Expression and cellular localization of mRNA encoding the "gastric" isoform of H(+)-K(+)-ATPase alpha-subunit in rat kidney 176. Am J Physiol 1995; 268(1 Pt 2):F99–109.
44. Codina J, Kone BC, Delmas-Mata JT, DuBose TD, Jr. Functional expression of the colonic H^+,K^+-ATPase α-subunit. Pharmacologic properties and assembly with X^+,K^+-ATPase β-subunits. J Biol Chem 1996; 271(47):29759–29763.
45. Zhou X, Wingo CS. H-K-ATPase enhancement of Rb efflux by cortical collecting duct. Am J Physiol 1992; 263:F43–F48.
46. Nakamura S, Wang Z, Galla JH, Soleimani M. K^+ depletion increases HCO_3^- reabsorption in OMCD by activation of colonic H^+-K^+-ATPase. Am J Physiol 1998; 43(4): F687–F692.
47. Ahn KY, Park KY, Kim KK, Kone BC. Chronic hypokalemia enhances expression of the H^+-K^+-ATPase α_2-subunit gene in renal medulla. Am J Physiol 1996; 271(Renal 40):F314–F321.
48. Kraut JA, Hiura J, Besancon M, Smolka A, Sachs G, Scott D. Effect of hypokalemia on the abundance of HKα_1 and HKα_2 protein in the rat kidney. Am J Physiol 1997; 272(Renal 41):F744–F750.
49. Sangan P, Rajendran VM, Mann AS, Kashgarian M, Binder HJ. Regulation of colonic H-K-ATPase in large intestine and kidney by dietary Na depletion and dietary K depletion. Am J Physiol 1997; 272(2 Pt 1):C685–C696.
50. DuBose TD, Jr., Codina J, Burges A, Pressley TA. Regulation of H^+,K^+-ATPase expression in kidney. Am J Physiol 1995; 269(Renal 38):F500–F507.
51. Wang Z, Baird N, Shumaker H, Soleimani M. Potassium depletion and acid-base transporters in rat kidney: Differential effect of hypophysectomy. Am J Physiol 1997; 41(6): F736–F743.
52. Verlander JW, Moudy RM, Campbell WG, Cain BD, Wingo CS. Immunohistochemical localization of H-K-ATPase α_{2c}-subunit in rabbit kidney. Am J Physiol Renal Physiol 2001; 281(2):F357–F365.
53. Fejes-Toth G, Rusvai E, Longo KA, Naray-Fejes-Toth A. Expression of colonic H-K-ATPase mRNA in cortical collecting duct: regulation by acid/base balance. Am J Physiol 1995; 38(4):F551–F557.
54. Silver RB, Soleimani M. H^+-K^+-ATPases: regulation and role in pathophysiological states. Am J Physiol 1999; 276(6 Pt 2):F799–F811.
55. Buffin-Meyer B, Younes-Ibrahim M, Barlet-Bas C, Cheval L, Marsy S, Doucet A. K depletion modifies the properties of Sch-28080-sensitive K-ATPase in rat collecting duct. Am J Physiol 1997; 272(Renal 41):F124–F131.
56. Younes-Ibrahim M, Bartlet-Bas C, Buffin-Meyer B, Cheval L, Rajerison R, Doucet A. Ouabain-sensitive and -insensitive K-ATPases in rat nephron: effect of K depletion. Am J Physiol 1995; 268(Renal 37):F1141–F1147.
57. Beltowski J, Wojcicka G. Spectrophotometric method for the determination of renal ouabain-sensitive H+,K+-ATPase activity. Acta Biochim Pol 2002; 49(2):515–527.
58. Lee J, Rajendran VM, Mann AS, Kashgarian M, Binder HJ. Functional Expression and Segmental Localization of Rat Colonic K-Adenosine Triphosphatase. J Clin Invest 1995; 96:2002–2008.
59. Fejes-Toth G, Naray-Fejes-Toth A. Immunohistochemical localization of colonic H-K-ATPase to the apical membrane of connecting tubule cells. Am J Physiol Renal Physiol 2001; 281(2):F318–F325.
60. Ahn KY, Turner PB, Madsen KM, Kone BC. Effects of chronic hypokalemia on renal expression of the "gastric" H^+-K^+-ATPase α-subunit gene. Am J Physiol 1996; 270(Renal 39):F557–F566.
61. Meneton P, Schultheis PJ, Greeb J, Nieman ML, Liu LH, Clarke LL et al. Increased sensitivity to K^+ deprivation in colonic H,K,ATPase-deficient mice. J Clin Invest 1998; 101(3):536–542.

62. Ahn KY, Park KY, Kim KK, Kone BC. Chronic hypokalemia enhances expression of the H(+)-K(+)-ATPase alpha 2-subunit gene in renal medulla. Am J Physiol 1996; 271(2 Pt 2):F314–F321.
63. Fejes-Toth G, Naray-Fejes-Toth A, Velazquez H. Intrarenal distribution of the colonic H,K-ATPase mRNA in rabbit. Kidney Int 1999; 56(3):1029–1036.
64. Modyanov NN, Petrukhin KE, Sverdlov VE, Grishin AV, Orlova MY, Kostina MB et al. The family of human Na,K-ATPase genes: *ATP1AL1* gene is transcriptionally competent and probably encodes the related ion transport ATPase. FEBS Lett 1991; 278:91–94.
65. Johansson M, Jansson T, Pestov NB, Powell TL. Non-gastric H+/K+ ATPase is present in the microvillous membrane of the human placental syncytiotrophoblast. Placenta 2004; 25(6):505–511.
66. Grishin AV, Sverdlov VE, Kostina MB, Modyanov NN. Cloning and characterization of the entire cDNA encoded by ATP1AL1 – A member of the human Na,K/H,K-ATPase gene family. FEBS Lett 1994; 349:144–150.
67. Sverdlov VE, Kostina MB, Modyanov NN. Genomic organization of the human ATP1AL1 gene encoding a ouabain-sensitive H,K-ATPase. Genomics 1996; 32(3): 317–327.
68. Pestov NB, Romanova LG, Korneenko TV, Egorov MV, Kostina MB, Sverdlov VE et al. Ouabain-sensitive H,K-ATPase: tissue-specific expression of the mammalian genes encoding the catalytic alpha subunit. FEBS Lett 1998; 440(3):320–324.
69. Kraut JA, Helander KG, Helander HF, Iroezi ND, Marcus EA, Sachs G. Detection and localization of H+-K+-ATPase isoforms in human kidney. Am J Physiol Renal Physiol 2001; 281(4):F763–F768.
70. Wingo CS, Smolka AJ. Function and structure of H,K-ATPase in the kidney. Am J Physiol 1995; 269(Renal 38):F1–F16.
71. Tsuruoka S, Schwartz GJ. Metabolic Acidosis Stimulates H^+ Secretion in the Rabbit Outer Medullary Collecting Duct (Inner Stripe) of the Kidney. J Clin Invest 1997; 99(6): 1420–1431.
72. Wingo CS. Potassium transport by the medullary collecting tubule of rabbit: effects of variation in K intake. Am J Physiol 1987; 253:F1136–F1141.
73. Wingo CS. Active proton secretion and potassium absorption in the rabbit outer medullary collecting duct – functional evidence for proton-potassium activated adenosine triphosphatase. J Clin Invest 1989; 84:361–365.
74. Guntupalli J, Onuigbo M, Wall S, Alpern RJ, DuBose TD, Jr. Adaptation to low-K^+ media increases H^+-K^+-ATPase but not H^+-ATPase-mediated pH_i recovery in $OMCD_i$ cells. Am J Physiol 1997; 273(Cell Physiol 42):C558–C571.
75. Codina J, Delmas-Mata JT, DuBose TD, Jr. Expression of HKalpha2 protein is increased selectively in renal medulla by chronic hypokalemia. Am J Physiol 1998; 275(3 Pt 2): F433–F440.
76. Weiner ID, Hamm LL. Use of fluorescent dye BCECF to measure intracellular pH in cortical collecting tubule. Am J Physiol 1989; 256:F957–F964.
77. Weiner ID, Frank AE, Wingo CS. Apical proton secretion by the inner stripe of the outer medullary collecting duct. Am J Physiol 1999; 276(4 Pt 2):F606–F613.
78. Yip KP, Tsuruoka S, Schwartz GJ, Kurtz I. Apical H(+)/base transporters mediating bicarbonate absorption and pH(i) regulation in the OMCD. Am J Physiol Renal Physiol 2002; 283(5):F1098-F1104.
79. Lynch IJ, Rudin A, Xia SL, Stow LR, Shull GE, Weiner ID et al. Impaired acid secretion in cortical collecting duct intercalated cells from H-K-ATPase-deficient mice: role of HKalpha isoforms. Am J Physiol Renal Physiol 2008; 294(3):F621–F627.
80. Knepper MA, Good DW, Burg MB. Ammonia and bicarbonate transport by rat cortical collecting ducts perfused in vitro. Am J Physiol 1985; 249:F870–F877.
81. Wall SM. NH_4^+ augments net acid secretion by a ouabain-sensitive mechanism in isolated perfused inner medullary collecting ducts. Am J Physiol 1996; 270:F432–F439.

82. Frank AE, Wingo CS, Weiner ID. Effects of ammonia on bicarbonate transport in the cortical collecting duct. Am J Physiol 2000; 278(2):F219–F226.
83. Frank AE, Wingo CS, Andrews PM, Ageloff S, Knepper MA, Weiner ID. Mechanisms through which ammonia regulates cortical collecting duct net proton secretion. Am J Physiol Renal Physiol 2002; 282(6):F1120–F1128.
84. Wang X, Wu J, Li L, Chen F, Wang R, Jiang C. Hypercapnic acidosis activates KATP channels in vascular smooth muscles. Circ Res 2003; 92(11):1225–1232.
85. Wang Z, Rabb H, Craig T, Burnham C, Shull GE, Soleimani M. Ischemic-reperfusion injury in the kidney: Overexpression of colonic H^+-K^+-ATPase and suppression of NHE-3. Kidney Int 1997; 51(4):1106–1115.
86. Schultheis PJ, Clarke LL, Meneton P, Miller ML, Soleimani M, Gawenis LR et al. Renal and intestinal absorptive defects in mice lacking the NHE3 Na+/H+ exchanger. Nat Genet 1998; 19(3):282–285.
87. Nakamura S, Amlal H, Schultheis PJ, Galla JH, Shull GE, Soleimani M. HCO_3^- reabsorption in renal collecting duct of NHE-3-deficient mouse: a compensatory response. Am J Physiol 1999; 276(6 Pt 2):F914–F921.
88. Wagner CA, Geibel JP. Acid-base transport in the collecting duct. J Nephrol 2002; 15 Suppl 5:S112–S127.
89. Stanton B, Pan L, Deetjen H, Guckian V, Giebisch G. Independent effects of aldosterone and potassium on induction of potassium adaptation in rat kidney. J Clin Invest 1987; 79: 198–206.
90. Koeppen BM, Helman SI. Acidification of luminal fluid by the rabbit cortical collecting tubule perfused in vitro. Am J Physiol 1982; 242:F521–F531.
91. Stone DK, Seldin DW, Kokko JP, Jacobson HR. Mineralocorticoid modulation of rabbit medullary collecting duct acidification. J Clin Invest 1983; 72:77–83.
92. Kuwahara M, Sasaki S, Marumo F. Mineralocorticoids and acidosis regulate $H^+/HCO^-{}_3$ transport of intercalated cells. J Clin Invest 1992; 89:1388–1394.
93. Weiner ID, Wingo CS, Hamm LL. Regulation of intracellular pH in two cell populations of inner stripe of rabbit outer medullary collecting duct. Am J Physiol 1993; 265(Renal 34):F406–F415.
94. DuBose TD, Jr., Caflisch CR. Effect of selective aldosterone deficiency on acidification in nephron segments of the rat inner medulla. J Clin Invest 1988; 82(5):1624–1632.
95. Khanna A, Simoni J, Hacker C, Duran MJ, Wesson DE. Increased endothelin activity mediates augmented distal nephron acidification induced by dietary protein. J Am Soc Nephrol 2004; 15(9):2266–2275.
96. Khanna A, Simoni J, Hacker C, Duran MJ, Wesson DE. Increased endothelin activity mediates augmented distal nephron acidification induced by dietary protein. Trans Am Clin Climatol Assoc 2005; 116:239–256.
97. Khanna A, Simoni J, Wesson DE. Endothelin-induced increased aldosterone activity mediates augmented distal nephron acidification as a result of dietary protein. J Am Soc Nephrol 2005; 16(7):1929–1935.
98. Laroche-Joubert N, Marsy S, Doucet A. Cellular origin and hormonal regulation of K(+)-ATPase activities sensitive to Sch-28080 in rat collecting duct. Am J Physiol Renal Physiol 2000; 279(6):F1053–F1059.
99. Laroche-Joubert N, Marsy S, Luriau S, Imbert-Teboul M, Doucet A. Mechanism of activation of ERK and H-K-ATPase by isoproterenol in rat cortical collecting duct. Am J Physiol Renal Physiol 2003; 284(5):F948–F954.
100. Laroche-Joubert N, Marsy S, Michelet S, Imbert-Teboul M, Doucet A. Protein kinase A-independent activation of ERK and H,K-ATPase by cAMP in native kidney cells: role of Epac I. J Biol Chem 2002; 277(21):18598–18604.
101. Beltowski J, Marciniak A, Wojcicka G, Gorny D. Regulation of renal Na(+),K(+)-ATPase and ouabain-sensitive H(+),K(+)-ATPase by the cyclic AMP-protein kinase A signal transduction pathway. Acta Biochim Pol 2003; 50(1):103–114.

102. Reinhardt J, Kosch M, Lerner M, Bertram H, Lemke D, Oberleithner H. Stimulation of protein kinase C pathway mediates endocytosis of human nongastric H+-K+-ATPase, ATP1AL1. Am J Physiol Renal Physiol 2002; 283(2):F335–F343.
103. Zhou X, Xia SL, Wingo CS. Chloride transport by the rabbit cortical collecting duct: dependence on H,K-ATPase. J Am Soc Nephrol 1998; 9(12):2194–2202.
104. Spicer Z, Clarke LL, Gawenis LR, Shull GE. Colonic H(+)-K(+)-ATPase in K(+) conservation and electrogenic Na(+) absorption during Na(+) restriction. Am J Physiol Gastrointest Liver Physiol 2001; 281(6):G1369–G1377.
105. Hebert SC, Desir G, Giebisch G, Wang W. Molecular diversity and regulation of renal potassium channels. Physiol Rev 2005; 85(1):319–371.
106. Lee MP, Ravenel JD, Hu RJ, Lustig LR, Tomaselli G, Berger RD et al. Targeted disruption of the Kvlqt1 gene causes deafness and gastric hyperplasia in mice. J Clin Invest 2000; 106(12):1447–1455.
107. Grahammer F, Herling AW, Lang HJ, Schmitt-Graff A, Wittekindt OH, Nitschke R et al. The cardiac K+ channel KCNQ1 is essential for gastric acid secretion. Gastroenterology 2001; 120(6):1363–1371.
108. Lambrecht NW, Yakubov I, Scott D, Sachs G. Identification of the K efflux channel coupled to the gastric H-K-ATPase during acid secretion. Physiol Genomics 2005; 21(1):81–91.
109. Zheng W, Verlander JW, Lynch IJ, Cash M, Shao J, Stow LR et al. Cellular distribution of the potassium channel KCNQ1 in normal mouse kidney. Am J Physiol Renal Physiol 2007; 292(1):F456–F466.
110. Spicer Z, Miller ML, Andringa A, Riddle TM, Duffy JJ, Doetschman T et al. Stomachs of mice lacking the gastric H,K-ATPase alpha – subunit have achlorhydria, abnormal parietal cells, and ciliated metaplasia. J Biol Chem 2000; 275(28):21555–21565.
111. Nakamura S. H+-ATPase activity in selective disruption of H+-K+-ATPase alpha 1 gene of mice under normal and K-depleted conditions. J Lab Clin Med 2006; 147(1):45–51.
112. Simpson AM, Schwartz GJ. Distal renal tubular acidosis with severe hypokalaemia, probably caused by colonic H(+)-K(+)-ATPase deficiency. Arch Dis Child 2001; 84(6):504–507.
113. Verlander JW, Moudy RM, Campbell WB, Cain BD, Wingo CS. Immunohistochemical localization of HKalpha2c in rabbit kidney. J Am Soc Nephrol 9(3), 13A. 1998. Ref Type: Abstract
114. Swarts HG, Klaassen CH, Schuurmans Stekhoven FM, De Pont JJ. Sodium acts as a potassium analog on gastric H,K-ATPase. J Biol Chem 1995; 270(14):7890–7895.

Sodium Transport Mechanisms in the Mammalian Nephron

Michelle L. Gumz, Lisa R. Stow, and Shen-Ling Xia

Abstract Na transport is highly coordinated and regulated along the length of the renal nephron. The proximal tubule reabsorbs the majority of filtered Na. The majority of Na transport is transcellular, through the action of Na-antiporters and Na-coupled cotransporters. The proximal tubule is responsible for the majority of Na reabsorption by the renal nephron. In the thick ascending limb of Henle, Na is cotransported with other ions. This nephron segment is critical for urine concentration. The distal tubule and collecting duct make the smallest contribution to renal Na reabsorption. However, Na transport in these segments is subject to stringent regulation and is critical for external Na balance under normal physiological conditions. The basic structure and function of the renal Na transporters are evaluated. Hormonal and signaling pathways that regulate Na transport in the nephron and collecting duct are also discussed.

Keywords Kidney · Nephron · Proximal tubule · Thick ascending limb · Distal tubule · Collecting duct · Ion transport · Ion exchange · Ion channel · Ion cotransport · Sodium · Aldosterone

1 Introduction

The mammalian kidney is a complex organ that filters the blood and is essential for the normal regulation of extracellular and blood fluid volume, acid/base and electrolyte balance. Consequently, the kidneys have a major role in determining the composition of body fluids and blood pressure. The blood plasma is filtered through the glomerulus, contained within Bowman's capsule. Together, the glomerulus and Bowman's capsule form the renal corpuscle. From there, the filtrate enters the functional unit of the kidney, the nephron. The nephron is the smallest functioning unit

---

M.L. Gumz (✉)
Wingo Laboratory, Veteran Affairs Medical Center, Gainesville, FL, 32608, USA
e-mail: michelle.gumz@medicine.ufl.edu

G.A. Gerencser (ed.), *Epithelial Transport Physiology*,
DOI 10.1007/978-1-60327-229-2_12,

of the kidney and comprises the renal corpuscle, the proximal nephron, and distal nephron, which coalesce into the collecting duct. The glomerulus of the renal corpuscle filters the blood to form an ultrafiltrate (~200 L/d) that ultimately is concentrated into ~1–2 L of urine. Both the proximal and distal nephrons are cylindrical structures and have several discrete segments that consist of a single epithelial layer of specialized cells. These cells tightly regulate the movement of water (H_2O), ions [including sodium (Na) and potassium (K)], and organic solutes that passes through the nephron (Fig.1). The ultrafiltrate enters successively the proximal tubule (PT), loop of Henle, the distal convoluted tubule, the connecting segment, and finally the collecting duct (CD) where two or more nephrons join. A typical human kidney consists of approximately 1.2 million nephrons.

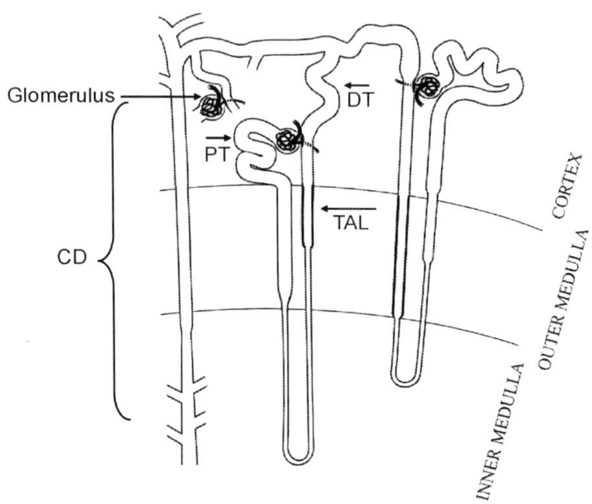

Fig. 1 The renal nephron. Two renal nephrons are pictured together with one collecting duct (CD). PT: proximal tubule, DT: distal tubule, TAL: thick ascending limb of Henle

Na is the principal extracellular cation making up over 90% of the osmotically active solute in the body, and under normal conditions Na content largely determines extracellular fluid (ECF) volume. Osmotic equilibrium exists between the extracellular and intracellular fluid. For example, K concentrations in the ECF are strictly maintained between 3.5 and 5 milliequivalents (mEq) per liter (L). Inside the cell, potassium levels are approximately 150 mEq/L. In contrast, the concentration of Na in the ECF is 145 mEq/L. Na concentration in the intracellular fluid is approximately 12 mEq/L. The reabsorption of H_2O and electrolytes like Na is a primary function of the nephron. Consequently, the reabsorption of Na and H_2O is essential to the maintenance of blood volume. The kidneys reabsorb approximately 25,000 mEq per day of Na and 179 L per day of H_2O. As Na is transported across the renal epithelium, its movement is followed by chloride (Cl) and H_2O, resulting in the net movement of NaCl and water. Na plays a vital role in maintaining ECF volume and therefore blood pressure.

The focus of this chapter is on Na transport in the nephron. There are four main segments in the nephron that contribute to Na reabsorption: the PT, the thick ascending limb of Henle (TAL), the distal tubule (DT), including the distal convoluted tubule and connecting segment, and the CD (Table 1). It should be noted that the relative contribution of each segment to Na transport is listed in Table 1 along with the transporter(s) responsible for apical (luminal) Na entry into the cell. The majority of Na reabsorption occurs in the PT through the action of the Na/H exchangers (NHE) and the Na/glucose cotransporter (SGLT). Twenty to thirty percent of Na reabsorption occurs in the TAL through the action of NHE and the Na, K, 2Cl cotransporter (NKCC). The DT and CD are responsible for the remaining Na reabsorption through the action of the Na/Cl cotransporter (NCC) and the epithelial Na channel (ENaC). In the following sections, the function and regulation of each of these transporters will be discussed.

Table 1 % Na reabsorption along the nephron. The contribution of each nephron segment to Na reabsorption is described together with the ion transporters that in each segment are responsible for the reabsorption of Na

Segment	Renal Na absorption (%)	Mechanism of apical Na entry
Proximal tubule	50–60	NHE3, SGLT, Na/organic anion
Thick ascending limb	20–30	NHE3, NKCC
Distal tubule[a]	5–10	NCC
Collecting duct	<5	ENaC

[a]Distal tubule includes the distal convoluted tubule and connecting segment.

The pathway for Na exit into the blood is common to all nephron segments: the ubiquitous Na, K-ATPase is located on the basolateral membrane of renal epithelial cells where it uses the energy from ATP hydrolysis to pump Na against its concentration gradient in exchange for K. The action of the Na, K-ATPase results in a transepithelial Na gradient that allows facilitated diffusion of Na across the apical membrane. The function [1], structure [2], accessory subunits [3, 4], and regulation [5, 6] of the Na, K-ATPase have been extensively studied and reviewed. The reader is directed to these several pertinent review articles for further information. Other basolateral Na exit pathways exist and will be discussed as they pertain to specific nephron segments.

2 Proximal Tubule

The PT reabsorbs the majority of Na, Cl, and bicarbonate (HCO_3) from the tubular fluid [7]. Na transport in the PT is coupled to the movement of organic solutes such as glucose as well as other ions, including protons (H), as depicted in Fig. 2. The basolateral Na, K-ATPase is responsible for active transport of Na into the blood.

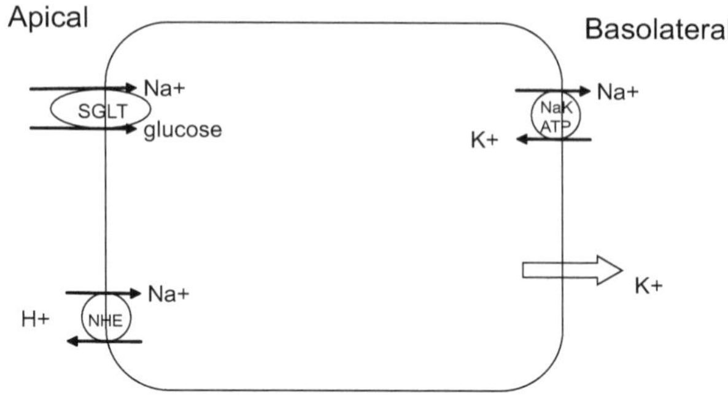

Fig. 2 Na transport in the proximal tubule. Na transport in the proximal tubule is mediated by the Na/glucose cotransporter (SGLT) and the Na/H exchanger (NHE) on the apical membrane and the Na, K ATPase on the basolateral membrane. A basolateral K channel is also pictured

2.1 Na/H Exchangers (NHE)

Nine Na/H exchanger isoforms have been identified. Na/H exchange regulates intracellular pH by extrusion of H in exchange for Na being brought into the cell. NHE proteins consist of an N-terminal domain that contains 12 transmembrane helices and a C-terminal domain that contains several regulatory domains targeted by protein kinases and other regulatory factors (Fig. 3) [8]. NHE1 is localized to the basolateral membrane of most renal tubule cells. NHE2 and NHE3 are expressed in the apical membrane of specific nephron segments. Specifically, NHE3 is expressed in the PT and TAL [9]. NHE3 is the primary pathway for Na reabsorption in the PT.

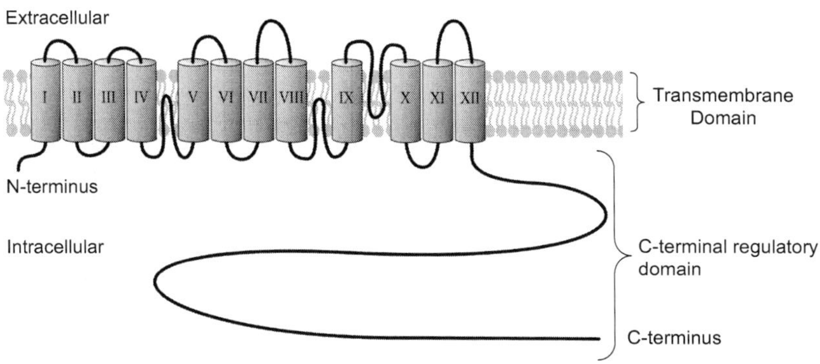

Fig. 3 Theoretical model of the Na/H Exchanger. A two dimensional model of the Na/H exchanger is pictured, based on [8]. Transmembrane helices are numbered I–XII. Both the N and C termini are located intracellularly

NHE3 works in parallel with an apical anion (Cl/base) exchanger to mediate the bulk of NaCl reabsorption [10, 11]. The importance of NHE3 is evident in NHE3-deficient mice. These animals exhibit negative Na balance and low blood pressure, which becomes lethal on a low salt diet (reviewed in [9]). On a normal diet, NHE3 null mice exhibit diarrhea and are mildly acidotic. The absorption of HCO_3 and fluid in the PT was decreased. Plasma aldosterone and renin levels were increased compared to wild-type animals. To compensate for the loss of NHE3 in these animals, the Na-phosphate cotransporter-2, ENaC-gamma and the anion exchanger AE1 were up-regualated. These studies in the NHE3 knockout mouse underscore the major role of NHE3 in mediating Na and H exchange in the kidney (as well as the intestine) as evidenced by perturbations in the acid/base balance and Na/fluid volume balance in these animals. In addition to its role in Na balance, NHE3 plays an important role in net acid excretion. NHE3 mediates bicarbonate reabsorption, ammonium secretion, and titration of filtered acids.

2.2 Na/Glucose Cotransporter (SGLT)

Two SGLT isoforms, SGLT1 and SGLT2, are present in the PT. They share 60% identity and 77% homology in their amino acid sequences, but differ in their affinity for glucose. SGLT1 is a high-affinity transporter consisting of 665 amino acids. The N-terminal region of SGLT1 contains the Na binding sites while the C-terminal region is responsible for glucose transport [12]. Figure 4 illustrates a possible topology model of SGLT1, which reportedly contains 14 transmembrane helices [12]. SGLT1 is expressed in renal cortex and outer medulla, specifically in renal proximal straight tubules. SGLT2 is a low-affinity/high capacity transporter consisting of 670 amino acids. SGLT2 is expressed in the apical membrane of convoluted proximal tubules. The role of SGLT2 is underscored by the effect of sergliflozin, a specific SGLT2 inhibitor and a possible new drug treatment for diabetes mellitus [13]. Blockade of SGLT2 by sergliflozin resulted in increased urinary glucose excretion as a consequence of inhibition of glucose reabsorption in the PT.

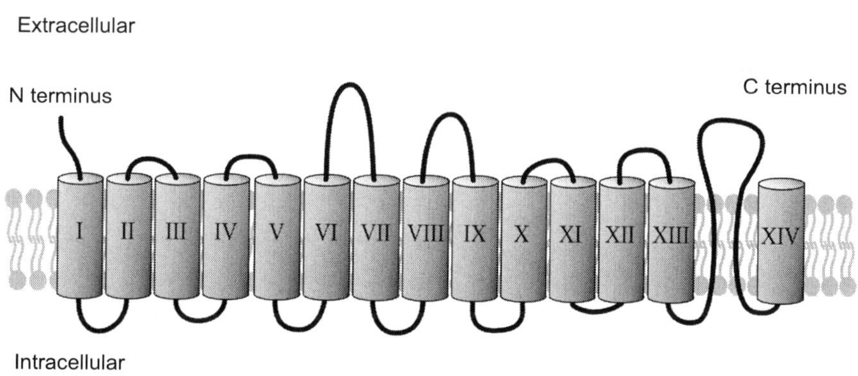

Fig. 4 Theoretical model of the Na/glucose cotransporter. A two-dimensional model of the Na/glucose cotransporter SGLT1 is pictured, based on [12]. The transmembrane helices are numbered I–XIV. The N and C termini are both extracellular

2.3 Other Na Transporters

The electrochemical gradient for Na favors Na reabsorption in the PT, therefore, Na is often used to cotransport other molecules. Transport of Na in the PT is coupled to organic ions including phosphate and sulfate. Na is also coupled to the movement of amino acids, and there are separate transporters for neutral, acidic, and basic amino acids. Na is also coupled to the transport of metabolites including lactate, acetoacetate, b-hydroxybutyrate, and tricarboxylic acid cycle intermediates such as a-ketoglutarate, succinate, and citrate.

2.4 Regulation of Na Transport in the PT

NHE3 is downregulated by parathyroid hormone, dopamine, and increased blood pressure. This downregulation results in saluresis and diuresis. Lack of downregulation of NHE3 by dopamine can cause hypertension due to retention of salt and water. NHE3 is upregulated by metabolic acidosis [10]. An increase in intracellular pH is sensed by Pyk2 that stimulates endothelin-1 (ET-1) production [14]. ET-1 then acts via the ET_B Receptor to stimulate NHE3 activity and facilitate proton secretion [15].

SGLT is downregulated by high glucose in the PT (reviewed in [16]). Protein kinase C (PKC) is activated by high glucose levels, resulting in the formation of reactive oxygen species. The action of reactive oxygen species leads to translocation of NF-κB into the nucleus which may downregulate expression of SGLT. In addition, PKC activates Ca^{2+}-dependent phospholipase A2 (cPLA2), resulting in release of arachidonic acid and the subsequent inhibition of SGLT activity.

SGLT1, SGLT2, and NHE are all regulated by angiotensin II (AngII). AngII is a critical effector of the renin-angiotensin-aldosterone system (RAAS), which is responsible for regulating Na transport and blood pressure. AngII has a biphasic effect on NHE-mediated Na transport in the PT [17, 18]. Picomolar levels of AngII induce Na transport while micromolar amounts inhibit AngII. Higher levels of AngII resulted in increased release of arachidonic acid which was associated with increased intracellular Ca^{2+} levels. The inhibitory effect of AngII on Na transport in PT cells was dependent on activation of a cPLA2. The resulting arachidonic acid acted through a cytochrome P450 epoxygenase pathway that involved cyclooxygenase.

Studies in cultured rabbit PT cells examined the effect of AngII on 14C-glucose uptake, as a measure of SGLT activity [19]. AngII caused a dose- and time-dependent decrease in SGLT protein levels. The inhibitory effect of AngII on SGLT activity was dependent on the AT1 receptor, and a pathway involving PKC, p44/42-MAPK, and cPLA2. Han et al. showed that ANG II activated the AT_1 receptor and PKC, resulting in activation of p44/42 MAPK. MAPK action led to the release of arachidonic acid and its subsequent metabolism by cyclooxygenase and cytochrome

P450. The breakdown products of arachidonic acid may result in the decreased SGLT protein expression and a resultant inhibition of SGLT activity.

The neurotransmitter dopamine plays a major role in the regulation of renal ion transport. Renal dopamine is formed mostly from PT cells. The dopamine precursor L-dopa is filtered freely from the glomerulus and is actively transported into the PT cells where L-dopa is converted to dopamine by decarboxylation. Once dopamine is synthesized, it is transported out of the cells where it can interact with dopamine receptors at each renal segment. Dopamine is not stable in a nonacidic environment and is therefore not converted to norepinephrine in renal tubules [20, 21].

Dopamine receptors have been classified into two families based on pharmacologic and molecular studies: D_1-like receptors (D_1 and D_5) and D_2-like receptors (D_2, D_3, and D_4) [22]; they are linked to the stimulation (via G_s protein) and inhibition (via G_i protein) of adenylyl cyclase, respectively. D_1-like receptors are generally connected to natriuretic and diuretic effects. Although D_2-like receptors have been suggested to produce anti-natriuretic effect, their role is still unclear. The activation of D_1-like receptors located on the PT causes inhibition of tubular Na reabsorption by inhibiting apical NHE and Na phosphate cotransporters, as well as the basolateral Na, K-ATPase. It is plausible that dopamine has direct effects on both apical and basolateral transporters. The process from the activation of dopamine receptors to the inhibition of Na, K-ATPase activity involves multiple cellular signaling pathways that, despite a great deal of work, appear to be complex and only partially understood [20, 21].

3 Thick Ascending Limb of Henle (TAL)

The H_2O-impermeable ascending limb of the loop of Henle functions to dilute the filtrate passing through the tubule. Ion transport in the thin ascending limb of Henle is passive, whereas active ion reabsorption occurs in the thick ascending limb. This segment of the nephron is critical for urine concentration. Figure 5 illustrates a typical cell in the thick ascending limb of Henle (TAL). The TAL absorbs 20–30% of

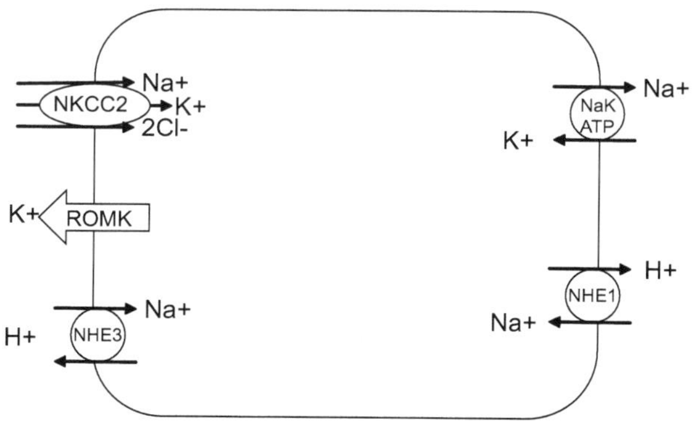

Fig. 5 Na transport in the thick ascending limb of Henle. Na transport in the TAL is mediated by the Na/H exchanger isoform 3 (NHE3) and the Na/K/2Cl cotransporter on the apical membrane and the Na, K ATPase and NHE1 on the basolateral membrane. The renal outer medullary K channel (ROMK) is pictured at the apical membrane

filtered NaCl load mostly through the action of apical NKCC2 [23] with some contribution of NHE3. Na exits basolaterally through the action of the Na, K-ATPase. Cl^- exits the cell through a Cl channel or via the action of the K/Cl cotransporter [24]. Furthermore, K recycling across the apical membrane via K channels contributes to the lumen positive voltage that promotes paracellular reabsorption of Na. Because the TAL is impermeable to H_2O, solute transport in the TAL creates the osmotic gradient that drives fluid reabsorption in the connecting segment and the collecting duct.

3.1 Na, K, 2Cl Cotransporter (NKCC)

There are two NKCC isoforms. NKCC1 consists of 1,205 amino acids and shares 66% identity and 80% homology with NKCC2, a 1,095 amino acid protein. The ubiquitously expressed NKCC1 is capable of forming a homo-oligomer due to the action of a self-interacting domain in the C-terminus [25]. NKCC1 is expressed in the basolateral membrane of cells in the CD [26]. NKCC2 contains 12 transmembrane helices and intracellular N and C termini (Fig. 6). Two glycosylation sites present in the extracellular loop between transmembrane helices 7 and 8 are important for trafficking of NKCC2 to the apical membrane [27] as mutagenesis of either or both residues resulted in decreased surface expression of the transporter.

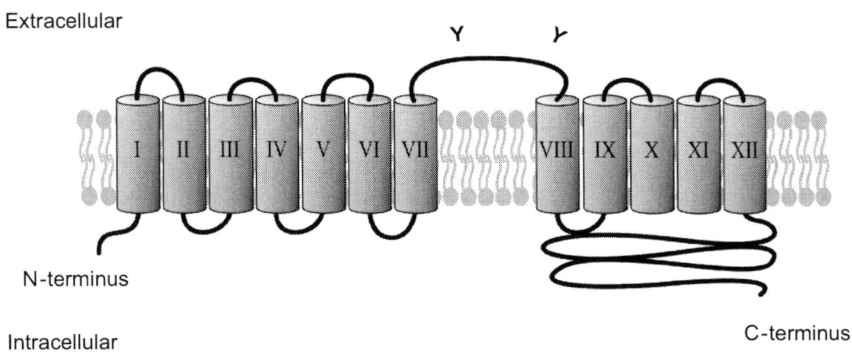

Fig. 6 Theoretical model of the Na/K/2Cl cotransporter. A two dimensional model of the Na/K/2Cl cotransporter (NKCC2) is pictured with the 12 transmembrane helices numbered. The N and C termini are both located inside the cell. On the extracellular face, important sites of glycosylation are indicated by "Y". The model is based on [42]

Studies in Xenopus oocytes demonstrated that the kidney-specific isoform, NKCC2, like NKCC1, is capable of forming a homo-oligomer [28]. Yeast-two-hybrid analysis suggested that a self-association domain was present in the C-terminus and was responsible for homo-oligomer formation [29]. Splice variants of NKCC2 have been identified and display distinct functional properties, affinities for transported ions, and distribution along the TAL [30–34]. Brunet et al. demonstrated that NKCC2 splice variants were differentially regulated by furosemide and a H_2O rich diet [29]. The amino acids responsible for some of the functional differences displayed by the splice variants have recently been identified [35].

The importance of NKCC2 to Na and volume regulation is underscored by studies showing that mutations or deletions of the NKCC2 gene are linked to salt-wasting and hypotension (reviewed in [36, 37]). These results are similar to the effect of pharmacological inhibition of NKCC2 using a loop diuretic such as furosemide. Conversely, increased NKCC2 activity is associated with genetic forms of hypertension in humans and mouse or rat models [38–40].

3.2 NHE Isoforms in the TAL

NHE1 and NHE3 were identified using RT-PCR analysis of mRNA from microdissected rat medullary TAL [41]. It was concluded and has been confirmed since that NHE1 is present at the basolateral membrane in TAL cells while NHE3 provides a pathway for apical Na entry (see Fig. 5). NHE3 action facilitates the reabsorption of HCO_3 in this segment. NHE1 and NHE3, due to their extrusion of protons from the cell, play a more important role in mediating acid/base balance in the TAL than Na reabsorption. In support of this, it has been shown that NKCC2 is the major pathway for Na entry in the TAL [42]. Interestingly, NKCC2 is also expressed on the macula densa cells and contributes to tubuloglomerular feedback, a mechanism by which the glomerular filtration rate is regulated [43]. The following section focuses on the regulation of NKCC2 in the TAL.

3.3 Regulation of Na Transport in the TAL

Na, K, 2Cl cotransporters play a role in regulating cell volume and its activity increases in response to cell shrinkage, resulting in a consequent increase in cell volume (reviewed in [44]). Because of its role in maintaining H_2O balance, NKCC2 is regulated by vasopressin [45] and inhibitors of prostaglandin synthesis [46], both of which resulted in increased levels of NKCC2 protein. NKCC2 can be regulated by phosphorylation as was demonstrated by Gimenez and Forbush using immunofluorescence [47]. The effect was rapid and induced by vasopressin in rat renal tubules. Specifically, three threonine residues in the intracellular N-terminal tail were linked to regulatory phosphorylation of NKCC2. Studies in mice with heterozygous disruption of the G protein signaling molecule Gαs displayed decreased NKCC2 activity, indicating that a G-protein coupled signaling pathway involving cAMP regulates this transporter as well [48].

NKCC2 is also regulated through modulation of trafficking to the plasma membrane. Gimenez and Forbush used electron microscopy to show that exposure to vasopressin caused increased translocation of NKCC2 to the apical membrane

[47]. Benziane et al. used a yeast-two-hybrid assay to identify novel binding partners for NKCC2 [42]. Aldolase B was found to specifically interact with the C-terminus of NKCC2, resulting in the decreased presence of NKCC2 at the apical membrane.

Nitric oxide (NO), a potent vasodilator, has a wide range of physiological effects in the body. In the kidney, NO has effects on hemodynamic regulation as well as Na transport. Stimulating NO production causes an increase in renal Na and water excretion, whereas blocking NO production reduces Na and water excretion. NO is produced from L-arginine by nitric oxide synthase (NOS). There are three NOS isoforms (NOS1, NOS2, and NOS3) and all are expressed in the kidney. The physiologically relevant isoform of NOS is still actively debated; however, mounting evidence supports the role of NOS3 in the TAL.

Studies have shown that ET-1 inhibits medullary TAL (mTAL) Cl transport via an ET_B receptor-mediated increase in NO synthase-3 (NOS3) expression and release of NO [49–52]. Furthermore, Herrera and colleagues demonstrated that a high salt diet increases medullary osmolality and stimulates NOS3 expression in the mTAL by an ET-1 and ET_B receptor mediated pathway [49]. Stimulation of NOS3 will increase levels of NO and inhibit Na transport.

ET-1 has well-defined effects to promote Na excretion in the kidney. Many components of the medullary ET-1 system are upregulated in response to a high salt diet, including ET-1, ET-converting enzyme -1 (ECE-1), and ET_BR [49, 53–57]. These observations provide direct evidence for a physiological role of ET-1-mediated Na excretion. In addition, hypertonicity also stimulates ET-1 synthesis and release from the inner medullary CD (IMCD) [54]. In healthy individuals, plasma ET-1 is not affected by dietary salt intake, but urinary ET-1, a marker of renal-derived ET-1 [58], positively correlates with Na excretion and urinary volume [59]. Taken together, these data suggest that renal ET-1 could mediate a physiological natriuresis in the presence of high salt intake in a healthy individual.

4 Distal Tubule and Collecting Duct

The DT, including the distal convoluted tubule and connecting segment, and CD comprise the nephron segment responsible for the least amount of Na reabsorption (see Table 1). However, the distal tubule and collecting duct represent the terminal site for reabsorption and is essential for fine-tuned regulation of Na transport in response to dietary intake. Accordingly, this segment plays a critical role in body salt and volume homeostasis [60–63]. Changes in dietary salt intake results in changes in Na transport in the CD [61] and pathophysiological states of Na retention occur due to increases in Na transport in the CD [64–66]. An extensive section below details the regulation of Na transport by the mineralocorticoid aldosterone. Two aldosterone target cells are pictured in Fig. 7. Panel A represents a DT cell wherein apical Na entry is through the Na/Cl cotransporter (NCC) as well as ENaC. In the principal cells of the CD, apical Na entry occurs via the ENaC.

Fig. 7 Na Transport in the Distal Nephron and Collecting Duct. **A**. Na transport in the early part of the distal tubule is mediated by the Na/Cl cotransporter at the apical membrane and the Na, K ATPase at the basolateral membrane. **B**. Na transport in the late distal tubule and collecting duct is mediated by the epithelial Na channel (ENaC) at the apical membrane and the Na, K ATPase at the basolateral membrane

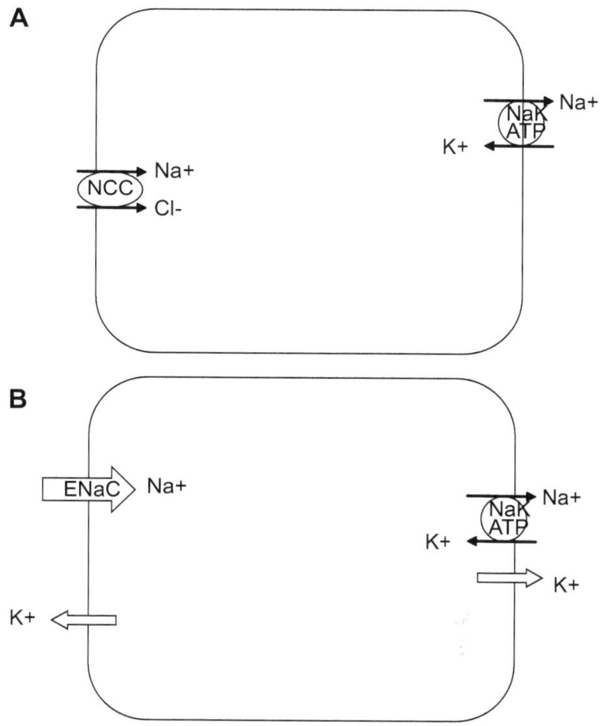

4.1 Na/Cl Cotransporter (NCC)

The Na/Cl cotransporter (NCC) is located at the apical membrane of cells in the distal convoluted tubule (DCT) where it provides the primary mode of entry for Na in that segment of the nephron. NCC is the target of thiazide diuretics, the first line of treatment for essential hypertension. Figure 8 shows the two-dimensional structure of NCC. NCC is a 1,002 amino acid protein containing 12 putative transmembrane helices, a short intracellular N-terminal tail, and a longer C-terminal intracellular tail. NCC shares approximately 50% identity with NKCC2 (reviewed in [37]). The extracellular loop between transmembrane helices 7 and 8 contains two glycosylation sites that appear to be necessary for proper function of the transporter [67]. Mutagenesis of the asparagine residues in the glycosylation site resulted in decreased surface expression and activity of NCC with a consequent increase in thiazide sensitivity, presumably because of increased access of the drug to its binding site. Structure-function studies of NCC indicate that the Cl binding site is located in the N terminal region, within transmembrane helices 1–7 while the thiazide binding site is located within transmembrane helices 8–12 [68]. Both the N-terminal and C-terminal transmembrane segments appear to be involved in Na binding.

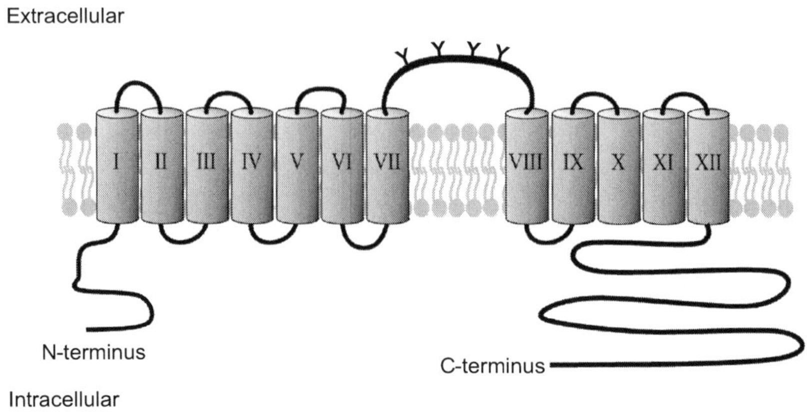

Fig. 8 Theoretical Model of the Na/Cl cotransporter. A two-dimensional model of the Na/Cl cotransporter (NCC) is pictured, based on [68]. The transmembrane helices are numbered I–XII. Four important sites of glycosylation are indicated on the extracellular face by "**Y**". Both the N and C termini are located in the cytoplasm

4.2 *Epithelial Sodium Channel (ENaC)*

The major mechanism of Na absorption in the CD is via ENaC [69]. ENaC is a heterotrimeric Na channel, thought to consist of two α subunits and one each of a β and γ subunit [70] (Fig. 9). Each subunit consists of two transmembrane helices, intracellular N and C termini, and a large extracellular loop containing several N-linked glycosylation sites. The glycosylation sites are thought to play a role in the maturation and processing of ENaC [71]. The importance of ENaC is emphasized by the discovery that activating mutations of α, β, or γ subunits of ENaC can produce severe hypertension [72].

ENaC function has mostly been assigned to the CD. In an intriguing study, Rubera et al. showed that ENaC is present in the distal convoluted tubule as well [73]. Even under a Na-restricted diet, CD-specific ENaCα subunit deficient transgenic mice did not display a Na wasting phenotype. Immunostaining showed the presence of the ENaCα subunit in the late DCT and connecting segment but not in the CD. ENaCα expression has been detected in the DCT and CNT previously [74].

4.3 *Regulation of Na Transport in the Distal Tubule and Collecting Duct*

4.3.1 Aldosterone

The renin-angiotensin-aldosterone system (RAAS) results in the release of aldosterone under conditions when ECF is low, Na levels are low or blood pressure has dropped. Indeed, every known form of Mendelian inherited hypertension is

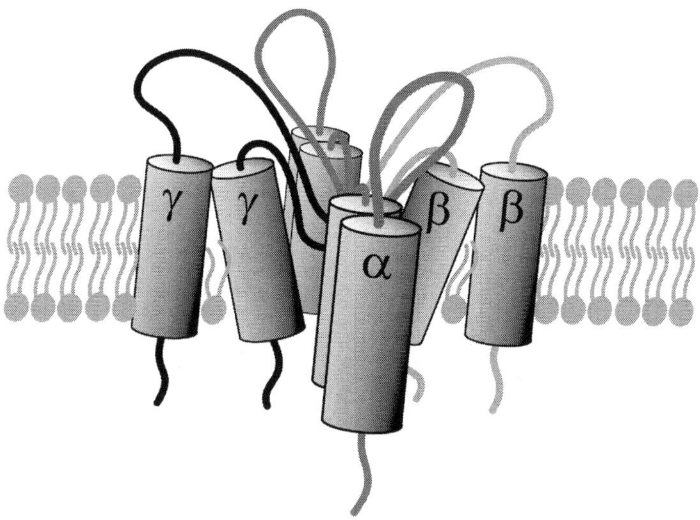

Fig. 9 Theoretical Model of the Epithelial Na Channel. The epithelial Na channel (ENaC) is thought to consist of two α subunits and one each of a β and γ subunit. The N and C termini of each subunit are located inside the cell. Tilting of membrane helices is arbitrary

the result of a defect in the downstream signaling targets of aldosterone [65]. Aldosterone acts directly on principal cells in the DT and CD to increase Na absorption, which in turn increases blood volume, vascular reactivity, and blood pressure. Aldosterone couples the movement of Na from the tubular fluid into the blood by increasing the activity and number of epithelial Na channels (i.e., ENaC) at the apical membrane and the Na, K-ATPase at the basolateral membrane. The transepithelial Na and H_2O reabsorption increases blood volume and vascular reactivity which increases systemic blood pressure. Thus, aldosterone contributes to the maintenance of blood pressure and renal and cardiac function [75, 76].

4.3.2 Mechanism of Aldosterone Action

The canonical receptor for aldosterone is the mineralocorticoid receptor (MR) [77]. Aldosterone binds MR with an affinity of 1.3 nM, under in vitro conditions [78]. The glucocorticoid receptor (GR) can also bind aldosterone with reported K_D values ranging from 0.5 to 3.0 nanomolar (nM). MR and GR are capable of forming hetero- and homodimers to drive transcription of target genes. Both hormone receptors recognize a 15 nucleotide hormone response element (HRE), an inverted palindrome repeat separated by several random nucleotides [78]. MR and GR are both able to stimulate Na transport in IMCD cells [79]. Mechanisms exist in aldosterone target cells to ensure that MR is activated specifically by aldosterone and not by glucocorticoids. In aldosterone-responsive tissues, the enzyme 11-β-hydroxysteroid dehydrogenase 2 (11βHSD2) inactivates glucocorticoid hormones so that MR activity is induced by aldosterone alone [80]. The importance of 11βHSD2 is underscored by the effects of loss of function mutations in this gene, which result in the syndrome of Apparent Mineralocorticoid Excess (AME) [81]. AME is characterized by severe hypertension.

4.3.3 Aldosterone Targets

Known aldosterone targets include the basolateral Na, K-ATPase [82] as well as the apical ENaC, both of which directly contribute to increase transcellular Na transport (see Fig. 1). The ENaCα subunit is specifically regulated by aldosterone in the kidney [83]. These effects of aldosterone occur in a late phase, typically 4–6 h or more following hormone exposure. However, the immediate transcriptional targets of aldosterone are still being characterized. One well-known aldosterone effector is the serum and glucocorticoid-regulated kinase, Sgk-1, which acts on ENaC [84–86]. Several studies have used microarray technology to identify these acute aldosterone targets. Mouse inner medullary collecting duct cells (mIMCD-3) were treated with aldosterone for 1 h followed by microarray analysis [87]. Among the novel aldosterone targets identified by Gumz et al. were preproendothelin-1, connective tissue growth factor, and period homolog; this study was validated by the appearance of Sgk-1 on the list of induced genes. Soundararajan et al. treated a mouse cortical CD (CCD) cell line with aldosterone for 1 or 6 h [88]. The microarray analysis showed that Sgk1 mRNA was upregulated more than ten-fold at both time points while the ENaCα subunit message was induced more than five-fold after 6 h of treatment. Two novel aldosterone targets were identified by this group: glucocorticoid-induced leucine zipper protein (GILZ) was upregulated at 1 and 6 h while FK506-binding protein 5 (FKBP5) was induced at 6 h. Subsequent experiments indicated that GILZ induced Na transport in Xenopus oocytes and renal epithelial cells.

Fakitsas et al. performed a microarray analysis on dissected CCD/connecting tubule segments from mice treated with aldosterone for 1 h [89]. This study was also validated by the detection of Sgk1 as an induced transcript. Three novel aldosterone-induced transcripts were identified and confirmed by independent methods: (1) protein-related to DAN and Cerberus, (2) activating transcription factor 3, and (3) ubiquitin-specific protease 2 (Usp2-45). Subsequent experiments showed that Usp2-45 stimulated Na current in Xenopus oocytes and renal epithelial cells. It was further demonstrated that Usp2-45 "de-represses" ENaC by removing the ubiquitin groups added by the ubiquitin ligase neural precursor cell-expressed developmentally downregulated (gene 4) protein (Nedd4-2).

"De-repression" of ENaC by the product of an aldosterone-induced gene is a recurring theme in this signaling pathway. The well-known aldosterone target Sgk1 phosphorylates Nedd4-2, a ubiquitin ligase that targets and degrades ENaC. Sgk1 phosphorylation of Nedd4-2 creates a docking site for 14-3-3 adaptor protein, resulting in Nedd4-2 inhibition and a subsequent increase in ENaC at the apical membrane due to decreased channel turnover [90]. 14-3-3 adaptor protein family members were identified as aldosterone targets in mIMCD-3 cells [87] and in a mouse CCD cell line [91]. Nagaki et al. recently showed that 14-3-3 inhibits the interaction between Nedd4-2 and ENaC in a manner which is dependent on Sgk1-mediated phosphorylation of Nedd4-2 [92]. During the acute phase of aldosterone action, Sgk1 inhibits Nedd4-2 via phosphorylation that results in association of Nedd4-2 with 14-3-3 (reviewed in [93]). The subsequent reduced interaction of Nedd4-2 with ENaC results in decreased ubiquitination of the Na channel, and thus increased stability of the channel. Consequently, ENaC activity at the apical membrane is increased by Sgk1 action in the presence of aldosterone.

4.3.4 Purinergic Signaling

It is becoming increasingly apparent that the regulation of Na transport in the CD is complex, involving multiple mechanisms. There is a growing appreciation for non-traditional regulatory mechanisms (in addition to aldosterone, insulin, and arginine vasopressin) that regulate salt and volume homeostasis. Purinergic signaling in the CD is emerging as one such potential mechanism [94, 95].

In the kidney, purinergic signaling has been suggested to play an important role in stimulating renal cell growth and division, developing polycystic kidney disease, controlling microvascular function, and in regulating epithelial ion transport [94, 96]. A recent in vivo study revealed that activation of apical P2 receptors by agonists inhibits amiloride-sensitive Na reabsorption in the distal nephron and CD [97]. It is consistent with previous reported in vitro results that ATP inhibits amiloride-sensitive short circuit current and amiloride-sensitive Na channel activity via activation of apical P2 receptors [98–101]. Li et al. [102] reported that nucleotides can downregulate Sgk1 through activation of P2 receptors in mIMCD-3 cells. The same study also demonstrated that aldosterone-stimulated increase of Sgk1 kinase activity can be suppressed by the activation of P2 receptors. Given the fact that renal tubule cells release ATP in physiological circumstances and in tissue ischemia or cell injury [103], it is likely that the purinergic signaling in the renal CD can provide an apical regulatory pathway to inhibit Na transport in both physiological and pathophysiological conditions.

4.3.5 Dopamine

In the CCD D_1-like and D_2-like receptors are colocalized and the effects of dopamine become more complex and are dependent on the conditions of Na balance and the presence of other agents [104, 20]. The CD has several dopamine subtype receptors. The D_1, D_3, and D_4 receptors have been demonstrated in the CCD [105–109]; the D_2 and D_4 receptors are present in the medullary CD (MCD) [108–111]. The animal data so far show that D_1-like receptors are only present in the CCD and D_2-like receptors are present in both the CCD and the MCD. However, D_1-like receptors are expressed in the MCD of human kidney [112].

Activation of D_1-like receptors produces inhibition in basolateral membrane Na, K-ATPase activity in rat CCD, via intracellular cAMP mediated PKA pathway and the phospholipase A2 pathway [113–115]. The action of this receptor could decrease Na reabsorption via inhibiting Na, K-ATPase activity up to 50%, an effect proportionally larger than in the PT [114].

Dopamine action to inhibit tubular Na transport at the CCD is different from the PT. In the CCD Na transport across the luminal membrane to a large extent occurs via Na channels. Although the effect of dopamine on Na channels in the CCD has not been formally studied, several groups of investigators have showed that there is "cross talk" between basolateral membrane Na, K-ATPase activity and apical cation channel activity [116, 117]. When basolateral membrane Na, K-ATPase activity is inhibited by ouabain, a decrease of apical Na and K channel activity is observed. The inhibition process is intracellular Ca^{2+}-dependent and may involve the PKC signaling pathway. Inhibition of basolateral membrane Na, K-ATPase activity by dopamine would be anticipated to decrease apical Na channel activity.

There is considerable precedent for the direct action of dopamine on Na channels in other tissues. In the central nervous system, D_1-like agonists can decrease the amplitude of Na current about 40% and the inhibition of Na current is mediated by cAMP-dependent phosphorylation of the α-subunit of the Na channel [118, 119]. Thus, activation of dopamine receptors in the CCD may reduce both basolateral membrane Na, K-ATPase activity and apical cation channel activity. This would allow the CCD to decrease the reabsorption of Na with minimum changes of intracellular Na concentration.

Whereas dopamine receptor subtypes and dopamine action signaling pathways to Na transport in the CCD have been extensively studied, few studies have shown the action of dopamine in the MCD. Although the dopamine receptor subtypes have been reported in the MCD, functional studies regarding dopamine action to basolateral Na, K-ATPase activity and to apical cation channel conductance are less well documented. The outer MCD (OMCD) of the outer stripe has been shown to be able to reabsorb Na and to secrete K but at rates that are less than that in the CCD [120–122]. However, one study has suggested that in normal rabbits, basal Na, K-ATPase activity in the MCD is about the same as in the CCD [123]. Previous studies demonstrated the presence of mRNA for a 28-pS cation channel in mouse OMCD of the inner stripe [124] and conductance of a 23-pS cation channel in rabbit OMCD of the inner stripe [125]. These data demonstrate that the OMCD has the ability to reabsorb Na, which provides a possible target for dopamine action via its receptors (probably D_2-like subtype). There is evidence that the IMCD cells may also produce dopamine locally [126].

4.3.6 ET-1 Mediated Natriuresis

In 1993, Tomita et al. demonstrated that ET-1 reversibly inhibited water and chloride transport in isolated CCD [127]. Mounting evidence exists that ET_B receptors promote natriuresis by direct inhibition of ENaC activity through a mechanism mediated, at least in part, by NO and SRC kinases [128–130]. Using a patch-clamp technique, Gallego et al. demonstrated that ET-1 inhibited amiloride-sensitive Na transport in an ET_BR-dependent manner [130]. In the IMCD, ET-1 is able to stimulate NOS3 [131], NO, cGMP [132, 133], and inhibit Na, K-ATPase via cyclooxygenase metabolites [134].

4.3.7 Renal ET-1 Mediated Natriuresis in Experimental Models

Medullary ET-1 is reduced in several models of experimental hypertension including spontaneously hypertensive, Dahl S, and Prague hypertensive rats [135]. However, the importance of ET_B-mediated natriuresis was originally revealed in studies conducted on spotting lethal (*sl*) ET_BR deficient rats. *Sl/sl* rats have a naturally occurring 301 bp deletion in the ET_BR gene that leads to aganglionic megacolon and mortality shortly after birth. Gariepy and colleagues rescued the fatal phenotype in the enteric system by a dopamine hydroxylase promoter [136]. Rescued-*sl/sl* rats exhibit low renin salt-sensitive hypertension and do not demonstrate an acute depressor response to systemic ET-1 injection [136]. These rats also have higher levels of circulating ET-1 as a result of decreased ET_BR clearance. In the absence of ET_BR, high circulating levels of ET-1 can activate ET_A receptors and lead to ET_AR-mediated vasoconstriction. In addition, s*l/sl* rats also have enhanced ET_AR-mediated sympathetic tone that may contribute to the increase in blood pressure [137]. However, treating *sl/sl* rats with a selective ET_AR antagonist only partially decreased blood pressure; whereas, treatment with the ENaC inhibitor, amiloride, leads to complete normalization of blood pressure. These data support the role of ET_BR-mediated inhibition of ENaC.

However, the most definitive evidence of ET-1/ET_BR-mediated natriuresis comes from collecting duct ET-1, ET_AR, and ET_BR knockout mice. Collecting duct ET-1 knockout mice are hypertensive on a normal Na diet [129]. A high Na intake exacerbates the hypertension due to an inability to excrete the Na load. Treatment with amiloride corrects inappropriate Na retention and normalizes blood pressure. Clearly, collecting duct-derived ET-1 plays an important role in inhibiting Na reabsorption. Collecting duct ET_BR knockout mice also develop hypertension due to impaired Na excretion, but to a lesser extent than collecting duct ET-1 knockout mice [138]. This may be due to surrounding ET_BR in the renal medulla that is able to compensate for the loss of ET_BR in the CD. In contrast to ET-1 and ET_BR, collecting duct ET_AR knockout mice exhibit normal Na excretion on both a normal and high salt diet [139]. Collectively, this data unambiguously establishes the role for collecting duct ET-1 and ET_BR in modulating transepithelial Na transport in the collecting duct.

There is also extensive literature demonstrating the functional regulation of the renal ET-1 system in response to mineralocorticoid-induced hypertension. Medullary ET-1, ECE-1, and ET_BR expression are upregulated in the prehypertensive and hypertensive deoxycorticosterone acetate (DOCA)-salt treated rat [140, 141]. Furthermore, blockade of ET_BR in DOCA-salt sensitive hypertensive animals exacerbates the increase in blood pressure and impairs Na excretion. These observations indicate that the upregulation of ET-1, ECE-1, and ET_BR is a compensatory mechanism that should reduce arterial pressure and promote Na excretion [141].

4.3.8 With-No-Lysine (K) (WNK) Kinases

The majority of this section has focused on regulation of Na transport via ENaC. A relatively new field of investigation has yielded exciting insight into regulation of NCC in the DCT via a family of serine-threonine kinases, WNK, that lack a lysine in the highly conserved active site common to kinases, hence the name with-no-lysine (WNK). Mutations in WNK kinases were linked to an inherited form of hypertension, or familial hyperkalemic hypertension (FHHt) [142]. Mutations in two WNK genes were identified. Intronic deletions in WNK1 resulted in increased WNK1 expression, while missense mutations in a highly conserved region of WNK4 were identified. Both WNK1 and WNK4 are localized to the cytoplasm of cells in the distal nephron and CD; WNK4 is also found in tight junctions. Numerous studies have subsequently been focused on identifying the signaling mechanisms of the WNKs and how they relate to Na transport mechanisms in the nephron. These studies are the subject of several excellent reviews [143–146]. The current prevailing paradigm is that WNK4 inhibits NCC activity; WNK4 itself is subject to inhibition by WNK1. Thus, mutations causing over-expression of WNK1 lead to inhibition of WNK4 and a consequent increase in NCC activity. Likewise, loss of function mutations in WNK4 result in relieved inhibition of NCC.

The role of WNK kinases in regulation of NCC has been well characterized. Mounting evidence suggests a role for WNKs in regulation of other renal ion transporters, including ENaC, the renal outer medullary K channel ROMK, and NKCC. This should be an interesting field of study in the future with wide implications for the treatment of renal transport disorders, including hypertension. In the final section, disorders of renal Na transport will be discussed along with a summary of Na transport regulatory mechanisms.

5 Summary

5.1 Disorders of Renal Na transport

Table 2 summarizes genetic Na transport disorders and the transporter with which the defect is associated. The autosomal recessive Bartter's syndrome phenotype includes hypokalemia, metabolic alkalosis, hyperaldosteronism, salt wasting, and hypotension. The disorder results from the inability of the TAL to reabsorb Na. Simon et al. [147] linked Bartter's syndrome to frameshift or nonsense mutations in NKCC2. Bartter's syndrome may also result from defects in K and Cl transport [148, 149].

Table 2 Disorders of renal Na transport. Known disorders of renal Na transport are listed together with the ion transporter that is dysfunctional, the nephron segment where the dysfunction occurs, and the resulting phenotype

Disorder	Transporter	Nephron segment	Phenotype
Bartter's syndrome	NKCC	TAL	Hypokalemia, metabolic alkalosis, hyperaldosteronism
Gitelman's syndrome	NCC	DCT	Hypomagnesemia, hypokalemic metabolic alkalosis, hypocalciuria, hypotension
Gordon syndrome	NCC	DCT	Hypertension, hyperkalemia
Liddle syndrome	ENaC β,γ	CD	Hypertension, decreased Na excretion
Pseudohypoaldosteronism type 1a	ENaC α, β,γ	CD	Hypotension, increased Na excretion

The autosomal recessive Gitelman's syndrome is caused by loss of function mutations in NCC in the DCT and results in hypomagnesemia, hypokalemic metabolic alkalosis, hypocalciuria, and hypotension. Eighty different mutations have been identified and appear to cause defective trafficking and decreased insertion of the transporter into the apical membrane [150, 151].

Gordon syndrome, also known as familial hyperkalemic hypertension (FHHt) or pseudohypoaldosteronism Type II (PHAII), is an autosomal dominant form of hypertension that is accompanied by hyperkalemia and renal tubular acidosis (reviewed in [144]). The condition is treatable with thiazide diuretics, which suggests the involvement of NCC. However, linkage analysis indicated that Gordon's disease is associated with the genes for the WNK family of kinases. Specifically, gain of function mutations in WNK1 and loss of function mutations in WNK4 result in increased activity of NCC.

Liddle's syndrome is an autosomal dominant disorder resulting from gain of function mutations in the β and γ subunits of ENaC that lead to hypertension. Pseudohypoaldosteronism type 1a is an autosomal recessive disorder caused by loss of function mutations in any of the three ENaC subunits. These mutations result in increased Na excretion and subsequent hypotension.

In summary, the transport of Na is highly coordinated and regulated along the length of the renal nephron. Through the action of SGLT and NHE3, the PT reabsorbs the great majority of Na from the tubular fluid. In the TAL, NKCC2 and NHE3 contribute to Na reabsorption as well. The DT and CD make the smallest contribution to Na reabsorption by the nephron. However, Na transport in the DT and CD is subject to the most stringent regulation, as evidenced by the complex hormonal and signaling pathways that regulate NCC and especially ENaC. The role of the kidney in regulating blood pressure via the control of Na transport is emphasized by the many genetic disorders affecting renal Na transporters.

Acknowledgments We would like to acknowledge the Medical Media Service of the Malcolm Randall Veterans Affairs Medical Center for their help with the illustrations in this chapter.

References

1. Rajasekaran SA, Barwe SP, Rajasekaran AK. Multiple functions of Na,K-ATPase in epithelial cells. Semin Nephrol 2005;25(5):328–34.
2. Morth JP, Pedersen BP, Toustrup-Jensen MS, Sorensen TL, Petersen J, et al. Crystal structure of the sodium-potassium pump. Nature 2007;450(7172):1043–9.
3. Geering K. FXYD proteins: new regulators of Na-K-ATPase. Am J Physiol Renal Physiol 2006;290(2):F241–50.
4. Garty H, Karlish SJ. FXYD proteins: tissue-specific regulators of the Na,K-ATPase. Semin Nephrol 2005;25(5):304–11.
5. Pedemonte CH, Efendiev R, Bertorello AM. Inhibition of Na,K-ATPase by dopamine in proximal tubule epithelial cells. Semin Nephrol 2005;25(5):322–7.
6. Vinciguerra M, Mordasini D, Vandewalle A, Feraille E. Hormonal and nonhormonal mechanisms of regulation of the NA,K-pump in collecting duct principal cells. Semin Nephrol 2005;25(5):312–21.
7. Rector FC, Jr. Sodium, bicarbonate, and chloride absorption by the proximal tubule. Am J Physiol 1983;244(5):F461–71.
8. Slepkov ER, Rainey JK, Sykes BD, Fliegel L. Structural and functional analysis of the Na+/H+ exchanger. Biochem J 2007;401(3):623–33.
9. Burckhardt G, Di Sole F, Helmle-Kolb C. The Na+/H+ exchanger gene family. J Nephrol 2002;15 Suppl 5:S3–21.
10. Aronson PS. Ion exchangers mediating NaCl transport in the proximal tubule. Wien Klin Wochenschr 1997;109(12–13):435–40.
11. Aronson PS, Giebisch G. Mechanisms of chloride transport in the proximal tubule. Am J Physiol 1997;273(2 Pt 2):F179–92.
12. Puntheeranurak T, Kasch M, Xia X, Hinterdorfer P, Kinne RK. Three surface subdomains form the vestibule of the Na+/glucose cotransporter SGLT1. J Biol Chem 2007;282(35):25222–30.
13. Katsuno K, Fujimori Y, Takemura Y, Hiratochi M, Itoh F, et al. Sergliflozin, a novel selective inhibitor of low-affinity sodium glucose cotransporter (SGLT2), validates the critical role of SGLT2 in renal glucose reabsorption and modulates plasma glucose level. J Pharmacol Exp Ther 2007;320(1):323–30.
14. Preisig PA. The acid-activated signaling pathway: starting with Pyk2 and ending with increased NHE3 activity. Kidney Int 2007;72(11):1324–9.
15. Laghmani K, Preisig PA, Moe OW, Yanagisawa M, Alpern RJ. Endothelin-1/endothelin-B receptor-mediated increases in NHE3 activity in chronic metabolic acidosis. J Clin Invest 2001;107(12):1563–9.
16. Lee YJ, Lee YJ, Han HJ. Regulatory mechanisms of Na(+)/glucose cotransporters in renal proximal tubule cells. Kidney Int Suppl 2007(106):S27–35.
17. Schuster VL, Kokko JP, Jacobson HR. Angiotensin II directly stimulates sodium transport in rabbit proximal convoluted tubules. J Clin Invest 1984;73(2):507–15.
18. Han HJ, Park SH, Koh HJ, Taub M. Mechanism of regulation of Na+ transport by angiotensin II in primary renal cells. Kidney Int 2000;57(6):2457–67.
19. Han HJ, Park SH, Lee YJ. Signaling cascade of ANG II-induced inhibition of alpha-MG uptake in renal proximal tubule cells. Am J Physiol Renal Physiol 2004;286(4):F634–42.
20. Hussain T, Lokhandwala MF. Renal dopamine receptor function in hypertension. Hypertension 1998;32(2):187–97.
21. Aperia AC. Intrarenal dopamine: a key signal in the interactive regulation of sodium metabolism. Annu Rev Physiol 2000;62:621–47.

22. Jose PA, Raymond JR, Bates MD, Aperia A, Felder RA, Carey RM. The renal dopamine receptors. J Am Soc Nephrol 1992;2(8):1265–78.
23. Herrera M, Ortiz PA, Garvin JL. Regulation of thick ascending limb transport: role of nitric oxide. Am J Physiol Renal Physiol 2006;290(6):F1279–84.
24. Greger R, Schlatter E. Properties of the basolateral membrane of the cortical thick ascending limb of Henle's loop of rabbit kidney. A model for secondary active chloride transport. Pflugers Arch 1983;396(4):325–34.
25. Simard CF, Brunet GM, Daigle ND, Montminy V, Caron L, Isenring P. Self-interacting domains in the C terminus of a cation-Cl- cotransporter described for the first time. J Biol Chem 2004;279(39):40769–77.
26. Wall SM, Knepper MA, Hassell KA, Fischer MP, Shodeinde A, et al. Hypotension in NKCC1 null mice: role of the kidneys. Am J Physiol Renal Physiol 2006;290(2):F409–16.
27. Paredes A, Plata C, Rivera M, Moreno E, Vazquez N, et al. Activity of the renal Na+-K+-2Cl- cotransporter is reduced by mutagenesis of N-glycosylation sites: role for protein surface charge in Cl- transport. Am J Physiol Renal Physiol 2006;290(5):F1094–102.
28. Starremans PG, Kersten FF, Knoers NV, van den Heuvel LP, Bindels RJ. Mutations in the human Na-K-2Cl cotransporter (NKCC2) identified in Bartter syndrome type I consistently result in nonfunctional transporters. J Am Soc Nephrol 2003;14(6):1419–26.
29. Brunet GM, Gagnon E, Simard CF, Daigle ND, Caron L, et al. Novel insights regarding the operational characteristics and teleological purpose of the renal Na+-K+-Cl2 cotransporter (NKCC2s) splice variants. J Gen Physiol 2005;126 (4):325–37.
30. Payne JA, Forbush B, 3rd. Alternatively spliced isoforms of the putative renal Na-K-Cl cotransporter are differentially distributed within the rabbit kidney. Proc Natl Acad Sci USA 1994;91 (10):4544–8.
31. Igarashi P, Vanden Heuvel GB, Payne JA, Forbush B, 3rd. Cloning, embryonic expression, and alternative splicing of a murine kidney-specific Na-K-Cl cotransporter. Am J Physiol 1995;269 (3 Pt 2):F405–18.
32. Gimenez I, Isenring P, Forbush B. Spatially distributed alternative splice variants of the renal Na-K-Cl cotransporter exhibit dramatically different affinities for the transported ions. J Biol Chem 2002;277 (11):8767–70.
33. Plata C, Meade P, Vazquez N, Hebert SC, Gamba G. Functional properties of the apical Na+-K+-2Cl- cotransporter isoforms. J Biol Chem 2002;277 (13):11004–12.
34. Gagnon E, Forbush B, Caron L, Isenring P. Functional comparison of renal Na-K-Cl cotransporters between distant species. Am J Physiol Cell Physiol 2003;284 (2):C365–70.
35. Gimenez I, Forbush B. The residues determining differences in ion affinities among the alternative splice variants F, A, and B of the mammalian renal Na-K-Cl cotransporter (NKCC2). J Biol Chem 2007;282 (9):6540–7.
36. Gimenez I. Molecular mechanisms and regulation of furosemide-sensitive Na-K-Cl cotransporters. Curr Opin Nephrol Hypertens 2006;15 (5):517–23.
37. Gamba G. Molecular physiology and pathophysiology of electroneutral cation-chloride cotransporters. Physiol Rev 2005;85(2):423–93.
38. Lang F, Capasso G, Schwab M, Waldegger S. Renal tubular transport and the genetic basis of hypertensive disease. Clin Exp Nephrol 2005;9(2):91–9.
39. Capasso G, Cantone A, Evangelista C, Zacchia M, Trepiccione F, et al. Channels, carriers, and pumps in the pathogenesis of sodium-sensitive hypertension. Semin Nephrol 2005;25(6):419–24.
40. Capasso G, Rizzo M, Evangelista C, Ferrari P, Geelen G, et al. Altered expression of renal apical plasma membrane Na+ transporters in the early phase of genetic hypertension. Am J Physiol Renal Physiol 2005;288(6):F1173–82.
41. Borensztein P, Froissart M, Laghmani K, Bichara M, Paillard M. RT-PCR analysis of Na+/H+ exchanger mRNAs in rat medullary thick ascending limb. Am J Physiol 1995;268(6 Pt 2):F1224–8.

42. Benziane B, Demaretz S, Defontaine N, Zaarour N, Cheval L, et al. NKCC2 surface expression in mammalian cells: down-regulation by novel interaction with aldolase B. J Biol Chem 2007; 282(46):33817–30.
43. Takahashi N, Chernavvsky DR, Gomez RA, Igarashi P, Gitelman HJ, Smithies O. Uncompensated polyuria in a mouse model of Bartter's syndrome. Proc Natl Acad Sci USA 2000;97(10):5434–9.
44. Haas M. Properties and diversity of (Na-K-Cl) cotransporters. Annu Rev Physiol 1989;51:443–57.
45. Kim GH, Ecelbarger CA, Mitchell C, Packer RK, Wade JB, Knepper MA. Vasopressin increases Na-K-2Cl cotransporter expression in thick ascending limb of Henle's loop. Am J Physiol 1999;276(1 Pt 2):F96–F103.
46. Fernandez-Llama P, Ecelbarger CA, Ware JA, Andrews P, Lee AJ, et al. Cyclooxygenase inhibitors increase Na-K-2Cl cotransporter abundance in thick ascending limb of Henle's loop. Am J Physiol 1999;277(2 Pt 2):F219–26.
47. Gimenez I, Forbush B. Short-term stimulation of the renal Na-K-Cl cotransporter (NKCC2) by vasopressin involves phosphorylation and membrane translocation of the protein. J Biol Chem 2003;278 (29):26946-51.
48. Ecelbarger CA, Yu S, Lee AJ, Weinstein LS, Knepper MA. Decreased renal Na-K-2Cl cotransporter abundance in mice with heterozygous disruption of the G(s)alpha gene. Am J Physiol 1999;277 (2 Pt 2):F235–44.
49. Herrera M, Garvin JL. A high-salt diet stimulates thick ascending limb eNOS expression by raising medullary osmolality and increasing release of endothelin-1. Am J Physiol 2005;288 (1):F58–64.
50. Plato CF, Pollock DM, Garvin JL. Endothelin inhibits thick ascending limb chloride flux via ET(B) receptor-mediated NO release. Am J Physiol 2000;279(2): F326–33.
51. de Jesus Ferreira MC, Bailly C. Luminal and basolateral endothelin inhibit chloride reabsorption in the mouse thick ascending limb via a Ca(2+)-independent pathway. J Physiol 1997;505 (Pt 3):749–58.
52. Herrera M, Garvin JL. Endothelin stimulates endothelial nitric oxide synthase expression in the thick ascending limb. Am J Physiol 2004;287 (2):F231–5.
53. Fattal I, Abassi Z, Ovcharenko E, Shimada K, Takahashi M, et al. Effect of dietary sodium intake on the expression of endothelin-converting enzyme in the renal medulla. Nephron Physiol 2004;98 (4):p89–96.
54. Kohan DE, Padilla E. Osmolar regulation of endothelin-1 production by rat inner medullary collecting duct. J Clin Invest 1993;91(3):1235–40.
55. Taylor TA, Gariepy CE, Pollock DM, Pollock JS. Gender differences in ET and NOS systems in ETB receptor-deficient rats: effect of a high salt diet. Hypertension 2003;41(3 Pt 2): 657–62.
56. Vassileva I, Mountain C, Pollock DM. Functional role of ETB receptors in the renal medulla. Hypertension 2003;41(6):1359–63.
57. Melo LG, Veress AT, Chong CK, Pang SC, Flynn TG, Sonnenberg H. Salt-sensitive hypertension in ANP knockout mice: potential role of abnormal plasma renin activity. Am J Physiol 1998;274(1 Pt 2):R255–61.
58. Serneri GG, Modesti PA, Cecioni I, Biagini D, Migliorini A, et al. Plasma endothelin and renal endothelin are two distinct systems involved in volume homeostasis. Am J Physiol 1995;268(5 Pt 2):H1829–37.
59. Malatino LS, Bellanuova I, Cataliotti A, Cuzzola F, Mallamaci F, et al. Renal endothelin-1 is linked to changes in urinary salt and volume in essential hypertension. Salt sensitivity group of the Italian society of hypertension. J Nephrol 2000;13(3):178–84.
60. Stein JH, Osgood RW, Boonjarern S, Ferris TF. A comparison of the segmental analysis of sodium reabsorption during Ringer's and hyperoncotic albumin infusion in the rat. J Clin Invest 1973;52(9):2313–23.

61. Stein JH, Reineck HJ. The role of the collecting duct in the regulation of excretion of sodium and other electrolytes. Kidney Int 1974;6(1):1–9.
62. Stein JH, Osgood RW, Kunau RT, Jr. Direct measurement of papillary collecting duct sodium transport in the rat. Evidence for heterogeneity of nephron function during Ringer loading. J Clin Invest 1976;58(4):767–73.
63. Osgood RW, Reineck HJ, Stein JH. Further studies on segmental sodium transport in the rat kidney during expansion of the extracellular fluid volume. J Clin Invest 1978;62(2): 311–20.
64. Shimkets RA, Warnock DG, Bositis CM, Nelson-Williams C, Hansson JH, et al. Liddle's syndrome: heritable human hypertension caused by mutations in the beta subunit of the epithelial sodium channel. Cell 1994;79(3):407–14.
65. Booth RE, Johnson JP, Stockand JD. Aldosterone. Adv Physiol Educ 2002;26(1–4):8–20.
66. Stockand JD. New ideas about aldosterone signaling in epithelia. Am J Physiol Renal Physiol 2002;282(4):F559–76.
67. Hoover RS, Poch E, Monroy A, Vazquez N, Nishio T, et al. N-Glycosylation at two sites critically alters thiazide binding and activity of the rat thiazide-sensitive Na(+):Cl(−) cotransporter. J Am Soc Nephrol 2003;14(2):271–82.
68. Moreno E, Cristobal PS, Rivera M, Vazquez N, Bobadilla NA, Gamba G. Affinity-defining domains in the Na-Cl cotransporter: a different location for Cl- and thiazide binding. J Biol Chem 2006;281(25):17266–75.
69. Canessa CM, Schild L, Buell G, Thorens B, Gautschi I, et al. Amiloride-sensitive epithelial Na+ channel is made of three homologous subunits. Nature 1994;367(6462):463–7.
70. Firsov D, Gautschi I, Merillat AM, Rossier BC, Schild L. The heterotetrameric architecture of the epithelial sodium channel (ENaC). Embo J 1998;17(2):344–52.
71. Ergonul Z, Frindt G, Palmer LG. Regulation of maturation and processing of ENaC subunits in the rat kidney. Am J Physiol Renal Physiol 2006;291(3):F683–93.
72. Hummler E, Horisberger JD. Genetic disorders of membrane transport. V. The epithelial sodium channel and its implication in human diseases. Am J Physiol 1999;276(3 Pt 1):G567–71.
73. Rubera I, Loffing J, Palmer LG, Frindt G, Fowler-Jaeger N, et al. Collecting duct-specific gene inactivation of alphaENaC in the mouse kidney does not impair sodium and potassium balance. J Clin Invest 2003;112(4):554–65.
74. Loffing J, Pietri L, Aregger F, Bloch-Faure M, Ziegler U, et al. Differential subcellular localization of ENaC subunits in mouse kidney in response to high- and low-Na diets. Am J Physiol Renal Physiol 2000;279(2):F252–8.
75. Goodfriend TL. Aldosterone – a hormone of cardiovascular adaptation and maladaptation. J Clin Hypertens (Greenwich) 2006;8(2):133–9.
76. Pitt B. Effect of aldosterone blockade in patients with systolic left ventricular dysfunction: implications of the RALES and EPHESUS studies. Mol Cell Endocrinol 2004;217(1–2): 53–8.
77. Arriza JL, Weinberger C, Cerelli G, Glaser TM, Handelin BL, et al. Cloning of human mineralocorticoid receptor complementary DNA: structural and functional kinship with the glucocorticoid receptor. Science 1987;237(4812):268–75.
78. Funder JW. Glucocorticoid and mineralocorticoid receptors: biology and clinical relevance. Annu Rev Med 1997;48:231–40.
79. Husted RF, Laplace JR, Stokes JB. Enhancement of electrogenic Na+ transport across rat inner medullary collecting duct by glucocorticoid and by mineralocorticoid hormones. J Clin Invest 1990;86(2):498–506.
80. Funder JW, Pearce PT, Smith R, Smith AI. Mineralocorticoid action: target tissue specificity is enzyme, not receptor, mediated. Science 1988;242(4878):583–5.
81. Melander O, Orho-Melander M, Bengtsson K, Lindblad U, Rastam L, et al. Association between a variant in the 11 beta-hydroxysteroid dehydrogenase type 2 gene and primary hypertension. J Hum Hypertens 2000;14(12):819–23.

82. Verrey F, Kraehenbuhl, J.P., Rossier, B.C. Aldosterone Induces a Rapid Increase in the Rate of Na,K-ATPase Gene Transcription in Cultured Kidney Cells. Molecular Endocrinology 1989; 3:1369–76.
83. Escoubet B, Coureau C, Bonvalet JP, Farman N. Noncoordinate regulation of epithelial Na channel and Na pump subunit mRNAs in kidney and colon by aldosterone. Am J Physiol 1997;272(5 Pt 1):C1482–91.
84. Alvarez de la Rosa D, Zhang P, Naray-Fejes-Toth A, Fejes-Toth G, Canessa CM. The serum and glucocorticoid kinase sgk increases the abundance of epithelial sodium channels in the plasma membrane of Xenopus oocytes. J Biol Chem 1999;274(53):37834–9.
85. Chen SYea. Epithelial sodium channel regulated by aldosterone-induced protein sgk. Proc Natl Acad Sci USA 1999;96:2514–9.
86. Naray-Fejes-Toth A, Canessa C, Cleaveland ES, Aldrich G, Fejes-Toth G. sgk is an aldosterone-induced kinase in the renal collecting duct. Effects on epithelial na+ channels. J Biol Chem 1999;274(24):16973–8.
87. Gumz ML, Popp MP, Wingo CS, Cain BD. Early transcriptional effects of aldosterone in a mouse inner medullary collecting duct cell line. Am J Physiol 2003;285(4):F664–73.
88. Soundararajan R, Zhang TT, Wang J, Vandewalle A, Pearce D. A novel role for glucocorticoid-induced leucine zipper protein in epithelial sodium channel-mediated sodium transport. J Biol Chem 2005;280(48):39970–81.
89. Fakitsas P, Adam G, Daidie D, van Bemmelen MX, Fouladkou F, et al. Early aldosterone-induced gene product regulates the epithelial sodium channel by deubiquitylation. J Am Soc Nephrol 2007;18(4):1084–92.
90. Pearce D. SGK1 regulation of epithelial sodium transport. Cell Physiol Biochem 2003;13(1):13–20.
91. Liang X, Peters KW, Butterworth MB, Frizzell RA. 14-3-3 isoforms are induced by aldosterone and participate in its regulation of epithelial sodium channels. J Biol Chem 2006;281(24):16323–32.
92. Nagaki K, Yamamura H, Shimada S, Saito T, Hisanaga S, et al. 14-3-3 Mediates phosphorylation-dependent inhibition of the interaction between the ubiquitin E3 ligase Nedd4-2 and epithelial Na+ channels. Biochemistry 2006;45(21):6733–40.
93. Lee IH, Campbell CR, Cook DI, Dinudom A. Regulation of epithelial Na+ channels by aldosterone: role of Sgk1. Clin Exp Pharmacol Physiol 2008;35(2):235–41.
94. Chan CM, Unwin RJ, Burnstock G. Potential functional roles of extracellular ATP in kidney and urinary tract. Exp Nephrol 1998;6(3):200–7.
95. Burnstock G. Pathophysiology and therapeutic potential of purinergic signaling. Pharmacol Rev 2006;58(1):58–86.
96. Leipziger J. Control of epithelial transport via luminal P2 receptors. Am J Physiol Renal Physiol 2003;284(3):F419–32.
97. Shirley DG, Bailey MA, Unwin RJ. In vivo stimulation of apical P2 receptors in collecting ducts: evidence for inhibition of sodium reabsorption. Am J Physiol Renal Physiol 2005;288(6):F1243–8.
98. McCoy DE, Taylor AL, Kudlow BA, Karlson K, Slattery MJ, et al. Nucleotides regulate NaCl transport in mIMCD-K2 cells via P2X and P2Y purinergic receptors. Am J Physiol 1999;277(4 Pt 2):F552–9.
99. Thomas J, Deetjen P, Ko WH, Jacobi C, Leipziger J. P2Y(2) receptor-mediated inhibition of amiloride-sensitive short circuit current in M-1 mouse cortical collecting duct cells. J Membr Biol 2001;183(2):115–24.
100. Ma HP, Li L, Zhou ZH, Eaton DC, Warnock DG. ATP masks stretch activation of epithelial sodium channels in A6 distal nephron cells. Am J Physiol Renal Physiol 2002;282(3): F501–5.
101. Lehrmann H, Thomas J, Kim SJ, Jacobi C, Leipziger J. Luminal P2Y2 receptor-mediated inhibition of Na+ absorption in isolated perfused mouse CCD. J Am Soc Nephrol 2002;13(1):10–8.

102. Li L, Wingo CS, Xia SL. Downregulation of SGK1 by nucleotides in renal tubular epithelial cells. Am J Physiol Renal Physiol 2007;293(5):F1751–7.
103. Schwiebert EM, Kishore BK. Extracellular nucleotide signaling along the renal epithelium. Am J Physiol Renal Physiol 2001;280(6):F945–63.
104. Jose PA, Asico LD, Eisner GM, Pocchiari F, Semeraro C, Felder RA. Effects of costimulation of dopamine D1- and D2-like receptors on renal function. Am J Physiol 1998;275(4 Pt 2):R986–94.
105. Felder RA, Kinoshita S, Sidhu A, Ohbu K, Kaskel FJ. A renal dopamine-1 receptor defect in two genetic models of hypertension. Am J Hypertens 1990;3(6 Pt 2):96S–9S.
106. Ohbu K, Felder RA. DA1 dopamine receptors in renal cortical collecting duct. Am J Physiol 1991;261(5 Pt 2):F890–5.
107. Takemoto F, Satoh T, Cohen HT, Katz AI. Localization of dopamine-1 receptors along the microdissected rat nephron. Pflugers Arch 1991;419(3–4):243–8.
108. Sun D, Wilborn TW, Schafer JA. Dopamine D4 receptor isoform mRNA and protein are expressed in the rat cortical collecting duct. Am J Physiol 1998;275(5 Pt 2): F742–51.
109. O'Connell DP, Vaughan CJ, Aherne AM, Botkin SJ, Wang ZQ, et al. Expression of the dopamine D3 receptor protein in the rat kidney. Hypertension 1998;32(5):886–95.
110. Huo T, Healy DP. [3H]domperidone binding to the kidney inner medullary collecting duct dopamine-2 K (DA2K) receptor. J Pharmacol Exp Ther 1991;258(2):424–8.
111. Sun D, Schafer JA. Dopamine inhibits AVP-dependent Na+ transport and water permeability in rat CCD via a D4-like receptor. Am J Physiol 1996;271(2 Pt 2):F391–400.
112. Ozono R, O'Connell DP, Wang ZQ, Moore AF, Sanada H, et al. Localization of the dopamine D1 receptor protein in the human heart and kidney. Hypertension 1997;30(3 Pt 2): 725–9.
113. Satoh T, Cohen HT, Katz AI. Intracellular signaling in the regulation of renal Na-K-ATPase. I. Role of cyclic AMP and phospholipase A2. J Clin Invest 1992;89(5):1496–500.
114. Takemoto F, Cohen HT, Satoh T, Katz AI. Dopamine inhibits Na/K-ATPase in single tubules and cultured cells from distal nephron. Pflugers Arch 1992;421(4):302–6.
115. Satoh T, Cohen HT, Katz AI. Different mechanisms of renal Na-K-ATPase regulation by protein kinases in proximal and distal nephron. Am J Physiol 1993;265(3 Pt 2):F399–405.
116. Silver RB, Frindt G, Windhager EE, Palmer LG. Feedback regulation of Na channels in rat CCT. I. Effects of inhibition of Na pump. Am J Physiol 1993;264(3 Pt 2):F557–64.
117. Wang WH, Geibel J, Giebisch G. Mechanism of apical K+ channel modulation in principal renal tubule cells. Effect of inhibition of basolateral Na(+)-K(+)-ATPase. J Gen Physiol 1993;101(5):673–94.
118. Schiffmann SN, Lledo PM, Vincent JD. Dopamine D1 receptor modulates the voltage-gated sodium current in rat striatal neurones through a protein kinase A. J Physiol 1995;483 (Pt 1):95–107.
119. Cantrell AR, Smith RD, Goldin AL, Scheuer T, Catterall WA. Dopaminergic modulation of sodium current in hippocampal neurons via cAMP-dependent phosphorylation of specific sites in the sodium channel alpha subunit. J Neurosci 1997;17(19):7330–8.
120. Stokes JB, Ingram MJ, Williams AD, Ingram D. Heterogeneity of the rabbit collecting tubule: localization of mineralocorticoid hormone action to the cortical portion. Kidney Int 1981;20(3):340–7.
121. Stokes JB. Na and K transport across the cortical and outer medullary collecting tubule of the rabbit: evidence for diffusion across the outer medullary portion. Am J Physiol 1982;242(5):F514–20.
122. Wingo CS. Active proton secretion and potassium absorption in the rabbit outer medullary collecting duct. Functional evidence for proton-potassium- activated adenosine triphosphatase. J Clin Invest 1989;84(1):361–5.
123. Garg LC, Knepper MA, Burg MB. Mineralocorticoid effects on Na-K-ATPase in individual nephron segments. Am J Physiol 1981;240(6):F536–44.

124. Ciampolillo F, McCoy DE, Green RB, Karlson KH, Dagenais A, et al. Cell-specific expression of amiloride-sensitive, Na(+)-conducting ion channels in the kidney. Am J Physiol 1996;271(4 Pt 1):C1303–15.
125. Xia SL, Noh SH, Verlander JW, Gelband CH, Wingo CS. Apical membrane of native OMCD(i) cells has nonselective cation channels. Am J Physiol Renal Physiol 2001;281(1):F48–55.
126. Huo T, Ye MQ, Healy DP. Characterization of a dopamine receptor (DA2K) in the kidney inner medulla. Proc Natl Acad Sci USA 1991;88(8):3170–4.
127. Tomita K, Nonoguchi H, Terada Y, Marumo F. Effects of ET-1 on water and chloride transport in cortical collecting ducts of the rat. Am J Physiol 1993;264(4 Pt 2):F690–6.
128. Gilmore ES, Stutts MJ, Milgram SL. SRC family kinases mediate epithelial Na+ channel inhibition by endothelin. J Biol Chem 2001;276(45):42610–7.
129. Ahn D, Ge Y, Stricklett PK, Gill P, Taylor D, et al. Collecting duct-specific knockout of endothelin-1 causes hypertension and sodium retention. J Clin Invest 2004;114(4):504–11.
130. Gallego MS, Ling BN. Regulation of amiloride-sensitive Na+ channels by endothelin-1 in distal nephron cells. Am J Physiol 1996;271(2 Pt 2):F451–60.
131. Ye Q, Chen S, Gardner DG. Endothelin inhibits NPR-A and stimulates eNOS gene expression in rat IMCD cells. Hypertension 2003;41(3 Pt 2):675–81.
132. Edwards RM, Pullen M, Nambi P. Activation of endothelin ETB receptors increases glomerular cGMP via an L-arginine-dependent pathway. Am J Physiol 1992;263(6 Pt 2):F1020–5.
133. Pollock DM. Renal endothelin in hypertension. Curr Opin Nephrol Hypertens 2000;9(2):157–64.
134. Zeidel ML, Brady HR, Kone BC, Gullans SR, Brenner BM. Endothelin, a peptide inhibitor of Na(+)-K(+)-ATPase in intact renal tubular epithelial cells. Am J Physiol 1989;257(6 Pt 1):C1101–7.
135. Kohan DE. The renal medullary endothelin system in control of sodium and water excretion and systemic blood pressure. Curr Opin Nephrol Hypertens 2006;15(1):34–40.
136. Gariepy CE, Ohuchi T, Williams SC, Richardson JA, Yanagisawa M. Salt-sensitive hypertension in endothelin-B receptor-deficient rats. J Clin Invest 2000;105(7):925–33.
137. Ohkita M, Wang Y, Nguyen ND, Tsai YH, Williams SC, et al. Extrarenal ETB plays a significant role in controlling cardiovascular responses to high dietary sodium in rats. Hypertension 2005;45(5):940–6.
138. Ge Y, Bagnall A, Stricklett PK, Strait K, Webb DJ, et al. Collecting duct-specific knockout of the endothelin B receptor causes hypertension and sodium retention. Am J Physiol Renal Physiol 2006;291(6):F1274–80.
139. Ge Y, Stricklett PK, Hughes AK, Yanagisawa M, Kohan DE. Collecting duct-specific knockout of the endothelin A receptor alters renal vasopressin responsiveness, but not sodium excretion or blood pressure. Am J Physiol 2005;289(4):F692–8.
140. Hsieh TJ, Lin SR, Lee YJ, Shin SJ, Lai YH, et al. Increased renal medullary endothelin-1 synthesis in prehypertensive DOCA- and salt-treated rats. Am J Physiol 2000;279(1):F112–21.
141. Pollock DM, Allcock GH, Krishnan A, Dayton BD, Pollock JS. Upregulation of endothelin B receptors in kidneys of DOCA-salt hypertensive rats. Am J Physiol 2000;278(2):F279–86.
142. Wilson FH, Disse-Nicodeme S, Choate KA, Ishikawa K, Nelson-Williams C, et al. Human hypertension caused by mutations in WNK kinases. Science 2001;293(5532):1107–12.
143. Gamba G. Role of WNK kinases in regulating tubular salt and potassium transport and in the development of hypertension. Am J Physiol Renal Physiol 2005;288(2):F245–52.
144. Flatman PW. Cotransporters, WNKs and hypertension: important leads from the study of monogenetic disorders of blood pressure regulation. Clin Sci (Lond) 2007;112(4):203–16.
145. Subramanya AR, Yang CL, McCormick JA, Ellison DH. WNK kinases regulate sodium chloride and potassium transport by the aldosterone-sensitive distal nephron. Kidney Int 2006;70(4):630–4.

146. Hadchouel J, Jeunemaitre X. Life and death of the distal nephron: WNK4 and NCC as major players. Cell Metab 2006;4(5):335–7.
147. Simon DB, Karet FE, Hamdan JM, DiPietro A, Sanjad SA, Lifton RP. Bartter's syndrome, hypokalaemic alkalosis with hypercalciuria, is caused by mutations in the Na-K-2Cl cotransporter NKCC2. Nat Genet 1996;13(2):183–8.
148. Simon DB, Karet FE, Rodriguez-Soriano J, Hamdan JH, DiPietro A, et al. Genetic heterogeneity of Bartter's syndrome revealed by mutations in the K+ channel, ROMK. Nat Genet 1996;14(2):152–6.
149. Simon DB, Nelson-Williams C, Bia MJ, Ellison D, Karet FE, et al. Gitelman's variant of Bartter's syndrome, inherited hypokalaemic alkalosis, is caused by mutations in the thiazide-sensitive Na-Cl cotransporter. Nat Genet 1996;12(1):24–30.
150. Kunchaparty S, Palcso M, Berkman J, Velazquez H, Desir GV, et al. Defective processing and expression of thiazide-sensitive Na-Cl cotransporter as a cause of Gitelman's syndrome. Am J Physiol 1999;277(4 Pt 2):F643–9.
151. De Jong JC, Van Der Vliet WA, Van Den Heuvel LP, Willems PH, Knoers NV, Bindels RJ. Functional expression of mutations in the human NaCl cotransporter: evidence for impaired routing mechanisms in Gitelman's syndrome. J Am Soc Nephrol 2002;13(6):1442–8.

Renal Acid–Base Regulation Via Ammonia Transport in Mammals

I. David Weiner

Abstract Maintenance of acid–base homeostasis is fundamental to all mammals. This is maintained through two mechanisms, respiratory control through CO_2 excretion and renal net acid excretion. Renal net acid excretion comprises titratable acid excretion, ammonia excretion, and bicarbonate excretion. Quantitatively, ammonia excretion is the most significant component of renal net acid excretion under most physiologic and pathophysiologic circumstances. Renal ammonia excretion includes the dual components of an intrarenal ammoniagenesis, followed by segment and transporter-specific transport of both molecular forms of ammonia, NH_4^+ and NH_3. Coordinated NH_4^+ and NH_3 transport results in exquisitely regulated renal ammonia excretion and thereby maintenance of system at pH homeostasis.

Keywords Acid–base · ammonia · proximal tubule · loop of Henle · collecting duct · Rh glycoprotei · Rh B Glycoprotein (Rhbg) · Rh C Glycoprotein (Rhcg) · NKCC2 · sodium hydrogen exchang · NHE3 · phosphoenolpyruvate carboxykinase · phosphate-dependent glutaminase · glutamine synthetase

1 Introduction

Acid–base regulation is a fundamental requirement for all living organisms. In the absence of balance between exogenous and endogenous acid production and excretion a wide variety of disease states can ensue, including failure to thrive in children, disorders of bone maturation and development, recurrent renal stone disease, skeletal muscle atrophy, increased cardiovascular susceptibility to arrhythmias and decreased cardiac sensitivity to circulating catecholamines [1].

I.D. Weiner (✉)
Division of Nephrology, Hypertension and Transplantation, University of Florida College of Medicine, Nephrology and Hypertension Section, North Florida/South Georgia Veterans Health System, Gainesville, FL, 32610-0224, USA
e-mail: david.weiner@medicine.ufl.edu

G.A. Gerencser (ed.), *Epithelial Transport Physiology*,
DOI 10.1007/978-1-60327-229-2_13,

Normal cellular metabolism of amino acids, carbohydrates, and lipids generates acids on a continuous and ongoing process. Amino acid metabolism, in contrast, results in ~0.8 mmol kg^{-1} d^{-1} of HCl and H_2SO_4 formation. Carbohydrate and lipid metabolism generates CO_2 in substantially greater quantities, ~15,000 mmol per day. Because CO_2 can be hydrated to form carbonic acid (H_2CO_3), which is a strong acid, it is considered an acid. However, because CO_2 is rapidly excreted by pulmonary respiration, it is considered a volatile acid. Because CO_2 is excreted by the lungs, and not the kidneys, it is not a subject for this review.

Acids produced from amino acid metabolism, and nonvolatile acids that result from either exogenous or endogenous sources, are rapidly buffered by intracellular and extracellular buffers. The most important of these buffer systems is the HCO_3^-/CO_2 buffer system: $H^+ + HCO_3^- \leftrightarrow H_2CO_3 \leftrightarrow H_2O + CO_2$. With this buffer system, H^+'s are rapidly buffered, converted to H_2CO_3, and then to H_2O and CO_2, and the CO_2 is rapidly excreted by respiration. A critical aspect of this buffer system is the necessity to replenish the HCO_3^- that is consumed in buffering acid and then being excreted as CO_2. Kidneys mediate a central role in acid–base homeostasis because they are the primary organ that generates HCO_3^-.

2 Renal Acid–Base Homeostasis

Renal acid–base homeostasis involves two fundamental processes, reabsorption of filtered bicarbonate and generation of new bicarbonate. In humans with normal renal function ~4,000 mEq per day of bicarbonate are filtered at the glomerulus; under normal conditions almost all of this filtered bicarbonate is completely reabsorbed by specific transport mechanisms in the renal nephron and collecting duct. Several excellent reviews of the mechanism of reabsorption of filtered bicarbonate, are available for the interested reader [2–4].

However, reabsorbing filtered bicarbonate is not sufficient to maintain acid–base homeostasis. Endogenous fixed acid production that results from amino acid metabolism averages ~0.8–1.0 mEq kg^{-1} day^{-1} and is buffered by an equivalent amount of base. To prevent progressive loss of this buffering capacity, new alkali equivalents must be formed. Kidneys, through the process of net acid excretion (NAE), generate these new alkali equivalents. NAE involves excretion of ammonia and titratable acid and is counter-balanced by bicarbonate excretion.

Ammonia metabolism is the major component of basal NAE and is also the primary component of the renal response to most acid–base disorders. Under basal conditions renal ammonia metabolism contributes 60–70% of NAE, titratable acid excretion accounts for only ~30–40%, and urinary bicarbonate excretion is essentially zero [5]. Metabolic acidosis increases NAE over a period of several days to restore acid–base homeostasis. Almost the entire NAE increase is due to increased ammonia excretion; titratable acid excretion increases only modestly (Fig. 1) [5, 7]. Because ammonia excretion represents new HCO_3^- generation, this is an adaptive response that contributes to correction of the metabolic acidosis.

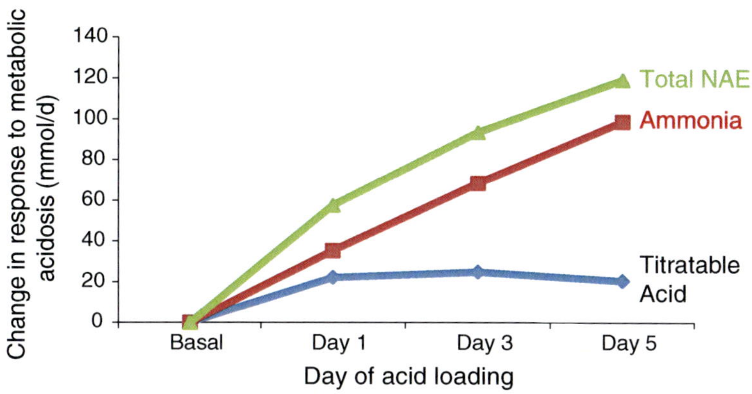

Fig. 1 Human renal net acid excretion response to metabolic acidosis. Induction of metabolic acidosis results in gradual increase in net acid excretion over several days. Changes in titratable acid are relatively minimal, whereas the majority of changes in net acid excretion result from increases in urinary ammonia excretion. Data recalculated from [6]

3 Ammonia Chemistry

Ammonia exists in two molecular forms in biological solutions, NH_3 and NH_4^+. The relative amounts of each are governed by the buffer reaction: $NH_3 + H^+ \leftrightarrow NH_4^+$. This reaction occurs nearly instantaneously and has a pKa under biologically relevant conditions of ~9.1. Accordingly, at pH 7.4 only ~1.7% of total ammonia is present as NH_3. Moreover, because most biological fluids exist at a pH substantially below the pKa of this buffer reaction, small changes in pH result in exponential changes in relative NH_3 concentration with almost no change in NH_4^+ concentration.

The two molecular forms of ammonia have very different biophysical characteristics. NH_3 is a relatively small, uncharged molecule, enabling it to diffuse across most lipid bilayers. However, its membrane permeability is not infinite, resulting in the possibility of transepithelial NH_3 gradients when there are high rates of either NH_4^+ or H^+ transport. Such gradients are known to exist in the kidney [8]. NH_4^+ has only a very limited permeability across lipid bilayers in the absence of specific transport proteins. However in aqueous solutions, NH_4^+ and K^+ have nearly identical biophysical characteristics [9], which enables NH_4^+ to be transported at the K^+-transport site of many proteins.

4 Renal Ammonia Metabolism

Renal ammonia metabolism and transport involves integrated responses of multiple portions of the kidney (Fig. 2). Only a minimal amount of urinary ammonia derives from glomerular filtration, making urinary ammonia excretion unique among the

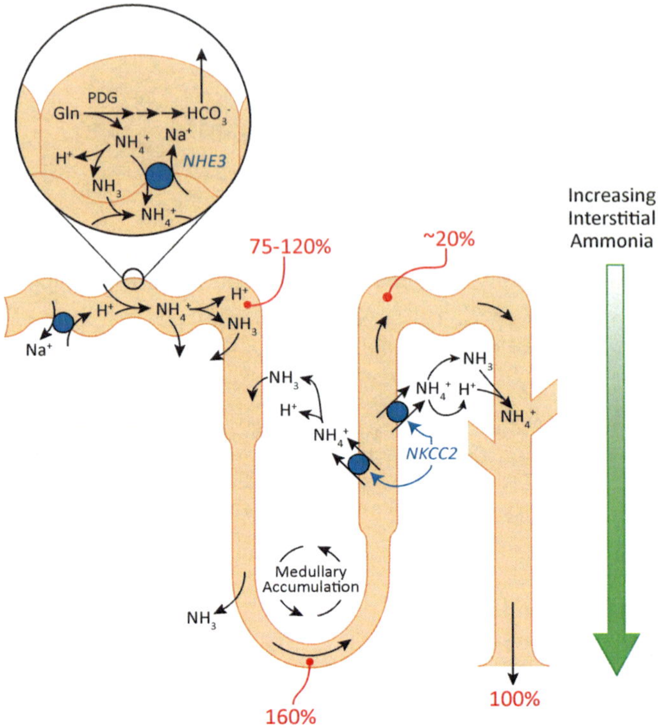

Fig. 2 Summary of renal ammonia metabolism. The proximal tubule produces ammonia, as NH4+, from glutamine. NH4+ is then secreted preferentially into the luminal fluid, primarily by NHE3. It is then reabsorbed by the TAL, primarily by NKCC2. This results in ammonia delivery to the distal nephron accounting for ~20% of final urinary ammonia; the remaining ~80% is secreted in the collecting duct through parallel NH3 and H+ transport. *Numbers in red* indicate proportion of total urinary ammonia present at the indicated sites under basal states. Modified from [9]

major compounds present in the urine. Instead, ammonia is produced by the kidney and is then selectively transported into either the urine or into the renal vein. Ammonia's selective transport into either the urine, where it can be excreted, or into the renal interstitium and then into the renal veins, resulting in return to the systemic circulation, determines its effect on acid–base homeostasis. Ammonia excreted into the urine as ammonium results in equimolar new bicarbonate formation. Ammonia returned to the systemic circulation is metabolized by the liver to urea and glutamine [10]. Under normal conditions, ~2/3 of hepatic ammonia metabolism results in urea formation in a process that utilizes equimolar amounts of bicarbonate, consuming the bicarbonate produced in the proximal tubule in ammoniagenesis [11]. The selective transport of ammonia into either the urine or the renal vein includes integrated transport mechanisms in the proximal tubule, thick ascending limb of the loop of Henle, and the collecting duct.

Proximal tubule segments metabolize the amino acid, glutamine, through a multistep enzymatic process that results in production of two NH_4^+ and two HCO_3^- molecules from each glutamine molecule. The biochemistry of ammonia production is reviewed in detail elsewhere [12, 13]. The proximal tubule then secretes ammonia preferentially into the tubule lumen, although there is some transport across the basolateral membrane (24–45% of ammonia produced depending on acid–base homeostasis) [14]. Ammonia secretion into the luminal fluid appears to be transported as the molecular species, NH_4^+, through a mechanism that appears primarily to involve via the apical Na^+/H^+ exchanger, NHE3 [15, 16], and to a lesser degree an apical Ba^{+2}-sensitive K^+ channel [16, 17]. In addition, a significant component (but still unresolved proportion) of NH_3 diffusion across the apical membrane occurs.

In the thick ascending limb of the loop of Henle, multiple proteins contribute to luminal ammonia reabsorption. The apical Na^+-K^+-$2Cl^-$ cotransporter, NKCC2, mediates the majority of ammonium reabsorption; an apical K^+/NH_4^+ antiporter and an amiloride-sensitive NH_4^+-conductance also contribute [18–20]. Some of the ammonia absorbed by the medullary thick ascending limb of the loop of Henle undergoes recycling into the thin descending limb of the loop of Henle. Ammonia recycling predominantly involves passive NH_3 diffusion, although there may be a small component of passive NH_4^+ transport [21, 22].

The net effect of ammonia absorption by the thick ascending limb of the loop of Henle and passive ammonia secretion into the thin descending limb of the loop of Henle is development of axial ammonia concentration in the medullary interstitium that parallels the hypertonicity gradient. Moreover, ammonia absorption by the medullary thick ascending limb results in ammonia delivery to the distal tubule averaging ~20% of final urinary ammonia content [5, 23].

Ammonia secretion by the collecting duct accounts for the majority, ~80%, of urinary ammonia content. Several studies have shown that collecting duct ammonia secretion involves parallel NH_3 and H^+ transport, with little-to-no NH_4^+ permeability in collecting duct segments [23, 24]. In the absence of luminal carbonic anhydrase activity in most segments of the collecting duct, a luminal disequilibrium pH amplifies the NH_3 gradient and increases the rate of transepithelial ammonia secretion [23, 24]. Thus, the fundamental mechanisms of ammonia transport differ in the collecting duct as compared to the proximal tubule and the thick ascending limb of the loop of Henle.

NH_3 movement across collecting duct segments could occur via either diffusive NH_3 transport or transporter-mediated NH_3 transport. One approach to differentiate between these two possibilities is to examine the correlation between extracellular ammonia concentration and ammonia transport; with diffusion the transport rate parallels the concentration gradient, whereas transporter-mediated movement exhibits saturable kinetics. This question was approached using cultured renal collecting duct epithelial cell line, mIMCD-3, grown on permeable support membranes to enable separate study of apical and basolateral transport mechanisms. Methylammonia (MA) was used as an ammonia analogue, enabling the use of the radiolabeled molecule analogue, $[^{14}C]$-MA, for quantification of transport.

Studies examining basolateral ammonia/methylammonia uptake showed that basolateral transport was a combination of both a saturable, transporter-mediated, component and a nonsaturable, diffusive component; the transporter-mediated component predominated at concentrations below 7 mM [25]. This process had an apparent affinity of ammonia of ~2 mM. Functional characterization of the saturable transport activity demonstrated that it did not involve extracellular Na^+ or K^+, was not mediated by Na^+-K^+-ATPase, Na^+-K^+-$2Cl^-$ cotransporter, NKCC-1 or NKCC-2, K^+ channels, or KCC proteins. Transport activity paralleled H^+ gradients and was not regulated by membrane voltage. Thus, a major mechanism of basolateral ammonia transport was via a basolateral, electroneutral NH_3 transporter [25]. Because these cells expressed basolateral Rhbg, this transport activity likely reflected Rhbg-mediated transport. However, Rhbg-specific inhibitors were unavailable and manipulation of Rhbg protein expression was not performed, preventing definitive proof of the role of Rhbg.

Similar studies examined apical ammonia transport mechanisms. Apical ammonia transport had characteristics similar to those of basolateral transport, with the exception of a lower apparent affinity for ammonia, ~7 mM [26]. Moreover, because apical transport exhibited trans stimulation by MA, it appeared to be mediated by an NH_4^+/H^+ exchange mechanism with affinity for NH_4^+ (MA^+) at both the cytosolic and extracellular binding sites rather than facilitated NH_3 transport [26]. These cells expressed apical Rhcg, suggesting Rhcg mediated the facilitated transport identified in these studies.

Although these studies [25, 26] establish facilitated ammonia transport as a major mechanism of both apical and basolateral ammonia transport, the relative extent of lipid solubility diffusion of NH_3 and protein mediated facilitated diffusion of NH_3 in both the apical and basolateral membranes of collecting duct cells in vivo remains to be determined.

Collecting duct ammonia secretion requires both H^+ secretion in parallel with NH_3 transport. A wide variety of studies have examined the cellular mechanisms of collecting duct H^+ secretion. H^+ are secreted across the apical plasma membrane by both H^+-ATPase and H^+-K^+-ATPase. Cytosolic H^+ are generated by carbonic anhydrase II-catalyzed hydration of CO_2 to H_2CO_3, which dissociates to H^+ and HCO_3^-. Cytosolic HCO_3^- exits the cell via basolateral Cl^-/HCO_3^- exchange, such as AE1 in intercalated cells. This HCO_3^- reacts with interstitial H^+ that derives from NH_4^+, to form H_2CO_3, which dehydrates to CO_2 and H_2O. This CO_2 is then able to diffuse across the basolateral plasma membrane into the cell. Interstitial NH_3 that results from dissociation of NH_4^+ to NH_3 and H^+, appears to be transported into the cell via either Rhbg or Rhcg, or via diffusive mechanisms, and is transported across the apical plasma membrane by Rhcg, and possibly by a component of diffusive mechanisms. In the IMCD the basolateral Na^+-K^+-ATPase contributes to basolateral NH_4^+ uptake. Overall, the current model of collecting duct ammonia secretion suggests that transepithelial ammonia transport involves the integrated function of a number of proteins (Fig. 3).

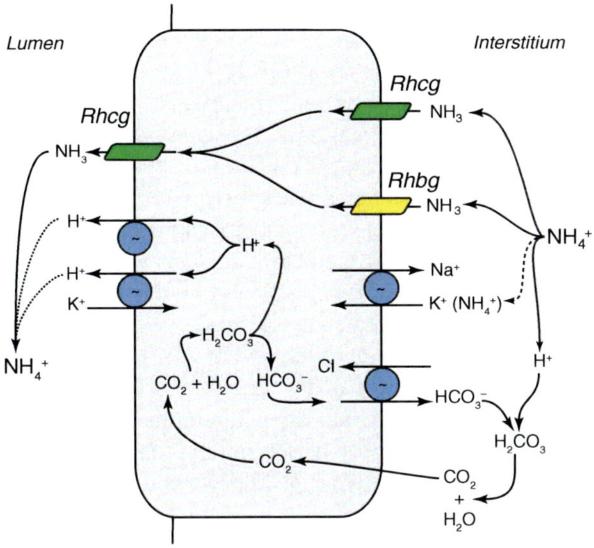

Fig. 3 Detailed model of collecting duct ammonia transport. Interstitial NH_4^+ is in equilibrium with NH_3 and H^+. NH_3 is transported across the basolateral membrane, predominantly by Rhcg, but also possibly partly by Rhbg. In the IMCD, basolateral Na^+-K^+-ATPase transports NH_4^+. Intracellular NH_3 is secreted across the apical membrane by apical Rhcg. H^+-ATPase and H^+-K^+-ATPase secrete H^+, which combines with luminal NH_3 to form NH_4^+, which is "trapped" in the lumen. Intracellular H^+ is generated by CA II-accelerated CO_2 hydration that forms carbonic acid, which dissociates to H^+ and HCO_3^-. Basolateral Cl^-/HCO_3^- exchange transports HCO_3^- across the basolateral membrane; HCO_3^- combines with H^+ released from NH_4^+, to form carbonic acid, which dissociates to CO_2 and water. This CO_2 can recycle into the cell, supplying the CO_2 used for cytosolic H^+ production. The net result is NH_4^+ transport from the peritubular space into the luminal fluid

5 Specific Proteins Involved in Renal Ammonia Transport

5.1 NHE3

Ammonium produced in the proximal tubule is transported preferentially across the apical membrane into the luminal fluid. In part, this preferential apical secretion is due to apical microvilli which results in amplification of the apical plasma membrane surface area. In addition, apical ammonia transport is inhibited by combining a low luminal Na^+ concentration with amiloride and its analogues, suggesting that it involves Na^+/NH_4^+ exchange via the apical Na^+/H^+ exchanger, NHE3 [27]. Consistent with this interpretation is that proximal tubule brush border vesicles exhibit NH_4^+/Na^+ exchange activity [28]. Although parallel Na^+/H^+ exchange and NH_3 secretion may account for some of this transport, luminal acidification to the pH generated by apical Na^+/H^+ exchange activity does not result in equivalent rates of net ammonia secretion [27], suggesting that diffusive NH_3 transport is not the primary mechanism of ammonia secretion.

NHE3 is also present in the apical membrane of the thick ascending limb of the loop of Henle. However, since this transporter secretes NH_4^+, and the thick ascending limb of the loop of Henle reabsorbs NH_4^+, NHE3 is unlikely to mediate an important role in loop of Henle ammonia transport.

Regulation of proximal tubule apical NHE3 appears to be important in the regulation of proximal tubule ammonia transport. Chronic metabolic acidosis, changes in extracellular potassium and angiotensin II each regulate proximal tubule ammonia secretion and appear to do so through changes in apical Na^+/H^+ exchange activity [29, 30] that, at least for chronic metabolic acidosis and hypokalemia, parallels changes in NHE3 expression [31, 32]. In the S3 segment, chronic metabolic acidosis increases angiotensin II (AII)-stimulated apical ammonia secretion via a mechanism that appears to involve increased transport by NHE3 [33]. Further studies show that inhibiting the angiotensin II, Type 1 (AT_1), receptor blunts the expected increase in urinary ammonia excretion, proximal tubule ammonia production and secretion and changes in NHE3 expression [34]. These observations suggest that AII, acting through the AT_1 receptor, may mediate an important role in regulating proximal tubule ammonia secretion through regulation of NHE3 expression. Other studies have shown that increased endothelin-1 expression with subsequent activation of the endothelin-B receptor mediates an important role in increasing NHE3 expression and subsequent renal ammonia excretion in metabolic acidosis [35].

5.2 Potassium Channels

At a molecular level, K^+ and NH_4^+ have nearly identical biophysical characteristics (Table 1). This molecular mimicry enables NH_4^+ to substitute for K^+ at the K^+ binding site of most K^+ transporters, including K^+ channels. The general characteristic of K^+ channels is that they are able to transport NH_4^+ has been shown for several different K^+ channel families, including strong and weak inward rectifying K^+ channels, voltage-gated, Ca^{+2}-activated, delayed rectifier and l-type transient K^+ channels [9, 36]. In general, the relative conductance of K^+ channels for NH_4^+ is 10–20% of that observed for K^+ [9, 36].

Table 1 Biophysical characteristics of common cations

Cation	Atomic weight	Ionic radius (Å)	Hydrodynamic radius (Å)	Mobility in H_2O ($10^{-4}\,cm^2\,s^{-1}\,V^{-1}$)	T_i^{H2O}
Li^+	6	0.060	1.73	4.01	0.33
Na^+	23	0.095	1.67	5.19	0.39
NH_4^+	18	0.133	1.14	7.60	0.49
K^+	39	0.143	1.14	7.62	0.49

The evidence that K^+ channels contribute to proximal tubule ammonia transport is based on in vitro microperfusion studies showing that a component of proximal tubule ammonia transport is blocked by barium, a nonspecific K^+ channel inhibitor

[16]. Multiple K^+ channels are present in the apical membrane of the proximal tubule, including KCNA10, TWIK-1, and KCNQ1 [37–41]; which of these mediate ammonia transport is not currently known. Because of the negative intracellular electrical potential, apical K^+ channels are more likely to contribute to the NH_4^+ absorption that can occur in the proximal straight tubule than to NH_4^+ secretion.

Apical plasma membrane K^+ channels in the medullary thick ascending limb of the loop of Henle are critical for normal Na^+, Cl^-, and Ca^{2+} transport. They can also contribute to apical plasma membrane NH_4^+ transport, particularly under conditions when apical Na^+-K^+-$2Cl^-$ cotransport is inhibited [20]. However, the observations that the transcellular component of thick ascending limb of the loop of Henle transepithelial ammonia reabsorption is nearly completed inhibited by NKCC-2 inhibitors, such as furosemide, suggests that apical K^+ channels are unlikely to mediate quantitatively important roles in thick ascending limb of the loop of Henle transepithelial ammonia transport [42].

Another possible role for K^+ channels is in collecting duct ammonia secretion. Under normal conditions, the electrochemical gradient for NH_4^+ transport by K^+ channels favors NH_4^+ uptake. Thus, it is possible that basolateral K^+ channels in the collecting duct may contribute to peritubular NH_4^+ uptake. At present, no studies have examined this possibility. However, the observation that ammonia incubation for as little as 10 min causes intracellular acidification of CCD intercalated cells [43] suggests there may be significant NH_4^+ transport, possibly due to K^+ channels.

5.3 *Na^+-K^+-$2Cl^-$ Cotransport*

Two isoforms of Na^+-K^+-$2Cl^-$ have been identified, NKCC-1 and NKCC-2 [44]. They have been identified in a wide variety of tissues where they mediate multiple physiologic functions, including ion transport, cell volume regulation, blood pressure regulation, saliva formation, and endolymph formation [44–46]. NKCC-1 is also known as BSC-2 or as the "secretory isoform," and it is expressed at the basolateral membrane of many cells involved in secretory functions. In the kidney, NKCC-1 is present almost exclusively in the basolateral membrane of A-type intercalated cells in the outer and inner medullary collecting ducts [47]. Because peritubular addition of bumetanide, a highly specific NKCC1 and NKCC-2 inhibitor, does not alter OMCD ammonia secretion NKCC1 appears unlikely to mediate a quantitatively important role in OMCD ammonia secretion [48]. In the IMCD, studies using cultured IMCD cells suggest that basolateral Na^+-K^+-$2Cl^-$ cotransport, presumably NKCC-1, contributes to basolateral NH_4^+ uptake [49, 50], but other studies using in vitro microperfused rat terminal IMCD segments have not confirmed this observation [51].

NKCC-2 is a kidney-specific Na^+-K^+-$2Cl^-$ cotransporter isoform specifically expressed in the apical plasma membrane of the thick ascending limb of the loop of Henle [52, 53]. It is the major mechanism for ammonia reabsorption in the thick ascending limb of the loop of Henle [42, 54]. Importantly, because NKCC-2 has a *Km* for ammonia of ~2 mM, luminal ammonia is able to effectively bind to the K^+-binding size, enabling effective competition of NH_4^+ with K^+ for transport [55]. Hyperkalemia suppresses renal ammonia excretion; this is in part due to increased luminal K^+ in the thick ascending limb of the loop of Henle, resulting in competitive inhibition of ammonia reabsorption [42].

Changes in thick ascending limb of the loop of Henle ammonia transport contribute to the increase in renal ammonia excretion with chronic metabolic acidosis [56]. In part, this involves increased NKCC-2 expression [57] that appears to be transcriptionally mediated [57] and may be related to the elevation in systemic glucocorticoids that occurs with chronic metabolic acidosis [58].

5.4 Na^+-K^+-ATPase

Na^+-K^+-ATPase is a family of heterodimeric proteins present in essentially all nucleated cells. The alpha subunit comprises the catalytic and ion transport domain. The function of the beta subunit is not completely clear, but may include directing membrane localization [59]. In the kidney, Na^+-K^+-ATPase is present in the basolateral plasma membrane of essentially all cells, and its expression is greatest in the medullary thick ascending limb of the loop of Henle, with lesser expression in the cortical thick ascending limb (CTAL), distal convoluted tubule (DCT), cortical collecting duct (CCD), medullary collecting duct (MCD), and the proximal tubule [60].

Ammonium can compete with K^+ at the K^+-binding site of Na^+-K^+-ATPase, thereby enabling net Na^+-NH_4^+ exchange energized by ATP hydrolysis [61, 62]. When examined, Na^+-K^+-ATPase has an affinity for NH_4^+ of $\sim 11 \pm 1.6$ mM, whereas the affinity for K^+ is ~1.9 mM [61, 62]. In the renal cortex, interstitial ammonia concentrations are less than 1 mM, suggesting that NH_4^+ is not likely to be transported to any significant extent by basolateral Na^+-K^+-ATPase [61]. Moreover, even in the presence of elevated peritubular ammonia concentrations, in vitro microperfused tubule studies show that ouabain, a Na^+-K^+-ATPase inhibitor, does not alter CCD ammonia secretion [63].

In contrast, interstitial ammonia concentrations are substantially higher in the inner medulla suggesting that Na^+-K^+-ATPase-mediated basolateral ammonia uptake contributes significantly to transepithelial ammonia secretion by the IMCD [61, 62]. Direct studies have shown that ammonium competes with K^+ as a substrate for hydrolytic activity, and that basolateral Na^+-K^+-ATPase mediates ammonium uptake, increases apical proton secretion, and is important for transepithelial ammonia secretion [51, 62, 64]. Mathematical analyses, using measured values for Na^+-K^+-ATPase K_m and $V_{\max}$ and assuming that vasa recta K^+ and ammonium concentrations are similar to interstitial concentrations, suggest that ammonium uptake via Na^+-K^+-ATPase is remarkably similar to measured rates of ammonium secretion [65]. Furthermore, physiologic changes in interstitial K^+ concentrations associated with hypokalemia predicted a two–three-fold increase in ammonium uptake [65]. Direct studies examining the effects of hypokalemia on Na^+-K^+-ATPase-mediated ammonia transport have confirmed this model [66]. Moreover, the increase in ammonia uptake was almost totally attributable to changes in interstitial K^+ concentration, and did not involve changes in Na^+-K^+-ATPase hydrolytic activity [66]. Thus, increased ammonia secretion during hypokalemia may be due, at least in part, to decreased interstitial K^+ concentration, which results in more effective NH_4^+ transport by basolateral Na^+-K^+-ATPase.

5.5 K^+/NH_4^+ (H^+) Exchange

An apical K^+/NH_4^+ (H^+) activity is present in the thick ascending limb of the loop of Henle that appears to be able to transport ammonia when NKCC-2 is inhibited [54]. It is electroneutral and is inhibited by barium and verapamil [19, 67]. The gene product and the protein which correlate with this transport activity have not yet been identified. However, the observation that NKCC-2 inhibitors nearly completely inhibit the transcellular component of thick ascending limb of the loop of Henle ammonia transport suggests that K^+/NH_4^+ (H^+) exchange activity may not mediate a major role in thick ascending limb of the loop of Henle ammonia reabsorption.

5.6 H^+-K^+-ATPase

H^+-K^+-ATPase proteins are other members of the P-type ATPase family that are able to transport NH_4^+. This has been demonstrated in heterologous expression systems, colonic apical membranes and inner medullary collecting ducts perfused in vitro [68–71]. Expression of the colonic H^+-K^+-ATPase is induced during potassium deficiency and has been postulated to facilitate increased ammonia excretion [70]. One possible mechanism by which this occurs may involve the observation that ammonia stimulates collecting duct H^+ secretion through mechanisms involving stimulation of H^+-K^+-ATPase [43], and that the increased luminal H^+ concentration facilitates ammonia secretion.

5.7 Aquaporins

The aquaporin family is an extended family of proteins that enable facilitated water transport [72, 73]. Because both H_2O and NH_3 have similar molecular sizes and charge distribution several studies have examined the role of aquaporins in NH_3 transport. The aquaporins, AQP1, AQP3, AQP8, and AQP9, have been most extensively studied.

The first study examining aquaporins as ammonia transporters examined AQP1. In these studies, several lines of evidence showed that AQP1 enabled facilitated NH_3 transport [74]. However, not all studies have confirmed the observation that

AQP1 can transport NH_3 [75, 76]. While it is tempting to speculate that AQP1 may contribute to NH_3 permeability in the thin descending limb of the loop of Henle, it is unknown whether there are alterations in thin descending limb of the loop of Henle ammonia transport in AQP1 knock-out mice.

AQP3 is present in the basolateral membrane of collecting duct principal cells [77] and when expressed in the Xenopus oocytes mediates NH_3 permeability [75]. Whether AQP3 contributes to renal principal cell basolateral NH_3 transport has not been determined.

Another aquaporin shown to transport ammonia is AQP8. Similar to other aquaporins, AQP8 appears to transport NH_3, not NH_4^+ [75]. In the kidney, AQP8 is weakly expressed in the cytoplasm of proximal tubule, CCD and OMCD, with no evidence of plasma membrane expression [78]. Genetic deletion of AQP8 does not alter serum chloride concentration, urine ammonia concentration, or urine pH either under basal conditions or in response to acid-loading, suggesting that AQP8 does not contribute to renal ammonia metabolism [79, 80]. However, AQP8 deletion does induce minor changes in hepatic ammonia accumulation, renal excretion of infused ammonia, and intrarenal ammonia concentrations [80]. Thus, the possibility that AQP8 contributes to intracellular ammonia transport in wild-type mice, and that compensatory changes in the expression of other ammonia transport mechanisms in AQP8 knockout mice maintain normal acid–base homeostasis cannot be excluded at present.

AQP9 is expressed in the liver, leukocytes, epididymis and in several other regions of the male reproductive tract including efferent ducts, vas deferens, prostate, and coagulating gland [81, 82]. Similar to AQP8, AQP9 expression in the Xenopus oocytes increases apparent NH_3 permeability and it increases permeability to the analogues, formamide and methylammonia [75]. However, AQP9 does not appear to be expressed in the kidney and is therefore unlikely to mediate an important role in renal ammonia metabolism.

The molecular specificity of ammonia transport by aquaporins may relate to specific amino acids in the aromatic/arginine constriction region of these proteins. This region is located below the channel mouth and may be narrower than the NPA constriction [83]. Site-directed mutagenesis has shown that specific amino acids, including Phe-56, His-180, and Arg-195 of AQP1, are important for ammonia permeability, but not for water permeability [76]. The central NPA constriction, while important for aquaporin water permeability, does not appear to be critical for aquaporin ammonia permeability [76].

5.8 Rh Glycoproteins

The most recent addition to our understanding of the molecular mechanisms of renal ammonia transport has been the identification of the Rh glycoprotein family. These proteins are mammalian orthologs of Mep/AMT proteins, the ammonia transporter families present in yeast, plants, bacteria, and many other organisms [84,

85]. Three mammalian Rh glycoproteins have been identified to date, Rh A glycoprotein (RhAG/Rhag), Rh B Glycoprotein (RhBG/Rhbg), and Rh C Glycoprotein (RhCG/Rhcg) [86, 87]. By convention, Rh A Glycoprotein is termed RhAG in human tissues and is termed Rhag in nonhumans; a similar terminology is used for RhBG/Rhbg and RhCG/Rhcg.

5.8.1 RhAG/Rhag

Rh A glycoprotein (RhAG/Rhag) is a component of the Rh complex, which is generally believed to consist of the nonglycosylated Rh proteins, RhD and RhCE in humans and Rh30 in nonhuman mammals, in association with RhAG/Rhag. The first suggestion that Rh proteins might function as ammonia transporters was the prediction in 1997 that Rh proteins have structural similarity to Mep and AMT proteins [88]. Subsequent studies utilizing yeast deficient in endogenous ammonia transporters (ΔMep1/ΔMep2/ΔMep3), which could not grow in media in which the only nitrogen source was a low concentration of ammonia, showed that RhAG expression restored growth [89]. Additionally, RhAG was able to mediate efflux of the ammonia analogue, methylammonia [89]. Further studies showed that RhAG transported the ammonia analogue, methylammonia, via an electroneutral, H^+-gradient coupled mechanism, with an EC_{50} of $\sim$1.6 mM, consistent with either facilitated NH_3 transport or with NH_4^+/H^+ exchange [89–91]. Finally, studies comparing erythrocytes from Rh_{null} individuals, who do not express RhAG, with erythrocytes with normal RhAG expression show that erythrocyte NH_3 transport parallels erythrocyte RhAG expression [92].

RhAG/Rhag appears to be an erythrocyte and erythroid-precursor specific protein [93, 94]. In particular, studies in the human kidney have not found evidence of renal epithelial cell RhAG expression [95]. At present, RhAG/Rhag appears unlikely to contribute to renal ammonia metabolism.

5.8.2 RhBG/Rhbg

RhBG/Rhbg is a second member of the mammalian Rh glycoprotein family. RhBG/Rhbg is expressed in a wide variety of organs involved in ammonia metabolism, including kidneys, liver, skin, stomach, and gastrointestinal tract [86, 96]. The kidneys express basolateral Rhbg immunoreactivity in the DCT, CCD, OMCD, and IMCD [97, 98]. In general, both intercalated and principal cells express Rhbg, with greater expression in intercalated cells than in principal cells. The two exceptions to this rule are the IMCD, where only intercalated cells express Rhbg, and the CCD B-type intercalated cell, which does not express Rhbg detectable by light microscopy [98]. Rhbg's basolateral expression appears to be due to basolateral stabilization through specific interactions of its cytoplasmic carboxy-terminus with ankyrin-G [99].

RhBG/Rhbg is expressed in multiple extrarenal sites involved in ammonia transport and metabolism. In the liver, Rhbg is expressed in the sinusoidal membrane of perivenous hepatocytes [100], the hepatocytes responsible for high-affinity hepatic ammonia metabolism [101]. In the skin, Rhbg mRNA is expressed in hair follicles and sweat glands [86]. Its role in sweat formation is undefined at present, but it is intriguing to note that sweat ammonia concentrations average ~3 mM and can exceed 11 mM [102, 103]. The gastrointestinal tract is another major site of ammonia transport and metabolism, and villous epithelial cells, the major site of ion transport, from the duodenum through the colon express basolateral Rhbg immunoreactivity [96].

Multiple studies demonstrate that RhBG/Rhbg can transport ammonia and the ammonia analogue, methylammonia. However, different studies have reached different conclusions regarding the exact molecular species, NH_3 or NH_4^+, transported. Some studies show that ammonia and methylammonia transport is electroneutral, coupled to H^+ gradients, Na^+- and K^+-independent, and unaffected by changes in membrane voltage, indicating electroneutral, cation-independent NH_3 transport [104–106]. However, other data suggests that Rhbg mediates electrogenic NH_4^+ transport [107]. In all of these studies, the affinity for ammonia is ~2–4 mM. The explanation underlying the seemingly differing molecular mechanisms in different studies is not defined at present.

Rhbg's specific role in renal ammonia metabolism has been examined in several studies. In chronic metabolic acidosis there were no detectable changes in Rhbg protein expression [108]. Thus, either Rhbg's role in renal ammonia metabolism is mediated through post-translation modifications not detectable by immunoblot assay and immunohistochemistry or it is not involved. In another study, genetic deletion of the B-type intercalated cell and non-A, non-B cell apical Cl^-/HCO_3^- exchanger, pendrin, decreased Rhbg expression [109]. This adaptation likely reflects the increased urinary acidification with pendrin deletion, which increases the gradient for NH_3 transport, thus decreasing the necessity for ammonia transport via Rhbg. Finally, an Rhbg knock-out mouse had normal basal acid–base parameters, normal responses to acid loading, and normal basolateral NH_3 and NH_4^+ transport in in vitro microperfused CCD segments [110]. These results suggest that either Rhbg is not involved in renal ammonia metabolism or that compensatory mechanisms activated in response to Rhbg deletion fully compensate for the lack of Rhbg [110]. However, there were no differences in Rhcg, H^+-ATPase or AE1 expression, likely candidate proteins if compensatory adaptation were to be present [110]. At present, RhBG/Rhbg's role in renal ammonia metabolism is unclear.

5.8.3 RhCG/Rhcg

RhCG/Rhcg is the third member of the mammalian Rh glycoprotein family. It is expressed in multiple ammonia transporting/metabolizing organs, including kidneys, CNS, testes, liver and throughout the gastrointestinal tract [87, 89, 96, 100, 111]. In the kidney, Rhcg expression parallels that of Rhbg, i.e., in the DCT, CNT, ICT, CCD, OMCD, and IMCD [98, 112]. Its cellular expression is similar to Rhbg, with intercalated cell expression exceeding principal cell expression in the DCT, CNT, ICT, CCD, and OMCD, and expression only in intercalated cells in the IMCD [98, 112].

However, Rhcg's subcellular location differs from the exclusive basolateral location observed for Rhbg. In the mouse kidney, Rhcg immunoreactivity is apical [98, 112], although recent preliminary reports indicate the presence of both apical and basolateral plasma membrane expression [113]. In the rat, both exclusively apical [112] and both apical and basolateral Rhcg immunoreactivity has been reported [108, 114–116], and in the human kidney both apical and basolateral Rhcg expression has been reported [117]. Quantitatively, basolateral plasma membrane Rhcg expression is ~20% of total cellular expression, at least in the rat OMCD in the inner stripe [116]. In addition, rat renal Rhcg, is also present in intracellular sites, including cytoplasmic vesicles, consistent with vesicular trafficking regulating apical plasma membrane Rhcg expression [116].

Similar to Rhbg, multiple studies have addressed the molecular ammonia species transported by RhCG/Rhcg. Some studies suggest that RhCG/Rhcg mediates electroneutral NH_3 transport [104–106], while others have reported both NH_3 and NH_4^+ transport [118] or only electrogenic NH_4^+ transport [119]. The explanation for these differing observations is unclear at present.

Multiple studies show that Rhcg's expression parallels changes in renal ammonia metabolism. Chronic metabolic acidosis significantly increases Rhcg protein expression in both the OMCD and the IMCD, but not in the cortex [108]. These changes appear to occur through a post-transcriptional mechanism, since Rhcg steady-state mRNA expression is not detectably altered. In reduced renal mass induced by a 5/6 ablation-infarction model there is a dramatic increase in single nephron ammonia metabolism that is associated with increased apical Rhcg expression in the CCD A-type intercalated cell, CCD principal cell, OMCD intercalated cell, and OMCD principal cell [114]. Basolateral Rhcg expression increases in the CCD principal cell, OMCD, intercalated cell and the OMCD principal cell [114]. In ischemia-reperfusion injury, where there is decreased renal ammonia excretion, there is selective damage to Rhcg-expressing OMCD intercalated cells, with induction of apoptosis and cellular extrusion [115]. Finally, chronic cyclosporine A nephropathy is associated with decreased Rhcg expression, which likely contributes to impaired ammonia excretion and to the development of metabolic acidosis [120].

Multiple mechanisms are involved in Rhcg's response to altered renal ammonia metabolism. Quantitative analysis of cell-specific Rhcg expression shows that metabolic acidosis increases cellular Rhcg expression in both the intercalated cell and the principal cell [116]. Although the absolute levels of Rhcg expression are greater in the intercalated cell than in the principal cell, the relative increase in cellular expression is substantially greater in the principal cell (↑;~240%) than in the intercalated cell (↑;~30%). A second regulatory mechanism appears to be changes in Rhcg's subcellular location. Under basal conditions, Rhcg is located in both the apical and basolateral plasma membrane and in subapical sites in both principal cell and intercalated cell. In response to chronic metabolic acidosis, particularly in the intercalated cell, there is a dramatic increase in apical plasma membrane expression and a decrease in cytoplasmic Rhcg expression [116]. These results suggest translocation of Rhcg from cytoplasmic to apical plasma membrane sites. A similar response occurs in the principal cell, although of substantially less magnitude.

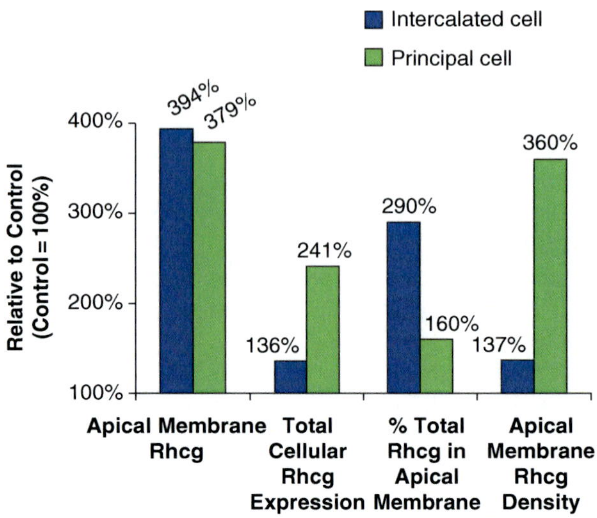

Fig. 4 Differential regulatory mechanisms of apical Rhcg expression in the intercalated cell and principal cell in metabolic acidosis. Cellular and subcellular Rhcg expression quantified using immunogold electron microscopy and quantitative morphometry. Results are relative to same cell under control conditions. Metabolic acidosis increases apical plasma membrane Rhcg expression ~4-fold in both the OMCD intercalated cell and principal cell. This is due to an increase in cell-specific Rhcg expression of ~241% in the principal cell and only ~30% in the intercalated cell. A second mechanism is altered subcellular distribution of Rhcg, with the proportion of cell-specific Rhcg expression in the apical plasma membrane increasing substantially more in the intercalated cell than in the principal cell. Furthermore, the apical plasma membrane Rhcg density (amount of Rhcg per unit length of apical plasma membrane) increases more in the principal cell than in the intercalated cell. Thus, there appear to be cell-specific differences in the relative role of these two regulatory mechanisms for Rhcg. Data recalculated from [116]

Changes in Rhcg's subcellular distribution also contribute to the response to reduced renal mass [114].

These observations suggest at least two mechanisms regulate apical plasma membrane Rhcg expression, changes in total cellular expression and changes in the subcellular distribution of Rhcg. Moreover, the relative roles of these two mechanisms differ in adjacent cell types, the principal cell and the intercalated cell (Fig. 4). These changes are also membrane specific, as the increase in principal cell basolateral plasma membrane Rhcg parallels changes in total cellular expression, and do not involve changes in the relative subcellular distribution of Rhcg. The mechanisms involved in the cell- and membrane-specific differences in Rhcg expression in response to specific physiologic stimuli are unknown at present.

Chronic metabolic acidosis also increases basolateral plasma membrane Rhcg expression [114, 116]. This increase paralleled the change in total cellular expression, and does not appear to involve redistribution from cytoplasmic sites to the basolateral plasma membrane.

5.8.4 Nonglycosylated Rh Proteins

The nonglycosylated Rh proteins, RhD and RhCE in humans and Rh30 in non-humans, are present in erythrocytes in a multimeric complex with RhAG. RhCE appears to neither transport ammonia or its analogue, methylammonia, nor alter transport by RhAG [90]. Moreover, structural models, using the *Escherichia coli* AmtB structure as a template, suggest that the arrangement of key amino acids is sufficiently different in RhCE and RhD that they either do not transport ammonia or that they do so by mechanisms that differ from that used by AmtB, RhAG, RhBG, and RhCG [121].

6 Significance

Normal renal ammonia metabolism is central to normal acid–base homeostasis, and abnormalities in ammonia metabolism lead to a variety of important clinical problems. These can include chronic metabolic acidosis, with subsequent osteoporosis and/or osteomalacia, skeletal muscle atrophy and nephrolithiasis, and hypokalemia-induced metabolic alkalosis, which may, in susceptible individuals with liver disease, lead to precipitation and/or worsening of hepatic encephalopathy. Understanding the molecular mechanisms regulating these proteins, and understanding the genetic polymorphisms which underlie susceptibility to these complications, is likely to be an important field for future study.

Acknowledgments The preparation of this review was supported by funds from the NIH (DK-45788) and the Research Service of the Malcolm Randall Veterans Affairs Medical Center.

References

1. Mitch WE. Metabolic and clinical consequences of metabolic acidosis. *J Nephrol* 2006; 19(Suppl 9):S70–S75.
2. Boron WF. Acid-base transport by the renal proximal tubule. *J Am Soc Nephrol* 2006; 17(9):2368–2382.
3. Capasso G, Unwin R, Rizzo M, Pica A, Giebisch G. Bicarbonate transport along the loop of Henle: molecular mechanisms and regulation. *J Nephrol* 2002; 15 Suppl 5:S88–S96.
4. Gluck SL, Iyori M, Holliday LS, Kostrominova T, Lee BS. Distal urinary acidification from Homer Smith to the present. *Kidney Int* 1996; 49(6):1660–1664.
5. Hamm LL, Simon EE. Roles and mechanisms of urinary buffer excretion. *Am J Physiol* 1987; 253:F595–F605.
6. Elkinton JR, Huth EJ, Webster GD, Jr., McCance RA. The renal excretion of hydrogen ion in renal tubular acidosis. *Am J Med* 1960; 36:554–575.
7. Sartorius OW, Roemmelt JC, Pitts RF, Calhoon D, Miner P. The renal regulation of acid-base balance in man. IV. The nature of the renal compensations in ammonium chloride acidosis. *J Clin Invest* 1949; 28(3):423–439.
8. Good DW, Caflisch CR, DuBose TD. Transepithelial ammonia concentration gradients in inner medulla of the rat. *Am J Physiol* 1987; 252:F491–F500.
9. Weiner ID, Hamm LL. Molecular mechanisms of renal ammonia transport. *Annu Rev Physiol* 2007; 69:317–340.

10. Haussinger D, Lamers WH, Moorman AF. Hepatocyte heterogeneity in the metabolism of amino acids and ammonia. *Enzyme* 1992; 46(1–3):72–93.
11. Haussinger D. Nitrogen metabolism in liver: structural and functional organization and physiological relevance. *Biochem J* 1990; 267(2):281–290.
12. Tannen RL, Sahai A. Biochemical pathways and modulators of renal ammoniagenesis. *Miner Electrolyte Metab* 1990; 16:249–258.
13. Tannen RL. Ammonia metabolism. *Am J Physiol Renal* 1978; 235(4):F265–F277.
14. Tizianello A, Deferrari G, Garibotto G, Robaudo C, Acquarone N, Ghiggeri GM. Renal ammoniagenesis in an early stage of metabolic acidosis in man. *J Clin Invest* 1982; 69(1):240–250.
15. Nagami GT. Ammonia production and secretion by the proximal tubule. *Am J Kidney Dis* 1989; 14:258–261.
16. Simon EE, Merli C, Herndon J, Cragoe EJ, Jr., Hamm LL. Effects of barium and 5-(*N*-ethyl-*N*-isopropyl)-amiloride on proximal tubule ammonia transport. *Am J Physiol* 1992; 262:F36–F39.
17. Hamm LL, Simon EE. Ammonia transport in the proximal tubule. *Miner Electrolyte Metab* 1990; 16:283–290.
18. Kikeri D, Sun A, Zeidel ML, Hebert SC. Cell membranes impermeable to NH_3. *Nature* 1989; 339:478–480.
19. Amlal H, Paillard M, Bichara M. NH_4^+ transport pathways in cells of medullary thick ascending limb of rat kidney. NH_4^+ conductance and $K^+/NH_4^+(H^+)$ antiport. *J Biol Chem* 1994; 269:21962–21971.
20. Attmane-Elakeb A, Amlal H, Bichara M. Ammonium carriers in medullary thick ascending limb. *Am J Physiol Renal Physiol* 2001; 280(1):F1–F9.
21. Flessner MF, Mejia R, Knepper MA. Ammonium and bicarbonate transport in isolated perfused rodent long-loop thin descending limbs. *Am J Physiol* 1993; 264(3 Pt 2):F388–F396.
22. Mejia R, Flessner MF, Knepper MA. Model of ammonium and bicarbonate transport along LDL: implications for alkalinization of luminal fluid. *Am J Physiol* 1993; 264(3 Pt 2):F397–F403.
23. DuBose TD, Good DW, Hamm LL, Wall SM. Ammonium transport in the kidney: new physiological concepts and their clinical implications. *J Am Soc Nephrol* 1991; 1:1193–1203.
24. Knepper MA. NH_4^+ transport in the kidney. *Kidney Int* 1991; 40:S95–S102.
25. Handlogten ME, Hong SP, Westhoff CM, Weiner ID. Basolateral ammonium transport by the mouse inner medullary collecting duct cell (mIMCD-3). *Am J Physiol Renal Physiol* 2004; 287(3):F628–F638.
26. Handlogten ME, Hong SP, Westhoff CM, Weiner ID. Apical ammonia transport by the mouse inner medullary collecting duct cell (mIMCD-3). *Am J Physiol Renal* 2005; 289(2):F347–F358.
27. Nagami GT. Luminal secretion of ammonia in the mouse proximal tubule perfused in vitro. *J Clin Invest* 1988; 81(1):159–164.
28. Kinsella JL, Aronson PS. Interaction of NH_4^+ and Li^+ with the renal microvillus membrane Na^+-H^+ exchanger. *Am J Physiol* 1981; 241:C220–C226.
29. Nagami GT. Effect of bath and luminal potassium concentration on ammonia production and secretion by mouse proximal tubules perfused in vitro. *J Clin Invest* 1990; 86(1):32–39.
30. Nagami GT, Sonu CM, Kurokawa K. Ammonia production by isolated mouse proximal tubules perfused in vitro: effect of metabolic acidosis. *J Clin Invest* 1986; 78:124–129.
31. Ambuhl PM, Amemiya M, Danczkay M et al. Chronic metabolic acidosis increases NHE3 protein abundance in rat kidney. *Am J Physiol* 1996; 271(4 Pt 2):F917–F925.
32. Elkjar ML, Kwon TH, Wang W et al. Altered expression of renal NHE3, TSC, BSC-1, and ENaC subunits in potassium-depleted rats. *Am J Physiol Renal* 2002; 283(6):F1376–F1388.
33. Nagami GT. Ammonia production and secretion by S3 proximal tubule segments from acidotic mice: role of ANG II. *Am J Physiol Renal* 2004; 287(4):F707–F712.

34. Nagami GT. Role of angiotensin II in the enhancement of ammonia production and secretion by the proximal tubule in metabolic acidosis. *Am J Physiol Renal Physiology* 2008; 294(4):F874–F880.
35. Laghmani K, Preisig PA, Moe OW, Yanagisawa M, Alpern RJ. Endothelin-1/endothelin-B receptor-mediated increases in NHE3 activity in chronic metabolic acidosis. *J Clin Invest* 2001; 107(12):1563–1569.
36. Choe H, Sackin H, Palmer LG. Permeation Properties of Inward-Rectifier Potassium Channels and Their Molecular Determinants. *J Gen Physiol* 2000; 115(4): 391–404.
37. Vallon V, Grahammer F, Volkl H et al. KCNQ1-dependent transport in renal and gastrointestinal epithelia. *Proc Natl Acad Sci USA* 2005; 102(49):17864–17869.
38. Nie X, Arrighi I, Kaissling B et al. Expression and insights on function of potassium channel TWIK-1 in mouse kidney. *Pflugers Arch* 2005; 451(3):479–488.
39. Merot J, Bidet M, Le MS, Tauc M, Poujeol P. Two types of K+ channels in the apical membrane of rabbit proximal tubule in primary culture. *Biochim Biophys Acta* 1989; 978(1):134–144.
40. Yao X, Tian S, Chan HY, Biemesderfer D, Desir GV. Expression of KCNA10, a voltage-gated K channel, in glomerular endothelium and at the apical membrane of the renal proximal tubule. *J Am Soc Nephrol* 2002; 13(12):2831–2839.
41. Cluzeaud F, Reyes R, Escoubet B et al. Expression of TWIK-1, a novel weakly inward rectifying potassium channel in rat kidney. *Am J Physiol* 1998; 275(6 Pt 1): C1602–C1609.
42. Good DW. Ammonium transport by the thick ascending limb of Henles loop. *Annu Rev Physiol* 1994; 56:623–647.
43. Frank AE, Wingo CS, Weiner ID. Effects of ammonia on bicarbonate transport in the cortical collecting duct. *Am J Physiol Renal* 2000; 278(2):F219–F226.
44. Haas M, Forbush B. The Na-K-Cl Cotransporters. *J Bioenerg Biomembr* 1998; 30(2): 161–172.
45. Wall SM, Knepper MA, Hassell KA et al. Hypotension in NKCC1 null mice: role of the kidneys. *Am J Physiol Renal* 2006; 290(2):F409–F416.
46. Meyer JW, Flagella M, Sutliff RL et al. Decreased blood pressure and vascular smooth muscle tone in mice lacking basolateral Na(+)-K(+)-2Cl(–) cotransporter. *Am J Physiol Heart Circ Physiol* 2002; 283(5):H1846–H1855.
47. Ginns SM, Knepper MA, Ecelbarger CA et al. Immunolocalization of the secretory isoform of Na-K-Cl cotransporter in rat renal intercalated cells. *J Am Soc Nephrol* 1996; 7(12): 2533–2542.
48. Wall SM, Fischer MP. Contribution of the Na(+)-K(+)-2Cl(–) cotransporter (NKCC1) to transepithelial transport of H(+), NH(4)(+), K(+), and Na(+) in rat outer medullary collecting duct. *J Am Soc Nephrol* 2002; 13(4):827–835.
49. Wall SM, Trinh HN, Woodward KE. Heterogeneity of NH_4^+ transport in mouse inner medullary collecting duct cells. *Am J Physiol* 1995; 269:F536–F544.
50. Glanville M, Kingscote S, Thwaites DT, Simmons NL. Expression and role of sodium, potassium, chloride cotransport (NKCC1) in mouse inner medullary collecting duct (mIMCD-K2) epithelial cells. *Pflugers Arch* 2001; 443(1):123–131.
51. Wall SM. Ouabain reduces net acid secretion and increases pHi by inhibiting NH4+ uptake on rat tIMCD Na(+)-K(+)-ATPase. *Am J Physiol* 1997; 273(6 Pt 2):F857–F868.
52. Ecelbarger CA, Terris J, Hoyer JR, Nielsen S, Wade JB, Knepper MA. Localization and regulation of the rat renal Na(+)-K(+)-2Cl- cotransporter, BSC-1. *Am J Physiol* 1996; 271 (3 Pt 2):F619–F628.
53. Plotkin MD, Kaplan MR, Verlander JW et al. Localization of the thiazide sensitive Na-Cl cotransporter, rTSC1 in the rat kidney. *Kidney Int* 1996; 50(1):174–183.
54. Karim Z, Attmane-Elakeb A, Bichara M. Renal handling of NH4+ in relation to the control of acid-base balance by the kidney. *J Nephrol* 2002; 15 Suppl 5:S128–S134.

55. Kinne R, Kinne-Saffran E, Schutz H, Schölermann B. Ammonium transport in medullary thick ascending limb of rabbit kidney: involvement of the Na^{+},K^{+},Cl^{-}-cotransporter. *J Membr Biol* 1986; 94(3):279–284.
56. Good DW. Adaptation of HCO3- and NH4+ transport in rat MTAL: effects of chronic metabolic acidosis and Na+ intake. *Am J Physiol* 1990; 258:F1345–F1353.
57. Attmane-Elakeb A, Mount DB, Sibella V, Vernimmen C, Hebert SC, Bichara M. Stimulation by in vivo and in vitro metabolic acidosis of expression of rBSC-1, the Na+-K+ (NH4+)-2Cl-cotransporter of the rat medullary thick ascending limb. *J Biol Chem* 1998; 273(50):33681–33691.
58. Attmane-Elakeb A, Sibella V, Vernimmen C, Belenfant X, Hebert SC, Bichara M. Regulation by glucocorticoids of expression and activity of rBSC1, the Na+-K+(NH4+)-2Cl-cotransporter of medullary thick ascending limb. *J Biol Chem* 2000; 275(43):33548–33553.
59. Martin DW. Structure-Function Relationships in the Na+,K+-Pump. *Semin Nephrol* 2005; 25(5):282–291.
60. Jorgensen PL. Sodium and potassium ion pump in kidney tubules. *Physiol Rev* 1980; 60:864–917.
61. Kurtz I, Balaban RS. Ammonium as a substrate for Na+-K+-ATPase in rabbit proximal tubules. *Am J Physiol* 1986; 250:F497–F502.
62. Wall SM, Koger LM. NH+4 transport mediated by Na(+)-K(+)-ATPase in rat inner medullary collecting duct. *Am J Physiol* 1994; 267:F660–F670.
63. Knepper MA, Good DW, Burg MB. Mechanism of ammonia secretion by cortical collecting ducts of rabbits. *Am J Physiol* 1984; 247:F729–F738.
64. Wall SM. NH4+ augments net acid-secretion by a ouabain-sensitive mechanism in isolated-perfused inner medullary collecting ducts. *Am J Physiol* 1996; 270(3):F432–F439.
65. Wall SM. Mechanisms of NH_4^+ and NH_3 transport during hypokalemia. *Acta Physiol Scand* 2003; 179(4):325–330.
66. Wall SM, Fischer MP, Kim GH, Nguyen BM, Hassell KA. In rat inner medullary collecting duct, NH4+ uptake by the Na,K-ATPase is increased during hypokalemia. *Am J Physiol Renal* 2002; 282(1):F91–102.
67. Attmane-Elakeb A, Boulanger H, Vernimmen C, Bichara M. Apical location and inhibition by arginine vasopressin of K+/H+ antiport of the medullary thick ascending limb of rat kidney. *J Biol Chem* 1997; 272(41):25668–25677.
68. Cougnon M, Bouyer P, Jaisser F, Edelman A, Planelles G. Ammonium transport by the colonic H+-K+-ATPase expressed in Xenopus oocytes. *Am J Physiol Cell Physiol* 1999; 277(2):C280–C287.
69. Codina J, Pressley TA, DuBose TD. The colonic H+,K+-ATPase functions as a Na+-dependent K+(NH4+)-ATPase in apical membranes from rat distal colon. *J Biol Chem* 1999; 274(28):19693–19698.
70. Nakamura S, Amlal H, Galla JH, Soleimani M. NH4+ secretion in inner medullary collecting duct in potassium deprivation: Role of colonic H+-K+-ATPase. *Kidney Int* 1999; 56(6):2160–2167.
71. Swarts HGP, Koenderink JB, Willems PHGM, De Pont JJ. The Non-gastric H,K-ATPase Is Oligomycin-sensitive and Can Function as an H+,NH4+-ATPase. *J Biol Chem* 2005; 280(39):33115–33122.
72. King LS, Kozono D, Agre P. From structure to disease: the evolving tale of aquaporin biology. *Nat Rev Mol Cell Biol* 2004; 5(9):687–698.
73. Knepper MA. The aquaporin family of molecular water channels. *Proc Natl Acad Sci USA* 1994; 91:6255–6258.
74. Nakhoul NL, Hering-Smith KS, Abdulnour-Nakhoul SM, Hamm LL. Transport of NH(3)/NH in oocytes expressing aquaporin-1. *Am J Physiol Renal Physiol* 2001; 281(2):F255–F263.
75. Holm LM, Jahn TP, Moller AL et al. NH3 and NH4+ permeability in aquaporin-expressing Xenopus oocytes. *Pflugers Arch* 2005; 450(6):415–428.

76. Beitz E, Wu B, Holm LM, Schultz JE, Zeuthen T. Point mutations in the aromatic/arginine region in aquaporin 1 allow passage of urea, glycerol, ammonia, and protons. *Proc Natl Acad Sci USA* 2006; 103(2):269–274.
77. Ma T, Song Y, Yang B et al. Nephrogenic diabetes insipidus in mice lacking aquaporin-3 water channels. *Proc Natl Acad Sci USA* 2000; 97(8):4386–4391.
78. Elkjar ML, Nejsum LN, Gresz V et al. Immunolocalization of aquaporin-8 in rat kidney, gastrointestinal tract, testis, and airways. *Am J Physiol Renal* 2001; 281(6):F1047–F1057.
79. Yang B, Song Y, Zhao D, Verkman AS. Phenotype analysis of aquaporin-8 null mice. *Am J Physiol Cell Physiol* 2005; 288(5):C1161–C1170.
80. Yang B, Zhao D, Solenov E, Verkman AS. Evidence from knockout mice against physiologically significant aquaporin-8 facilitated ammonia transport. *Am J Physiol Cell Physiol* 2006; 291:C417–C423, In press.
81. Takata K, Matsuzaki T, Tajika Y. Aquaporins: water channel proteins of the cell membrane. *Prog Histochem Cytochem* 2004; 39(1):1–83.
82. Pastor-Soler N, Bagnis C, Sabolic I et al. Aquaporin 9 Expression along the Male Reproductive Tract. *Biol Reprod* 2001; 65(2):384–393.
83. de Groot BL, Grubmuller H. Water Permeation Across Biological Membranes: Mechanism and Dynamics of Aquaporin-1 and GlpF. *Science* 2001; 294(5550):2353–2357.
84. von Wiren N, Gazzarrini S, Gojon A, Frommer WB. The molecular physiology of ammonium uptake and retrieval. *Curr Opin Plant Biol* 2000; 3(3):254–261.
85. Gazzarrini S, Lejay L, Gojon A, Ninnemann O, Frommer WB, von Wiren N. Three functional transporters for constitutive, diurnally regulated, and starvation-induced uptake of ammonium into Arabidopsis roots. *Plant Cell* 1999; 11(5):937–948.
86. Liu Z, Peng J, Mo R, Hui Cc, Huang CH. Rh type B glycoprotein is a new member of the Rh superfamily and a putative ammonia transporter in mammals. *J Biol Chem* 2001; 276(2):1424–1433.
87. Liu Z, Chen Y, Mo R et al. Characterization of human RhCG and mouse Rhcg as novel nonerythroid Rh glycoprotein homologues predominantly expressed in kidney and testis. *J Biol Chem* 2000; 275(33):25641–25651.
88. Marini AM, Urrestarazu A, Beauwens R, Andre B. The Rh (rhesus) blood group polypeptides are related to NH4+ transporters. *Trends Biochem Sci* 1997; 22(12): 460–461.
89. Marini AM, Matassi G, Raynal V, Andre B, Cartron JP, Cherif-Zahar B. The human Rhesus-associated RhAG protein and a kidney homologue promote ammonium transport in yeast. *Nat Genet* 2000; 26(3):341–344.
90. Westhoff CM, Ferreri-Jacobia M, Mak DO, Foskett JK. Identification of the erythrocyte Rh-blood group glycoprotein as a mammalian ammonium transporter. *J Biol Chem* 2002; 277(15):12499–12502.
91. Westhoff CM, Siegel DL, Burd CG, Foskett JK. Mechanism of genetic complementation of ammonium transport in yeast by human erythrocyte Rh-associated glycoprotein (RhAG). *J Biol Chem* 2004; 279(17):17443–17448.
92. Ripoche P, Bertrand O, Gane P, Birkenmeier C, Colin Y, Cartron JP. Human Rhesus-associated glycoprotein mediates facilitated transport of NH_3 into red blood cells. *Proc Natl Acad Sci USA* 2004; 101(49):17222–17227.
93. Cartron JP. RH blood group system and molecular basis of Rh-deficiency. *Baillieres Best Pract Res Clin Haematol* 1999; 12(4):655–689.
94. Liu Z, Huang CH. The mouse Rhl1 and Rhag genes: sequence, organization, expression, and chromosomal mapping. *Biochem Genet* 1999; 37(3–4):119–138.
95. Weiner ID. The Rh gene family and renal ammonium transport. *Curr Opin Nephrol Hyper* 2004; 13(5):533–540.
96. Handlogten ME, Hong SP, Zhang L et al. Expression of the ammonia transporter proteins, Rh B Glycoprotein and Rh C Glycoprotein, in the intestinal tract. *Am J Physiol Gastrointest* 2005; 288(5):G1036–G1047.

97. Quentin F, Eladari D, Cheval L et al. RhBG and RhCG, the putative ammonia transporters, are expressed in the same cells in the distal nephron. *J Am Soc Nephrol* 2003; 14(3): 545–554.
98. Verlander JW, Miller RT, Frank AE, Royaux IE, Kim YH, Weiner ID. Localization of the ammonium transporter proteins, Rh B Glycoprotein and Rh C Glycoprotein, in the mouse kidney. *Am J Physiol Renal* 2003; 284:F323–F337.
99. Lopez C, Metral S, Eladari D et al. The ammonium transporter RhBG : Requirement of a tyrosine-based signal and ankyrin-G for basolateral targeting and membrane anchorage in polarized kidney epithelial cells. *J Biol Chem* 2004; 280(9):8221–8228.
100. Weiner ID, Miller RT, Verlander JW. Localization of the ammonium transporters, Rh B Glycoprotein and Rh C Glycoprotein in the mouse liver. *Gastroenterology* 2003; 124(5):1432–1440.
101. Kaiser S, Gerok W, Haussinger D. Ammonia and glutamine metabolism in human liver slices: new aspects on the pathogenesis of hyperammonaemia in chronic liver disease. *Eur J Clin Invest* 1988; 18(5):535–542.
102. Czarnowski D, Gorski J. Sweat ammonia excretion during submaximal cycling exercise. *J Appl Physiol* 1991; 70(1):371–374.
103. Brusilow SW, Gordes EH. Ammonia secretion in sweat. *Am J Physiol* 1968; 214(3):513–517.
104. Mak DO, Dang B, Weiner ID, Foskett JK, Westhoff CM. Characterization of transport by the kidney Rh glycoproteins, RhBG and RhCG. *Am J Physiol Renal* 2006; 290(2): F297–F305.
105. Ludewig U. Electroneutral ammonium transport by basolateral Rhesus B glycoprotein. *J Physiol* 2004; 559(Pt 3):751–759.
106. Zidi-Yahiaoui N, Mouro-Chanteloup I, D′Ambrosio AM et al. Human Rhesus B and Rhesus C glycoproteins: properties of facilitated ammonium transport in recombinant kidney cells. *Biochem J* 2005; 391:33–40.
107. Nakhoul NL, DeJong H, Abdulnour-Nakhoul SM, Boulpaep EL, Hering-Smith K, Hamm LL. Characteristics of Renal Rhbg as an NH4+ Transporter. *Am J Physiol Renal Physiol* 2004; 288(1):F170–F181.
108. Seshadri RM, Klein JD, Kozlowski S et al. Renal expression of the ammonia transporters, Rhbg and Rhcg, in response to chronic metabolic acidosis. *Am J Physiol Renal* 2006; 290(2):F397–F408.
109. Kim YH, Verlander JW, Matthews SW et al. Intercalated cell H^+/OH^- transporter expression is reduced in *Slc26a4* null mice. *Am J Physiol Renal* 2005; 289(6):F1262–F1272.
110. Chambrey R, Goossens D, Bourgeois S et al. Genetic ablation of Rhbg in mouse does not impair renal ammonium excretion. *Am J Physiol Renal* 2005; 289(6):F1281–F1290.
111. Weiner ID. Expression of the non-erythroid Rh glycoproteins in mammalian tissues. *Transfus Clin Biol* 2006; 13(1–2):159–163.
112. Eladari D, Cheval L, Quentin F et al. Expression of RhCG, a new putative NH_3/NH_4^+ transporter, along the rat nephron. *J Am Soc Nephrol* 2002; 13(8):1999–2008.
113. Kim HY, Handlogten ME, Cain B, Guo H, Han KH, Verlander JW. Basolateral expression of the ammonia transporter family member, Rh C Glycoprotein, in the mouse kidney. *J Am Soc Nephrol* 2007; 18:589A. (Abstract).
114. Kim HY, Baylis C, Verlander JW et al. Effect of reduced renal mass on renal ammonia transporter family, Rh C glycoprotein and Rh B glycoprotein, expression. *Am J Physiol Renal Physiology* 2007; 293(4):F1238–F1247.
115. Han KH, Kim HY, Croker BP et al. Effects of Ischemia-reperfusion injury on renal ammonia metabolism and the collecting duct. *Am J Physiol Renal Physiology* 2007; 293(4):F1342–F1354.
116. Seshadri RM, Klein JD, Smith T et al. Changes in the subcellular distribution of the ammonia transporter Rhcg, in response to chronic metabolic acidosis. *Am J Physiol Renal* 2006; 290(6):F1443–F1452.

117. Han KH, Croker BP, Clapp WL et al. Expression of the ammonia transporter, Rh C Glycoprotein, in normal and neoplastic human kidney. *J Am Soc Nephrol* 2006; 17:2670–2679.
118. Bakouh N, Benjelloun F, Hulin P et al. NH_3 is involved in the NH_4^+ transport induced by the functional expression of the human Rh C glycoprotein. *J Biol Chem* 2004; 279(16):15975–15983.
119. Nakhoul NL, Palmer SA, Abdulnour-Nakhoul S, Hering-Smith K, Hamm LL. Ammonium transport in oocytes expressing Rhcg. *FASEB J* 2002; 16(4):A53. (Abstract).
120. Lim SW, Ahn KO, Kim WY et al. Expression of Ammonia Transporters, Rhbg and Rhcg, in Chronic Cyclosporine Nephropathy in Rats. *Nephron Exp Nephrol* 2008; 110(2):e49–e58.
121. Conroy MJ, Bullough PA, Merrick M, Avent ND. Modelling the human rhesus proteins: implications for structure and function. *Br J Haematol* 2005; 131(4):543–551.

Hexose Transport Across Mammalian Epithelia

Chris I. Cheeseman

Abstract Hexoses (sugars) are a major source of carbon and a key group of energy supplying nutrients for the majority of mammalian cell types. This chapter compares the mechanisms responsible for hexose transport across the epithelia of mammalian intestine, kidney and placenta. The small intestine and proximal convoluted tubule can move glucose and galactose against a concentration gradient from the lumen to the blood stream by coupling the transport of sodium to that of the sugars. In contrast, movement of glucose across the placenta appears to occur down a concentration gradient and so no energy coupling is required. There are two families of protein responsible for this transport, the SLC5 gene family which includes SGLT1 and 2, and the SLC2A, GLUT isoforms. There are significant differences between the three tissues as to which hexose transporters they express and in which membrane domains they function. Finally, all three tissues can change the expression levels of these proteins in response to both short- and long-term signals related to either physiological or pathological conditions.

Keywords Sodium-coupled glucose transport · Facilitated glucose transport · Intestine · Kidney · Placenta · Glucose · Fructose · Mannose · Galactose

1 Intestinal Hexose Transport

1.1 Background

The intestine was one of the first tissues in which the kinetics of carrier-mediated transport was determined [1] and the application of Michaelis–Menten-type kinetics was interpreted to indicate that transport involved both a binding site for substrates

C.I. Cheeseman (✉)
Membrane Protein Research Group, Department of Physiology, University of Alberta, Edmonton, AB T6G 2H7, Canada
e-mail: chris.cheeseman@ualberta.ca

G.A. Gerencser (ed.), *Epithelial Transport Physiology*,
DOI 10.1007/978-1-60327-229-2_14,

and some type of reorientation of that binding site (translocation) across the lipid bilayer. Evidence for the presence of sodium in the lumen influencing sugar absorption had existed since the turn of the twentieth century [2], but it was the work of Riklis and Quastel [3], Csáky [4] and finally Crane [5] and Schultz [6] which led to the development of the concept of sodium coupling. Crane hypothesized that sodium and glucose fluxes were linked on one transporter, such that the uphill movement of the hexose into the cell against its concentration gradient was driven by the movement of sodium ions down their electrochemical gradient. Kinetic experiments suggested that when the sodium became bound to the transporter, the affinity for glucose (or galactose) was increased and the subsequent de-binding of sodium on the cytoplasmic side resulted in a reduced affinity for glucose and hence its release [7]. Given that sodium concentrations are maintained at close to the plasma concentration within the intestinal lumen (145 mM), the electrochemical gradient for sodium across the apical membrane always favours the entry of this ion, providing the motive force for glucose and galactose uptake. This would result in the accumulation of these hexoses within the intestinal epithelial cells and hence their subsequent downhill release into the blood stream mediated by a facilitated hexose transporter. However, there is no evidence for sodium-coupled fructose uptake in the mammalian small intestine and it was only recently that the mechanisms of fructose uptake were defined.

1.2 Intestinal Morphology and Function

The surface of the small intestine is made up of a series of finger-like projections, villi, which extend several hundred microns into the lumen. Each of these is surrounded by 6–8 crypts which dip down into the underlying submucosa (Fig. 1). Despite its complex structure the intestinal epithelium is only one cell layer thick and represents the barrier between the lumen (outside world) and the capillaries and lymph ducts within the basal lamina. There are numerous cell types making up this epithelium with the absorptive cells (enterocytes) representing about 80% of the total while the remaining proportion is a mixture of mucus-secreting goblet cells, endocrine and immune cells. In mammals this entire epithelial layer is replaced about every 3–5 days from a population of stem cells, found near the base of the crypts, which are continually dividing. The resulting undifferentiated cells then migrate up the walls of the crypts onto the adjacent villi where they start to transform into epithelial cells. During this differentiation process, as they migrate toward the villus tip, enterocytes start to express hexose transporter proteins as well as carbohydrate digestive enzymes, disaccharidases. This process involves the insertion of new membrane into the microvilli covering their apical surface and so these microprojections elongate accommodating the additional embedded and surface proteins. Normally, enterocytes reach full maturity about half way to two thirds up the villi, but during some adaptive processes such as reduced food intake the dynamics of cell migration and differentiation can change significantly. Thus, under basal conditions

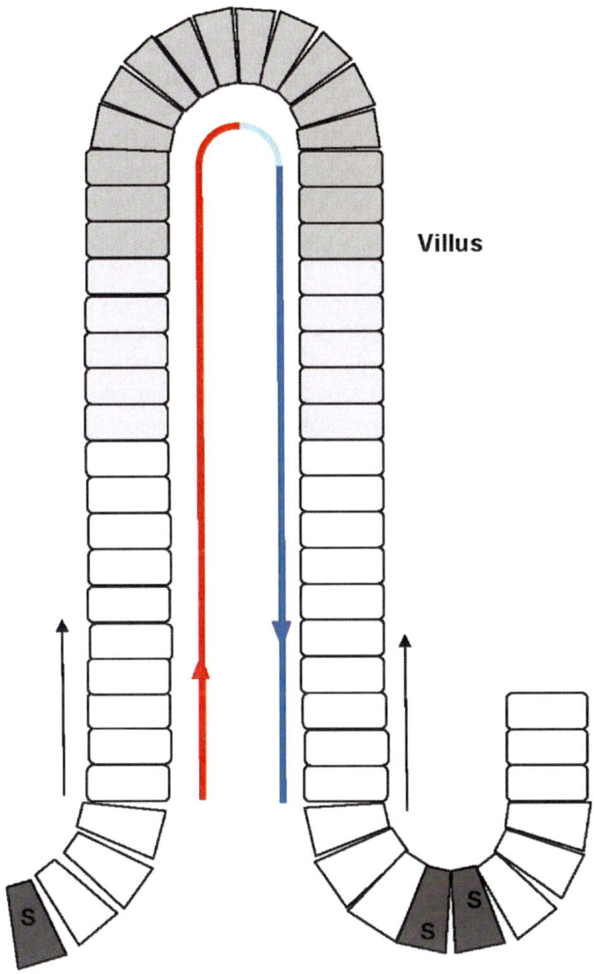

Fig. 1 Intestinal villus and associated crypts. Stem cells in the crypt (S) divide and produce progenitor cells which differentiate as they migrate up the crypt and villus. Shaded cells at the villus tip represent mature enterocytes with full transport and digestive capability. The blood supply flows up the core of the villi and nutrients leaving the enterocytes cross the endothelial cell layer of the capillaries to enter the portal blood

it is only the top half or less of the enterocytes population which contributes significantly to hexose absorption, but when stimulated more cells can be recruited into the transport pool (see regulation section below). Until recently, it had been assumed that epithelial cell loss was restricted to the tip of the villus and likely involved very rapid apoptotic events, presumably to maintain epithelial integrity and barrier function. However, recent vital microscopy studies indicate otherwise. It appears that in a number of mammalian species examined to date, including humans, enterocytes are lost all along the length of the villi [8–10]. What is currently not understood is why the cells are lost before reaching the villus tip, the stimulus or mechanism for this exfoliation or how the barrier function is maintained at the points of cell removal.

1.3 Uptake Across the Apical (Brush-Border Membrane)

The basic model of hexose transport across the intestinal epithelium became established in the 1960s and 1970s and consists of what has been termed an "input" system. The active transport component for glucose and galactose is found in the apical membrane and depends upon the sodium-coupled glucose transporter, SGLT1. In 1987 this protein was cloned in Ernest Wright's laboratory and they subsequently showed it coupled the movement of two sodium ions with one hexose molecule [11, 12]. The lumenal sodium concentration in mammals is in equilibrium with that of plasma because the first region of the small intestine after the stomach, the duodenum, is quite permeable to sodium [13]. This movement occurs primarily paracellularly as the so-called tight junction between the epithelial cells is actually permselective for sodium [14]. Enterocytes also express sodium-potassium ATPase only in their basolateral membrane [15] and this active transporter maintains the intracellular sodium concentration at about 50 mM, resulting in an activity of between 15 and 20 mM [16]. This electrogenic pump is also primarily responsible for generating the membrane potential (*Vm*) of these cells as it exchanges three sodium ions for two potassium ions, and so sodium entry across the brush-border membrane (BBM) is driven by an electrochemical gradient. While there are variations in the reported values for the BBM *Vm*, it appears to be in the range of between −40 and −50 mV [17] and microelectrode studies indicate that this increases as the enterocytes migrate toward the villus tip [18]. Sodium-coupled glucose entry across the BBM tends to initially depolarize the membrane and appears to have little effect on intracellular sodium activity, while the potassium activity is reduced slightly [19]. At the same time, the short circuit current across the tissue increases indicating that the activity of the basolateral Na^+/K^+ ATPase increases in response to the sodium entry [20]. The combined hexose and sodium movement into the enterocytes appears to promote osmotically driven water uptake resulting in cell swelling, which is in turn compensated for by the increased gating of basolateral potassium channels [21].

1.3.1 Glucose and Galactose Uptake – SGLT1

The sodium-coupled hexose transporter, SGLT1, was identified using expression cloning with rabbit intestinal cDNA and determined to be a 664 residue protein with a single N-glycosylation site. The sequence similarity of this protein across species is well conserved and numerous examples have now been confirmed in the intestines of many mammalian and other species. The gene family is distinct from that of facilitated hexose transporters and has been termed SLC5A (solute carrier gene family 5A). Heterologous expression of this protein in *Xenopus* oocytes has confirmed that this protein mediates the movement of glucose and galactose (not fructose) in an electrogenic fashion with a coupling ratio of two sodium ions to one hexose [22, 23]. The kinetic analyses using the current measurements associated with hexose fluxes indicate a *Km* for glucose of about 0.5 mM and a half maximal binding constant for sodium of about 2.5 mM [24]. Despite 20 years of very intensive study there is still no atomic structure of this protein available and there are many questions outstanding regarding just how the movement of sodium and hexoses are coupled. The most recent studies employing different sugar analogues and site-directed mutagenesis have identified a number of residues within the protein that likely form part of the hexose binding pocket [25–27]. However, it is still not clear exactly which of the protein's membrane spanning domains (TMs) form the lining of the aqueous channel through the membrane nor how the binding of sodium changes the shape of the hexose binding site [28]. Ernest Wright's group have argued in favour of the last 5 TMs of SGLT1 forming the hexose translocation pore with some of the N-terminal TMs providing the sodium binding site and the energy coupling mechanism. This model is based on the initial observations that chimeric proteins and truncation of the WT protein to the final five C-terminal helices, (TM10 – 14) still provided for hexose translocation, but with no sodium coupling [29, 30]. Since then, a series of mutational studies combined with cysteine scanning mutagenesis and interaction with sulphydryl reacting compounds have helped to identify a number of residues within TMs 10, 11, 12 and 13 which form part of the hexose binding pocket. Recently, much of this data was combined into a model suggesting that these TMs form an aqueous pore for the translocation of glucose and galactose [31].

However, other groups have suggested that the structure may be more complex and that the orientation of TM 14 is not entirely clear. The Wright model proposes a very long intracellular loop between TMs 13 and 14 with the C-terminal end of the protein facing the outside of the membrane. In contrast, Kinne's group has proposed that this long loop may form a re-entrant structure with TM 14 oriented with its C-terminal end facing the cytoplasm [32], Fig. 2. Further, they propose that residues in this loop combine with amino acids in the extracellular loop between TMs 6 and 7 and several more in the loop between TMs 8 and 9 to form a hexose binding pocket [33]. Such a model suggests components of the protein closer to the N-terminus also contribute to the hexose translocation mechanism.

There does seem to be more consensus as to the position of key amino acid residues which form part of the sodium binding domain of SGLT1. Both Silverman's group and that of Wright have identified residues within the region of TMs 3 and 4 which contribute to sodium binding suggesting that this region of the protein plays an important role in the initial interaction with sodium [34–36]. However, in the absence of a crystal structure, it is currently difficult to imagine how the binding of hexose and sodium result in their coupled translocation across the membrane.

1.3.2 Fructose Uptake

Numerous studies had indicated that fructose uptake by the intestine is not sodium-dependent, but appears to be carrier mediated because it was taken up quickly [37, 38]. However, isolated BBM vesicle experiments showed an almost linear rate of uptake in relation to fructose concentration which was unaffected by the presence of sodium [39]. The identification of a series of facilitated non-sodium dependent hexose transporters (gene family SLC2A) of the type expressed in red cells and the liver (GLUTs) resulted in the isolation of an intestinal clone, GLUT5 [40]. At first

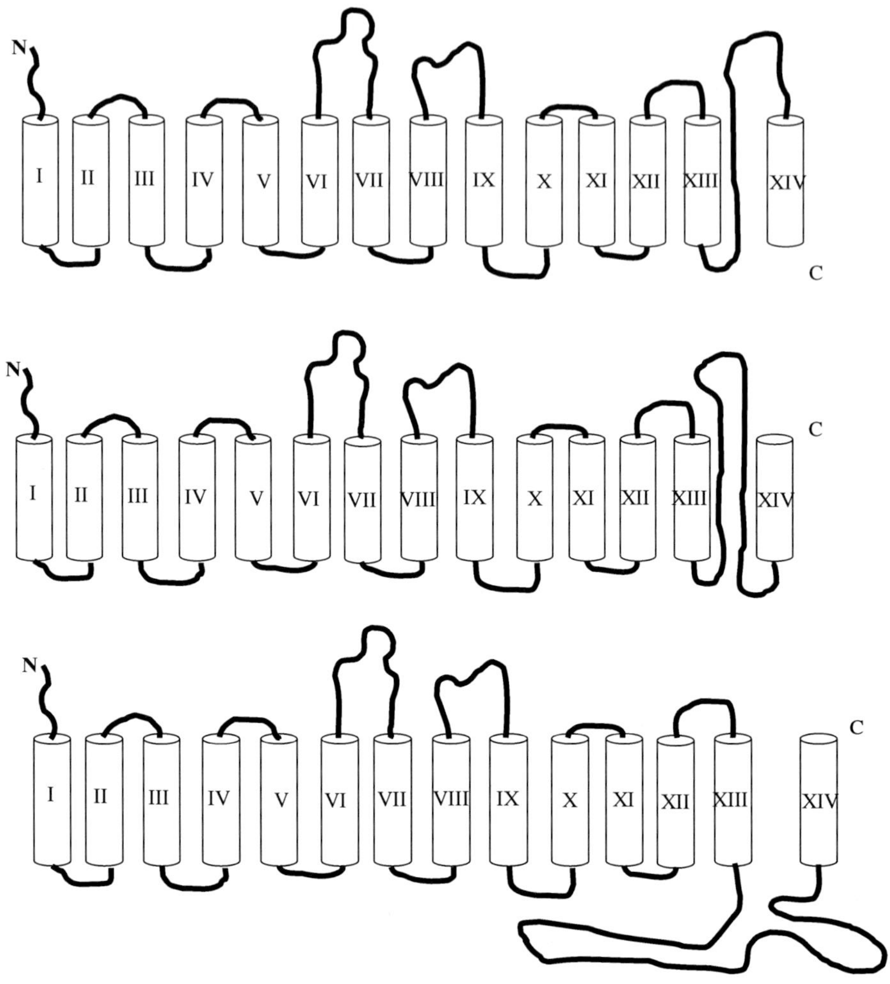

Fig. 2 Putative structures of SGLT1. All have 14 TMs, but Panel A shows the C terminus of TM 14 being intracellular while B and C have the C terminus extracellular. Further the linkage between TMs 13 and 14 is re-entrant in models A and B, while it is totally intracellular in model C. Modified from references [31, 32]

this transporter was believed to mediate primarily glucose movement, although poorly, until it was demonstrated that a major substrate for this protein is fructose which it readily transported with a *Km* of about 5–6 mM [41, 42]. GLUT5 is expressed in the intestinal BBM with a similar longitudinal distribution to that of SGLT1 and until recently was believed to be the sole route of entry for fructose into the epithelium. Interestingly, there does not appear to be a need for an active transporter for the uptake of fructose because plasma concentrations of this hexose are kept low by the rapid uptake into the liver and its conversion to glucose [43].

1.3.3 Role of Apical GLUT2

The second mammalian facilitated hexose transporter to be cloned was the so-called "liver" glucose transporter (GLUT2) which is also expressed in the kidney, pancreatic beta cells and the small intestine [44]. This protein mediates the movement of all three dietary hexoses, glucose, galactose and fructose with kinetics indicating a low affinity and high capacity for all three sugars. Initially, evidence was only found for expression of this transporter in the BLM of the intestine (see below) [45, 46]. However, recently significant evidence has accumulated to show that this protein is transiently expressed in the BBM in response to the presence of hexoses in the intestinal lumen. The first indication that GLUT2 might appear in the BBM was found in the diabetic rats [47], but the physiological significance of this observation was difficult to interpret. Subsequently, Kellett found that during perfusion of the intestinal lumen with hexoses what had been interpreted to be a diffusive component uptake was actually mediated by the low affinity GLUT2 in the BBM [48, 49]. Thus, at the initial part of a meal when lumenal free hexose concentrations can reach >50 mM [50] the rapid insertion of apical GLUT2 allows for fast and efficient downhill movement of hexoses. However, as the lumenal concentration falls GLUT2 re-internalizes preventing subsequent backflux of substrate from the blood to the lumen. The complex signalling pathways involved in promoting GLUT2 insertion into the BBM are covered in the section below on regulation of hexose transport.

1.3.4 Possible Role for Apical GLUT7

An additional member of the SLC2A family has recently been shown to be expressed in the BBM of the jejunum and ileum [51]. This facilitated hexose transporter takes up glucose and fructose, but not galactose, with very high affinity, $Km \sim 300\,\mu M$, which may relate to it being expressed in the distal intestine where lumenal hexose concentrations are low. The transport capacity of this protein for hexoses is low suggesting that its physiological substrate may not be glucose or fructose. So, while definitive evidence for a role of this protein in the uptake of hexoses in the distal intestine is currently lacking, it does need to be kept in mind as another possible route of entry.

1.3.5 Mannose Uptake

Mannose is a hexose making up a substantial component of protein glycosylation and therefore is a significant component in most diets. The normal blood circulating levels are however about 1% of that of glucose [52]. Vesicle experiments had clearly indicated that mannose is taken up across the apical membrane of both the intestine and the PCT of the kidney (also see section below) [53, 54]. However, the lack of the ability for SGLT1 to transport this hexose strongly suggested that another transporter must handle this hexose. An in silico screen of the human intestinal cDNA libraries using SGLT1 as the template uncovered a novel sequence which was termed SGLT4 (SLC5A4) [55]. Primers were used to clone the mRNA and expression of the protein in COS-7 cells induced sodium-dependent α-methyl-D-glucoside transport which was significantly inhibited by mannose, less so by glucose or fructose, but not by 3-O-methyl-D-glucose or galactose. Thus, the transport activity of this protein is quite different from that of SGLT1 of SGLT2. Quantification of mRNA levels in different tissues using PCR indicated that SGLT4 is expressed at about one tenth of the level of SGLT1 in the intestine, but at about the same level in the kidney. Thus, this transporter likely is responsible primarily for sodium-dependent mannose uptake in the intestine and kidney. However, it may also contribute to some sodium-coupled glucose and fructose uptake in both tissues as well.

1.4 Exit Across the Basolateral Membrane

The efflux of hexoses across the basolateral membrane is sodium-independent and initial characterization suggested that the *Km* for glucose transport was much higher than for transport across the red cell membrane [56–59]. Subsequent cloning of the facilitated hexose transporters led to the determination that it was GLUT2 expressed in the BLM of the small intestine [60]. To date, this is the only hexose transporter that has been found to be expressed at significant levels in the intestinal BLM [61]. The expression of this protein, which can mediate the exit of all three dietary sugars, seems to be sufficient to explain transepithelial hexose absorption using the original Crane model. Thus, the simplest explanation is that after entry into the remodeling all three hexoses move down their concentration gradient into the extracellular fluid, across the capillary endothelium and ultimately into the plasma. In the case of galactose and fructose the concentration gradient from the cytoplasm to the plasma is always favourable for exit, as both are rapidly cleared from the circulation by the liver. However, there are a number of reports which suggest that GLUT2 may not be the only route of exit for glucose. Studies of a GLUT2 knock out (KO) mouse model showed that if the expression of GLUT2 in the pancreatic beta cells was rescued and thus signalling for insulin secretion maintained the animals survived. Of interest was that they appeared to be able to eat carbohydrate and absorb hexoses without difficulty, although renal glucose reabsorption was somewhat impaired [62]. Introduction of small volume boluses of very high hexose concentration (4 M) into the lumen of vascularly perfused small intestine in these GLUT2 KO animals indicated that glucose and fructose transepithelial movement was apparently normal [63]. However, the vascular appearance of the non-metabolized hexose analogue, 3-O-methyl-D-glucose, was almost completely abolished. Further, the application of a specific glucose-6-phosphate translocase inhibitor, S4048, significantly reduced the flux of glucose in a dose-dependent manner in the GLUT2 KO animals, but had no effect on glucose permeation in the Wild Type (WT) animals. Also, the inhibitor had no effect on fructose appearance in either the WT or KO animals. The authors concluded from this study that an exocytosis mechanism might be working under these conditions much as they had proposed for the efflux of glucose from the GLUT2$^{-/-}$ liver [64].

In vivo human studies have also been performed in patients with Fanconi–Bickel (FB) syndrome, a genetic disease involving mutations in their SLC2A gene resulting in a failure to properly express GLUT2. This results in several metabolic disturbances, including a large build up of glycogen in the liver, poor renal reabsorption of glucose leading to glucosuria (see renal section below) and symptoms resulting possibly from carbohydrate malabsorption. Certainly, there are several reports of children with the disease having difficulty in absorbing carbohydrate but in later life the problem seems to be less severe [65]. One recent report, using the hydrogen breath test to determine if glucose absorption was compromised, in a FB patient indicated that hexose absorption was normal. Interestingly, comparison with a patient who did not express the glucose-6-phosphate transporter, found in the endoplasmic reticulum (ER), indicated that this individual did have a compromised ability to absorb hexoses [66]. Therefore, they argued in favour of the model proposed by Thorens that glucose could exit across the BLM via some type of exocytotic mechanism which involved the phosphorylation-dephosphorylation pathway in the endoplasmic reticulum. However, it should be borne in mind that with the recent expansion of the SLC2A gene family there could well be additional facilitated hexose transporters expressed in the BLM. Thus, in the absence of GLUT2 one or more additional proteins may increase their expression to compensate. There is already some evidence suggesting that GLUT5 can be expressed in the BLM and this could account for the normal fructose absorption in the GLUT2$^{-/-}$ mice. It is also important to note that an analysis using a gene chip array indicated that there are significant differences in the mRNA levels for hexose transporters between mouse, rat and human duodenum [67]. It should be noted that in most mammals hexose absorption is predominantly achieved in the jejunum, not the duodenum, however these differences do suggest that while a generic model for hexose transport across mammalian small intestine can be used as a starting point, there may be species differences.

1.5 Regulation of Transport

Sporadic reports over the last 20 years have suggested that the capacity for hexose absorption in the small intestine can be rapidly altered to match the lumenal load during a single meal or feeding period. For example, the facilitated hexose transport rate in the BLM has been shown to increase during hyperglycemia and lumenal hexose perfusion in the ileum can increase the absorptive capacity of the jejunum [68]. There is also evidence that the enteric peptide glucagon-like peptide 2 (GLP-2) can promote the insertion of apical SGLT1 and upregulate BLM transport [69]. However, recently it has become clear that there exists a complex set of sensing and signalling pathways which control hexose transporter expression and activity in the small intestine. In this section I will focus on rapid regulatory mechanisms which operate within the time course of a single meal and the subsequent section will deal with adaptive changes in protein expression which anticipate a more prolonged change in meal composition.

In 1988 Cásky showed that in rats 4 h of hyperglycemia induced an elevated hexose transport in the jejunum of 3-O-methyl-D-glucose (3MG), but not α-O-methyl-D-glucopyranoside (αMG) [70]. He concluded that this response must involve GLUT2 as 3MG is not a substrate for SGLT1, while αMG can be transported by this protein. At that time GLUT2 was only known to be expressed in the BLM and so he speculated that the upregulation occurred on the exit side of the epithelium. Subsequently, it was shown that 30 min of hyperglycemia or 30 min of high glucose in the intestinal lumen could result in a three-fold upregulation of BLM glucose transport, apparently mediated by GLUT2 [62]. Lumenal hexoses can promote the release of the enteric peptides GIP, GLP-1 and GLP-2 [71–73] and vascular infusion of GIP or GLP-2 causes a series of rapid responses including, insertion of SGLT1 and GLUT2 into the BBM and upregulation of BLM GLUT2 activity [74–76]. Only the fructose transporter GLUT5 appears to be unaffected by these hormones [77]. Cholecystokinin (CCK) can induce a rapid reduction of SGLT1 in the apical membrane in a dose-dependent manner well within physiological plasma concentrations (2–8 pM) [78]. While it was possible through pharmacological intervention to show that this effect is mediated by CCK_A receptors their location was not identified. There is currently no evidence for the expression of these receptors on the BLM of the enterocytes, which prompts speculation that they could be on submucosal neurons. Thus, the inhibitory effect is likely mediated through the release of other transmitters.

These two sets of opposing effects presumably form a sensitive integrated response to the nutrient load within the intestinal lumen. But what is (are) the signal(s) for these responses and how is the presence of lumenal carbohydrate detected? Recent findings indicate that there are several interacting pathways which lead to these responses. When glucose is taken up by SGLT1 the resulting depolarization of the apical membrane gates the L-type calcium channel $Ca_V1.3$ promoting calcium entry [79–81]. This results in a contraction of the cytoskeleton (the terminal web) beneath the apical membrane and this remodeling appears to be essential for additional transporter insertion into the BBM. However, this alone is not enough to promote the insertion of GLUT2 into the apical membrane unless the lumenal hexose concentration is above 20 mM indicating a second essential component for signalling. Within the mammalian epithelium are "brush cells" which appear to be solitary chemosensing units having structures very similar to those of taste cells in the tongue and which have been speculated to perform a similar function. Taste cells express several sweet receptor proteins which appear to heterodimerize to form a functional unit. The primary unit is a dimer of T1R1 and T1R3 which in turn interact with the G-proteins alpha-gustducin ($G_{\alpha gust}$) and transducin [82–84]. Screening of the intestinal epithelium for these proteins showed no less than three locations of their expression: (1) In the apical membrane of rat jejunal enterocytes, (2) within SCC cells distributed throughout the epithelium, and (3) in Paneth cells found in the crypts [85]. Their location in the two endocrine cell types has led to the speculation that the receptors may play a role in GIP and GLP1 release and by extension could also include the control of GLP2 secretion, which has been shown along with GIP to upregulate GLUT2 and SGLT1 activity [70, 71].

Direct evidence for a role of taste receptors inducing apical GLUT2 insertion comes from the work of Kellett who showed that low concentrations of chemical sweeteners (e.g. 1 mM sucralose) can promote the lumenal insertion of GLUT2 in the presence of a permissive 20 mM concentration of glucose. He proposes that the minimum 20 mM glucose is required for the SGLT1-mediated calcium channel opening and the elevated intracellular calcium then somehow interacts with the PKCβ2 signalling stimulated by the T1R receptors [85]. This would explain why the SGLT1 inhibitor phloridzin blocks the ability of lumenal glucose to promote apical GLUT2 insertion. It also indicates that there are multiple regulatory pathways which can integrate the regulatory response to match the carbohydrate load and composition to the required transport capacity of the tissue.

1.6 Adaptation to Altered Nutrient Load

It has long been known that, unlike most tissues, the mammalian small intestine increases its expression of hexose transporter proteins as well as surface digestive enzymes when exposed to increased dietary carbohydrate for a period of several days [86,87]. Ligand binding studies showed that the increased expression of SGLT1 is programmed into new enterocytes as they leave the crypts, thus the temporal pattern of expression parallels the repopulation of the villus epithelium, taking 3–5 days for completion [88–90]. Similarly, downregulation of SGLT1 expression when the diet is low in carbohydrates follows a similar time course. This suggests that expression of SGLT1 may involve monitoring of dietary carbohydrate by cells within the villus crypt at the point of epithelial cell birth, but does not indicate the polarity of that signal which could be either lumenal or from the blood supplying those cells [91]. Two recent studies have addressed how the promoters for SGLT1 expression are controlled in enterocytes. In some cell systems metabolism plays a key role in hexose transporter expression, yet there is evidence that non-metabolized and non-transported hexoses such as 3-O-methyl-D-glucose and xylose, respectively, can increase SGLT1 expression [92]. Mice in which either the T1R3 or the $G_{\alpha gust}$gene had been knocked out failed to alter SGLT1 transcription in response to dietary carbohydrate strongly implicating these receptors in the response [93]. Interestingly, this study focused on the duodenum where they found T1R3 in the SCC cells and the crypt, but not in the apical membrane of the enterocytes. This is in contrast to Kellett's finding where the taste receptor complexes were also expressed in the BBM of the jejunal enterocytes. This indicates there is a spatial gradient along the length of the gut which appears to match with the known responses to lumenal glucose. A second study using Caco2/T7 cells also found that the upregulation of SGLT1 and GLUT5 expression involves stimulation of taste receptors, an effect which can be at least partially blocked by the sweet taste inhibitor lactisole [94]. In contrast, the control of GLUT2 expression appears to use a different pathway/s and while both glucose and fructose can induce increased levels of GLUT2 mRNA the effect appears to be predominantly mediated by hexoses coming from the blood side of the epithelium. However, it must be kept in mind that this data comes from cultured cells and still needs to be confirmed in the native tissue.

2 Renal Hexose Transport

2.1 Structure and Function of the Renal Nephron or Tubule

The proximal convoluted tubule (PCT) is the first region of the nephron after the filtration apparatus and is responsible, in large part, for the recapture of essential metabolic solutes lost into the urine. Under normal metabolic conditions the reabsorption of glucose from the urine and its release back into the blood draining from the kidney is close to 100% efficient. In vivo studies showed that as plasma glucose concentrations rise the ability to reabsorb the hexose from the filtered urine becomes saturated leading to the concept of *Tm* or maximal reabsorption rate [95]. The basic model for transporting glucose from the urine to the plasma against a concentration gradient is very similar to that established for the small intestine [96–98]. There are however, some key differences between the two tissues.

2.2 Uptake Across the Apical Membrane

2.3 Sodium-Coupled Hexose Transport

Hexose transport studies in the late 1970s and early 1980s using isolated renal tubules and membrane vesicles showed that were significant differences in the rate of glucose fluxes and the kinetics of transport between the three morphologically distinct regions or segments of the proximal convoluted tubule. Just below the Bowman's capsule, in segment S1, fluxes were high but showed low affinity, while by segment S3 transport rates were slower, paracellular permeability much lower and the affinity much higher [99]. A large proportion of the glucose uptake was found to be sodium-dependent and the kinetic differences were postulated to be the result of either one transporter showing some type of cooperativity [100] or two distinct sodium-coupled transporters [101]. The cloning of SGLT1 rapidly led to the resolution that this protein was responsible for the high affinity transport in segment S3 [102], while the subsequent identification of the lower affinity SGLT2 explained the uptake in segment S1 [103].

2.3.1 SGLT1

Northern blotting and hybridization experiments indicated expression of SGLT1 primarily in segment S3 of the PCT and immunohistochemistry localized the protein to the brush-border membrane. As described in the intestinal section, SGLT1 couples the movement of two sodium ions to one glucose/ galactose molecule resulting in a very efficient transport of hexoses against their concentration gradient. This protein is therefore responsible for the late stages of glucose and galactose recapture from the filtered urine ensuring that almost no hexose is lost from the body under normal conditions.

2.3.2 SGLT2

The existence of this transporter was the subject of considerable speculation and debate regarding the validity of the apparent low affinity component for sodium-dependent glucose transport in the PCT. However, low stringency screening with a human SGLT1 probe of a human cDNA library quickly led to the identification of a second sodium-coupled glucose transporter which was initially termed Hu14 and which had 59% identity with SGLT1 [104]. This protein was found to be clearly expressed in the S1 segment of the PCT [105]. Subsequent characterization of this protein in *Xenopus* oocytes showed that this transporter had a lower affinity for glucose with a Km of $\sim$1.6 mM and a much lower affinity for sodium, $Km > 100$ mM. Most importantly the coupling ratio between sodium and substrate was determined to be 1:1 and unlike SGLT1 this protein could not transport galactose [106]. All of these features fitted with the low affinity renal S1 transporter and it was then renamed SGLT2. In the same paper in situ hybridization using a cRNA probe for this transporter with rat renal sections showed a clear expression of message for SGLT2 in S1 of the PCT, but not S2 or S3.

2.3.3 Renal Transport of Mannose

Mannose is an important circulating hexose used in the glycosylation of membrane proteins. Most cell types in the body have mannose transporters which need to have a high affinity as the plasma concentrations are usually in the 50–70 μM range. Isolated PCT BBM vesicles show high affinity sodium-dependent mannose transport which is inhibited by the hexoses glucose and fructose and by the pharmacological agents phloridzin and phloretin [107, 108]. Further, isolation of renal mRNA and injection into *Xenopus* oocytes induces the expression of sodium-dependent mannose transport with the same characteristics [109]. However, this was later reported to result from a renal protein inducing the activity of an endogenous oocyte protein which mediated the mannose uptake [110]. More recently, an SGLT4 was cloned from the human small intestine and found to be a sodium-dependent mannose, glucose and fructose transporter [50]. This transporter appears to fit with the physiological characterization for mannose uptake in the PCT BBM, but gene knock out and immunohistochemistry is required to confirm its role.

2.3.4 Facilitated Transporters

Once GLUT2 had been cloned from liver cells, Northern blotting and subsequent immunohistochemistry localized renal expression of GLUT2 to the BLM of the PCT [111]. This pattern of expression confirmed the models of transepithelial glucose movement in which apical uptake was mediated by the sodium-coupled transporters followed by movement down a concentration gradient and facilitated exit via GLUT2. However, subsequent in situ hybridization screening of renal tissue during development indicated that other GLUTs, including 1, 4 and 5, were also expressed in regions including the PCT [112, 113]. The mature rat kidney appeared to express GLUT1 and GLUT5 in the proximal straight tubule as well as GLUT2 suggesting these transporters might also play a role in renal hexose handling. GLUT5 is primarily a fructose transporter and several studies using isolated BBM from the PCT showed that fructose uptake in S3 was sodium-independent and insensitive to phloridzin or CB inhibition [114]. Further, only fructose could inhibit its own uptake, whereas neither glucose nor galactose had any effect. Finally, kinetic measurements indicated a Km of about 5 mM. All of these features fitted with the known characteristics of GLUT5-mediated fructose transport.

Most recently, GLUT9, a class II GLUT, has also been found in the renal cortex. This protein has several splice variants which appear to affect the membrane domain of the epithelial cells in which it is expressed, but here variation occurs between species. Kelle Moley's group has shown that GLUT9 is expressed in the BLM of either the distal convoluted tubule or collecting duct of the mouse kidney [115]. They were not able to further refine the localization but did clearly show that it did not co-localize with GLUT2. Also, they found a significant increase in GLUT9 expression in diabetic animals suggesting again that under altered metabolic conditions this high affinity facilitative glucose and fructose transporter may have an important role. However, in the human kidney the expression pattern of GLUT9a and b is quite different. Both splice variants are found in the PCT and whereas GLUT9b is again expressed in the BLM the full length variant is trafficked to the apical membrane [116]. Both proteins have been shown to be high affinity glucose and fructose transporters suggesting that they also play some kind of role in hexose recovery from urine. However, a series of recent population genetic studies have found that elevated plasma levels of urate correlate with SNP's in the SLC2A9 gene. This implicated the GLUT9 transporter in the renal handling of urate and two studies have found that it actually has a high capacity for this substrate rather than hexoses [117]. Interestingly, in humans urate is an important plasma anti-oxidant, but because primates lack the enzyme uricase they cannot then excrete this metabolic product of purine breakdown as the more soluble allantoin. Therefore, the human kidney has both absorptive and secretory systems for urate, but while the transporters responsible for reabsorption have largely been defined at the molecular level those responsible for secretion are poorly understood. Thus, the correlation between SNP's in the GLUT9 gene, its location in both poles of the human PCT epithelium and its ability to mediate urate fluxes, implicate this protein as part of the urate secretory mechanism.

2.4 Regulation of Transepithelial Hexose Flux

Recently Debnam looked at the effects of diabetes and overnight fasting on renal transporter expression and function in the rat. He found that GLUT1 is constitutively expressed at low levels in the BBM of the PCT, however, surprisingly he also found GLUT2 in the BBM as well [118]. The hyperglycemia of diabetes had no effect on the expression of GLUT1, but significantly increased the abundance of GLUT2 in the BBM and vesicle flux experiments indicated that it was functionally active. It was possible to separate out the GLUT2 activity from that of GLUT5 by their difference in CB sensitivity. Interestingly, an overnight fast of the diabetic animals led to a significant drop in the BBM expression of GLUT2 suggesting that there is a regulatory process for the apical insertion or removal of GLUT2 similar to that seen in the intestine [119]. GLUT1 abundance was unaffected by either diabetes or fasting emphasizing the specificity of the effect. Finally, they confirmed the presence of GLUT5 in the BBM of control animals and again this expression was increased in diabetes, but unaffected by fasting. The authors concluded that in conditions which induce hyperglycemia the recovery of glucose from the urine was promoted by the expression of these facilitated hexose transporters in the apical membrane. Interestingly, under these conditions they found no change in the functional activity or expression of the sodium-coupled transporters.

The kidney has a rich sympathetic innervation and there are both α and β adrenergic receptors expressed throughout the organ. At least one primary cell culture study has shown that epinephrine induces a reinternalization of SGLT1 and SGLT2 from the apical membrane in the PCT and an inhibition of α-methyl-D-glucoside transport [120]. Interestingly, the transport inhibition was greater than the reduction in transporter abundance suggesting two levels of regulation. An additional study on the expression of GLUT2 in the kidney of diabetic animals has indicated that sympathetic input increases the level of GLUT2 abundance in the apical membrane [121]. In contrast, a related study showed that apical GLUT5 activity and fructose uptake in the PCT was inhibited by epinephrine [122]. These apparently conflicting data clearly indicate the possibility of sympathetic control of glucose and fructose reabsorption in the PCT, but more work is needed to fit these observations into a homeostatic model.

2.5 Gluconeogenesis

It has been known for a very long time that the kidney not only reabsorbs glucose from the urine but it is also a very important organ metabolically. The renal medulla uses a lot of energy in pumping sodium to generate the osmotic gradient necessary for water reabsorption in the collecting ducts. However, despite estimates that at rest in the human this activity may utilize up to 25% of the body's total metabolism, numerous studies indicate no real arteriovenous difference for glucose in the kidney [123]. That is the renal arterial and venous concentrations of glucose are essentially the same. How can that be given the very high rates of metabolism in the renal medulla? In fact, under some conditions the concentration of glucose may actually be higher in the blood draining from the kidney. This can only be explained if some part of the kidney is involved in gluconeogenesis, i.e. converting other substrates into glucose and releasing it into the circulation. To do this, just as occurs in the liver, the renal nephrons need to express glucose-6-phosphatase and phosphoenolpyruvate carboxykinase (PEPCK-C). A recent immunohistochemical study has shown that both enzymes are indeed expressed in the PCT which indicates that indeed the molecular pathways needed for gluconeogenesis do exist [124]. However, just as for the interesting debate about the role of GLUT2 in the intestinal release of glucose across the BLM, the same issues apply for the kidney.

3 The Placenta

3.1 Structure and Function

The placenta is the epithelium which forms the barrier between foetus and mother allowing for the uptake of nutrients and removal of waste products. It also forms a key immunological barrier between the host and the offspring stopping the otherwise inevitable attack and rejection of the developing "non-self" individual. This tissue develops after implantation of the blastocyst when the cells from the outer cell layer, the trophoblast, form a syncytium (syncytiotrophoblast, STB) by the fusion of cytotrophoblast cells. These secrete metalloproteinases and other enzymes to penetrate the endometrial epithelium of the uterine wall. Blood vessels from the uterus then gradually invade the syncytiotrophoblast and the placenta continues to grow progressively along with the foetus. Cytotrophoblasts still remain as single nucleated cells and these divide and contribute to the expansion of the STB throughout pregnancy [125]. The final structure of the placenta varies enormously between different mammalian species and so I will focus on those of the human, rat and mouse.

3.2 Human Placenta

When fully developed the syncytiotrophoblast is organized into highly branched villus-like structures which are surrounded by maternal blood, Fig. 3. Exchange of materials occurs across a single layer of STB cells encasing the endothelium of the foetal blood supply running up the centre of the villi. Glucose moves down a concentration gradient from the maternal blood into the foetal circulation and has to cross the STB and the endothelial cells. In vitro perfusion studies have confirmed that indeed glucose does readily pass down its concentration gradient from the maternal to the foetal blood with no accumulation within the tissue [126]. D-glucose, D-galactose, 2-deoxyglucose and 3-O-methyl-glucose are all transported

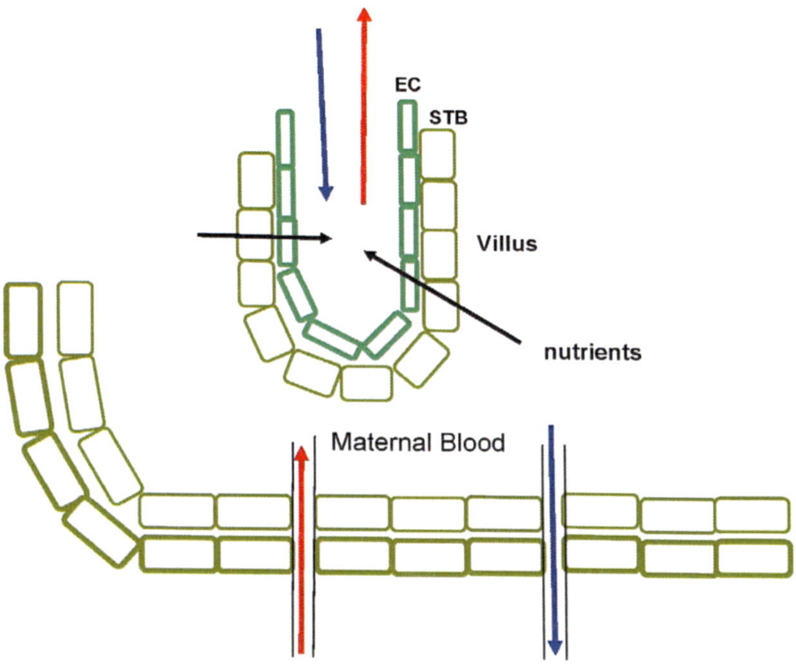

Fig. 3 Human placenta. Maternal arterial blood enters the cavities via spiral arteries. The villi formed from projections of the syncytiotrophoblast (STB) are bathed in the maternal blood and nutrients have to cross the single STB cell layer and the endothelial cells (EC) to reach the foetal blood

equally rapidly, while L-glucose and fructose move very slowly by simple diffusion [127, 128]. Studies employing isolated apical or basolateral membranes also show that the transport is saturable, highly temperature sensitive, sodium independent and inhibited by cytochalasin B and phloretin. Molecular studies have determined that the major hexose transporter expressed in the microvilli of the STB facing the maternal blood is GLUT1 [129–131]. Hybridization studies have detected the mRNA for GLUT1 in the STB and immunohistochemistry and Western blotting of isolated membranes have clearly shown the presence of the protein in the apical membrane. However, unlike the intestinal and renal epithelium there is no evidence for the presence of a sodium-coupled transporter in either pole of these cells. Neither SGLT1/2 message or protein have been detected in the STB which confirms the lack of sodium-dependence seen in the physiological studies of glucose fluxes. It has also been reported that the mRNA for the higher affinity GLUT3 is present in the STB and there is some indication that the protein may also be present in the apical membrane [132]. However, this has been difficult to verify between studies using Western blotting or immunohistochemistry and further work is needed to clarify what role, if any GLUT3 plays in mediating glucose uptake into the human foetus. The same techniques have also shown the presence of GLUT1 in the basolateral membrane of the STB, although at lower abundance. When membrane surface area of the apical and BLMs is taken into account it appears that there is approximately 15–20-fold more GLUT1 at the apical surface. Studies of cultured BeWO cells which are a choriocarcinoma-derived cell line show a similar, but less pronounced, differential expression of GLUT1 in their apical and BLM's [133]. The effect of increasing concentrations of pCMBS, which inactivates GLUTs, in either the apical or basolateral solution indicated that the rate limiting step for glucose fluxes across this epithelium is exit via the BLM. Thus, given the much larger differential GLUT expression in the STB it is very likely that exit is the rate limiting step in the human placenta. However, this raises important questions as to how glucose actually traverses the cytoplasm of the syncytium. This tissue has a high rate of metabolism and presumably some of the glucose taken up from the maternal circulation will be used to satisfy these needs. There is good evidence that the STB expresses hexokinase which would rapidly phosphorylate the entering glucose and trap it within the cytoplasm, as glucose-6-phosphate cannot be handled by any of the known GLUT isoforms. So just as the intestinal and renal epithelia this still requires the resolution as to how glucose crosses the BLM of the STB to reach the foetal circulation. The STB does express glucose-6-phosphatase, an ER enzyme found primarily in tissues where gluconeogenesis occurs, for example the liver [134–136]. But, it is unknown how this would occur in relation to the ability of the released glucose to access the GLUT1 in the BLM of the STB. Arguing against a role for glucose phosphorylation and de-phosphorylation during the transepithelial fluxes are the studies which show that non-metabolized glucose analogues, such as 2-D-O-G or 3-O-MG, move just as rapidly across the perfused placenta as does glucose [137]. Another possibility is that much of the glucose entering across the apical membrane is in some way metabolically compartmentalized and remains as free glucose until it leaves on the other side. Interestingly, this same problem applies to the transepithelial flux of glucose in both the small intestine and the renal nephron. Thick EM sections of proximal convoluted tubules show that the endoplasmic reticulum appears to form channels connecting the apical and BLM's which could act as a conduit through the cytoplasm for some solutes [138]. (See the intestinal section of this chapter for additional evidence for G-6-phosphatase playing a role in transepithelial hexose fluxes).

In the human placenta the cytotrophoblasts lie between the STB and the endothelial cells of the blood vessels. Here they can still divide and fuse with the syncytium allowing for growth to parallel that of the foetus. They too express primarily GLUT1 presumably for their own energy requirements as they are not positioned to contribute to transepithelial hexose fluxes.

The endothelial cells of the blood vessels within the villi also express GLUT1 and possibly GLUT3, so glucose may also traverse this cell barrier by facilitated diffusion. However, there are also significant fenestrations which may well allow the glucose to pass rapidly into the foetal blood. The high affinity GLUT3 in these cells could be argued to help lower the glucose concentration and promote substrate flux from the maternal blood [139].

3.2.1 Expression of Additional GLUT Isoforms

There is good evidence that the human placenta expresses insulin receptors, yet numerous studies have failed to find any effect of insulin on glucose fluxes, metabolism or the expression of GLUT1 or GLUT3 [140]. The known coupling between insulin action and the expression and trafficking of GLUT4 in insulin-sensitive tissues prompted a search for this hexose transporter in the placenta but this isoform is not expressed in the STB, nor the endothelium of this tissue. The lack of an ability of the placenta to transport fructose also strongly argues against the expression of either GLUT2 or GLUT5 and there is no molecular evidence for their presence in this tissue. The recent expansion of the GLUT family has prompted a screening for the presence of mRNA for other isoforms which has detected messages for GLUTs 8, 9, 10, 11 and 12 [143–146]. To date the protein distribution in the placenta has only been determined for GLUT12 and found to be present in the trophoblast, cytotrophoblast and STB in the first trimester [147]. However, to date little is known regarding the physiological role of this protein in any tissue. It is a class III hexose transporter and very little has been published on their functional characteristics. GLUT12 has internalization targeting motifs like GLUT4 which probably means that the protein is retained within an intracellular compartment [148]. However, signals for the trafficking of this protein have yet to be identified. GLUTs 8 and 9 have been found in the pre-implantation blastocyst and there is evidence that they play a key role in metabolism at this stage of development, but it is not clear how they might contribute to placental transfer of glucose [149, 150].

Thus, the current model of glucose handling by the human placenta is facilitated transport via GLUT1 (and possibly GLUT3) into the STB followed by a similar movement down its concentration gradient to the endothelial cells and on into the foetal circulation. At this time it is not known what role intracellular metabolism plays in this transfer and whether glucose-6-phosphate is an intermediate.

3.3 *Rat and Mouse Placenta*

The rodent placenta is more complex than that of the human, having two layers of STB after implantation. The first layer is similar to that found in the human, but below there is a second layer oriented the opposite way so that the basolateral membrane is next to that of the first layer and the second microvillus surface faces the endothelial cells of the foetal blood vessels. Gap junctions connect the two sets of basolateral membranes and may well provide a route for the passage of glucose from the first layer to the second [151, 152] (Fig. 4). Again, the major GLUT isoform found in both sets of microvilli is GLUT1 and this appears to provide both the route of uptake from the maternal blood and release to the foetal endothelial cells which are highly fenestrated [153]. However, recent evidence from the mouse indicates that GLUT3 is expressed in the trophoblast cells of the blastocyst prior to implantation and that this high affinity transporter is very important in ensuring sufficient energy supplies prior to and during implantation [153, 154].

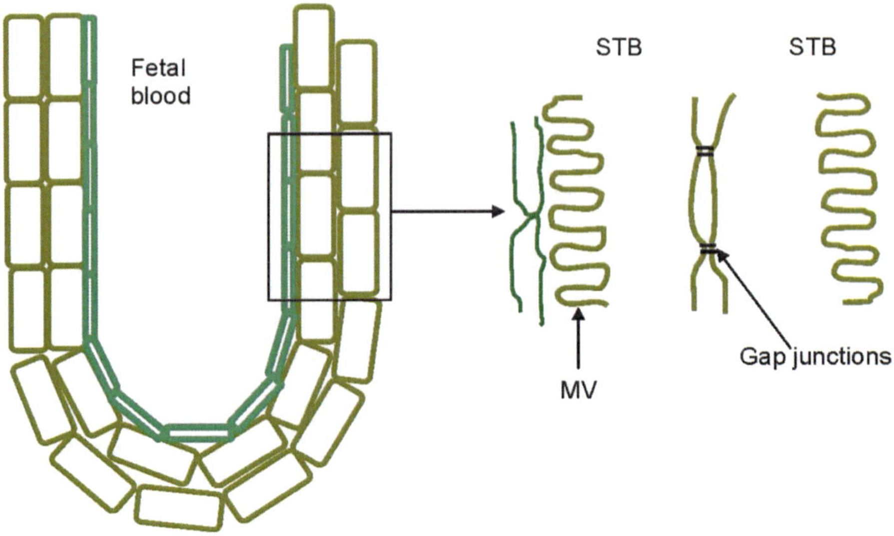

Fig. 4 Rodent placenta. Structure of rat or mouse placental villus showing double layer of syncytiotrophoblasts (STB) overlying a single layer of endothelial cells. Maternal nutrients cross the apical membrane (MV) of the first STB cell layer which is connected to the underlying STBs by gap junctions. The nutrients then cross the second STB cell layer and exit across the apical membrane to reach endothelial cells and then the foetal blood

There is only one report suggesting that SGLT1 might be expressed in the placenta and that was for rabbit [155]. Thus, the overall model for transplacental hexose movement would appear to employ facilitated hexose transporters with a downhill flux of glucose, and possibly fructose, from the maternal circulation to that of the foetus.

4 Inherited Diseases Related to Epithelial Hexose Transporters

There are relatively few reports of significant genetic mutations involving hexose transporters, which emphasizes the essential nature of glucose and fructose as metabolic fuels. However, there are a couple of rare syndromes which are of some significance clinically.

4.1 Fanconi–Bickel Syndrome

This syndrome was first described in 1949 as a rare autosomal metabolic disease which resulted in growth retardation, low fasting blood glucose, glycogen accumulation in the liver and kidney and glucose galactose malabsorption [156]. Since then 109 cases have been studied and found to have one of 34 different mutations in the

SLC2A2 gene which codes for GLUT2 [157]. In all cases tested this resulted in either a trafficking or functional defect in this facilitated hexose transporter.

4.2 Glucose-Galactose Malabsorption

Although this disease had been described in the literature from the 1960s as a rare inherited disease in which individuals could not tolerate glucose or galactose in the diet the genetic defect was not identified until almost 20 years later. The first report that it was related to a loss of sodium-coupled glucose transport in the intestinal apical membrane appeared in 1988 [158]. However, it took another 5 years for the first mutation in SGLT1 to be shown to produce a trafficking defect resulting in the failure to express the protein in the apical membrane [159, 160]. Since then over 23 separate mutation sin SGLT1 have been identified, 22 of which appear to induce trafficking defects and one of which does make it to the cell membrane but it is non-functional [161] Treatment for this condition is simply to remove glucose and galactose from the diet and diarrhoea induced by the hexoses reaching the colon is then resolved.

5 Summary

The recent rapid advance in molecular techniques has added a significant number of new hexose transporter isoforms to those which had been characterized over the previous 50 years. However, the basic models for transepithelial hexose movement have proven to be quite robust. In the intestine, the addition of a GLUT2-mediated pathway in the apical membrane has certainly provided an explanation for the low affinity / high capacity pathway well described using in vivo techniques. In addition to the adaptive responses of the intestine to protracted changes in the diet, it is now appreciated how dynamic the epithelium is with respect to the sensing and responses to the lumenal contents during an individual meal. The basic model for renal hexose absorption has also stood the test of time with a sodium-dependent apical uptake mediated by a low affinity and a high affinity transporter (SGLT2 and SGLT1). However, the addition of GLUT2 and GLUT5 to the apical membrane is still not explained physiologically and awaits further study. Finally, the placenta does not appear to need more than a facilitated mediation of hexose movement which appears to be managed through GLUT1 and GLUT3.

However, what is missing from all of these models is a satisfactory explanation of how glucose or fructose moves through these epithelial cells. In most cases these substrates appear to be rapidly phosphorylated upon entry into the cells, yet the majority of release on the other side is in the form of the free hexose. GLUT knock-out experiments have helped to identify a key role for glucose-6-phosphatase in the movement of glucose, yet why this is so important has not been elucidated. Clearly, the next area of focus is going to be on just how glucose and fructose cross the epithelial cell. Is it through the endoplasmic reticulum in the form of glucose-6-phosphate, to be released adjacent to the BLM in the free form? New fluorescent imaging techniques are being developed to follow the flux of glucose through cells and perhaps this type of approach may well help to provide the answer [162].

Acknowledgements Chris Cheeseman is supported by grants from the Canadian Breast Cancer Foundation and the Canadian Institutes for Health Research.

References

1. Fisher RB, Parson DS. Glucose movements across the wall of the rat small intestine. J Physiol. 1953;119:210–223.
2. Reid EW. On intestinal absorption, especially on the absorption of serum, peptone, and glucose. Phil Trans Roy Soc B. 1900;192:211–227.
3. Riklis E, Quastel JH. Effects of cations on sugar absorption by isolated surviving guinea pig intestine. Can J Biochem Physiol. 1958;36:347–362.
4. Casky TZ Significance of sodium ions in active intestinal transport of nonelectrolytes. Am J Physiol. 1961;201:999–1001.
5. Crane RK. Na^+ -dependent transport in the intestine and other animal tissues. Fed Proc. 1965;24(5):1000–1006.
6. Schultz SG, Hudson RL, Lapointe JY. Electrophysiological studies of sodium cotransport in epithelia: toward a cellular model. Ann NY Acad Sci. 1985;456:127–135.
7. Crane RK, Forstner G, Eichholz A. Studies on the mechanism of the intestinal absorption of sugars. X. An effect of Na+ concentration on the apparent Michaelis constants for intestinal sugar transport, in vitro. Biochim Biophys Acta. 1965;109(2):467–477.
8. Watson AJ, Chu S, Sieck L, Gerasimenko O, Bullen T, Campbell F, McKenna M, Rose T, Montrose MH. Epithelial barrier function in vivo is sustained despite gaps in epithelial layers. Gastroenterology. 2005;129(3):902–912.
9. Bullen TF, Forrest S, Campbell F, Dodson AR, Hershman MJ, Pritchard DM, Turner JR, Montrose MH, Watson AJ. Characterization of epithelial cell shedding from human small intestine. Lab Invest. 2006;86(10):1052–1063.
10. Kiesslich R, Goetz M, Angus EM, Hu Q, Guan Y, Potten C, Allen T, Neurath MF, Shroyer NF, Montrose MH, Watson AJ. Identification of epithelial gaps in human small and large intestine by confocal endomicroscopy. Gastroenterology. 2007;133(6):1769–1778.
11. Hediger MA, Coady MJ, Ikeda TS, Wright EM. Expression cloning and cDNA sequencing of the Na+/glucose co-transporter. Nature. 1987;330(6146):379–381.
12. Ikeda TS, Hwang ES, Hirayama BA, Hediger MA, Wright EM. Characterization of a Na+/glucose cotransporter cloned from rabbit small intestine. J Membr Biol. 1989;110(1):87–95
13. Soergel KH, Whalen GE, Harris JA. Passive movement of water and sodium across the human small intestinal mucosa. J Appl Physiol. 1968;24(1):40–48.
14. Simmons NL, Naftalin RJ. Bidirectional sodium ion movements via the paracellular and transcellular routes across short-circuited rabbit ileum. Biochim Biophys Acta. 1976;19:448(3):426–450.
15. Wright EM, Harms V, Mircheff AK, van Os CH. Transport properties of intestinal basolateral membranes. Ann NY Acad Sci. 1981;372:626–636.
16. Vanthanouvong V, Gman MH, Roomans GM. In Vitro and In Situ Experimental Model for X-Ray Microanalysis of Intestinal Epithelium. Micros Res Tech. 2003;62:211–217.
17. Debnam ES, Thompson CS. The effect of fasting on the potential difference across the brush-border membrane of enterocytes in rat small intestine. J Physiol. 1984;355:449–456.
18. Cremaschi D, James PS, Meyer G, peacock MA, Smith MW. Membrane potentials of differentiating enterocytes. Biochim Biophys Acta. 1982;688(1):271–274.

19. Hudson RL, Schultz SG. Sodium-coupled sugar transport: effects on intracellular sodium activities and sodium-pump activity. Science. 1984;224(4654):1237–1239.
20. Gunter-Smith PJ, Grasset E, Schultz SG. Sodium-coupled amino acid and sugar transport by Necturus small intestine. An equivalent electrical circuit analysis of a rheogenic co-transport system. J Membr Biol. 1982;66(1):25–39.
21. Schultz SG, Dubinsky WP. Sodium absorption, volume control and potassium channels. J. Mem Biol. 2001;184:255–261.
22. Ikeda TS, Hwang ES, Coady MJ, Hirayama BA, Hediger MA, Wright EM. Characterization of a Na+/glucose cotransporter cloned from rabbit small intestine. J Membr Biol. 1989;110(1):87–95.
23. Chen XZ, Coady MJ, Jackson F, Berteloot A, Lapointe JY. Thermodynamic determination of the Na+: Glucose coupling ratio for the human SGLT1 cotransporter. Biophys J. 1995;69:2405–2414.
24. Birnir B, Loo DD, Wright EM. Voltage-clamp studies of the Na+/glucose cotransporter cloned from rabbit small intestine. Pflugers Arch. 1991;418(1–2):79–85.
25. Wright EM, Loo DD, Panayotova-Heiermann M, Hirayama BA, Turk E, Eskandari S, Lam JT. Structure and function of the Na+/glucose cotransporter. Acta Physiol Scand Suppl. 1998 Aug;643:257–264.
26. Lo B, Silverman M. Cysteine scanning mutagenesis of the segment between putative transmembrane helices IV and V of the high affinity Na+/Glucose cotransporter SGLT1. Evidence that this region participates in the Na+ and voltage dependence of the transporter. J Biol Chem. 1998;273(45):29341–29351.
27. Gagnon DG, Frindel C, Lapointe JY. Voltage-clamp fluorometry in the local environment of the C255–C511 disulfide bridge of the Na+/glucose cotransporter. Biophys J. 2007;92(7):2403–2411.
28. Wright EM, Hirayama BA, Loo DF. Active sugar transport in health and disease. J Intern Med. 2007;261(1):32–43.
29. Panayotova-Heiermann M, Loo DDF, Kong CT, Lever JT, Wright EM. Sugar binding to Na+/glucose cotransporters is determined by the carboxyl-terminal half of the protein, J Biol Chem1996;271: 10029–10034.
30. Panayotova-Heiermann M, Eskandari S, Turk E, Zampighi GA, Wright EM. Five transmembrane helices form the sugar pathway through the Na+/glucose cotransporter, J Biol Chem 1997;272: 20324–20327.
31. Hirayama BA, Loo DD, Díez-Sampedro A, Leung DW, Meinild AK, Lai-Bing M, Turk E, Wright EM. Sodium-dependent reorganization of the sugar-binding site of SGLT1. Biochemistry. 2007;46(46):13391–13406.
32. Kumar A, Tyagi NK, Arevalo E, Miller KW, Kinne RK. A proteomic study of sodium/D-glucose cotransporter 1 (SGLT1): topology of loop 13 and coverage of other functionally important domains. Biochim Biophys Acta. 2007;1774(8):968–974.
33. Puntheeranurak T, Kasch M, Xia X, Hinterdorfer P, Kinne RK. Three surface subdomains form the vestibule of the Na+/glucose cotransporter SGLT1. J Biol Chem. 2007;282(35):25222–25230.
34. Lo B, Silverman M. Cysteine scanning mutagenesis of the segment between putative transmembrane helices IV and V of the high affinity Na+/Glucose cotransporter SGLT1. Evidence that this region participates in the Na+ and voltage dependence of the transporter. J Biol Chem. 1998;273(45):29341–29351.
35. Huntley SA, Krofchick D, Silverman M. Position 170 of rabbit Na+/glucose cotransporter (rSGLT1) Lies in the Na1 pathway; modulation of polarity/charge at this site regulates charge transfer and carrier turnover. Biophys J 2004;87:295–310.
36. Meinild AK, Loo DD, Hirayama BA, Gallardo E, Wright EM. Evidence for the involvement of Ala 166 in coupling Na(+) to sugar transport through the human Na(+)/glucose cotransporter. Biochemistry. 2001;40(39):11897–11904.
37. Holdsworth CD, Dawson AM. Glucose and fructose absorption in idiopathic steatorrhoea. Gut. 1965;6(4):387–391.

38. Schultz SG, Strecker CK. Fructose influx across the brush border of rabbit ileum. Biochim Biophys Acta. 1970;211(3):586–588.
39. Sigrist-Nelson K, Hopfer U. A distinct D-fructose transport system in isolated brush border membrane. Biochim Biophys Acta. 1974;367(2):247–254.
40. Kayano T, Burant CF, Fukumoto H, Gould GW, Fan YS, Eddy RL, Byers MG, Shows TB, Seino S, Bell GI. Human facilitative glucose transporters. Isolation, functional characterization, and gene localization of cDNAs encoding an isoform (GLUT5) expressed in small intestine, kidney, muscle, and adipose tissue and an unusual glucose transporter pseudogene-like sequence (GLUT6). J Biol Chem. 1990;265(22):13276–13282.
41. Burant CF, Takeda J, Brot-Laroche E, Bell GI, Davidson NO. Fructose transporter in human spermatozoa and small intestine is GLUT5. J Biol Chem. 1992;267(21):14523–14526.
42. K Miyamoto K, Tatsumi S, Morimoto A, Minami H, Yamamoto H, Sone K, Taketani Y, Nakabou Y, Oka T, Takeda E. Characterization of the rabbit intestinal fructose transporter (GLUT5). Biochem J. 1994; 303(Pt 3): 877–883.
43. Tappy L, Randin JP, Felber JP, Chiolero R, Simonson DC, Jequier E, DeFronzo RA. Comparison of thermogenic effect of fructose and glucose in normal humans. Am J Physiol. 1986;250(6):E718–E724.
44. Thorens B, Sarkar HK, Kaback HR, Lodish HF. Cloning and functional expression in bacteria of a novel glucose transporter present in liver, intestine, kidney, and beta-pancreatic islet cells. Cell. 1988;55(2):281–290.
45. Thorens B, Cheng ZQ, Brown D, Lodish HF. Liver glucose transporter: a basolateral protein in hepatocytes and intestine and kidney cells. Am J Physiol. 1990;259(6 Pt 1):C279–C285.
46. Cheeseman CI. GLUT2 is the transporter for fructose across the rat intestinal basolateral membrane. Gastroenterology. 1993;105(4):1050–1056.
47. Corpe CP, Basaleh MM, Affleck J, Gould G, Jess TJ, Kellett GL. The regulation of GLUT5 and GLUT2 activity in the adaptation of intestinal brush-border fructose transport in diabetes. Pflugers Arch. 1996;432(2):192–201.
48. Kellett GL, Helliwell PA. The diffusive component of intestinal glucose absorption is mediated by the glucose-induced recruitment of GLUT2 to the brush-border membrane. Biochem J. 2000;350(1):155–162.
49. Helliwell PA, Richardson M, Affleck J, Kellett GL. Stimulation of fructose transport across the intestinal brush-border membrane by PMA is mediated by GLUT2 and dynamically regulated by protein kinase C. Biochem J. 2000;350(1):149–154.
50. Ferraris RP, Yasharpour S, Lloyd KC, Mirzayan R, Diamond JM. Luminal glucose concentrations in the gut under normal conditions. Am J Physiol. 1990;259(5 Pt 1):G822–G837.
51. Li Q, Manolescu A, Ritzel M, Yao S, Slugoski M, Young JD, Chen XZ, Cheeseman CI. Cloning and functional characterization of the human GLUT7 isoform (SLC2A7) from the small intestine. Am J Physiol. 2004;287(1):G236–G242.
52. Pitkanen E, Mannose, mannitol, fructose and 1,5-anhydroglucitol concentrations measured by gas chromatography/mass spectrometry in blood plasma of diabetic patients. Clinica Chimica Acta. 1996;251:91–103.
53. Cano M, Calonge ML, Peral MJ, Ilundain, AA. A Na+-dependent D-mannose transporter in the apical membrane of chicken small intestine epithelial cells. Pflügers Archiv Eur J Physiol. 2001;441:686–691.
54. De la Horra MC, Cano M, Peral MJ, Garcia-Delgado M, Duran JM, Calonge ML, Ilundain AA. Na(+)- dependent D-mannose transport at the apical membrane of rat small intestine and kidney cortex. Biochimica et Biophysica Acta 2001;1512, 225–230.
55. Tazawa S, Yamato T, Fujikura H, Hiratochi M, Itoh F, Tomae M, Takemura Y, Maruyama H, Sugiyama T, Wakamatsu A, Isogai T, Isaji M. LC5A9/SGLT4, a new Na+-dependent glucose transporter, is an essential transporter for mannose, 1,5-anhydro-D-glucitol, and fructose Life Sci. 2005;76(9):1039–1050.
56. Hopfer U. Intestinal sugar transport: studies with isolated plasma membranes. Ann NY Acad Sci. 1975;264:414–427.

57. Murer H, Hopfer U, Kinne-Saffran E, Kinne R. Glucose transport in isolated brush-border and lateral-basal plasma-membrane vesicles from intestinal epithelial cells. Biochim Biophys Acta. 1974;345(2):170–179.
58. Wright EM, van Os CH, Mircheff AK. Sugar uptake by intestinal basolateral membrane vesicles. Biochim Biophys Acta. 1980;597(1):112–124.
59. Maenz DD, Cheeseman CI. The Na+-independent D-glucose transporter in the enterocyte membrane: orientation and cytochalasin B binding characteristics. J Membrane Biol 1987;97:259–266.
60. Thorens B, Sarkar HK, Kaback HR, Lodish HF. Cloning and functional expression in bacteria of a novel glucose transporter present in liver, intestine, kidney, and beta-pancreatic islet cells. Cell. 1988;55(2):281–290.
61. Cheeseman CI. GLUT2 is the transporter for fructose across the rat intestinal basolateral membrane. Gastroenterology 1993;105:1050–1056.
62. Thorens B, Guillam MT, Beermann F, Burcelin R, Jaquet M. Transgenic reexpression of GLUT1 or GLUT2 in pancreatic beta cells rescues GLUT2-null mice from early death and restores normal glucose-stimulated insulin secretion. J Biol Chem. 2000;275(31): 23751–23758
63. Stümpel F, Burcelin R, Jungermann K, Thorens B. Normal kinetics of intestinal glucose absorption in the absence of GLUT2: evidence for a transport pathway requiring glucose phosphorylation and transfer into the endoplasmic reticulum. Proc Natl Acad Sci USA. 2001;98(20):11330–511333.
64. Burcelin R, del Carmen Muñoz M, Guillam MT, Thorens B. Liver hyperplasia and paradoxical regulation of glycogen metabolism and glucose-sensitive gene expression in GLUT2-null hepatocytes. Further evidence for the existence of a membrane-based glucose release pathway. J Biol Chem. 2000;275(15):10930–10936
65. Santer R, Steinmann B, Schaub J. Fanconi-Bickel syndrome – a congenital defect of facilitative glucose transport. Curr Mol Med. 2002;2(2):213–227.
66. Santer R, Hillebrand G, Steinmann B, Schaub J. Intestinal glucose transport: evidence for a membrane traffic-based pathway in humans. Gastroenterology. 2003;124(1):34–39.
67. Kim HR, Park SW, Cho HJ, Chae KA, Sung JM, Kim JS, Landowski CP, Sun D, Abd El-Aty AM, Amidon GL, Shin HC. Comparative gene expression profiles of intestinal transporters in mice, rats and humans. Pharmacol Res. 2007;56(3):224–236.
68. Tsang R, Ao Z, Cheeseman CI. The influence of luminal and vascular hexoses on intestinal basolateral glucose transport. Can J Physiol Pharmacol. 1994;72: 317–326.
69. Cheeseman CI, Tsang R. The effect of gastric inhibitory polypeptide and glucagon-like peptides on intestinal hexose transport. Am J Physiol. 1996;271: G477–G482.
70. Csaky TZ. Membrane-transport of sugars in diabetes mellitus. Prog Clin Biol Res. 1988;258:37–42
71. Schulz TB, Burhol PG, Jorde R, Waldum HL. Gastric inhibitory polypeptide release into the portal vein in response to intraduodenal glucose loads in anesthetized rats. Scand J Gastroenterol. 1981;16(8):1061–1065.
72. Falko JM, Crockett SE, Cataland S, O'Dorisio TM, Kramer W, Mazzaferri EL. The effect of increasing doses of ingested glucose on insulin and gastric inhibitory polypeptide (GIP) concentrations in man. Clin Endocrinol. 1980;13(6):587–593.
73. Orskov C, Holst JJ, Knuhtsen S, Baldissera FG, Poulsen SS, Nielsen OV. Glucagon-like peptides GLP-1 and GLP-2, predicted products of the glucagon gene, are secreted separately from pig small intestine but not pancreas. Endocrinology. 1986;119(4):1467–1475.
74. Au A, Gupta A, Schembri P, Cheeseman CI. Rapid insertion of GLUT2 into the Rat Jejunal Brush-Border Membrane Promoted by Glucagon-Like Peptide 2. Biochem J 2002; 367: 254.
75. Cheeseman CI. Upregulation of SGLT-1 transport activity in the rat jejunum induced by GLP-2 infusion in vivo. Am J Physiol. 1997;273: R1965–R1971.

76. Cheeseman CI, O'Neill D. Basolateral D-glucose transport activity along the crypt villus axis in rat jejunum and upregulation induced by GIP and GLP-2. Exp Physiol. 1998;83: 605–616.
77. Ferraris RP. Dietary and developmental regulation of intestinal sugar transport. Biochem J. 2001;360(Pt 2):265–276.
78. Hirsh AJ, Cheeseman CI. Cholecystokinin decreases intestinal hexose absorption by a parallel reduction in SGLT1 abundance in the brush border membrane. J Biol Chem 1998;273: 14545–14549.
79. Helliwell PA, Richardson M, Affleck J, Kellett GL. Regulation of GLUT5, GLUT2 and intestinal brush-border fructose absorption by the extracellular signal-regulated kinase, p38 mitogen-activated kinase and phosphatidylinositol 3-kinase intracellular signalling pathways: implications for adaptation to diabetes. Biochem J. 2000;350 Pt 1:163–169.
80. Mace OJ, Morgan EL, Affleck JA, Lister N, Kellett GL. Calcium absorption by Cav1.3 induces terminal web myosin II phosphorylation and apical GLUT2 insertion in rat intestine. J Physiol. 2007;580(Pt. 2):605–616.
81. Morgan EL, Mace OJ, Affleck J, Kellett GL. Apical GLUT2 and Cav1.3: regulation of rat intestinal glucose and calcium absorption. J Physiol. 2007;580(Pt. 2):593–604.
82. Nelson G, Hoon MA, Chandrashekar J, Zhang Y, Ryba NJ, Zuker CS. Mammalian sweet taste receptors. Cell. 2001;106(3):381–390.
83. Nie Y, Vigues S, Hobbs JR, Conn GL, Munger SD. Distinct contributions of T1R2 and T1R3 taste receptor subunits to the detection of sweet stimuli. Curr Biol. 2005;15(21): 1948–1952.
84. Zhao GQ, Zhang Y, Hoon MA, Chandrashekar J, Erlenbach I, Ryba NJ, Zuker CS. The receptors for mammalian sweet and umami taste. Cell 2003;115:255–266.
85. Mace OJ, Affleck J, Patel N, Kellett GL. Sweet taste receptors in rat small intestine stimulate glucose absorption through apical GLUT2. J Physiol. 2007;582(Pt 1):379–392.
86. Rosenweig NS, Herman RH. Time response of jejunal sucrase and maltase activity to a high sucrose diet in normal man. Gastroenterology. 1969;56(3):500–505
87. Diamond JM, Karasov WH, Cary C, Enders D, Yung R. Effect of dietary carbohydrate on monosaccharide uptake by mouse small intestine in vitro. J Physiol. 1984;349:419–440.
88. Ferraris RP, Diamond JM. Use of phlorizin binding to demonstrate induction of intestinal glucose transporters. J Membr Biol. 1986;94(1):77–82.
89. Ferraris RP, Villenas SA, Hirayama BA, Diamond J. Effect of diet on glucose transporter site density along the intestinal crypt-villus axis. Am J Physiol. 1992;262(6 Pt 1):G1060–1068.
90. Ferraris RP, Diamond JM. Specific regulation of intestinal nutrient transporters by their dietary substrates. Annu Rev Physiol. 1989;51:125–141.
91. Ferraris RP, Diamond JM. Crypt/villus site of substrate-dependent regulation of mouse intestinal glucose transporters. Proc Natl Acad Sci USA. 1993;90(12):5868–5872.
92. Miyamoto K, Hase K, Takagi T, Fujii T, Taketani Y, Minami H, Oka T, Nakabou Y. Differential responses of intestinal glucose transporter mRNA transcripts to levels of dietary sugars. Biochem J. 1993;295:211–215.
93. Margolskee RF, Dyer J, Kokrashvili Z, Salmon KS, Ilegems E, Daly K, Maillet EL, Ninomiya Y, Mosinger B, Shirazi-Beechey SP. T1R3 and gustducin in gut sense sugars to regulate expression of Na+-glucose cotransporter 1. Proc Natl Acad Sci USA. 2007;104(38):15075–15080.
94. Le Gall M, Tobin V, Stolarczyk E, Dalet V, Leturque A, Brot-Laroche E. Sugar sensing by enterocytes combines polarity, membrane bound detectors and sugar metabolism. J Cell Physiol. 2007;213(3):834–843.
95. Liu HY, Anderson GJ, Tsao MU, Moore BF, Giday Z. Tm glucose in a case of congenital intestinal and renal malabsorption of monosaccharides. Pediatr Res. 1967;1(5):386–394
96. Silverman M. Glucose reabsorption in the kidney. Can J Physiol Pharmacol. 1981;59(3):209–224.

97. Lee SH, Pritchard JB. Role of the electrochemical gradient for Na+ in D-glucose transport by mullet kidney. Am J Physiol. 1983;244(3):F278–288.
98. Turner RJ, Silverman M. Sugar uptake into brush border vesicles from normal human kidney. Proc Natl Acad Sci USA. 1977;74(7):2825–2829.
99. Quamme GA, Freeman H. Evidence for a high-affinity sodium-dependent D-glucose transport system in the kidney. Am. J. Physiol. 1987;253:Fl51–F157.
100. Oulianova N, Berteloot A. Sugar transport heterogeneity in the kidney: two independent transporters or different transport modes through an oligomeric Protein? 1. Glucose transport studies. J Membr Biol. 1996;153(3):181–194.
101. Turner RJ and Moran A. Heterogeneity of sodium-dependent D-glucose transport sites along the proximal tubule: evidence form vesicle studies. Am J Physiol Renal Fluid Electrol Physiol. 1982;242: F406–F411.
102. Lee WS, Kanai Y, Wells RG, Hediger MA. The high affinity Na+/glucose cotransporter. Re-evaluation of function and distribution of expression. J Biol Chem. 1994;269(16):12032–12039.
103. Kanai Y, Lee WS; You G; Brown D; Hediger MA. The human kidney low affinity Nal/glucose cotransporter SGLT2. Delineation of the major renal reabsorptive mechanism for D-glucose. J. Clin. Invest. 1994;93:397–404.
104. Wells RG, Pajor AM, Kanai Y, Turk E, Wright EM, Hediger MA. Cloning of a human kidney cDNA with similarity to the sodium-glucose cotransporter. Am J Physiol. 1992;263(3 Pt 2):F459–F465.
105. Silverman M, Speight P, Ho L. Identification of two unique polypeptides from dog kidney outer cortex and outer medulla that exhibit different Na+/D-glucose cotransport functional properties. Biochim Biophys Acta. 1993;1153(1):43–52
106. Mackenzie B, Loo DD, Panayotova-Heiermann M, Wright EM. Biophysical characteristics of the pig kidney Na+/glucose cotransporter SGLT2 reveal a common mechanism for SGLT1 and SGLT2. J Biol Chem. 1996;271(51):32678–3283.
107. Mendelssohn DC, Silverman M. A D-mannose transport system in renal brush-border membranes. Am J Physiol. 1989;257(6 Pt 2):F1100–F1107.
108. De la Horra MC, Cano M, Peral MJ, García-Delgado M, Durán JM, Calonge ML, Ilundáin AA. Na(+)-dependent D-mannose transport at the apical membrane of rat small intestine and kidney cortex. Biochim Biophys Acta. 2001;1512(2):225–230.
109. Blasco T, Aramayona JJ, Alcalde AI, Halaihel N, Sarasa M, Sorribas V. Expression and molecular characterization of rat renal D-mannose transport in Xenopus oocytes. J Membr Biol. 2000;178(2):127–135.
110. Blasco T, Aramayona JJ, Alcalde AI, Catalán J, Sarasa M, Sorribas V. D. Rat kidney MAP17 induces cotransport of Na-mannose and Na-glucose in Xenopus laevis oocytes. Am J Physiol Renal Physiol. 2003;285(4):F799–F810.
111. Dominguez JH, Camp K, Maianu L, Garvey WT. Glucose transporters of rat proximal tubule: differential expression and subcellular distribution. Am J Physiol. 1992;262(5 Pt 2):F807–F812.
112. Sugawara-Yokoo M, Suzuki T, Matsuzaki T, Naruse T, Takata K. Presence of fructose transporter GLUT5 in the S3 proximal tubules in the rat kidney. Kidney Int. 1999;56(3):1022–1028.
113. Suganuma N, Segade F, Matsuzu K, Bowden DW. Differential expression of facilitativc glucose transporters in normal and tumour kidney tissues. BJU Int. 2007;99(5):1143–1149.
114. Mate A, de la Hermosa MA, Barfull A, Planas JM, Vázquez CM. Characterization of D-fructose transport by rat kidney brush-border membrane vesicles: changes in hypertensive rats. Cell Mol Life Sci. 2001;58(12–13):1961–1967.
115. Keembiyehetty C, Augustin R, Carayannopoulos MO, Steer S, Manolescu A, Cheeseman CI, Moley KH. Mouse glucose transporter 9 splice variants are expressed in adult liver and kidney and are up-regulated in diabetes. Mol Endocrinol. 2006;20(3):686–697.

116. Augustin R, Carayannopoulos MO, Dowd LO, Phay JE, Moley JF, Moley KH. Identification and characterization of human glucose transporter-like protein-9 (GLUT9): alternative splicing alters trafficking. J Biol Chem. 2004;279(16):16229–16236.
117. Vitart V, et al. SLC2A9 is a newly identified urate transporter influencing serum urate concentration, urate excretion and gout. Nature Genet. 2008; e pub.
118. Marks J, Carvou NJ, Debnam ES, Srai SK, Unwin RJ. Diabetes increases facilitative glucose uptake and GLUT2 expression at the rat proximal tubule brush border membrane. J Physiol. 2003;553(Pt 1):137–145.
119. Goestemeyer AK, Marks J, Srai SK, Debnam ES, Unwin RJ. GLUT2 protein at the rat proximal tubule brush border membrane correlates with protein kinase C (PKC)-β1 and plasma glucose concentration. Diabetologia. 2007;50(10):2209–2217.
120. Kim EJ, Lee YJ, Lee JH, Han HJ. Effect of epinephrine on alpha-methyl-D-glucopyranoside uptake in renal proximal tubule cells. Cell Physiol Biochem. 2004;14(4–6):395–406.
121. Schaan BD, Irigoyen MC, Lacchini S, Moreira ED, Schmid H, Machado UF. Sympathetic modulation of the renal glucose transporter GLUT2 in diabetic rats. Auton Neurosci. 2005;117(1):54–61.
122. Park SH, Lee YJ, Lim MJ, Kim EJ, Lee JH, Han HJ. High glucose inhibits fructose uptake in renal proximal tubule cells: involvement of cAMP, PLC/PKC, p44/42 MAPK, and cPLA2. J Cell Physiol. 2004;200(3):407–416.
123. Meyer C, Gerich J: Role of the human kidney in glucose homeostasis. Curr Opin Endocrinol Diabetes. 2000; 7:19–24.
124. Gerich JE, Meyer C, Woerle HJ, Stumvoll M. Renal gluconeogenesis: its importance in human glucose homeostasis. Diabetes Care. 2001;24(2):382–391.
125. Benirschke K. Placentation. J Exp Zool. 1983;228(2):385–389.
126. Quraishi AN, Illsley NP. Transport of sugars across human placental membranes measured by light scattering. Placenta. 1999;20(2–3):167–174.
127. Rice PA, Rourke JE, Nesbitt REL (1976) In vitro perfusion studies of the human placenta, IV. Some characteristics of the glucose transport system in the human placenta. Gynecol Invest. 1976;7: 213–221.
128. Schneider H, Reiber W, Sager R, Malek A. Asymmetrical transport of glucose across the in vitro perfused human placenta. Placenta. 2003;24(1):27–33.
129. Barros LF, Yudilevich DL, Jarvis SM, Beaumont N, Baldwin SA. Quantitation and immunolocalization of glucose transporters in the human placenta. Placenta. 1995;16(7):623–633.
130. Shin BC, Suzuki T, Matsuzaki T, Tanaka S, Kuraoka A, Shibata Y, Takata K. Immunolocalization of GLUT1 and connexin 26 in the rat placenta. Cell Tissue Res. 1996;285(1):83–89.
131. Takata K. Glucose transporters in the transepithelial transport of glucose. J Electron Microsc (Tokyo). 1996;45(4):275–284.
132. Zhou J, Bondy CA. Placental glucose transporter gene expression and metabolism in the rat. J Clin Invest. 1993;91(3):845–852.
133. Vardhana PA, Illsley NP. Transepithelial glucose transport and metabolism in BeWo choriocarcinoma cells. Placenta. 2002;23(8–9):653–660.
134. Barash V, Erlich T, Bashan N. Microsomal hexose-6-phosphate and 6-phosphogluconate dehydrogenases in extrahepatic tissues: human placenta and pig kidney cortex. Biochem Int. 1990;20(2):267–274.
135. Matsubara S, Takizawa T, Sato I. Glucose-6-phosphatase is present in normal and pre-eclamptic placental trophoblasts: ultrastructural enzyme-histochemical evidence. Placenta. 1999;20(1):81–85.
136. Leonce J, Brockton N, Robinson S, Venkatesan S, Bannister P, Raman V, Murphy K, Parker K, Pavitt D, Teoh TG, Regan L, Burchell A, Steer P, Johnston DG. Glucose production in the human placenta. Placenta. 2006;27 Suppl A:S103–108.
137. Hay WW Jr, Molina RA, DiGiacomo JE, Meschia G. Model of placental glucose consumption and glucose transfer. Am J Physiol. 1990;258(3 Pt 2):R569–577.

138. Thiéry G, Bernier J, Bergeron M. A new method based on cobalt for histochemical and cytochemical demonstration of glucose-6-phosphatase activity. J Histochem Cytochem. 1990;38(10):1503–1509.
139. Hauguel-De Mouzon S, Challier J, Kacemi A, Cauzac M, Malek A, Girard J. The GLUT3 glucose transporter isoform is differentially expressed within human placental cell types. J Clin Endocr Metab. 1997;82:2689–2694.
140. Desoye G, Hartmann M, Blaschitz A, Dohr G, Hahn T, Kohnen G, Kaufmann P. Insulin receptors in syncytiotrophoblast and fetal endothelium of human placenta. Immunohistochemical evidence for developmental changes in distribution pattern. Histochem. 1994;101: 277–285.
141. Limesand SW, Regnault TR, Hay WW Jr. Characterization of glucose transporter 8 (GLUT8) in the ovine placenta of normal and growth restricted fetuses. Placenta. 2004;25(1):70–77.
142. Illsley NP. Glucose transporters in the human placenta. Placenta. 2000;21(1):14–22.
143. Augustin R, Carayannopoulos MO, Dowd LO, Phay JE, Moley JF, Moley KH. Identification and characterization of human glucose transporter-like protein-9 (GLUT9): alternative splicing alters trafficking. J Biol Chem. 2004;16:279(16):16229–16236.
144. Dawson PA, Mychaleckyj JC, Fossey SC, Mihic SJ, Craddock AL, Bowden DW. Sequence and functional analysis of GLUT10: a glucose transporter in the Type 2 diabetes-linked region of chromosome 20q12–13.1. Mol Genet Metab. 2001;74(1–2):186–199.
145. Scheepers A, SchMidt S, Manolescu A, Cheeseman CI, Bell A, Zahn C, Joost HG, Schürmann A. Characterization of the human SLC2A11 (GLUT11) gene: alternative promoter usage, function, expression, and subcellular distribution of three isoforms, and lack of mouse orthologue. Mol Membr Biol. 2005;22(4):339–351.
146. Gude NM, Stevenson JL, Murthi P, Rogers S, Best JD, Kalionis B, King RG. Expression of GLUT12 in the fetal membranes of the human placenta. Placenta. 2005;26(1):67–72.
147. Gude NM, Stevenson JL, Rogers S, Best JD, Kalionis B, Huisman MA, Erwich JJ, Timmer A, King RG. GLUT12 expression in human placenta in first trimester and term. Placenta. 2003;24(5):566–570.
148. Macheda ML, Kelly DJ, Best JD, Rogers S. Expression during rat fetal development of GLUT12 – a member of the class III hexose transporter family. Anat Embryol (Berl). 2002;205(5–6):441–452.
149. Carayannopoulos MO, Chi MM, Cui Y, Pingsterhaus JM, McKnight RA, Mueckler M, Devaskar SU, Moley KH. GLUT8 is a glucose transporter responsible for insulin-stimulated glucose uptake in the blastocyst. Proc Natl Acad Sci USA. 2000;97(13):7313–7318.
150. Pinto AB, Carayannopoulos MO, Hoehn A, Dowd L, Moley KH. Glucose transporter 8 expression and translocation are critical for murine blastocyst survival. Biol Reprod. 2002;66(6):1729–1733.
151. Forssmann WG, Metz J, Heinrich D 1975 Gap junctions in the hemotrichorial placenta of the rat. J Ultrastruct Res. 1975;53:374–381.
152. Takata K, Fujikura K, Shin B-C Ultrastructure of the rodent placental labyrinth: a site of barrier and transport. J Reprod Develop. 1997;43:13–24.
153. Shin BC, Fujikura K, Suzuki T, Tanaka S, Takata K. Glucose transporter GLUT3 in the rat placental barrier: a possible machinery for the transplacental transfer of glucose. Endocrinology. 1997;138(9):3997–4004.
154. Ganguly A, McKnight RA, Raychaudhuri S, Shin BC, Ma Z, Moley K, Devaskar SU. Glucose transporter isoform-3 mutations cause early pregnancy loss and fetal growth restriction. Am J Physiol Endocrinol Metab. 2007;292(5):E1241–1255.

155. Kevorkova O, Ethier-Chiasson M, Lafond J. Differential expression of glucose transporters in rabbit placenta: effect of hypercholesterolemia in dams. Biol Reprod. 2007;76(3): 487–495.
156. Fanconi G, Bickel H. Die chronische Aminoacidurie (Aminosäurediabetes oder nephrotisch-glukosurischer Zwergwuchs) bei der Glykogenose und der Cystinkrankheit. Helv Paediatr Acta 1949;4:359–396.
157. Santer R, Groth S, Kinner M, Dombrowski A, Berry GT, Brodehl J, Leonard JV, Moses S, Norgren S, Skovby F, Schneppenheim R, Steinmann B, Schaub J. The mutation spectrum of the facilitative glucose transporter gene SLC2A2 (GLUT2) in patients with Fanconi-Bickel syndrome. Hum Genet. 2002;110(1):21–29.
158. Booth IW, Patel PB, Sule D, Brown GA, Buick R, Beyreiss K. Glucose-galactose malabsorption: demonstration of specific jejunal brush border membrane defect. Gut. 1988;29(12):1661–1665.
159. Desjeux JF, Wright EM. Thirty years of research on congenital glucose and galactose malabsorption: from phenotype to genotype. Bull Acad Natl Med. 1993;177(1):125–131.
160. Lam JT, Martín MG, Turk E, Hirayama BA, Bosshard NU, Steinmann B, Wright EM. Missense mutations in SGLT1 cause glucose-galactose malabsorption by trafficking defects. Biochim Biophys Acta. 1999;1453(2):297–303.
161. Wright EM, Turk E, Martin MG. Molecular basis for glucose-galactose malabsorption. Cell Biochem Biophys. 2002;36(2–3):115–121.
162. Fehr M, Takanaga H, Ehrhardt DW, Frommer WB. Evidence for high-capacity bidirectional glucose transport across the endoplasmic reticulum membrane by genetically encoded fluorescence resonance energy transfer nanosensors. Mol Cell Biol. 2005;25(24):11102–11112.

Amino Acid Transport by Epithelial Membranes

Bruce R. Stevens

Abstract This review summarizes the current view of amino acid transport by epithelial cells of vertebrates. A wide variety of transporter proteins are expressed in apical and basolateral membranes and collectively play complex interactive roles in controlling the entire organism's overall metabolism of amino acids. Regulation of the transport systems can be manifested at many levels, including gene splicing and promoter regulation, interactions between requisite subunits of oligomers, thermodynamic electrochemical gradients contributed by ion exchangers, overlap of substrate specificity, selective tissue distribution, and specific spatial distribution of transporters leading to net vectorial flow of the amino acids. The next frontier for workers in this field is to uncover a comprehensive molecular understanding of the manner by which epithelial cells signal gene expression of transporters as triggered by substrates, hormones or other triggers, in order to further understand the trafficking and interactions among multimeric transport system proteins, to extend discoveries of novel small drug substrates for oral and ocular delivery, and to examine gene therapy or nanotherapy of diseases using small molecules delivered via amino acid transporters.

Keywords Membrane Transporter · Amino acid · Epithelial · Epithelium

1 Text

Transporters of epithelial membranes arguably represent a vertebrate organism's most significant collective array of regulatory mechanisms that control the intermediary metabolism of amino acids. Such transporters control metabolic reactions by regulating throughput of substrates and products, and creating steady state pools of metabolites. Epithelial membrane transporters modulate metabolic flows and

B.R. Stevens (✉)
Department of Physiology and Functional Genomics, College of Medicine, University of Florida, Gainesville, FL 32610-0274, USA
e-mail: stevensb@ufl.edu

G.A. Gerencser (ed.), *Epithelial Transport Physiology*,
DOI 10.1007/978-1-60327-229-2_15,

pools of both the proteinogenic and the nonproteinogenic amino acids. Working in conjunction with one another, amino acid transporters maintain coordination and integrity of metabolic events within tissues and even between organs.

This chapter reviews the current understanding of amino acid transport systems and transporters in basolateral membranes and apical membranes of epithelial cells in humans and other vertebrates. The chapter content is focused on transporters and transport systems that serve amino acids which are either nonproteinogenic or proteinogenic nutrients and metabolites, although some transporter genes and polypeptides are technically subsets categorized within larger groupings that are structurally and mechanistically related to the neurotransmitter transporters [1]. Also, many apparently unrelated so-called transporter "orphan" genes have been historically clustered as a single branch off the SLC6 neurotransmitter superfamily, although they are now emerging as actually being epithelial amino acid transporters related in structure, function, and probably oligomeric association in the membrane [2, 3].

This chapter attempts to catalog the body of literature encompassing the epithelial barriers of the kidney, gastrointestinal tract, lung, cornea, liver, gallbladder, and other tissues. The reader is directed to several fine reviews [4–8] that delve into other details of renal and intestinal absorptive epithelial membranes in physiological processes or their involvement in absorptive pharmacokinetics [9] of monomer amino acids or natural dipeptides and drug analogues [10, 11].

The concept of membrane transport physiology involves physical entities of discrete transporter genes and expressed polypeptides, in addition to the interactions and functionality of these units. "Transporter" and "transport system" are colloquially – although incorrectly – thought of as one and the same entity. Although a transporter's functional activity can indeed occur by the action of a single species of monomeric transporter protein, there are many instances of amino acid transport functional activity that are catalyzed only by the concerted interaction of multiple subunit proteins in the membrane. This latter scenario is termed a "transport system," which is comprised of multimeric arrangement of transporter proteins, usually as heterodimers.

2 Monomeric Transporter Proteins, Genes, and Transport Systems

Table 1 summarizes the known monomeric transporter proteins involved in epithelial regulation of amino acid movement across apical and basolateral membranes. Although most work has been done for small intestinal and renal transport, representation occurs for colon, liver, lung, cornea, adipose, testis, stomach, choroid plexus, and placenta. There is evidence for some SLC6 family members to form oligomers/homodimers [12, 13] in nonepithelial membranes, but this has not been confirmed for epithelial cell membranes. Further studies in this arena may indeed uncover that all SLC6 members form oligomers.

Table 1 Monomeric amino acid transporters and transport systems in epithelial membranes. Transporters are grouped by SLC (solute carrier) gene families [1], utilizing Human Genome Organization (HUGO) Nomenclature Committee classifications [14]. Functional transport systems are named using the Christensen naming plan [15]. Expression and localization derived from the Swedish Human Proteome Resource [69]. There is evidence for some SLC6 family members to form oligomers/homodimers [12, 13] in nonepithelial membranes, but this has not been established for epithelial cells

Transport "system" functional name	Monomer protein common alias	Gene	Human gene locus	Sequence accession ID	Representative substrates	Ion dependency	Tissue epithelium	Membrane location
SLC1 family								
X_{AG}^-	EAAT3	SLC1A1	9q24	NM_004170	L-glutamate, D/L-aspartate, cystine (SS)	H^+, Na^+, K^+	Kidney (K), Intestine (I), liver	Apical
X_{AG}^-	EAAT2	SLC1A2	11p13–p12	NM_004171	L-glutamate, D/L-aspartate, cystine (SS)	H^+, Na^+, K^+	Liver	Apical
ASC	ASCT1	SLC1A4	2p13–p15	NM_003038	Alanine, serine, threonine, cysteine, glutamine	Na^+	Lung, stomach, K, I, cornea	Apical
ASC	ASCT2 or ATB^0	SLC1A5	19q13.3	NM_005628	Alanine, serine, threonine, cysteine, glutamine, branched neutrals	Na^+	K, I, lung, colon	Apical
X_{AG}^-	EAAT5	SLC1A7	1p32.3	NM_006671	L-Glutamate, D/L-Aspartate	H^+, Na^+, K^+	Retina	Apical
SLC6 family								
5-HTT	SERT	SLC6A4	17q11.1–q12	NM_001045	Serotonin, carnitine, organic cations	Na+, Cl^-, K^+		

Table 1 (continued)

Transport "system" functional name	Monomer protein common alias	Gene	Human gene locus	Sequence accession ID	Representative substrates	Ion dependency	Tissue epithelium	Membrane location
β (beta)	TAUT	SLC6A6	3p25–p24	NM_003043	Taruine, β-alanine, GABA	Na^+, Cl^-	K, retina, placenta	Basolateral
Creatine	CRTR	SLC6A8	Xq28	NM_005629	Creatine	Na^+, Cl^-	K, I, retina	Apical
GLY	GLYT1	SLC6A9	1p33	NM_201649	Glycine	Na^+, Cl^-	Lung, stomach retina	Basolateral
GAT2	GAT2	SLC6A13	12p13.3	NM_016615	GABA	Na^+, Cl^-	K, retina, choroid plexus	
$B^{0,+}$	$ATB^{0,+}$	SLC6A14	Xq23–q24	NM_007231	Neutrals and dibasics, arginine, D-serine	Na^+, Cl^-	Colon	Apical
GLY	XTRP2	SLC6A18	5p15.33	NM_182632	Glycine	Na^+, Cl^-	K	Apical
B^0 (or B)	B^0AT1	SLC6A19	5p15.33	NM_001003841	Neutrals, glutamine	Na^+	K, I, placenta	Apical
IMINO	SIT1	SLC6A20	3p21.6	NM_020208	Proline, sarcosine, pipecolate	Na^+	K, I, choroid plexus, stomach, colon	Apical
SLC7 family								
y^+	CAT-1	SLC7A1	13q12–q14	NM_003045	Arginine, ornithine, lysine, histidine, dibasics	None	Universal	Basolateral

Table 1 (continued)

Transport "system" functional name	Monomer protein common alias	Gene	Human gene locus	Sequence accession ID	Representative substrates	Ion dependency	Tissue epithelium	Membrane location
SLC15 family								
Pept1	PEPT1	SLC15A1	13q33–q34	NM_005073	Dipeptides and tripeptides; carnosine, β-lactam antibiotics, angiotensin converting enzyme inhibitors	H^+ with NHE3	I, K	Apical
Pept2	PEPT2	SLC15A2	3q13.3–q21	NM_021082	Dipeptides and tripeptides	H^+ with NHE3	I, K, lung, mammary gland	Apical
PTR4	PHT1	SLC15A4	12q24.32	NM_145648	Histidine, di-, tri-peptides	H^+	Retina	Apical
SLC16 family								
T	TAT1	SLC16A10	6q21–q22	NM_018593	Aromatics, L-DOPA	–	K, I, placenta	Basolateral
SLC22 family								
OCTN2VT	OCTN2	SLC22A5	5q23.3	NM_003060	L-carnitine, acetyl-L-carnitine	Na^+	I, K, lung, liver	Apical

Table 1 (continued)

Transport "system" functional name	Monomer protein common alias	Gene	Human gene locus	Sequence accession ID	Representative substrates	Ion dependency	Tissue epithelium	Membrane location
SLC36 family								
Iminoacid	PAT1	SLC36A1	5q33.1	AF516142	Proline, glycine, b-alanine, GABA, taurine, D-serine	H^+ with NHE3	K, I, liver, lung, colon	Apical
Iminoacid	PAT2	SLC36A2	5q33.1	AY162214	Proline, glycine, L-alanine	H^+ with NHE3	K, Lung	Apical
SLC38 family								
A	SNAT1	SLC38A1	12q12–q13.11	NM_0030674	Glutamine alanine, asparagine, cysteine, histidine, serine	Na^+	Placenta, retina	Apical
A	SNAT2	SLC38A2	12q	NM_018976	Alanine, asparagine, cysteine, glutamine, glycine, histidine, methionine, proline, serine	Na^+	I, K, placenta, lung, liver	Basolateral
N	SNAT3, SN1	SLC38A3	3p21.3	NM_006841	Glutamine, histidine	Na^+, H^+	K, retina	Basolateral

Table 1 (continued)

Transport "system" functional name	Monomer protein common alias	Gene	Human gene locus	Sequence accession ID	Representative substrates	Ion dependency	Tissue epithelium	Membrane location
A	SNAT4	SLC38A4	12q13	NM_018018	Alanine, asparagine, cysteine, glycine, threonine	Na^+	K, liver, placenta	Basolateral
N	SNAT5	SLC38A5	Xp11.23	NM_033518	Glutamine, histidine, serine, asparagine, alanine	Na^+ , H^+	I	
SLC43 family								
LAT3	POV1	SLC43A1	11p11.2	NM_003627	Branched chain amino acids, phenylalanine	None	K	
LAT4	LAT4	SLC43A2	17p13.3	NM_152346	Branched chain amino acids, phenylalaninine	None	I, K, placenta	Basolateral

The components of Table 1 are based on the Human Genome Organization (HUGO) Nomenclature Committee classification nomenclature [14] for transporters as SLC families, with gene family refinement by Hediger et al. [1] Transport "system" functional names are derived from the original nomenclature of Christensen [15] as extended by many laboratories over the past several decades. Note that SLC genes can be grouped by the function of their expression product. In Table 1 a given physiologically functional transporting system results from the activity of a single polypeptide expression product trafficked to the appropriate membrane.

So-called historically "orphan transporters" are now emerging as actually being clustered related amino acid transporters of the SLC6 family. The dendogram in Fig. 1 shows the mouse SLC6 series of transporter proteins. The human creatine transporter-2 (SLC6A10) apparently does not have an ortholog correlate in mouse.

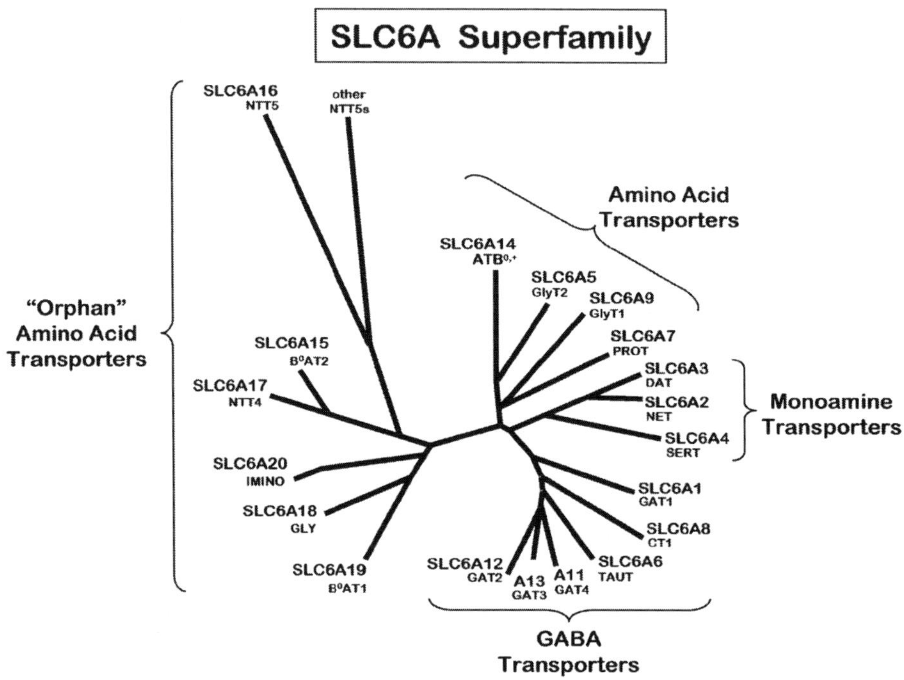

Fig. 1 SLC6 family clustering of amino acid monomer transporters. Note that many apparently unrelated transporters have been historically clustered as a major branch of "orphans," although their transporter polypeptide structures and possible oligomeric relationships are currently being uncovered as related. In this dendogram, the human gene SLC number is listed above the common name for each polypeptide expression product. Adapted from laboratories of Broer [2] and Verrey [3]

One of the "orphan" family genes, SLC6A20, encodes the SIT1 transporter serving proline and imino analogues via the sodium-dependent "IMINO" system in the intestine (IMINO has been characterized in[16]). In contrast, however, in the kidney and lung these substrates are served by the proton-activated "PAT/iminoacid"

Fig. 2 Genomic exon/intron organization of the genes SLC36A1 (encoding PAT1) and SLC36A2 (encoding PAT2) on human chromosome 5q33.1. These genes are flanked by genes *FAT2* and *GM2A*, shown for orientation. The mouse orthologs are located on chromosome 11B1.3. Putative exons are represented by *vertical lines*. Adapted from Boll et al. [66]

system which exhibits the activity profiles of two expressed transporter proteins PAT1 and PAT2, as encoded by the SLC36 family genes SLC36A1 and SLC36A2 (see Table 1). The latter genes reside on chromosome 5q33.1 in humans or 11B1.3 in mice, as shown in Fig. 2. The human and mouse open reading frames are each encoded by ten exons. This figure shows an example of the genomic exon/intron organization of the SLC36 genes responsible for tissue-dependent differential expression of the transporters.

3 Heterodimer Regulation of Transport

For several transporting systems, a multimeric arrangement of polypeptides is required for physiological activity. Table 2 summarizes the well-established heterodimers required for these cases. Note that 4F2hc (alternative name CD98) is utilized as a subunit in many of these systems.

Subunit polypeptides of amino acid transporters play multiple roles in nonepithelial cell types, and it is likely that in epithelial membranes they also serve multiple roles. For example, CD98/4F2hc at the basolateral membrane of intestinal epithelium contains a PDZ class II-binding domain [17] regulating integrin signaling, serving in such diverse roles as inflammation modulation, cell growth via mTOR [4, 18], cell adhesion, migration, and binding rotavirus enterotoxin [19]. 4F2hc (CD98) forms a heterodimer with amino acid transporter subunits in basolateral membranes – asc, LAT1, LAT-2, y^+LAT1, y^+LAT-2, and xCT – thereby resulting in six different functional transport systems (Table 2). Burdo et al. [20] reported evidence for xCT and 4F2hc in apical membranes of mouse and primate duodenum and renal tubules, as well as in choroid plexus although the particular membrane aspect was not specified for this brain tissue. In situ heterodimerization of y^+LAT1 with 4F2hc has been visualized using acceptor photobleaching FRET microscopy [21]. Figure 3 shows the diverse ways by which the 4F2hc subunit plays a role in modulating amino acid transport at the basolateral membrane.

In addition to these 4F2hc basolateral systems, a separate dimerization coupling occurs at the apical surface which involves $b^{0,+}$AT plus rBAT (Table 2). Extensive descriptions of models of the apical membrane heterodimer comprised of $b^{0,+}$AT plus rBAT are presented elsewhere [5, 6, 22–26].

Table 2 Heterodimeric amino acid transporters and transport systems in epithelial membranes. Transporter SLC (solute carrier) gene family names [1] are shown with Human Genome Organization (HUGO) Nomenclature Committee classifications [14]. Functional transport systems are named using the Christensen naming plan [15]. Expression and localization derived from the Swedish Human Proteome Resource [69]. The heavy chain 4F2hc subunit is synonymously called CD98

Transport "System" functional unit name	Heterodimer subunits; protein common aliases	Subunit genes	Human gene locus	Sequence accession ID	Representative substrates	Ion dependency	Tissue epithelium	Membrane location
Asc	asc1 plus 4F2hc	SLC7A10 plus SLC3A2	19q12–13.1 11q13	NM_019849 NM_002394	Small neutral L- and D-amino acids, alanine, D-serine, cyst(e)ine, glycine	None		Basolateral
$b^{0,+}$	$b^{0,+}$AT plus rBAT	SLC7A9 plus SLC3A1	19q13.1 2p16.3–p21	NM_014270 NM_000341	Dibasics, arginine, cystine, large neutrals (exchange extracellular dibasics with intracellular neutrals)	None	I, K, placenta	Apical
L	LAT1 plus 4F2hc	SLC7A5 plus SLC3A2	16q24.3 11q13	NM_003486 NM_002394	Branched chain neutrals, L-DOPA	None	I, placenta	Basolateral
L	LAT2 plus 4F2hc	SLC7A8 Plus SLC3A2	14q11.2 11q13	NM_012244 NM_002394	Branched chain neutrals (small and large)	None	I, K, placenta, stomach	Basolateral

Table 2 (continued)

Transport "System" functional unit name	Heterodimer subunits; protein common aliases	Subunit genes	Human gene locus	Sequence accession ID	Representative substrates	Ion dependency	Tissue epithelium	Membrane location
x_c^-	xCT plus 4F2hc	SLC7A11 plus SLC3A2	4q28–q32 11q13	NM_014331 NM_002394	Cystine/glutamate exchange	None	Choroid plexus, intestine, kidney	Basolateral (evidence in apical)
y^+L	y^+LAT1 plus 4F2hc	SLC7A7 plus SLC3A2	14q11.2 11q13	NM_003982 NM_002394	Lysine, arginine, dibasics, neutrals	None for dibasics; Na^+ for large neutrals	I, K, lung	Basolateral
y^+L	y^+LAT2 plus 4F2hc	SLC7A6 plus SLC3A2	16q22.1–22.2 11q13	NM_003983 NM_002394	Lysine, arginine, dibasics, neutrals	None for dibasics; Na^+ for large neutrals	I, K, lung	Basolateral

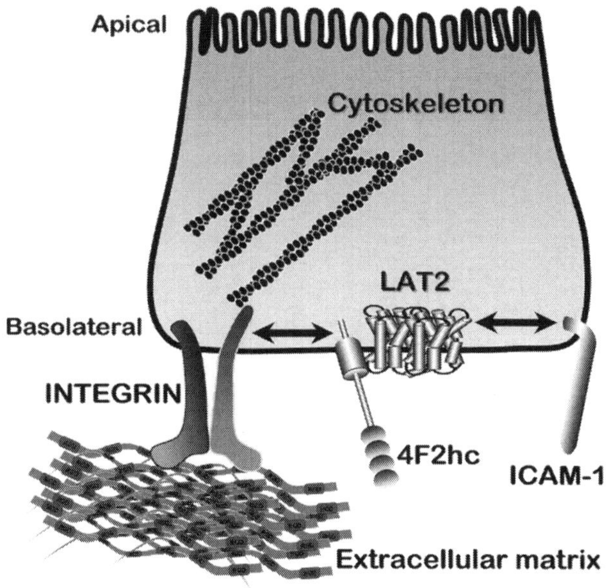

Fig. 3 Interactions of subunit regulation of amino acid transport in intestinal basolateral membranes. Various example complexes are shown, including heterodimerization with LAT2, β_1-integrins, and ICAM-1. 4F2hc (synonymous with CD98) modulates β_1-integrin activation as well as activity of amino acid transport via LAT2. Furthermore, another polypeptide, ICAM-1, could independently modulate amino-acid transport activities via LAT2 at its extracellular C-terminal domain that contains a PDZ-binding domain. Adapted from Yan et al. [17, 67]

Mineralocorticoid receptors regulate at least some intestinal transporters. Work from Soares-da-Silva's laboratory [27] demonstrates that chronic systemic aldosterone administration, acting through a spironolactone-sensitive mechanism, regulates rat intestinal LAT2 and 4F2hc subunit mRNA and protein expression as well as ASCT2 and LAT1 protein levels.

4 Dipeptide Transporters

Transepithelial movement of amino acids occurs by highly efficient transport of di- and tri-peptides across the apical membrane, with cytosolic hydrolysis by aminopeptidases then final basolateral transport of the constituent free amino acids. Proton-dependent secondary active transporters PEPT1 (SLC15A1) and PEPT2 (SLC15A2) are expressed in epithelia of small intestine, renal tubules, liver, and gallbladder. These transporters serve natural dipeptides, and are increasingly exploited for their ability to absorb certain oral pharmaceuticals as substrates [10]. Some examples of substrates accommodated by PEPT1 are shown in Table 3. Brandsch and colleagues [10] have extensively described peptidomimetic pharmaceutical substrates of PEPT1 and PEPT1.

Table 3 Natural peptides and pharmaceuticals interacting with PEPT1. Affinity constants K_i were obtained in Caco-2 cell uptake competition assays. Adapted from Brandsch et al. [10]

PEPT1 substrate or inhibitor	K_i(mM)	Affinity range (relative K_i)
Lys[ZNO_2)]–Val	0.002	High affinity (<0.5 mM)
Alafosfalin	0.19	
Ala-Lys	0.21	
Ceftibuten	0.34	
Valaciclovir	0.49	
Gly-Sar	1.1	Medium affinity (0.5–5 mM)
Pro–Pro	1.2	
δ-Aminolevulinic acid	1.5	
Cloxacillin	3.0	
Lys–Lys	3.4	
D-Ala-Lys	7.0	Low affinity (5–15 mM)
Cefadroxil	7.2	
Pro-Ala	9.5	
4-Aminophenylacetic acid	14	
Cephalexin	14	

It has become increasingly apparent that natural biological systems, PEPT1 and PEPT2 play roles other than simply nitrogen absorption for the host organism. For example, although PEPT1 is not normally expressed in colon, its expression is induced at the transcriptional level in pathological inflamed states of colonic epithelial cells in inflammatory bowel disorders such as ulcerative colitis or Crohn′s disease [28]. In this condition, bacterial metabolite dipeptides provide PEPT1 substrates absorbed intact into colonic cells and subsequently into the colonic lamina propria where immune cells reside [28]. *Escherichia coli* pathogenic peptide metabolites include Ac-muramyl-Ala-Glu and N-formylmethionyl-leucyl-phenylalanine. This triggers an inflammatory response by induction of MHC I molecules and AP1 and NFκB-related pathways leading to cytokines and macrophage infiltration. In the inflamed state, the sustained luminal acidic state further drives colonic transepithelial transport of the bacterial peptides. The distribution of PEPT1 in various regions along the gut is shown in Fig. 4.

5 Transepithelial Transport of Amino Acids

Transepithelial movement of amino acids requires communication between events at the apical and basolateral membrane of a given tissue epithelium (Fig. 5). The lion's share of amino acid transport occurs in the intestine and the renal nephron. Apical membrane transporters of small intestine and large intestine handle the fundamental interface between the entire internal milieu and the external world of nutrient nitrogen, while the kidney rescues filtered monomer amino acids as well as di- and tri-peptides in the form of the peptides' constituent free amino acids.

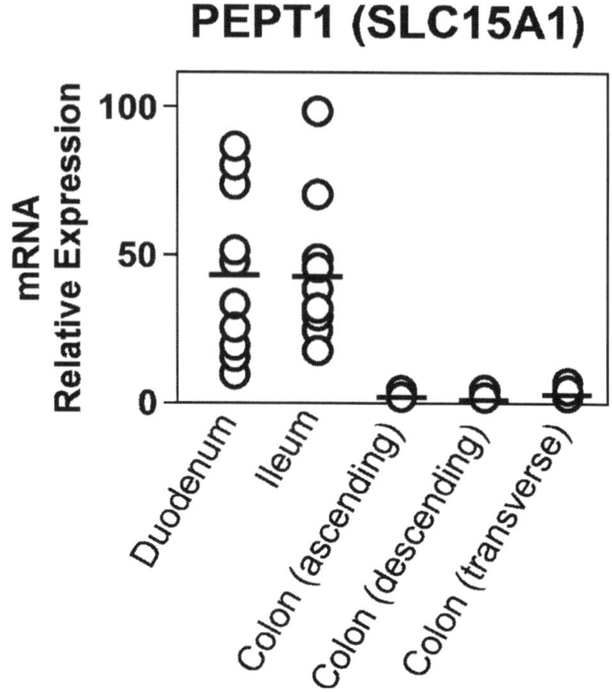

Fig. 4 Relative mRNA expression of PEPT1 (SLC15A1) in various regions of the gastrointestinal tract. Data points from individual patients. Means are shown as *horizontal bars* for each intestinal region. Adapted from data of Meier et al. [68]

Epithelial membrane systems of other tissues throughout the body provide a means of communication linking metabolic events among various organ systems.

6 Thermodynamic Coupling

Secondary active transport of amino acids exploit epithelial transmembrane electrochemical gradients of Na^+ and H^+, and K^+. As demonstrated by examples in Fig. 6, the activation can occur by Na^+ alone, H^+ alone, Na^+ plus H^+, or by coordinated coupling of Na^+ and H^+ gradient by two different transport systems. In the case of indirect H^+ coupling, apical NHE sodium/proton exchangers are the predominant players. In all cases, the Na^+/K^+ ATPase regulates Na^+, K^+, and electrical gradients sensed by the apical and basolateral sides. Gerencser and Stevens [29, 30] provide a quantitative explanation of thermodynamic coupling regarding amino acid transport.

7 Coordinated Thermodynamic and Metabolic Communication Among Amino Acid Transport Systems

Transport systems cooperate in a concerted manner to bring about net transepithelial transport, in addition to modulating the entire metabolism of amino acids.

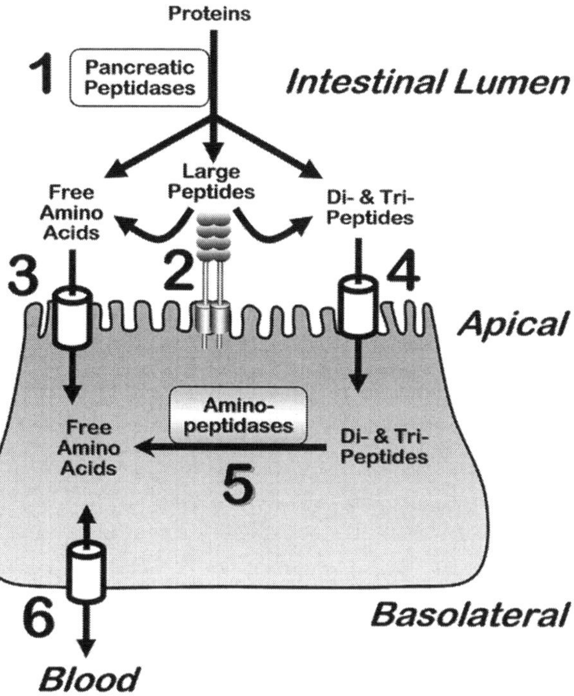

Fig. 5 The fundamental concept of nutrient nitrogen digestion and absorption by small intestinal epithelium enterocytes. Luminal pancreatic peptidases (1) generate products of free amino acids, small di- and tri-peptide fragments, and residual large polypeptides. Enterocyte apical membrane-bound aminopeptidases (2) further hydrolyze large peptides to yield di- and tri-peptides. A variety of amino acid transport systems (3) move free amino acids across the apical membrane into the enterocyte cytosol. Certain di- and tri-peptides are absorbed intact into the cytosol via apical membrane transporters (4). Subsequently, cytosolic aminopeptidases (5) hydrolyze these peptides to generate free amino acids. The pool of free amino acids is then transported (6) across the basolateral membrane to the serosal side of the epithelium, and subsequently into the portal blood. In the interdigestive postprandial state, amino acids are taken up by the basolateral membrane from the blood. A variety of transporters are responsible for steps 3, 4, and 6, as detailed throughout the text and tables. Analogous events occur in the kidney lumen relating to degradation of peptide hormones, rescue of amino acids, and whole body nitrogen regulation

Membrane ion exchangers, primarily H^+/Na^+ electroneutral exchanger NHE3 and Na^+/K^+ ATPase, interject their important roles in modulating the activity of secondary active transporters of amino acids. Figure 7 demonstrates the coordination of selected transporter proteins and their functional transport systems in epithelial metabolism and movement of amino acids concerning the cytosol, and apical and basolateral membranes. In Fig. 7 note the coordination of TAT1 with LAT2/4F2hc which allows efflux of small neutral amino acids at the expense of cycling of aromatic amino acids [31]. Also note the roles of the many transporter subunits involved in coupling and regulating movements of cationic and neutral amino acids. Table 4

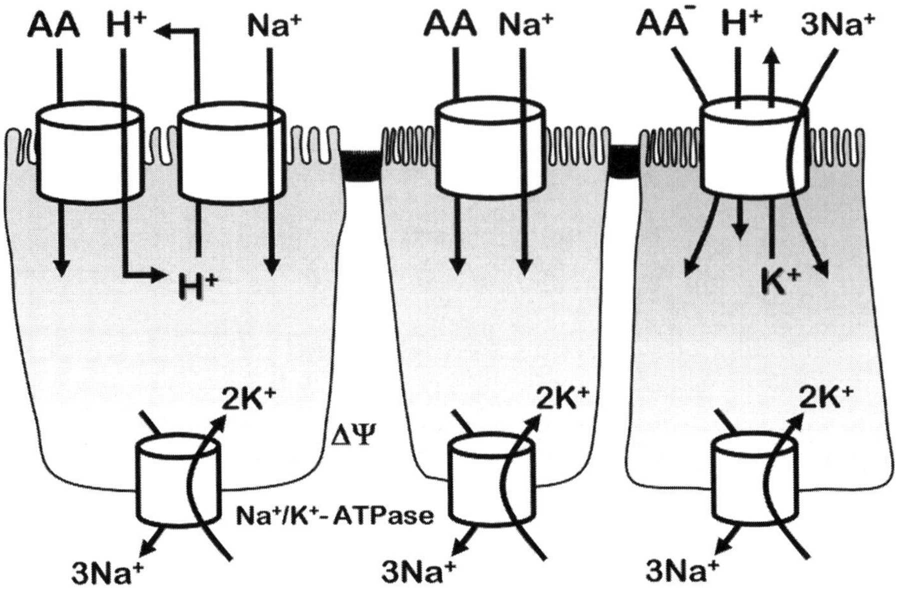

Fig. 6 Schematic examples of ion-dependent coupling in amino acid transport at the apical membrane of an epithelium. *Left cell* shows a generic proton-dependent amino acid uptake coupled with Na^+/H^+ exchanger NHE3 (SLC9A3). The *center cell* shows a generic transporter coupling amino acid influx with Na^+ influx, activated by the Na^+ electrochemical potential across the apical membrane. The *right cell* shows active transport of anionic amino acids (e.g., glutamate or aspartate) via EAAT3 (SLC1A1) coupling to electrochemical gradients of Na^+, H^+, and K^+. In all cases, transepithelial and epithelial cell membrane ion gradients and electrical potentials ($\Delta\Psi$) ultimately depend on the basolateral membrane Na^+/K^+-ATPase activity

describes the manner by which the selectivities of all the cationic/dibasic amino acid transport systems are determined experimentally.

8 Nonproteinogenic Creatine in the Intestinal Epithelium

Creatine is an important amino acid that serves several critical roles in intestinal epithelial transport physiology and intermediary metabolism. A metabolic interplay exists between creatine absorption, arginine and leucine absorption, and gross control of epithelial absorption of other nutrients in these tissues [32–35]. A putative model of creatine (Cr^+) absorption in renal or intestinal absorptive epithelium is shown in Fig. 8. Na^+ and Cl^--dependent creatine transport occurs across apical membranes and possibly basolateral membranes. In addition to transepithelial absorption, creatine enhances transepithelial absorption of other nutrients (glucose, other amino acids, etc) via villus contraction and pedundular motility through the actions of enterocyte actin-myosin as the result of entyerocyte creatine kinase phosphorylation. Intracelluar trafficking of SLC6A8 proteins sequested in cytosolic endosomal vesicles is intiated by mammalian targeting of the rapamycin (mTOR)

Table 4 Determination of epithelial membrane cationic/dibasic amino acid transport systems' functional activities. Unlabeled amino acids in assay media are at concentrations >100X concentration of [^{3}H]L-arginine. Modified from Bae et al. [70]

Transport "system" functional activity name	Assay conditions; uptake of [^{3}H]L-arginine	Parallel transport systems potentially involved with assay conditions	Calculating [^{3}H]L-arginine uptake via single target transport system	Protein subunits giving target transport system activity in epithelium	Genes encoding subunits
$B^{0,+}$	(a) Na^+ , Cl^-	y^+ ; y^+L ; $b^{0,+}$; $B^{0,+}$	$B^{0,+} = (a) - (b)$	$ATB^{0,+}$	SLC6A14
$b^{0,+}$	(b) Na^+ free	y^+ ; y^+L ; $b^{0,+}$	$b^{0,+} = (b) - (c)$	$b^{0,+}AT$ plus rBAT	SLC7A9 plus SLC3A1
y^+L	(c) Na^+ free plus L-alanine	y^+ ; y^+L	$y^+L = (c) - (d)$	y^+LAT1 plus 4F2hc; or y^+LAT2 plus 4F2hc	SLC7A6 plus SLC3A2;or SLC7A7 plus SLC3A2
y^+	(d) Na^+ plus L-leucine	y^+	$y^+ = (d)$	CAT-1	SLC7A1

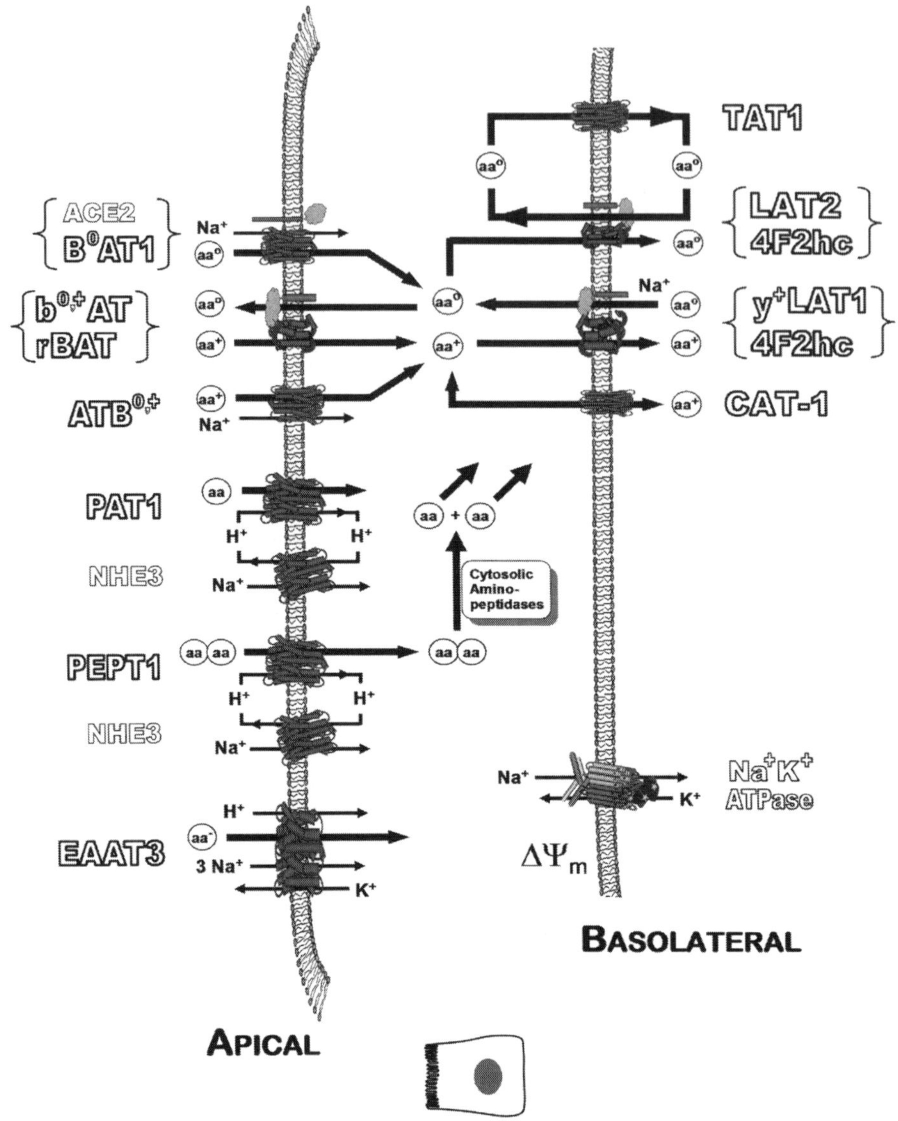

Fig. 7 Coordinated thermodynamic and metabolic communication among selected amino acid transport systems in epithelial membranes. Amino acids are shown as cationic (AA^+), zwitterion neutral (AA^O), anionic (AA^-), or generic (AA). Ion exchangers interject their important roles in modulating the secondary active transporters of amino acids. Various mechanisms maintain pH < 7 at the immediate apical membrane surface in various tissues. This figure is a generic representation of an epithelial cell, and not all shown transporters are in every tissue type. The role of ACE2 is described in the main text

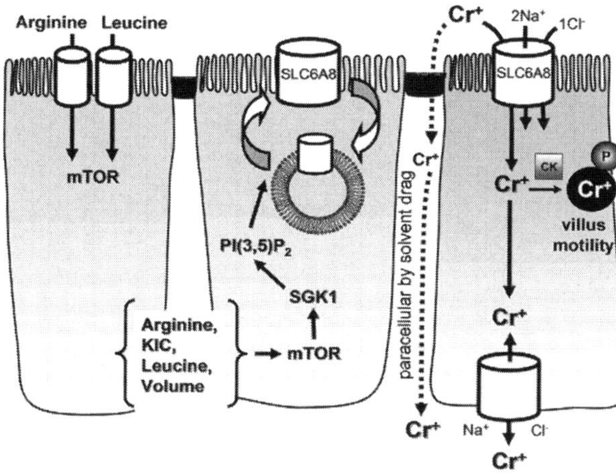

Fig. 8 Participation of nonproteinogenic creatine in intestinal epithelium physiology. Putative model of creatine (Cr^+) absorption in renal or intestinal absorptive epithelium. The branched chain amino acid L-leucine or its keto analogue α-ketoisocaproic acid (KIC), or L-arginine, can initiate mTOR-dependent SLC6A8 trafficking

signaling pathway [36]. The branched chain amino acid L-leucine or its keto analogue α-ketoisocaproic acid (KIC), or L-arginine, can initiate mTOR-dependent SLC6A8 trafficking.

9 A Major Transporter: SLC6A19 (B^0AT1)

A major functional transport system serving the nutrient zwitterionic amino acids, initially called System NBB (Neutral Brush Border) was originally described by Stevens [37, 38] using rabbit small intestinal brush border membrane vesicles. Stevens then renamed NBB as B^0 in order to be consistent with the pattern of competitive inhibition observed for Van Winkle's "b" and "B" series of systems in blastocyte membranes [39]. However, H. Christensen interjected (personal communications), requesting that the "^{0}" superscript be dropped, with the argument that it was redundant in naming a sodium-dependent transport system which served only zwitterion substrates. Thus, the designation "B" appeared briefly in the literature [40]. The mRNA was subsequently cloned as SLC6A19 by Kekuda and colleagues [41–43], and the literature now alternatively refers to the functional properties of expressed SLC6A19 as either B^0 or B^0AT1 (sodium-dependent cloned B^0 Amino acid Transporter number 1).

B^0AT1 is expressed prominently in apical membranes of renal proximal tubule and small intestine, as well as other nutrient absorptive epithelia (Table 3). Actually, B^0AT1 is expressed all along the gastrointestinal tract intestine [44], with dominance in jejunum and ileum (Fig. 9).

The B^0AT1 functional transport system represents an apical membrane polypeptide monomer activated by a sodium-dependent Cl^- independent mechanism.

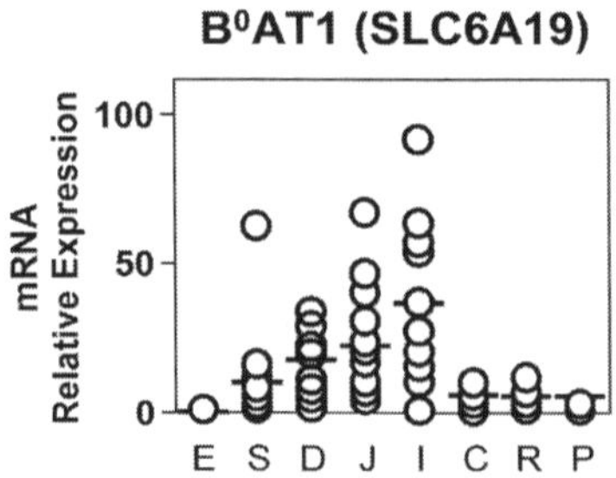

Fig. 9 Relative mRNA expression of system B^0 (B^0AT1 polypeptide monomer transporter from SLC6A19 gene) in various regions of the gastrointestinal tract: esophagus (E), stomach (S), duodenum (D), jejunum (J), ileum (I), colon (C), rectum (R), as well as pancreas (P). Real-time PCR forward primer was GTGTGGACAGGTTCAATAAGGACAT; reverse primer was CCACGTGACTTGCCAGAAGAT. Data points from individual patients redrawn from figure originally erroneously labeled in Terada et al. [44]. Means are shown as *horizontal bars* for each intestinal region

A putative structure-function relationship has been determined by Broer and colleagues [6, 45] , and a putative kinetic mechanism (see Fig. 10) has been worked out by the laboratories of Broer and Verrey [46–48].

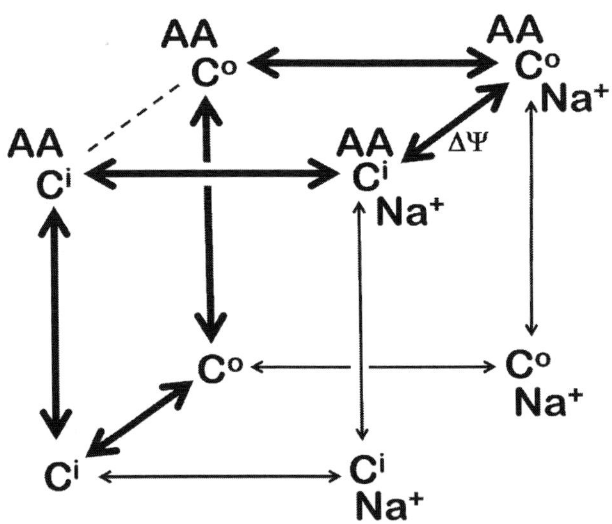

Fig. 10 Kinetic model of B^0AT1 transporter. Apical membrane carrier (C^i intracellular; C^o extracellular) first binds substrate amino acid AA then Na^+ activator, as influenced by membrane potential $\Delta\Psi$. Preferred pathway shown with *bold arrows*. On the basis of models from Bröer's group [46, 47] and Verrey's group [48]

10 Regulation of Amino Acid Transport by ACE2 and Its Homolog Collectrin

Figure 7 above shows the requirement of ACE2 for apical membrane functioning of B^0AT1. Therefore, ACE2 is also necessary for cooperative functioning of the other transporters that interact with B^0AT1, such as the $b^{0,+}$AT/rBAT heterodimer complex.

Many years ago our laboratory demonstrated that human intestinal apical membranes expressed angiotensin converting enzyme (ACE) [49–51], which is a peptidyl dipeptidase that removes a dipeptide from the C-terminus of peptides; we suggested that the function of this nonaminopeptidase was to hydrolyze peptides for subsequent absorption by transporters. However, ACE2, which is a carboxypeptidase, was subsequently discovered in apical membranes in kidney [52] and intestine [53–55]. ACE2 fortuitously is recognized as a binding receptor by a severe acute respiratory syndrome (SARS) coronavirus [56] (SARS-CoV), consistent with clinical observations that SARS-CoV causes severe gastrointestinal complications. A presumed intended biological function of ACE2 is proper trafficking and functioning of B^0AT1, and then membrane removal of the C-terminal single neutral amino acid from nutrient peptides. These particular amino acid products of ACE2 hydrolysis are the very amino acids that are B^0AT1 substrates [45]. A homologue of ACE2 found in kidney, collectrin, was shown to be a mediator of neutral amino acid transport activity and regulator of blood pressure [45, 57–59] via the transport interplay of arginine and neutral amino acids through cooperation of B^0AT1 and the $b^{0,+}$AT/rBAT transporters, as shown in Fig. 7. Collectrin knock out mice demonstrated reduced expression of apical membrane B^0AT1, rBAT, and $b^{0,+}$AT, as well as impaired intracellular trafficking of EAAC1, with the residual activity of B^0AT1 attributable to ACE2. Broer's laboratory demonstrated that the R240 moiety of B^0AT1 interacts with intestinal ACE2 permitting proper apical membrane trafficking and expression of this sodium-dependent amino acid transporter [45].

SLC6A19 genomic DNA is composed of 19 exons and 18 introns on human chromosome 5 at location 5p15.33. Mutations in SLC6A19 on more and more alleles of patients with Hartnup disorder gives the aminoaciduria phenotype manifested in both intestine and kidney [3, 45, 60–63]. Dysfunctional SLC6A19 mutants of B^0AT1 expressed in renal proximal tubule, arising from a variable number of tandem repeat minisatellite polymorphisms, are likely to be a risk factor for hypertension [64]. Figure 11 shows genomic intron/exon relationships concerning known SLC6A19 mutations, and a particular minisatellite repeat unit in intron 9 of SLC6A19 encodes a dysfunctional B^0AT1 associated with essential hypertension in humans [64]. Interestingly, SLC6A18 encoding another "orphan" transporter, XTRP2, which serves glycine reabsorption in kidney, is adjacent to the SLC6A19 gene sequence, although SLC6A18 minisatellites are not involved in hypertension [65].

Taken together, these elegant studies suggest that epithelial cell regulation of amino acid transport activity likely involves other undiscovered modulation factors

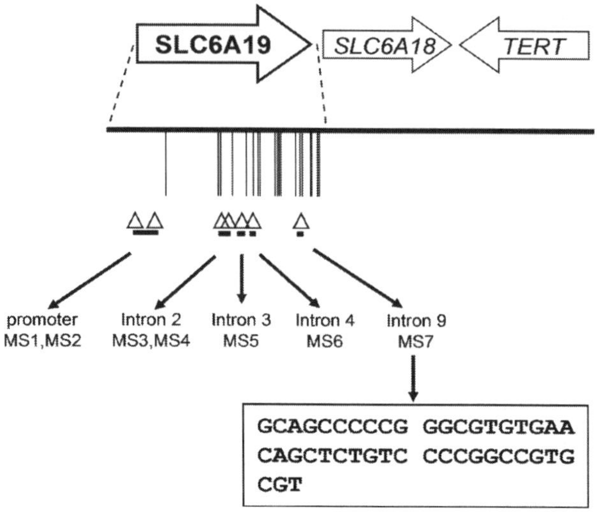

Fig. 11 Allelic structure of human genomic DNA near the SLC6A19 gene. Monomeric transporter B^0AT1 polypeptide is encoded by 19 exons of SLC6A19 (location 5p15.33) indicated by *vertical bars*. Minisatellite repeat (MS) positions are indicated by *triangles*, and introns are indicated by *short horizontal bars* under *triangles*. Minisatellite repeat unit MS7 in intron 9 of SLC6A19 encodes a dysfunctional B^0AT1 associated with hypertension. The sequence of MS7 is shown. TERT (telomerase reverse transcriptase) and SLC6A18 appear adjacent upstream at 5p15.33. Adapted from Seol et al. [64]

that are affective either through trafficking or perhaps other mechanisms. One overall physiological relevance might be that collectrin in renal tubule membranes or allelic mutations could affect blood pressure [59, 64].

11 Summary

This review summarized the current view of amino acid transport by epithelial cells of vertebrates. A wide variety of transporter proteins are expressed in apical and basolateral membranes and collectively play complex interactive roles in controlling the entire organism's overall metabolism of amino acids. Regulation of the transport systems can be manifested at many levels, including gene splicing and promoter regulation, interactions between requisite subunits of oligomers, thermodynamic electrochemical gradients contributed by ion exchangers, overlap of substrate specificity, selective tissue distribution, and specific spatial distribution of transporters leading to net vectorial flow of the amino acids. The next frontier for workers in this field is to uncover a comprehensive molecular understanding of the manner by which epithelial cells signal gene expression of transporters as triggered by substrates, hormones or other triggers, to further understand the trafficking and interactions among multimeric transport system proteins, to extend discoveries of novel small drug substrates for oral and ocular delivery, and to examine gene therapy or nanotherapy of diseases using small molecules delivered via amino acid transporters.

References

1. Hediger MA, Romero MF, Peng JB, Rolfs A, Takanaga H, Bruford EA. The ABCs of solute carriers: physiological, pathological and therapeutic implications of human membrane transport proteins: Introduction. Pflugers Arch 2004;447:465–8.
2. Broer S. The SLC6 orphans are forming a family of amino acid transporters. Neurochem Int 2006;48:559–67.
3. Romeo E, Dave MH, Bacic D, et al. Luminal kidney and intestine SLC6 amino acid transporters of B0AT-cluster and their tissue distribution in Mus musculus. Am J Physiol Renal Physiol 2006;290:F376–83.
4. Boyd CA. Facts, fantasies and fun in epithelial physiology. Exp Physiol 2008;93:303–14.
5. Broer S. Amino acid transport across mammalian intestinal and renal epithelia. Physiol Rev 2008;88:249–86.
6. Broer S. Apical transporters for neutral amino acids: physiology and pathophysiology. Physiology (Bethesda) 2008;23:95–103.
7. Verrey F, Ristic Z, Romeo E, et al. Novel renal amino acid transporters. Annu Rev Physiol 2005;67:557–72.
8. Cynober LA, ed. Metabolic & Therapeutic Aspects of Amino Acids in Clinical Nutrition, Second Edition: CRC Press; 2003.
9. Del Amo EM, Urtti A, Yliperttula M. Pharmacokinetic role of L-type amino acid transporters LAT1 and LAT2. Eur J Pharm Sci 2008;35:161–74.
10. Brandsch M, Knutter I, Bosse-Doenecke E. Pharmaceutical and pharmacological importance of peptide transporters. J Pharm Pharmacol 2008;60:543–85.
11. Brandsch M. Transport of L-proline, L-proline-containing peptides and related drugs at mammalian epithelial cell membranes. Amino Acids 2006;31:119–36.
12. Bartholomaus I, Milan-Lobo L, Nicke A, et al. Glycine transporter dimers: evidence for occurrence in the plasma membrane. J Biol Chem 2008;283:10978–91.
13. Hastrup H, Sen N, Javitch JA. The human dopamine transporter forms a tetramer in the plasma membrane: cross-linking of a cysteine in the fourth transmembrane segment is sensitive to cocaine analogs. J Biol Chem 2003;278:45045–8.
14. Eyre TA, Ducluzeau F, Sneddon TP, Povey S, Bruford EA, Lush MJ. The HUGO gene nomenclature database, 2006 updates. Nucleic Acids Res 2006;34:D319–21.
15. Christensen HN. Naming plan for membrane transport systems for amino acids. Neurochem Res 1984;9:1757–8.
16. Kowalczuk S, Broer A, Munzinger M, Tietze N, Klingel K, Broer S. Molecular cloning of the mouse IMINO system: an Na+- and Cl—dependent proline transporter. Biochem J 2005;386:417–22.
17. Yan Y, Vasudevan S, Nguyen H, Bork U, Sitaraman S, Merlin D. Extracellular interaction between hCD98 and the PDZ class II domain of hCASK in intestinal epithelia. J Membr Biol 2007;215:15–26.
18. Reynolds B, Laynes R, Ogmundsdottir MH, Boyd CA, Goberdhan DC. Amino acid transporters and nutrient-sensing mechanisms: new targets for treating insulin-linked disorders? Biochem Soc Trans 2007;35:1215–7.
19. Seo NS, Zeng CQ, Hyser JM, et al. Inaugural Article: Integrins {alpha}1{beta}1 and {alpha}2{beta}1 are receptors for the rotavirus enterotoxin. Proc Natl Acad Sci USA 2008.
20. Burdo J, Dargusch R, Schubert D. Distribution of the cystine/glutamate antiporter system xc- in the brain, kidney, and duodenum. J Histochem Cytochem 2006;54:549–57.

21. Kleemola M, Toivonen M, Mykkanen J, Simell O, Huoponen K, Heiskanen KM. Heterodimerization of y(+)LAT-1 and 4F2hc visualized by acceptor photobleaching FRET microscopy. Biochim Biophys Acta 2007;1768:2345–54.
22. Wagner CA, Lang F, Broer S. Function and structure of heterodimeric amino acid transporters. Am J Physiol Cell Physiol 2001;281:C1077–93.
23. Verrey F, Closs EI, Wagner CA, Palacin M, Endou H, Kanai Y. CATs and HATs: the SLC7 family of amino acid transporters. Pflugers Arch 2004;447:532–42.
24. Fort J, de la Ballina LR, Burghardt HE, et al. The structure of human 4F2hc ectodomain provides a model for homodimerization and electrostatic interaction with plasma membrane. J Biol Chem 2007;282:31444–52.
25. Palacin M, Kanai Y. The ancillary proteins of HATs: SLC3 family of amino acid transporters. Pflugers Arch 2004;447:490–4.
26. Fernandez E, Jimenez-Vidal M, Calvo M, et al. The structural and functional units of heteromeric amino acid transporters. The heavy subunit rBAT dictates oligomerization of the heteromeric amino acid transporters. J Biol Chem 2006;281: 26552–61.
27. Amaral JS, Pinho MJ, Soares-da-Silva P. Genomic regulation of intestinal amino acid transporters by aldosterone. Mol Cell Biochem 2008;313:1–10.
28. Charrier L, Merlin D. The oligopeptide transporter hPepT1: gateway to the innate immune response. Lab Invest 2006;86:538–46.
29. Gerencser GA, Stevens BR. Thermodynamics of symport and antiport catalyzed by cloned or native transporters. J Exp Biol 1994;196:59–75.
30. Gerencser GA, Stevens BR. Energetics of sodium-coupled active transport mechanisms in invertebrate epithelia. Am J Physiol 1989;257:R461–72.
31. Ramadan T, Camargo SM, Herzog B, Bordin M, Pos KM, Verrey F. Recycling of aromatic amino acids via TAT1 allows efflux of neutral amino acids via LAT2-4F2hc exchanger. Pflugers Arch 2007;454:507–16.
32. Garcia-Delgado M, Peral MJ, Cano M, Calonge ML, Ilundain AA. Creatine transport in brush-border membrane vesicles isolated from rat kidney cortex. J Am Soc Nephrol 2001;12: 1819–25.
33. Garcia-Delgado M, Garcia-Miranda P, Peral MJ, Calonge ML, Ilundain AA. Ontogeny up-regulates renal Na(+)/Cl(–)/creatine transporter in rat. Biochim Biophys Acta 2007;1768:2841–8.
34. Peral MJ, Garcia-Delgado M, Calonge ML, et al. Human, rat and chicken small intestinal Na+ – Cl– -creatine transporter: functional, molecular characterization and localization. J Physiol 2002;545:133–44.
35. Tosco M, Faelli A, Sironi C, Gastaldi G, Orsenigo MN. A creatine transporter is operative at the brush border level of the rat jejunal enterocyte. J Membr Biol 2004;202: 85–95.
36. Strutz-Seebohm N, Shojaiefard M, Christie D, Tavare J, Seebohm G, Lang F. PIKfyve in the SGK1 mediated regulation of the creatine transporter SLC6A8. Cell Physiol Biochem 2007;20:729–34.
37. Stevens BR, Ross HJ, Wright EM. Multiple transport pathways for neutral amino acids in rabbit jejunal brush border vesicles. J Membr Biol 1982;66:213–25.
38. Stevens BR, Kaunitz JD, Wright EM. Intestinal transport of amino acids and sugars: advances using membrane vesicles. Annu Rev Physiol 1984;46:417–33.
39. Van Winkle LJ. Amino acid transport in developing animal oocytes and early conceptuses. Biochim Biophys Acta 1988;947:173–208.
40. Stevens B. Amino acid transport in intestine. In: Kilberg M, Haussinger D, eds. Mammalian Amino Acid Transport. New York: Plenum Press; 1992:149–63.
41. Kekuda R, Prasad PD, Fei YJ, et al. Cloning of the sodium-dependent, broad-scope, neutral amino acid transporter Bo from a human placental choriocarcinoma cell line. J Biol Chem 1996;271:18657–61.

42. Kekuda R, Torres-Zamorano V, Fei YJ, et al. Molecular and functional characterization of intestinal Na(+)-dependent neutral amino acid transporter B0. Am J Physiol 1997;272:G1463–72.
43. Talukder JR, Kekuda R, Saha P, Arthur S, Sundaram U. Identification and characterization of rabbit small intestinal villus cell brush border membrane Na-glutamine cotransporter. Am J Physiol Gastrointest Liver Physiol 2008;295:G7–G15.
44. Terada T, Shimada Y, Pan X, et al. Expression profiles of various transporters for oligopeptides, amino acids and organic ions along the human digestive tract. Biochem Pharmacol 2005;70:1756–63.
45. Kowalczuk S, Broer A, Tietze N, Vanslambrouck JM, Rasko JE, Broer S. A protein complex in the brush-border membrane explains a Hartnup disorder allele. Faseb J 2008;22:2880–7.
46. Bohmer C, Broer A, Munzinger M, et al. Characterization of mouse amino acid transporter B0AT1 (slc6a19). Biochem J 2005;389:745–51.
47. O′Mara M, Oakley A, Broer S. Mechanism and putative structure of B(0)-like neutral amino acid transporters. J Membr Biol 2006;213:111–8.
48. Camargo SM, Makrides V, Virkki LV, Forster IC, Verrey F. Steady-state kinetic characterization of the mouse B(0)AT1 sodium-dependent neutral amino acid transporter. Pflugers Arch 2005;451:338–48.
49. Stevens BR, Fernandez A, Kneer C, Cerda JJ, Phillips MI, Woodward ER. Human intestinal brush border angiotensin-converting enzyme activity and its inhibition by antihypertensive Ramipril. Gastroenterology 1988;94:942–7.
50. Stevens BR, Fernandez A, Martinez del Rio C. Angiotensin converting enzyme in brush-border membranes of avian small intestine. J Exp Biol 1988;135:1–8.
51. Stevens BR, Phillips MI, Fernandez A. Ramipril inhibition of rabbit (Oryctolagus cuniculus) small intestinal brush border membrane angiotensin converting enzyme. Comp Biochem Physiol C 1988;91:493–7.
52. Warner FJ, Lew RA, Smith AI, Lambert DW, Hooper NM, Turner AJ. Angiotensin-converting enzyme 2 (ACE2), but not ACE, is preferentially localized to the apical surface of polarized kidney cells. J Biol Chem 2005;280:39353–62.
53. Chan PK, To KF, Lo AW, et al. Persistent infection of SARS coronavirus in colonic cells in vitro. J Med Virol 2004;74:1–7.
54. Gembardt F, Sterner-Kock A, Imboden H, et al. Organ-specific distribution of ACE2 mRNA and correlating peptidase activity in rodents. Peptides 2005;26:1270–7.
55. Hamming I, Timens W, Bulthuis ML, Lely AT, Navis GJ, van Goor H. Tissue distribution of ACE2 protein, the functional receptor for SARS coronavirus. A first step in understanding SARS pathogenesis. J Pathol 2004;203:631–7.
56. Li W, Moore MJ, Vasilieva N, et al. Angiotensin-converting enzyme 2 is a functional receptor for the SARS coronavirus. Nature 2003;426:450–4.
57. Malakauskas SM, Quan H, Fields TA, et al. Aminoaciduria and altered renal expression of luminal amino acid transporters in mice lacking novel gene collectrin. Am J Physiol Renal Physiol 2007;292:F533–44.
58. Danilczyk U, Sarao R, Remy C, et al. Essential role for collectrin in renal amino acid transport. Nature 2006;444:1088–91.
59. Verrey F. Does kidney amino acid transport have something to do with blood pressure? Nephrol Dial Transplant 2007;22:2449–51.
60. Seow HF, Broer S, Broer A, et al. Hartnup disorder is caused by mutations in the gene encoding the neutral amino acid transporter SLC6A19. Nat Genet 2004;36:1003–7.
61. Kleta R, Romeo E, Ristic Z, et al. Mutations in SLC6A19, encoding B0AT1, cause Hartnup disorder. Nat Genet 2004;36:999–1002.
62. Azmanov DN, Kowalczuk S, Rodgers H, et al. Further evidence for allelic heterogeneity in Hartnup disorder. Hum Mutat 2008.

63. Ristic Z, Camargo SM, Romeo E, et al. Neutral amino acid transport mediated by ortholog of imino acid transporter SIT1/SLC6A20 in opossum kidney cells. Am J Physiol Renal Physiol 2006;290:F880–7.
64. Seol SY, Lee SY, Kim YD, et al. Minisatellite polymorphisms of the SLC6A19: susceptibility in hypertension. Biochem Biophys Res Commun 2008;374:714–9.
65. Yoon YH, Seol SY, Heo J, Chung CN, Park IH, Leem SH. Analysis of VNTRs in the Solute Carrier Family 6, Member 18 (SLC6A18) and Lack of Association with Hypertension. DNA Cell Biol 2008;27(10):559–567.
66. Boll M, Foltz M, Rubio-Aliaga I, Daniel H. A cluster of proton/amino acid transporter genes in the human and mouse genomes. Genomics 2003;82:47–56.
67. Yan Y, Vasudevan S, Nguyen HT, Merlin D. Intestinal epithelial CD98: An oligomeric and multifunctional protein. Biochim Biophys Acta 2008;1780:1087–92.
68. Meier Y, Eloranta JJ, Darimont J, et al. Regional distribution of solute carrier mRNA expression along the human intestinal tract. Drug Metab Dispos 2007;35:590–4.
69. Resource SHP. Human Protein Atlas, Transporters, http://www.proteinatlas.org/search_advanced.php?proteinclass=Tr. In; 2008.
70. Bae SY, Xu Q, Hutchinson D, Colton CA. y+ and y+ L arginine transporters in neuronal cells expressing tyrosine hydroxylase. Biochim Biophys Acta 2005;1745:65–73.

Molecular Ontology of Amino Acid Transport

Dmitri Y. Boudko

Abstract This review focuses on the comparative physiology and phylogeny of plasma membrane transporters that absorb and redistribute amino acids in organisms. The first section briefly summarizes the life history of the environmental flux and metabolism of amino acids. It reveals a set of geological and biological events that may have shaped amino acid transport mechanisms, which evolved under everlasting antagonism of environmental availability, endogenous synthesis, and metabolic consumption of proteinogenic amino acids. The second section addresses the phylogenetic and physiological diversity of experimentally and theoretically defined amino acid transporters. It reveals a set of gene duplications, expansions, and extinctions in the phylogenetic tree of the amino acid transporters, which correlate with the rapid acquisitions of new transport phenotypes and assured remarkable adaptive plasticity of the amino acid transport network. Specific emphasis in this review is placed on the Excitatory Amino Acid Transporters and the Sodium Neurotransmitter symporter families (SLC1 and SLC6, respectively). The early evolution of these cation-coupled transporters compensating the anabolism of proteinogenic amino acids may have simultaneously driven the expansion of heterotrophy and the extinction of principal metabolic pathways (e.g. nitrogen fixation in prokaryotes and essential amino acid synthesis cascades in Metazoans). Furthermore, the extant physiological functions of these transporters – including the balance of dispensable and essential amino acids, cellular signaling, and neurochemical communication – are critical for the metabolic integrity and health of metazoan organisms. Molecular, genetic, and structural analyses of amino acid transporters have emphasized this point, and continue to provide us with an expanding knowledge base that will ultimately lead to new biomedical technologies for curing metabolic disorders and controlling pathogenic and pest organisms.

D.Y. Boudko (✉)
Department of Physiology and Biophysics, Chicago Medical School, Rosalind Franklin University of Medicine and Science, North Chicago, IL 60064, USA
e-mail: dmitri.boudko@rosalindfranklin.edu

G.A. Gerencser (ed.), *Epithelial Transport Physiology*,
DOI 10.1007/978-1-60327-229-2_16,

Keywords Essential · Amino acid · Transport · System · Metabolism · Evolution · SLC · TCDB

1 Introduction

1.1 Transport Domains of Life

The study of amino acid transport may provide a key to one of the most intriguing mysteries in the history of Earth, the origin and evolution of metazoan organisms. The integrated activity of amino acid transporters is essential for normal amino acid metabolism and brain function in humans and is involved in metabolic, neurodegenerative and mental disorders, including Alzheimer's disease, anxiety, drug abuse, obesity, and distinct forms of neutral aminoacidurias. On the other hand, amino acid transporters are attractive targets for selective and safe control of specified eukaryotes. The comparative study of genotype and phenotype interplay is critical to elucidate the structure-function relationships of membrane transporters and could raise possibilities for in silico predictions and the molecular engineering of new transport phenotypes, substrates, and drugs for correcting metabolic and neuronal disorders.

Life tames the energy of the expanding universe to transport and organize matter. To withstand entropy, life uses an array of membrane transport mechanisms including metabolic or auxiliary energy-driven ion pumps, ion channels, and electrochemical gradient-driven secondary transporters that act in synergy to support homeostasis of dividing cells and organisms (Fig. 1). The first two domains, pumps and ion channels generate and regulate sub-membrane pools and electrochemical gradients of major inorganic ions. They perform a variety of physiological functions by interacting with a narrow set of inorganic ions, predominantly Na^+, K^+, H^+, Cl^-, $HCO_3{}^-$, Ca^{++}, and Mg^{++}. Most pumps (a.k.a. primary transport mechanisms, Fig. 1.1;) represent ancient and exceptionally conserved mechanisms evolved to convert a metabolic energy, usually accumulated in the form of ATP, to the potential energy of transmembrane ion gradients. For example, P-type and V-type ATPases generate canonical Na^+/K^+ and H^+ electrochemical gradients by hydrolyzing ATP molecules. These gradients are subsequently employed by secondary transporters and ion channels (Figs. 1, 2 and 3), which together with pumps maintain homeostasis of the cell. By definition the term pump also includes ATP-Binding Cassette (ABC) transporters, which may directly couple the translocation of metabolites to ATP hydrolysis. Compared to secondary transporters, ABC pumps are massive and slow mechanisms. They evolved to support specific metabolic demands with small capacity and high affinity transport processes. They perform a variety of functions in prokaryotes but have been generalized as xenobiotics/metabolite-export mechanisms and ATP-regulated channels (i.e. CFTR) in eukaryotes. The secondary transporters comprise a phylogenetically separate domain of membrane-spanning mechanisms that use electrochemical energy of the major transmembrane

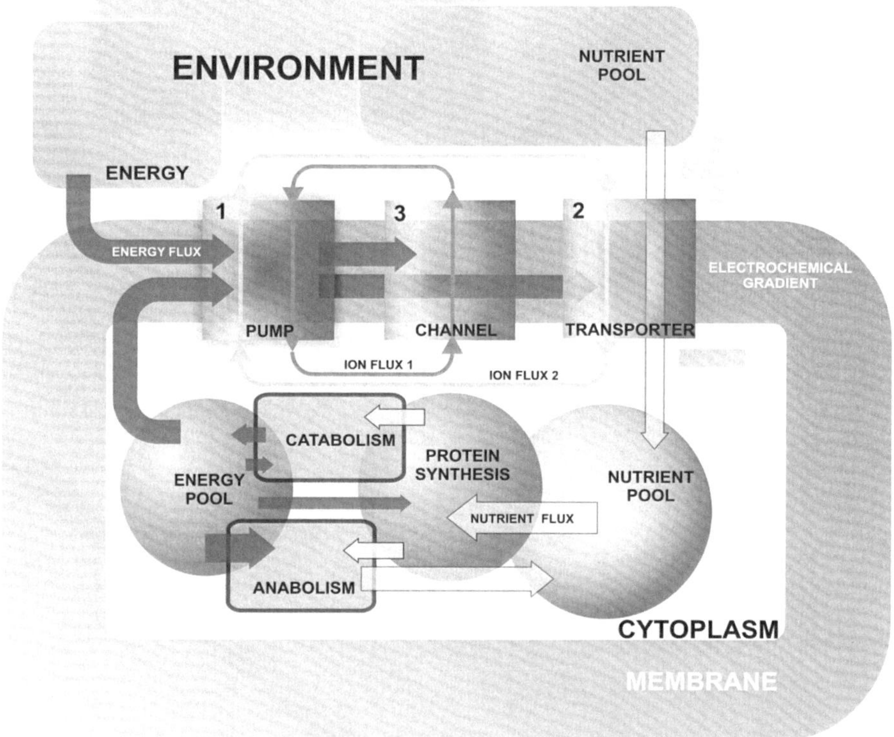

Fig. 1 The synergetic action of pump (*1*), transporter (*2*), and channel (*3*) at the membrane interface between Universe and intracellular Microcosm. The *arrows* represent specific fluxes of energy (*dark*), nutrients (*empty*), and mineral ions (*thin*)

ion gradients for substrate translocations. The substrate specificity of secondary transporters ranges from inorganic ions to polypeptide and oligonucleotides, but the functional consensus of all secondary transporters is to move substrates against unfavorable gradients or phase transition barriers of biological membranes. One subset of secondary transporters comprises passive transport mechanisms (permeases or uniporters) that act as substrate-selective channels catalyzing transmembrane substrate forward down their chemical gradients. The uniporters likely represent a primitive ancestral modality of all membrane transport proteins [1]. A large group of secondary transporters has evolved capacities to utilize motive forces of major inorganic ions Na^+, K^+, H^+, Cl^- and $HCO3^-$, and less frequently omnipresent organic solutes, whose transmembrane gradients may energize the translocations of different substrates. This group comprises active secondary transporters. Secondary transporters are divided into symporters and antiporters based on their translocation of their substrates in the same or opposite directions, respectively. The ion channels and secondary transporters that are driven by the transmembrane voltage are often

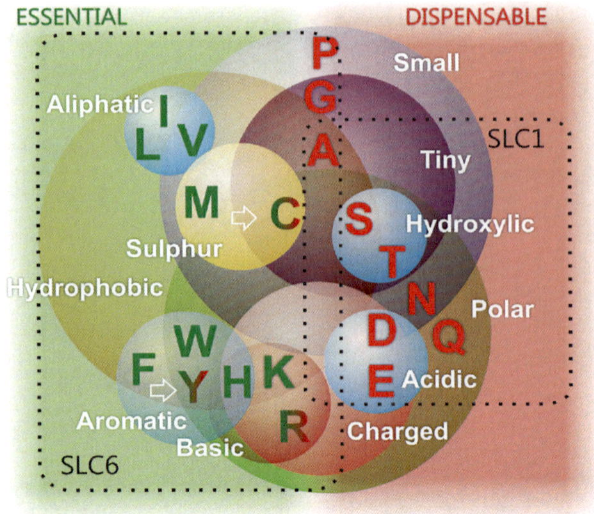

Fig. 2 The summary of physicochemical characteristics of essential and dispensable amino acids. Spheres outline different physicochemical groups of amino acids. *Red* and *green fonts* and *background colors* indicate dispensable and essential amino acids, respectively. *White empty arrows* indicate metabolic links between conditionally essential amino acids and their precursors. *Dotted lines* outline preferable substrates of SLC1 and SLC6 families as shown in the *upper left and lower right* corners, respectively

called electrogenic. However, both these groups use net transmembrane energy, which does do not induce, but rather reduces the transmembrane voltage; so they are "electrophoretic," whereas the ion pumps that generate transmembrane voltage are properly called "electrogenic." It is conceivable that a secondary transporter could use the gradient of a neutral solute to transport an ion against its electrochemical gradient, thus generating a transmembrane voltage, however the biological significance of such a mechanism remains to be demonstrated. The secondary transporters utilizing ion motive forces and organic solute motive forces can be comprehensively delineated as ionophoretic and organophoretic mechanisms, because some of these mechanisms rely on a chemical gradient and do not use electrical gradients.

1.2 Roles of Amino Acids

Amino acids comprise the foundation of existing life. They carry biologically accessible nitrogen and carbon between cells and tissues (within organisms), and between organisms in ecological nutrient chains. A large number of amino acids are generated in abiotic and biotic processes. However the majority of such amino acids, a.k.a. uncommon or noncanonical amino acids play minor, usually unique biological roles (for examples see [2, 3]). For elusive reasons only 18 L-amino acids plus

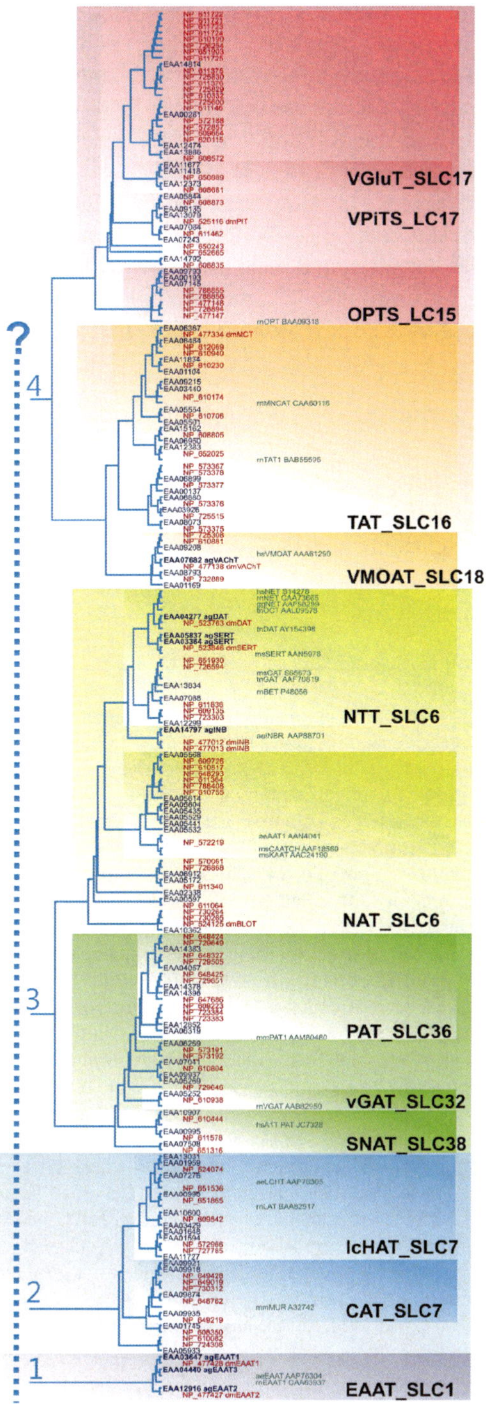

Fig. 3 (continued)

the achiral glycine and the imino acid, proline are included in the structures of naturally occurring proteins (a.k.a. standard or common amino acids [4]). The extension of that set may include selenocysteine (Sec), pyrrolysine (Pyl) [5, 6], several post-translational modifications, and the use of D-amino acids in some secretory peptides [7, 8] and proteins [9] with specific functions. Amino acids are precursors in a variety of catabolic and anabolic pathways in which amino acid transporters serve as rate-limiting components [1]. In addition to the rate-limiting substrate role in the syntheses of proteins and other biopolymers, amino acids serve as messengers of intra- and intercellular signaling and metabolic precursors of neurotransmitters and hormones, which act as osmolites [10]; and as auxiliary, usually mission-critical or starvation-induced resources of metabolic fuel [11, 12].

2 Ontology of the Amino Acid Pool

Amino acids undergo transport and recycling in the cell, organism, ecological system, and Biosphere. The mechanisms for assimilation and transport of amino acids form relatively stable nutrient chains that are crucial for the circulation and balance of bioavailable nitrogen in organisms and biocenoses. Nevertheless, a significant

Fig. 3 (continued) A phylogenetic relationship of selected dipteran proteins justifying ten SLC families and four upper phylogenetic divisions of secondary amino acid transporters. This phylogenetic analysis includes 193 dipteran sequences (Smith-Waterman all-against-all cutoff $< 10^{-5}$) that comprise members of ten solute carrier families (SoLute Carrier numbers; SLC with Arabic integers) and 15 identified phenotypes of amino acid transporters (*large font* on the transparent ribbon). This tree includes the oligopeptide transporters family (SLC15) since it shares common ancestry with few families of amino acid transporters and plays important roles in the nutrient amino acid balance. The recently identified SLC43 family is not present in the tree because it is absent in dipteran insects. Sequences are represented by nonredundant subsets of gene products' accession numbers that can be used to retrieve these sequences from NCBI protein databases (http://www.ncbi.nlm.nih.gov). The first row from the left (*blue font*) shows 85 putative gene products identified in *A. gambiae*, each of which is represented by a single splicoform; the second row (*red font*) shows 108 gene products found in *D. melanogaster*, which are encoded by 88 genes, but some are represented by several established splicoforms; the third row (*green font*) represents 25 characterized transporters from various species which are used to reference tree clusters. Common transporter phenotype abbreviations are used e.g.: EAAT, Excitatory Amino Acid Transporter; CAT, Cationic Amino acid Transporter; lcHAT, light chain subunit of Heterodimeric Amino acid Transporter; SNAT, Sodium Neutral Amino acid Transporter; vGAT, vesicular GAba Transporter; PAT, Proton Amino acid Transporter; SNF, Sodium Neurotransmitter Transporter; NAT, Nutrient Amino acid Transporter; NTT, NeuroTransmitter Transporter; vMAT, vesicular Monoamine transporter; TAT, T-system Amino acid Transporter; OPT, OligoPeptide Transporter; PiT, inorganic phosphate (Pi) Transporter; vGluT vesicular Glutamate Transporter. The *scale bar* represents evolutionary distances in mutations per site. *Leave colors* represent different taxons: *A. aegypti*, green; *A. gambiae*, blue; *D. melanogaster*, red; *H. sapiens*, black. Tree reconstruction was made from the first draft sequences as they were deposited in the NCBI database, so be aware that some *A. gambiae* sequences may have incorrect open reading frames as well as predictions of 3′ and 5′ ends, which could produce tree distortion

portion of amino acids may be added to – and removed from – the biological and ecological circulations via numerous biological and abiotic processes. Among those, a metabolic decomposition (deamination) and photosynthesis-coupled accumulation appear to be the most notable. These two processes predetermine a relatively steady pool and fluxes of amino acids that are reciprocally important for the evolution, composition, and existence of individual organisms and the Biosphere.

2.1 Natural History of Amino Acid Pool

Descending into the past of amino acid transport mechanisms, we have to consider that the Earth's pool and flux of amino acids are changing in space and time dimensions. In addition to relatively slow gradual trends, the local fluxes of amino acid have experienced several dramatic changes that coincide with major events in the course of geological and biological maturations of the planet. Some of these changes can be clearly identified in the geological, paleontological, or phylogenetic records.

2.2 Abiotic Stage

Oparin–Haldane hypothesis [13] proposes that hypoxic and/or reducing conditions occupying naive Earth between 4.0 and 3.8 billion years ago favored a chemical condensation of inorganic components to various forms of organic matter including simple amino acids, carbohydrates, and nucleotides. The efficiency of prebiotic chemistry is low, but the size of the reactor is enormous. One half of the proteinogenic amino acids can be produced upon abiotic interaction of inorganic precursors [CHON] (Table 1, Gray background) with an agitation by different physicochemical factors (a.k.a. Miller–Urey-type experiments [14–16]; reviewed in [17, 18]). In addition to terrestrial synthesis, a large-scale accumulation of organic matter may occur upon gravitational collapse of interstellar medium and formations of carriers with a cold surface, such as interstellar dust particles (IDPs), comets, meteorites, and planets [19]. The ongoing quest for extraterrestrial organic matter including analytical screening of meteorite samples [2, 20–22], spectral tests of interstellar clouds [23–26] and analytical tests of comet dust [21, 27] together with laboratory and computer simulation of abiotic reactions [24, 28] provides increasing support for an understanding of a large-scale extraterrestrial accumulation of abiotic organic matter – including simple amino acids or their precursors (reviewed in [29]). A high IDP flux and frequent introductions of large amounts of extraterrestrial matter carrying intact amino acids and amino acid precursors [30, 31] likely supplied a significant amount of abiotic amino acids, especially during a primordial stage of the planet, which is coincidental with energized and reduced or mildly oxidized atmosphere [32]. Another possible massive source of terrestrial amino acids and primitive peptides is linked to high energy and element-liberating terrestrial geochemistry – including volcanic eruptions [33–35], geothermal activity, and large-scale mineral catalysis at deep geothermal vents [36–41].

Table 1 Selected properties of canonical amino acids

		Precursors	ATP	NADH	NADPH	EQ	ST	pI	Vr	Mass	[CHON]	F%	Dr inf	ad
W	**Trp**	pep, eryP, penP	5	−2	3	79	12	5.89	163	204.23	C11H12N2O2	1.3	13	4
FF	**Phe**	pep	1	0	2	63	9	5.48	135	165.19	C9H11NO2	4	69	14
Y	**Tyr**	eryP, pep, pep	1	−1	2	57	9	5.66	141	181.19	C9H11NO3	3.3	F*	
H	**His**	penP	6	−3	1	33	1	7.47	118	155.16	C6H9N3O2	2.9	10	3
L	**Leu**	pyr, pyr, acCoA	0	−1	2	33	7	5.98	124	131.17	C6H13NO2	7.6	73	14
V	**Val**	pyr, pyr	0	0	2	25	4	5.96	105	117.15	C5H11NO2	6.8	38	10
I	**Isol**	pyr, oaa	2	0	5	20	11	5.94	124	131.17	C6H13NO2	3.8	31	10
M	**Met**	oaa	7	0	8	19	9	5.74	124	149.21	C5H11NO2S	1.8	27	13
K	**Lys**	oaa, pyr	2	0	4	19	10	9.59	135	146.19	C6H14N2O2	7.2	64	12
C	**Cys**	gly-3p	4	−1	5	25	9	5.02	86	121.16	C3H7NO2S	3.3	M*	
R	**Arg**	αkg	7	−1	4	18	10	11.2	148	174.20	C6H14N4O2	4.2	uc*	
S	**Ser**	gly-3p	−1	0	1	15	3	5.68	73	105.10	C3H7NO3	8.1		
G	**Gly**	gly-3p	0	−1	1	14	4	5.97	48	75.07	C2H5NO2	7.4		
A	**Ala**	pyr	0	0	1	13	1	6.00	67	89.09	C3H7NO2	7.4		
P	**Pro**	αkg	1	0	3	12	4	6.30	90	115.13	C5H9NO2	5		
Q	**Glt**	αkg	1	0	1	9	2	5.65	114	146.15	C5H10N2O3	3.7		
E	**Glu**	αkg	0	0	1	8	1	3.22	109	147.13	C5H9NO4	5.8		
T	**Thr**	0	2	0	3	6	6	5.64	93	119.12	C4H9NO3	6.2		
N	**Asn**	oaa	3	0	1	4	1	5.41	96	132.12	C4H8N2O3	4.4		
D	**Asp**	oaa	0	0	1	1	1	2.77	91	133.10	C4H7NO4	5.9		

Notes: Remodeled from [91, 454]; for precursors and synthesis cost references see [455]. Amino acid precursor abbreviations αkg, α-ketogluterate (9.5 ATPs); acCoA, acetyl-CoA (12 ATPs); eryP, erythrose 4-phosphate (28 ATPs); gly-3p, glycerate-3-phosphate (14.5 ATPs); oaa, oxaloacetate (2 ATPs); penP, ribose 5-phosphate (30 ATPs); pep, phosphenyl pyruvate (14.5 ATPs); pyr, pyruvate (13.5 ATPs). Amino acids are shown as a single letter code and complete name in ionized form. Abbreviation: ATP, NAD, NADPH, represent use of specific energo-carriers, respectively; EQ, total energy of synthesis in ATPs equivalent; ST, enzymatic steps; pI, isoelectric point; Vr, residual volume in cubic Angstroms; Mms, molecular mass; [CHON] atomic composition; F%, published overage frequency per protein sequence residue; inf, ad, dietary requirements in mg/kg/day infant and adult human subject, respectively (WHO). Essential and conditionally essential amino acids are highlighted with dark and light green background; Prebiotically plausible amino acids are highlighted with dark and light gray background inferred from a probability and yield of the abiotic reactions. *F, M, uc,- Phenylalanine, Methionine and a metabolite of urea cycle precursors.

The eclectic view recognizes many geophysical factors, including IDPs flux, meteorite impact, composition and temperature of atmosphere, condensation and mineralization of the Ocean, and volcanism and hydrothermal activities contributing to the formation of a prebiotic pool of amino acids, which may ignite and support earlier heterotrophic life [42]. The gaseous and aquatic simulations of [CHON] interactions yield from 5 up to 14 proteinogenic amino acids (reviewed by [18]); however, only a fraction of amino acids in these sets, obviously the most simple ones, with a higher synthesis probability and lower decomposition probability have yield/stability factors satisfying a prebiotic accumulation [43]. This fraction was necessary and sufficient for the origin of life according to the Oparin–Haldane model [13, 44]. It also may have been essential for an anticipated transition from Gilbert's [45] RNA World [46–48] or primitive GADV-protein World [49–52] to the extent RNA/DNA/20AA-Protein World somewhere at the demarcation between the Hadean and Archean eons (approx. 3.8 Bya). In contrast, the accumulation of the second part of the proteinogenic amino acids set comprises large aliphatic amino acids, whose production requires an accumulation of unstable precursors, is problematic during an abiotic stage. This second fraction was very limited in prebiotic environments [43], and was under-represented among substrates and metabolites of naive heterotrophic and/or primitive metabolic life (see Table 1 and Fig. 2). Hence, this fraction may not be essential for the sparks and initial steps of life on the planet. The phylogenetic analysis of amino acid gain and loss trend in protein sequences suggests that the expansion of initially under-represented amino acids began approximately 3.4 billion years ago [53, 54] and continues up to this day [55]. The acquisition of those amino acids in the protein synthesis framework coincides with the origin and sophistication of amino acid anabolism and protein synthesis cascades [56, 57].

2.3 Biotic Stage

Abiotic contributions to the global and local pools of amino acids declined in the path of geophysical evolution of the Solar system and Earth [32]. In contrast, biotic contributions emerged with life and graduallyincreased upon sophistications of energy-producing catabolism and amino acid-producing anabolism, as well as the spatial and ecological annexing of life. For example, the majority of existing amino acids are produced by efficient and structurally organized enzymatic pathways involving nitrogen fixation, amination of α-ketoglutarate (yield glutamate), transamination of glutamate (yield alanine and aspartate) and catalysis of the glutamate, alanine and aspartate, followed by specific amino acid synthesis cascades [58, 59]. In addition, glutamate interaction with ammonia catalyzed by glutamine synthase yields a two-amino-group glutamine that may supply a significant portion of fixed nitrogen to the plant and animal anabolism of amino acids [60].

2.4 Evolution of Anabolism

Considering a low yield and high degradation of large amino acids upon prebiotic chemistry the plausible concentrations of these substrates in the environment were disproportionally low and should decay rapidly with escalation of our heterotrophic ancestor and increased protein synthesis. Hence, biogenic compensations for decaying environmental amino acids are essential and it should evolve in parallel with protein synthesis. According to Horowitz's retro-evolution scenario: "... (the) first living entity was a completely heterotrophic unit, reproducing itself at the expense of prefabricated organic molecules in its environment" [61]. Such units acquired amino acids by primitive membrane transport systems or by fusion with other units. Some units receive selective advantages through an evolved biosynthesis (anabolism) of the decaying substrates. During "retro evolution," enzymes of an anabolic pathway would evolve in a reverse order to compensate for gradual extinction of abiotic substrates. On the other hand, the evolution of amino acid synthesis is the most significant biotic factor modifying pool and flux of free and biologically mobilized amino acids. Organisms synthesizing energetically expensive amino acids also can gain an advantage by evolving more efficient catabolic frameworks for breaking down prebiotic substrates and the extraction of an additional fraction of chemical energy. In catabolic pathways, the first enzyme to evolve would be the first in the pathways, i.e. the "forward evolution" scenario [62]. The systemic integration of the primitive anabolic and catabolic frameworks may provide structural and energetic bases in the evolution of branched aliphatic amino acid anabolism, which require unstable intermediates and cannot be described in terms of retro/forward evolution of a pre-existing acetolactate pathway [63]. The accumulation of large neutral amino acids is virtually impossible upon prebiotic synthesis and primitive biosynthesis. It is most likely that synthesis of those amino acids evolved as a random patchwork expansion of an earlier metabolic network [64]. Under the patchwork model, an accidental recruitment of enzymes with an opportunistically broad substrate spectra may have enhanced metabolic fitness via acquisition of new chemistries, as well as via generating new products [65]. The pool and flux of new amino acids would increase in parallel with the expansion of proteinogenic amino acid anabolism [56] and evolution of the genetic code [43, 66, 67]. The changes of amino acid pool, fluxes, and membrane transport during the biotic stage reflect the generalized evolution of metabolic complicity. However, a few metabolic inventions of earlier life were especially dramatic.

2.5 Diazotrophy

The origin of biological fixation of extremely inert atmospheric nitrogen N_2 (approx. 3.3 billion years ago [68, 69]) enabled earlier hemitropic and phototrophic prokaryotes to be independent from prebiotic sources of essential nitrogen species. Diazotrophs had launched new ecosystems and added new dimensions to the biotic accumulation and flux of amino acids. The azote fixation (diazotrophy) is catalyzed by a conserved family of nitrogenases, an ancient enzymatic complex with ultimate reducing power, which may supply three electron donors to cleave the triple bond of N_2, and subsequently catalyze binding of hydrogen atoms to nitrogen [70]. However, such unique reducing power nitrogenases are susceptible to oxidation. Their action requires strictly anaerobic conditions since they are irreversibly destroyed by molecular oxygen or other reactive species [68]. Diazotrophy is an anticipated part of "Last Universal Common Ancestor" (LUCA) metabolism. Perhaps initially it was supported by a primitive anaerobic metabolism [71]. Nitrogen fixation enzymes are very conserved in diazotrophs but present in limited numbers of extent prokaryotes including cyanobacteria, heterotrophic nitrogen-fixing symbionts of plants, some free-living soil bacteria, and archaea. It is extinct in the majority of prokaryotes and all eukaryotes. Such a broad metabolic reductionism is a possible outcome of global oxygenation. However, the oxygenation problem was successfully solved in earlier diazotrophs and remains effective in present diazotrophs despite the long-term oxygenation of the atmosphere. It is more likely that the high energy requirement for cleaving the excessively stable triple bond of N_2 (eight electron transfers and 16 ATPs $\Delta G_0 = 945$ kJ/mol, physiologically approx. 20–30 ATPs), along with the appearance of alternatives – such as increasing possibilities for the environmental acquisition of fixed nitrogen in a form of symbiotically produced ammonium cation (like in plants) or "much better" amino acids (like in animals) were the most notable factors behind the broad extinction of nitrogen fixation. Plants are still paying a high energy tax for fixed nitrogen and its conversion to amino acids (i.e., 20% of pea ATP energy budget is consumed by the symbiotic diazotrophs [58]), and animals need to evolve and carry on a set of adaptive mechanisms to obtain amino acids (e.g. mobility, digestion, plasma membrane absorption, and symbiotic interactions). Nevertheless, the increased environmental fluxes of fixed nitrogen and amino acids along with the evolution of specific membrane transport systems appear to be plausibly central to the broad compensation and extinction of Diazotrophy.

2.6 Energization Metabolism

Earlier amplification and spatial expansions of amino acid fluxes may have coincided with inventions of exotic redox reactions, supporting noncanonical lithotrophic and organotrophic cascades that explore specific combinations of inorganic and organic electron donors/acceptors and attempt connections to light harvesting energetics. For example, both phylogenetic and geological records suggest that waves of metabolic inventions corresponding with sulfate reduction in mesophilic bacteria (approx. 2.45 Bya) have induced dramatic geochemical changes that lead to additional origination events coinciding with photosynthesis, global oxygenation, and aerobic metabolism (2.2–2.3 Bya) [72]. The addition of nitrogen fixation to the nitrogen cycle also may have had a dramatic impact on the large-scale sequestration of N_2 and CO_2 in the Ocean, especially in the deep Ocean, that leads to global changes in the distribution of carbon and fixed carbon/nitrogen reservoirs [73]. A considerable amount of amino acids is produced by heterotrophic nitrogen-fixing bacteria [74], such as the free-living aerobic soil bacterium, *Azotobacter* [75] and nitrogen-fixing plant-symbiont, *Rhizobium* [76], which resolve incompatibility of aerobic metabolism and anaerobic nitrogen fixation by protecting nitrogenases with an oxygen-limiting diffusion barrier, rapid metabolic consumption, and intracellular buffering of oxygen. However, the majority of existing diazotrophs have "just enough" capacity to produce amino acids upon generic limitation of an approximate 18-times lesser energyyield of anaerobic vs. aerobic metabolism. Yet, the lion's share of present natural amino acids are produced by plants that combine efficient light energy-harvesting photosynthesis, supporting CO_2 fixation and aerobic metabolism, and perform assimilation of reduced nitrogen produced by free soil and symbiotic bacteria or are introduced in form of fertilizers (Haber–Bosch synthesis [77]). The cyanobacteria are the most important diazotrophs that actually can produce oxygen as a byproduct of photosynthesis and have evolved capacities to maintain anaerobic nitrogen fixation in the presence of molecular oxygen [78]. It is likely that the systemic coupling of oxygenic photosynthesis and nitrogen fixation played a key role in the earlier explosive evolution of cyanobacteria ancestors leading to the saturation of the atmosphere with molecular oxygen [78] and the biosphere with relatively stable and easy-metabolized carriers of fixed nitrogen in the form of free and protein-bound amino acids.

2.7 Mobilization Fluxes

Considering essential roles of the plasma membrane [79] and environmental amino acids in LUCA metabolism, it is reasonable to propose that the first transmembrane amino acid transport system translocating amino acids across purely permeable cellular membrane coincided with the membrane encapsulation of life. Hence, amino acid transport of abiotic amino acids may emerge before amino acid biogenesis and long before the culmination of life with a complete set of proteinogenic amino acids relaying on the existence of sophisticated metabolic pathways. Some popular scenarios for the origin of life suggest only a minor requirement for a prebioticamino acid pool in the event (see for example "Metabolism first" [40, 80] and "RNA word" [46–48]. Nevertheless, it is difficult to discard that the transition to the "existing cellular word" of protein-based metabolism and membrane-based bioenergetics would: (i) increase a net flux of amino acids on the planet; (ii) introduce new species of amino acids which were absent in prebiotic environments; (iii) make fluxes of amino acids more redundant by displacing a prebiotic "variety" with biotic "consistency"; (iv) increase environmental aggregation of amino acids via mobilization of amino acids in a proteins–cells–organism framework; (v) and provide deliberate environmental fluxes of amino acids inside specific nutrient chains and networks of organisms. Biological mobilization of amino acids in proteins and cells would also reduce a pool of dissolved amino acids that is anticipated for the prebiotic and earlier biotic environment. Life actually cooked a "real chunky soup" from the diluted prebiotic "bouillon." The collateral evolution of amino acid anabolism and saturation of the environment with sources of mobilized amino acids provides a basis for the dichotomy of autotrophic vs. heterotrophic forms that generalize in synthesis vs. environmental acquisition of amino acids. The biotic clearance of dissolved amino acids together with biological deconvolution of amino acid fluxes in the nutrient chains induces a selection of mechanisms which may facilitate the acquisition of amino acids from the nutrient chains including sensing, taxis, and digestion which may emerge many times in prokaryotes and become dominant traits of eukaryotic heterotrophs such as Protozoa and Metazoa, with mild exceptions for ultimate symbiosis adaptations. The origin of ecologically stable sources leads to the selection of transport-based vs. metabolism-based acquisition of proteinogenic amino acids, starting from the most metabolically expensive species.

2.8 Transport vs. Anabolism

Prokaryotes evolved and largely retained endogenous synthesis of all proteinogenic amino acids. However, a common expression: "plant and bacteria are able to produce all amino acids and mammals unable to synthesize some of them" has a largely ignored drawback. First, there are a number of prokaryotes with an unconditional or conditional depression of specific amino acid synthesis pathways. For example, exogenous amino acids can strongly facilitate bacterial growth and may be critical for symbiotic forms [81, 82]. Many prokaryotes depend on the environmental acquisition of certain amino acids, including lactic bacteria [83] and some archean thermophiles [84]. Second, about half of the amino acid synthesis cascades have been lost not only in mammals but also in other animals (Metazoa) as well as in many Protozoa, supported by the evidences of amino acid synthesis enzyme extinction in complete genomes of selected organisms [85]. The situation with fungal amino acid synthesis and requirement remain to be analyzed, but it has been shown by minimal media assay that many fungi species prefer fixed nitrogen in the form of amino acids vs. inorganic ions [86–88].

In spite of apparently independent extinctions of amino acid synthesis cascades in different heterotrophs [85], the set of essential amino acids is virtually identical across analyzed species. It includes ten of the most "expensive" to synthesize large neutral amino acids with some additions to that set as a result of ultimate symbiotic adaptations, i.e. intracellular parasites may have more essential amino acids. The elevated protein synthesis in the developing organisms or some physiological processes involving protein secretion may sharply increase demand for amino acids ultimately limiting such a secretion at the anabolic capacity of the cell. The amino acid synthesis represents a great portion of the overall energy budget in prokaryotes [89] and obviously in eukaryotes. In addition, the energy used in synthesis of different amino acids differs more than by an order of magnitude (Table 1). For example, approximately 8 vs. 79 high-energy phosphate bonds are used to synthesize a single molecule of glutamate vs. tryptophan [90, 91]. The "energy cost" has a clearly identifiable correlation with an amino acid "usage" in proteins of all three domains of life [92]. For example, such a correlation is virtually identical in *Caenorhabditis elegance* and *Homo sapiens*, which have ten essential amino acids, and it is very similar to that found in *Bacillus subtilis*, which synthesizes all essential amino acids [92].

The evolution of metabolism develops, energizes and mobilizes environmental fluxes of amino acids, which may support selection of transport versus anabolism. The facts of explosive evolution of Metazoa (e.g. Cambrian explosion) with ten essential amino acids support the notion that the replacement of synthesis by transport of aliphatic amino acids may dramatically benefit multicellular organisms. The high energetic footprint is only a fraction of the aliphatic amino acid anabolism problem. The synthesis of large aliphatic amino acids requires an increased number of enzymatic steps; for example glutamate synthesis requires one enzymatic step, whereas the tryptophan syntheses involve about 11 additional enzymatic reactions [91]. Such an extension of the integrated cascade leads to the dramatic increase in the execution time and housekeeping maintenance of the cascades that include recycling of enzymatic framework and accumulation of reactive intermediates. In addition, longer enzymatic cascades with larger genetic footprints appear to be at higher chance of knockdown upon random mutations and physiological stress. The enzymatic complicity and metabolic costs are increasing in the set of large neutral amino acids including the most expensive aromatic amino acids (Fig. 1, Table 1). The penalty for the escalation of metabolic and genetic footprints is extinction. Hence, for billions of years amino acid anabolism evolved under assaults from a diversity of biological mechanisms providing alternatives to extensive anabolism of amino acids. This includes frequency reduction of energetically expensive amino acids in protein sequences [92], ultimate reduction of metabolic rates [93, 94], symbiotic adaptations [82], and, perhaps one of the most successful attacks leading to the extinction of a half of amino acid synthesis cascades in Metazoa, active acquisition of neutral amino acids [95].

Obviously, the syntheses of branched amino acids with tremendous energetic and genetic footprints are the first tradeoffs upon heterotrophic adaptation to a steady-state external supply of these amino acids. Compared to the expensive synthesis, the cost of environmental acquisition via membrane transport mechanisms is small and uniform in the set of proteinogenic amino acids. For example, transport of glutamate and tryptophan by distinct sodium ion gradient coupled transporters, for example specific members of SLC1 and SLC6 families have similar energy price tags. The members of these families may utilize similar but varied electrochemical mechanisms including different ion-motive forces and substrate-ion-coupling stoichiometries (e.g. distinct combinations of Na^+, K^+, H^+, and Cl^- ions). They may adapt to act against different organic substrate gradients (e.g. an order difference in an electrochemical substrate gradient may require just one additional sodium ion per translocation). Nevertheless, the overall cost of a single translocation event which is mediated by the active secondary transporters would range from 0.5 up to 3 ATP bonds whose hydrolysis may be sufficient to recover electrochemical gradient of Na^+, K^+, H^+, and Cl^- discharging by the event. For example, P-type Na/K ATPase moves 3 Na^+ and 2 K^+ against their electrochemical gradients and plasma membrane H^+V-ATPase recovers 2 H^+ by hydrolyzing a single ATP molecule. Auxiliary mechanisms of ion homeostasis may be involved without significant increase in overall energy cost. In addition, assembling and maintaining of the amino acid transport network is far simpler vs. that required for synthesis for each amino acid in each individual cell. It may be limited to generic synthesis and traffic of a single transport protein for the accumulation of the proteins for translocation of a specified amino acid, given that metabolic – electrochemicalenergy-coupling pumps are present. Some auxiliary processes of transport assembling such as homo- and heteromerization, interaction with an enzymatic complex, and post-translational modifications add too little complicityto reverse selection of transport vs. anabolism.

2.9 Autotrophy vs. Heterotrophy

Plants comprise an example of generalized autotrophy. Plants declared independence from the environmental pool of proteinogenic amino acids upon invention of light energy-harvesting photosynthesis (approx. 2.5 Bya) that energizes the assimilation of fixed nitrogen and CO_2, and supplies the anabolism of proteinogenic amino acids. In plants, as in all other organisms, amino acids are central carriers of fixed nitrogen [96]. Amino acid redistribution is critical for metabolic integration of plant organism, because the capacities to produce amino acids are limited in many plant tissues [97]. For example, roots of plants assimilate fixed nitrogen; however, the protein synthesis in the root tissues depends on the systemic delivery of free amino acids and/or metabolic fuel that would energize the amino acid anabolism. Leaves produce metabolic fuel by means of photosynthesis and CO_2 fixation but leaf tissues would also require the delivery of either fixed nitrogen or free amino acids for protein synthesis. The role of the transport acquisition of amino acids is even more explicit in other plant tissues. A high demand for transport of amino acids is recognized in specific physiological processes, including plant development and regeneration. For example, early development and germination mainly rely on membrane transport mechanisms. The majority of amino acids are transported via the vascular system from the site of primary production or storage to other tissues with limited ability to produce amino acids. A specific network of plasma membrane transporters performs the systemic allocation and intracellular partitioning of amino acids [97]. However, plants may also express transporters for environmental assimilation of free amino acids [98] from decomposing or nitrogen-fixing bacteria, parasitic interface with a host organism [99], and captured prey digestion in some carnivorous plants [100, 101]. Several comprehensive reviews on the molecular basis and regulation of membrane transporters in plants have been published during the last few years providing a guideline for the future development of plant transport biology [97, 102–104]. In addition to transporters identified within plastids and mitochondria, the analysis of the complete genome of *Arabidopsis thaliana* [105] revealed over 50 amino acid transporters [97, 103] comprising two previously identified superfamilies [97, 106, 107]: the Polyamine and Choline Transporter (PCT) superfamily (the plant APCcluster is a counterpart of the mammalian SLC7 family), with 14 members; and the amino Acid Transporter Family (the plant ATF cluster is a counterpart of the mammalian SLC38) with 46 identified members. Comparative phylogeny of plant, fungi, and animal amino acid transporters has been based on the available genomes of selected representatives of the Kingdoms [97, 102–104]. The characterized plant amino acid transporters differ by substrate specificity and affinity, spatial and developmental expression profiles, and environmental control of expression and activity (reviewed in [97]). Photosynthesis played a key role in the conservation of comprehensive amino acid synthesis and simplified amino acid transport in a major part of the plant Kingdom. However, the balance of these functions may experience remarkable shift in the plants exploring active heterotrophy. Heterotrophy in plants has emerged at least 12 times [108], and it evidently produced a dramatic impact on metabolism; i.e. loss of photosynthetic capacity in the parasitic genus *Cuscuta* (dodder) [109], and acquisition of a sensory modality in *Cuscuta pentagona* using volatile cues for host location [110]. These adaptations reflect a micro-scenario of metazoan evolution that drives the substitution of essential amino acid synthesis by environmental acquisition and leads to the expansion of transport mechanisms performing absorption and allocation of essential amino acids in organisms. We also may expect differences in the physiology and genetics of amino acid transport in unicellular alga dominating the aquatic and terrestrial ecological scenes. Recent sequencing and annotation of the *Chlamydomonas reinhardtii* genome revealed that this organism largely retained the transporter diversity present in the common plant-animal ancestor [111], however the diversity of amino acid transporters in this practically and theoretically important model organism remains to be analyzed.

Protozoa and Metazoa represent two major steps in the evolution of heterotrophy. One half of the proteinogenic set of amino acids is essential in mammals and insects, as well as presently analyzed metazoan and protozoan organisms with a unique published exception [112]. Together with apparently independent extinction of essential amino acid synthesis cascades in different organisms [85], it may suggest that the evolution of amino acid essentiality, expressed as a selection of the environmental (nutrient) acquisition vs. synthesis for a particular set of amino acids (Fig. 2), is convergent in heterotrophs. The synchronized knockdown of endogenous syntheses of a half of the proteinogenic amino acids would be a devastating roadblock, hardly possible to bypass. So, it is reasonable to assume that the expansion of the essential amino acid set was a gradual process in which environmental acquisition via emerging membrane transporters replaced one by one anabolism of intracellular amino acids. Such a process may arise in complex bacterial communities that may share the costs of amino acid synthesis, thus restraining circulation of biologically accessible nitrogen within an interspecific amino acid recycling network. The steady commensalism may be sufficient to reduce the requirement of interacting members of the quorum to maintain synthesis of all proteinogenic amino acids and provide an opportunity for the specialization of the interacting species in the synthesis and transport of specified proteinogenic amino acids. It may support a collapse of specific amino acid synthesis cascades along with expansion of amino acid transport systems. It requires a hierarchy of metabolism and membrane transport, demanding sensory signaling and coordination of the metabolic and transport labor from individuals of the quorum. Similar metabolic hierarchy and integration were proposed as key factors in the origin of the last universal metazoan ancestor (LUMA) [113] and LUCA, which was anticipated to be "a complex community of protoeukaryotes with a RNA genome," adapted to a broad range of moderate temperatures, genetically redundant, morphologically and metabolically diverse [114].

To acquire amino acids, metazoan organisms evolved sophisticated sensory navigation, active taxis, and effective enzymatic extraction and cleavage of dietary proteins, digestion. In addition, virtually all metazoans acquire and support some symbiotic organisms or systems of organisms as amino acid providers that may notably reduce or completely eliminate the requirement of external essential amino acids. In addition, many protozoan and metazoan organisms are able to remain intact as plant cellsor chloroplasts to support demanding requirements of essential amino acids (for example [115]). Obviously, symbiotic compensations of the environmental acquisition do not reduce the generic requirement for epithelial absorption and cellular redistribution of essential amino acids.

Heterotrophic acquisition of amino acids is the major trade-in in metazoan evolution, dramatically reducing time and energy devoted to amino acid synthesis and, respectively, development (Sect. 2.8, Transport vs. Anabolism). However, it is collateral with the major metazoan trade-off, the loss of a half of amino acid synthesis cascades. Hence, the transport vs. anabolism of essential amino acids represents the key metazoan alternative fitting a plausible scenario of metazoan evolution. The origin and subsequent optimizations of light-harvesting systems over a last half of the Proterozoic Eon [116–118] culminates with dramatic spatial expansion of new biocenoses beyond a constrain of the geothermal "Goldilocks" layer (not too hot – not too cold), leading to explosive saturation of biospheres by the "difficult to build" biotic amino acids (Table 2). The increasing availability would normally correlate with elevated incorporation of the "difficult" vs. "prebiotic" amino acids in concomitant protein sequences, which is supported by the expansion of genetic code [119] and amino acid frequency trends in reconstructed ancestral proteomes [53, 54]. An increased flux of amino acids generates a layer of new opportunities for heterotrophs. Transport-mediated acquisition of the metabolically extensive amino acids yields remarkable energy equivalents and drastically reduces time of proliferation. On the one hand it provides obvious selective advantages to organisms with a best framework for the absorption and redistribution of environmental amino acids. On the other hand, it drives the extinction of amino acid synthesis cascades and demands ecological specialization leading to the divergence of a metazoan domain and formation of specific nutrient chains. The key metazoan modality, sensation and mobility evolved to access essential amino acid fluxes. It is plausible that the brain evolved to coordinate these modalities.

Table 2 Families and systems of amino acid transporter

Type	TCDB #	Mammalian SLC members	Insect protein	Transport systems	Substrate selectivity (weak substrate)	TMDs	Carrier type	Apparent affinity	Stoichiometry ↓ absorption ↑ emission	References
EAAT	2A23.2	SLC1A1, 2(EAAT3,2)	EAAT2 (dEAAT2)	X^{-}_{AG}	E,D ($K_m{}^{D} << K_m{}^{E}$ in dEAAT2)	8	S	H	↓$3Na^{+}Sb^{-}H^{+}$↑K^{+}	[169]
		SLC1A3,6, 7(EAAT1,4,5)	–	X^{-}_{AG}	E,D	8	S	H	↓$3Na^{+}Sb^{-}H^{+}$↑K^{+}	
ASCT	2A23.3	SLC1A4,5		ASC	A,S,C, (T,Q)	8	S E	L,M L,M	↓$Na^{+}Sb[Cl^{-}]$ ↓$Na^{+}Sb$↑Sb	
		insect specific	EAAT1 (dEAAT1)	X^{-}_{AG}	E,D	8	S	H	↓2–$3Na^{+}Sb^{-}H^{+}$↑K^{+}	
			EAAT3 (AeaEAAT)	X^{-}_{AG}	E,D	8	S	H	↓2–$3Na^{+}Sb^{-}H^{+}$↑K^{+}	
GABAT	2A22.3	SLC6A1	–	β	GABA, neuronal high affinity	12	S	H	↓2–$3Na^{+}GABA[Cl^{-}]$	[229]
DAT, NET	2A22.1	SLC6A3,2	tnOCT, tnDAT, dDAT		DA, OA, NE	12	S	H	↓$2Na^{+}Sb$↓$K^{+}[Cl–]$	
SERT	2A22.1	SLC6A4	ms5HTT, dSERT		5HT	12	S	H	↓$2Na^{+}5HT$↑$K^{+}[Cl–]$	
GlyT, ProT	2A22.1	SLC6A5, 7,9	–	Gly	G,P	12	S	H	↓2–$3Na^{+}Sb[Cl–]$	
		SLS6A8, 6,11.12,13	–	GABA/Bet, CT	GABA, betaine, creatine	12	S	H	↓$2Na^{+}Sb[Cl–]$	
		SLC6A14	–	B^{0}+ (ATB0+)	K,R,A,S, C,T,N,Q,H, M,I,L,V,F, Y,W	12	S	M	↓$2Na^{+}Sb[Cl–]$	[456]
IMINO SIT1	2A22.6	SLC6A20	–	IMINO	P, HO-P	12	S	M	↓$Na^{+}SB[Cl–]$	[330, 331]
B^{0}AT1	2A22.6	SLC6A19	–	B^{0}	F,C,H,A, S,M,I,Y, T,G,N,P	12	S	L	↓$Na^{+}SB[Cl–]$	[136]

Table 2 (continued)

Type	TCDB #	Mammalian SLC members	Insect protein	Transport systems	Substrate selectivity (weak substrate)	TMDs	Carrier type	Apparent affinity	Stoichiometry ↓ absorption ↑ emission	References
B^0AT2		SLC6A15	–	B^0	P,L,V, I,M	12	S	H	↓Na^+SB[Cl–]	[137]
NAT	2A22.2	insect specific	DmNAT1	B^0-like	F,C,H,A,S, M,I,Y,T, G,N,P (D&L AA)	12	S	M	↓Na^+Sb[Cl–]; ↓K^+Sb[Cl–]	[342]
			MsKAAT1		F,C,H,A, S,M,I, Y,T,G, N,P	12	S	M	↓2Na^+Sb[Cl–]; ↓2 K^+Sb[Cl-]	[304]
			MsCAATCH1		F,C,H,A,S, M,I,Y,T, G,N,P	12	S	M	↓2Na^+Sb[Cl–]; ↓2 K^+Sb[Cl-]	[317]
			AeAAT1		F,C,H,A, S,M,I,Y, T,G,N,P	12	S	M	↓2Na^+Sb[Cl–]; ↓2 K^+Sb[Cl-]	[95]
			AgAAT6		W,Y, 5-HT,F,A	12	S	H	↓2Na^+Sb[Cl–]	[336]
			AgAAT8		Y,F,L-DOPA,W,5-HT	12	S	H	↓2Na^+Sb[Cl–]	[126]
CAT	2A3.3	SLC7A1	–	y+	R,K,H	14	U,E	L,M	↓Sb; ↓Sb↑Sb^+	[327]
		SLC7A2	–	y+	R,K,H	14	U	L	↓Sb	
		SLC7A3	–	y+	R,K	14	U	L	↓Sb	
		SLC7A4	–		–	14	–	–	–	
HAT	2A3.8.	SLC7A6	–	y+L, y+LAT2	K,R,Q,H, M,L,A,C	12+AP^1	E	M,H	↓Sb^+↑Sb; ↓Na^+Sb↑Sb^{+0}	
		SLC7A7	–	y+L, y+LAT1	K,R,Q, H,M,L	12+AP^1	E	M,H	↓Na^+Sb↑Sb^+	
LAT1	2A3.8	SLC7A5	–	L, LAT1	H,M,L,I, V,F,Y,W, (Q)	12+AP^1	E	↓H↑L	↓Sb↑Sb	
LAT2		SLC7A8	–	L, LAT2	A,S,C,T, N,Q,R,H, M,L,I,V,F, Y,W	12+AP^1	E	M	↓Sb↑Sb	
		SLC7A10	–	Asc1	G,A,S,C,T (D & L isoforms)	12+AP^1	E	–	↓Sb↑Sb	

Table 2 (continued)

Type	TCDB #	Mammalian SLC members	Insect protein	Transport systems	Substrate selectivity (weak substrate)	TMDs	Carrier type	Apparent affinity	Stoichiometry ↓ absorption ↑ emission	References
		SLC7A11	–	x^{-}_{c}(xCT)	E, C– (substrate for GSH)	12+[AP[1]]	E	H	$\downarrow C^- \uparrow E^-$	
		SLC7A9	–	rBAT, $b^{0,+}$	K,R,A,S, C,T,N,Q, H,M,I,L, V,F,Y,W	12+AP[1]	E	M	$\downarrow Sb^{+0} \uparrow Sb$	
PEPT	2A17.4	SLC15A1	–	–	di-, tree-peptides	12	S	L	$\downarrow 2H^+Sb^0$; $\downarrow 1–2H^+Sb^{+/-}$	[384]
		SLC15A2	–	–	di-, tree-peptides	12	S	H	$\downarrow 2H^+Sb^0$; $\downarrow 1–2H^+Sb^{+/-}$	
PHT1,2	2A17.4	SLC15A1,2	–	–	histidine, di-, tree-peptides	12	S	–	$\downarrow 2H^+Sb^0$; $\downarrow 1–2H^+Sb^{+/-}$	
MCT	2A1.13	SLC16A1,7,8,3	–	MCT	lactate, pyruvate, ketone bodies	12+AP[2]	SE	LM	$\downarrow H^+Sb$; $\downarrow Sb \uparrow Sb$	[401]
		SLC16A4, 5,6,11, 12,13	–		–	12	–	–	–	
		SLC16A2	–	T3,T4	D,L aromatic amino acids	12	U	ML	$\downarrow Sb$	
	2A1.13.2	SLC16A10	–	TAT	D,L aromatic amino acids	12	U	ML	$\downarrow Sb$	
vGutT	2A1.14	SLC176,7,8			Glu		E	H	$\downarrow H^+ \uparrow Sb$	[412]
vMOAT	2A1.2	SLC18A1,2			5-HT, DA, NA, HA	10–12	E	H	$\downarrow H^+ \uparrow Sb$	[457]
		SLC18A3			ACh	10–12	E	H	$\downarrow H^+ \uparrow Sb$	
vGAT	2A18.5	SLC32			GABA	10	E	L	$\downarrow H^+Sb$	[444]
PAT	2A18.8	SLC36A1	–	PAT	P,G,A, GABA, β-Ala	11(9)	S	L	$\uparrow SbH^+$ (lys); $\downarrow SbH^+$ (ep)	[458]

Table 2 (continued)

Type	TCDB #	Mammalian SLC members	Insect protein	Transport systems	Substrate selectivity (weak substrate)	TMDs	Carrier type	Apparent affinity	Stoichiometry ↓ absorption ↑ emission	References
		SLC36A2	–	Tramdorin 1	P,G,A	11(9)	S	M	$\uparrow SbH^+$ (lys); $\downarrow SbH^+$ (ep)	
		SLC36A3,4	–		–	11(9)	–	–	–	
SNAT	2A18.6	SCL38A1	–	A1, ATA1	Q, N,H,S, G,A,C,M	11	S	L	$\downarrow Na^+Sb$	[445]
		SCL38A2	–	A2, ATA2	Q, N,H,S, G,A,C,M,P	11	S	L	$\downarrow Na^+Sb$	
		SCL38A3	–	N1 , SN1	Q,N,H	11	SE	L	$\downarrow Na^+Sb\uparrow H^+$	
		SCL38A5	–	N2, SN2	Q,N,H, S,G	11	SE	L	$\downarrow Na^+Sb\uparrow H^+$	
		SCL38A4	–	A3, ATA3	N,S,G, A,C,M,P	11	S	L	$\downarrow Na^+Sb$	
LAT4	2A1.44	SLC43A1	lost in insect	L	L,I,M,F	12	U	L	↑S	[459]
		SLC43A2		L	L,I,M,F	12	U	L	↑S	

Abbreviations and special signs: + and –, positive and negative charges; AP^1, essential auxiliary subunit of SLC7 transporters (4F2hc/CD98 or rBAT); AP^2, ancillary protein (CD147); U, uniporters (facilitated diffusion), E, exchanger (antiporters); S, symporters (co-transporter); C, channels (high throughput tunneling of inorganic ions); Sb, substrate; H, M, L, apparent substrate affinity high< 10^{-5}, medium 10^{-3}–10^{-4}, and low > 10^{-3} respectively; lys, lysosomal membrane; ep, epithelial membrane; ves, vesicular membrane; na, not available. *Ligands*: a, some amino acid (specific amino acids are shown according to a single capital letter nomenclature); HA, histamine; DA, dopamine; 5HT, serotonin; NE, norepinephrine (noradrenalin); EP, epinephrine (adrenaline); ACh, acetylcholine; Ch, choline; GABA, gamma amino butyric acid; DOPA. EAAT, excitatory amino acid.

The extinction of essential amino acid synthesis in the protozoan heterotrophs may coincide with the invention of endocytosis, vacuolar digestion, and vacuolar transport delivering amino acids from the digestive vacuoles to cytoplasm. Hence the vacuolar digestion and transport of amino acids are secondary adaptations that may derive from the plasma membrane-coupled digestion and transport of earlier heterotrophs. Such an adaptation may evolve under pressure of the declining flux of dissolved carriers of fixed nitrogen and increasing density of mobilized amino acid sources (see Sect. 2.7, Mobilization Fluxes). The internalization of source provides more efficient acquisition of nutrients, and drives the origin of cellular organelles, and also changes the environment of secondary transporters. The secondary transporters of intracellular organelles have to deal with steady-state low intracellular sodium and elevated pH gradients. Indeed, many secondary protozoan transporters explore proton motive forces while some families of sodium ion-dependent secondary transporters are expired and may be entirely extinct in protozoa that specialize in endocytosis and vacuolar digestion of nutrients.

Metazoan organisms minimize the use of endocytosis and intracellular digestion to a set of specific processes, cell types, and tissues. Some primitive and parasitic forms can use endocytosis as a primary nutrition mode, i.e. flat worms and ticks. However, in the majority of metazoan organisms nutrient amino acid absorption occurs in the alimentary canal, which is evolved to extract nutrients from solid-state sources. The alimentary canal conducts extracellular digestion and absorption of nutrients and can deal with a larger subject compared to that accessible by cellular endocytosis and vacuolar digestion. The primary absorption of digested amino acids from the lumen of the alimentary canal to the body cavity and from the body cavity or circulatory system to the individual cell normally occurs against electrochemical gradients. Conceptually, amino acid transporters with specific selectivity, affinity, electrochemical dependency, and spatial expression patterns form a comprehensive transport network coupling the environment and metabolism of each individual cell, tissue, and organs of a metazoan organism. Similar to plants, different cells and tissues of metazoan organisms have different demands for amino acids. In contrast to plants, animals are unable to synthesize a half of the proteinogenic amino acids; so, the transport of these amino acids is the ultimate requirement of protein synthesis of each metazoan cell. Metazoan anabolism of dispensable amino acids is also limited by the accessibility of the metabolic energy pool that is not uniform over the animal tissues and relies exclusively on the catabolism of organic nutrients; so, the transport of dispensable amino acids may also be an important limiting factor in protein synthesis and other metabolic cascades. This implies that each essential amino acid, as well as each conditionally essential dispensable amino acid, requires an active delivery system. The system for essential amino acids has to include at list three components: apical and basal transport in the epithelial tissues, and plasma membrane transport of each individual cell. The system for conditionally essential amino acids may not require epithelial transport components but need an additional transporter for the efflux of amino acid from synthesizing cells.

A short-term disorder of the essential amino acid transport can be lethal for an organism because storage and reuse of these amino acids in the animal cells is very limited. It is generally accepted that a lion's share of amino acids is absorbed via broad substrate spectra low affinity and selectivity peptide transporters, which act in synergy with low-specificity peptidases. The theoretical drawback of such a system is the limited ability to deliver specific and especially under-represented amino acids on demand. Consider your reaction if you ordered a beefsteak and received a substitute combination that includes three cups of coffee, a glass of lemonade and a cup of tea, just about the same energy yield. In addition, the peptide transporters (see SLC15) utilize proton motive forces and rely on a pool of available protons, which can be very limited in the neutral and mild basic conditions existing in the major absorptive regions of the alimentary canal as well as in the internal circulation system (e.g. concentration of 10^{-7}–10^{-9} M of OH_3^+ vs. 10^{-2}–10^{-3} for Na^+). So, a proton-dependent peptide transport system may play a certain role in bulk amino acid absorption but cannot substitute specific amino acid transport systems that are crucial for the alimentary absorption and specific redistribution of amino acids between cells and tissues. Even after the recognition that di- and tri-peptides can be absorbed by specific apical transporters, we still do not know the quantitative importance of these mechanisms in overall amino acid absorption [120].

3 Phylogenetic and Functional Diversity of Amino Acid Transport

The diversity of transport mechanisms increases with the evolution of metabolic and cellular complexity of organisms; however, strong selection and extinction of inefficient transport systems also occurs. Unsurprisingly, different organisms may use distinct transport systems to balance the same intracellular substrates and metabolites. Prokaryotic domains comprise the greatest diversity of amino acid transport mechanisms including ABC pumps [121], secondary transporters, and permeases. The phylogenetic aspect as well as occurrence, function, and biotechnological value of prokaryotic amino acid transporters have been summarized in recent reviews [122, 123]. Remarkably, most amino acid transporters in prokaryotes are secondary transporters that utilize ion motive forces. For example, only 27% (13 of 48) of prokaryotic ABC families aid in the transport, predominantly efflux, of amino acids; whereas 90% (19 of 21 families) of secondary transporters mediate transport of amino acids [122]. Some of these transporters share common phylogenetic roots across life kingdoms [103]. The indispensable role of amino acids in protein synthesis and nitrogen metabolism anticipates that most amino acid transporters have to emerge earlier in the history of life to maintain particular nodes of the amino acid transport network. Indeed, 9 of 13 families of eukaryotic amino acid transporters share common ancestors among 19 families of prokaryotic transporters [122]. In contrast to the apparent reduction of phylogenetic diversity, virtually every family of eukaryotic amino acid transporters undergoes notable expansions associated with specialization and amplification of the transport functions. It becomes increasingly clear that amino acid transporters represent a very dynamic entity, with a remarkable degree of physiological and evolutionary plasticity [95, 124, 125]. Analysis of recently available genomes suggests that evolution of structural complicitycorrelates with metabolic specialization of cells and expansion of membrane transport mechanisms rather than the invention of new metabolic cascades. Amino acid transport is critical for regulation of systemic fluxes and local concentrations of amino acids and it must work in precise coordination with energy conversion between metabolic storage and electrochemical gradients as well as the mineral ion balance of cells and organisms. Each amino acid requires specific molecular carriers to pass lipid bilayers of cellular membranes. All essential and most dispensable amino acids require active transport against electrochemical gradients because their intracellular concentrations are higher vs. extracellular concentrations, except for amino acids excessively synthesized in the cell. The transport requirement differs in different cells, tissues and organisms, depending on the local (A)nabolism, (C)atabolism, and (T)ransport ratios, e.g. $A–C+T>=0$. For example, if $A > C$ then $T < 0$ requiring efflux of amino acids. In contrast, if $A < C$ or A equals 0 (requirements for the dispensable amino acid often exceed biosynthesis and the essential amino acid anabolism equals 0, Table 1, Fig. 2) then T is > 0, indicating the requirement of active, energy-coupled transport of the amino acid. Hence, we may expect the existence of mechanisms that enable apical uptake (1) and basal efflux, (2) in epithelial tissues associated with amino acid absorption in the alimentary canal, as well as plasma membrane uptake, (3) for each essential amino acid. Ionophoretic mechanisms must be present in the position 1 and 3, while in position 2 it may be selective permeases and/or an organophoretic transporter, an amino acid coupling exchanger. The properties of amino acid systems together with recent studies of cloned transporters showed that single transporter genes may encode carriers for several amino acids with similar properties (i.e. broad substrate spectra transporters); however several more selective transporters (narrow spectra transporters) are necessary to reduce competition of amino acids for binding sites and to increase transport efficiency [125, 126]. A high affinity and selectivity uptake is obligatory for the absorption of under-represented substrates from a mixture, implying that a large number of transporter phenotypes may exist for essential amino acids. Dispensable amino acids: glycine, alanine, serine, proline, aspartate, glutamate, tyrosine, cysteine and their derivatives can be synthesized de novo in organisms but the capacity to produce dispensable amino acids depends on many metabolic factors and appears to be very different in specialized cells and tissues. Hence, essential roles of transporters for distribution of nonessential amino acids have evolved and have become more prominent with increased histological complexity, cellular specialization, and developmental complicity of organisms. In addition to the population of plasma membrane amino acid transporters, a number of specific transporters may be necessary for translocation of amino acids between different intracellular domains. Moreover, many metazoan organisms evolved with radical physiological and metabolic transitions between developing and developed stages (e.g. larvae and adult). These transitions imply dramatic differences in amino acid requirements, which, together with certain requirements for genomic and phenotypic plasticity of amino acid transport mechanisms, may result in an increase of amino acid transporters being involved in organism homeostasis. Some of these stage-dependent transporters may remain silent in the genome until activated.

3.1 Systematics of Amino Acid Transporters

3.1.1 Amino Acid Transport Systems

Studies by Christensen and colleagues during the last few decades of the twentieth century led to the discovery of several types of amino acid transport activities which can be identified in different cells, tissues and organisms based on substrate specificity and major electrochemical characteristics [127, 128]. The identifiable functional specificity and electrochemical profile of membrane transport was designated as a specific "transport system" and funded the Amino acid Transport Systems classification (ATS) [129, 130]. Further refinements in system discrimination may require more intensive comparative analysis of kinetic, thermodynamic, physiologic, and genetic properties of the transport mechanisms [131]. It was postulated that individual ATS might serve as transport for several substrates with a consensus of physicochemical characteristics (e.g. acidic, basic, etc., see Fig. 2). The selectivity toward a particular physicochemical amino acid group has been used as a major integer in the name of specific ATS. In addition, capital vs. small case and some other modifiers were used to reflect Na^+ dependency vs. independency and some other distinctive properties of a particular transport system [130, 132]. Current ATS nomenclature differentiates from 16 to 18 [103, 133, 134] up to 21 [135] amino acid transport systems (Table 2). It also has been recently recognized that ATS signature may reflect a mutual activity of a transport complex where properties of individual components are difficult to discriminate. Hence, at a molecular level, ATS typically represent a single transporter, but also may represent a mutual activity of a few spatially colocalized, most likely phylogenetically related but sometimes unrelated, transport mechanisms that require additional effort to discriminate in vivo or in vitro assay systems. For example, two recently cloned B^01 [136] and B^02 [137] transporters comprise a broad substrate spectra (B^0) system transporter [138]. On the other hand, phylogenetically related transporters might represent different ATS. The original ATS nomenclature was based on the study of mammalian cells, and ATS activity can be traced across different tissues with a mild variety in the phenotypes across different mammalian model organisms [131, 134]. In addition, ATS-like activities can be recognized in invertebrates, fungi, plants, archaea, and bacteria despite reduced similarity of the electrochemical properties [103]. The phylogenetic relationship between newly cloned representatives from mammalian and nonmammalian ATS supports common ancestry of such correlates rather than convergence of the phenotypes. Hence, it also suggests an earlier origin and conservation of phenotypes in a major group of amino acid transporters. At the same time, cloning of new transporters reveals that ATS nomenclature may be insufficient to accommodate a variety of newly characterized phenotypes from nonmammalian and even recently diverged mammalian species. It is becoming increasingly difficult to interpolate the ATS systematically across biological species. The differences in the genetic and functional aspects of homologous amino acid transport mechanisms are increased with phylogenetic distances. It may involve alterations in substrate spectra, substrate affinity, and even ion dependency. It may also involve variations of gene numbers comprising a particular system-like unity. In summary, although ATS classification represents a very important step in the understanding of amino acid transport and has given insight into the parallel evolution of amino acid transport mechanisms, it has limited capacity to integrate new dimensions of postgenomic transport biology and provide a cross-specific classification along with functional inference of individual amino acid transporters.

The increasing number of cloned and characterized transporters from different species, along with the possibility to reveal sequence-function homology patterns in the growing databases of complete genomes, raises the possibility to establish a phylogenetic relationship between all existing transport phenotypes. Attempts to review genomic varieties of amino acid transporters naturally lead to the evolutionary-based universal systematics of transport mechanisms across transporter groups and biological species. Building such a scheme is absolutely critical for understanding an integrative and evolutionary basis for membrane transport. There are two equally important efforts of phylogenetically derived classifications of secondary transporters: the universal Transporter Classification Data Base (TCDB [139–141]; and SoLute Carrier family classification (SLC; originally proposed by the Human Genome Organization (HUGO) Nomenclature Committee Database). Both of these schemes include an amino acid transporter classification as a significant portion of their systematics.

3.1.2 Transporter Classification Data Base, TCDB

The TCDB classification aims to provide a universal nomenclature for transport mechanisms across a variety of biological species [142, 143]. It was developed and updated by Saier's laboratory Bioinformatics Group. It combines characterized phenotypes linked to over 10,000 relevant membrane mechanisms and ranges from the most structurally simple ion permeases to the most complex ion channel receptors, electrochemically driven transporters, and chemical energy-coupled pumps. In another dimension it incorporates many representative transporters cloned and characterized from different prokaryotic and eukaryotic model organisms. The TSDB is freely accessible via http://www.tcdb.org/ (see also http://www.chem.qmul.ac.uk/iubmb/mtp/ and http://en.wikipedia.org/wiki/TCDB). Presently it includes about 550 transporter families differentiating approximately 3,000 sequenced proteins, for which the annotation is supported by a representative set of cloned and expression-characterized transporters and literature references [144]. The TCDB used a classification scheme similar to that of enzyme nomenclature (a.k.a. Enzyme Commission number, EC). A classification key represents a 5-position string (see [142] and http://www.tcdb.org/). The first number is a transporter class (from 1 to 7), transport-coupled auxiliary components (8) or incompletely characterized presumable transporter (9); the second letter represents subclasses inferred from the energy coupling; the third number is a phylogenetic family; the fourth number is the subfamily (author remark: the fourth number often represents distinct families or phenotypes instead of a subfamily); and the last number represents a substrate specificity. The TSDB framework is useful for a large-scale revision of transport mechanisms and phylogenetic inferences of transporter phenotypes to genomic annotations, which however should be used with special care, especially for the inference of paralogous and orthologous mechanisms. The amino acid transport aspect of TCBD has been reviewed [140, 145]. All together TCDB recognizes 11 families of metazoan amino acid transporters that belong to six superfamilies (Table 2).

3.1.3 SoLute Carrier Family (SLC)

The SLC classification is focused on human transporters, which are sorted as SLC#a# members, where the first and second # represent particular family and family member numbers, respectively. Notably, it includes comprehensive referencing beyond the human transporters, for example rat and mouse, and is a good starting point for comparative phylogenetic analysis of transporter families in other model organisms with complete genomes as well as individually cloned transporters. It represents a very convenient experimental framework to use for systematic and comparative revision of eukaryotic, especially metazoan, transport mechanisms (e.g. [103, 146], and it is broadly used in the phylogenetic comparison of newly cloned transporters from mammalian and invertebrate model organisms). The SLC table was originally prepared by the authors of the SLC mini-review series of Pflügers Archives [147] (see also http://en.wikipedia.org/wiki/Solute_carrier_family for the sorted references of the review articles). The tabulated version of the SLC families and their members as well as relevant hyperlinks to genetic and proteomic databases and key references in the literature is maintained at http://www.bioparadigms.org/slc/menu.asp. The phylogenetic family of transporters are easy to justify using simple phylogenetic analysis, BLAST, by retrieving proteins with an average sequence identity >15% (i.e. cutoff $< 10^{-9}$; the value can be higher for interspecific analysis, or lower to retrieve evolutionary related families), or the BLINK tool (NCBI). At the time of this review the SLC classification has been virtually completed. It includes 47 families comprising 362 human transporters. The SLC series does not include members of transport protein families which have previously been classified by other widely accepted nomenclature systems including: primary active FF, P, and ABC pump coupling transport to ATP hydrolysis; electron transport coupled membrane mechanisms; and ion channels and aquaporins.

Present SLC classification differentiates 11 families whose members all represent or include at least one amino acid transporter (SLCs 1, 6, 7, 15, 16, 17, 18, 32, 36, 38, 43; Table 2). The incremental understanding of molecular and integrative physiology of amino acid transport families expanded by complete genome annotations and molecular cloning-analysis of new representatives has been the subject of several recent reviews [103, 133, 134, 146]. The diversity of mammalian transporters also was a subject of several comprehensive reviews focusing on the molecular basis of inherited transport disorders [134, 148, 149] and transport-mediated toxicity [150]. Evolutionary and comparative aspects of the amino acid transporter families are reviewed in the following sections. The example of a comparative genome-wide phylogenetic aspect of secondary amino acid transporters from model dipterans, *Drosophila melanogaster* and *Anopheles gambiae*, is presented in Fig. 3.

3.2 SLC1 Family, Small Neutral and Acidic (Excitatory) Amino Acid Transporters (ASC, EAAT Families)

SLC1 comprises sodium and proton-dependent transporters for small neutral amino acids (alanine, serine, conditionally essential cysteine and threonine; ASC or ASCT system) and acidic amino acids (L-Glu and L/D-Asp; EAAT or X^{-}_{AG} system) (A, S, C, T and D, E in Fig. 2). On the basis of the currently available genomes, the SLC1 family exists in animals, bacteria, and archaea (Fig. 4), but apparently is extinct in protozoans, fungi, red algae, and angiosperm plants. Interestingly, the genome of single-celled heterotrophic (facultative autotrophic organism) motile green alga with well-developed phototaxis, *C. reinhardtii,* remains one SLC1 member with high homology to HsSLC1a2 (Fig. 4). Representative archean genomes include a single SLC1 member, while bacterial genomes may have expansions of up to four members of the family. For example, four SLC1 genes of the *B. subtilis* represent: one C_4-dicarboxylic acid transporter (DctP), two glutamate transporters (GltP and GltT), and an L-cystine transporter (YhcL) [151–154]. C_4-dicarboxylates are required for symbiotic nitrogen fixation, which correlates with the apparent expansion of DctP group transporters in the representative nitrogen-fixing bacteria (e.g. *Azotobacter vinelandii*, Az in Fig. 4).

Mammalian genomes have the largest SLC1 family (Fig. 4, HsSLC1). This includes seven SLC1 genes that are orthologous between different mammalian species, which are clearly divided into two subfamilies: a two-gene subfamily representing mammalian ASC systems, which encode transporters for small neutral amino acids, and a five-gene subfamily representing X^{-}_{AG} systems, which include glutamate transporters of the EAAC1, GLT, and EAATs types. These X^{-}_{AG}-types differ in expression and electrochemical properties, including their substrate affinities [155]. The comparative phylogenetic analysis (Fig. 4) suggests that the X^{-}_{AG} cluster may be divided into three additional subfamilies based on the proximity of genes representing different phyla: the EAAT2-type that also includes a four-gene expansion of Ce transporters, a sole representative of the algal transporters, and a single Ci transporter; the insect-specific EAATs that only include the dipteran transporters; the EAAT5-type that includes one mammalian and two Ci transporters; and the EAAT1,3,4-type that includes a three-gene expansion of mammalian EAATs routed with one Ce gene. The broad spectrum of EAATs' functions have been associated with a signaling role of glutamate while terminal expansions of SLC1 in

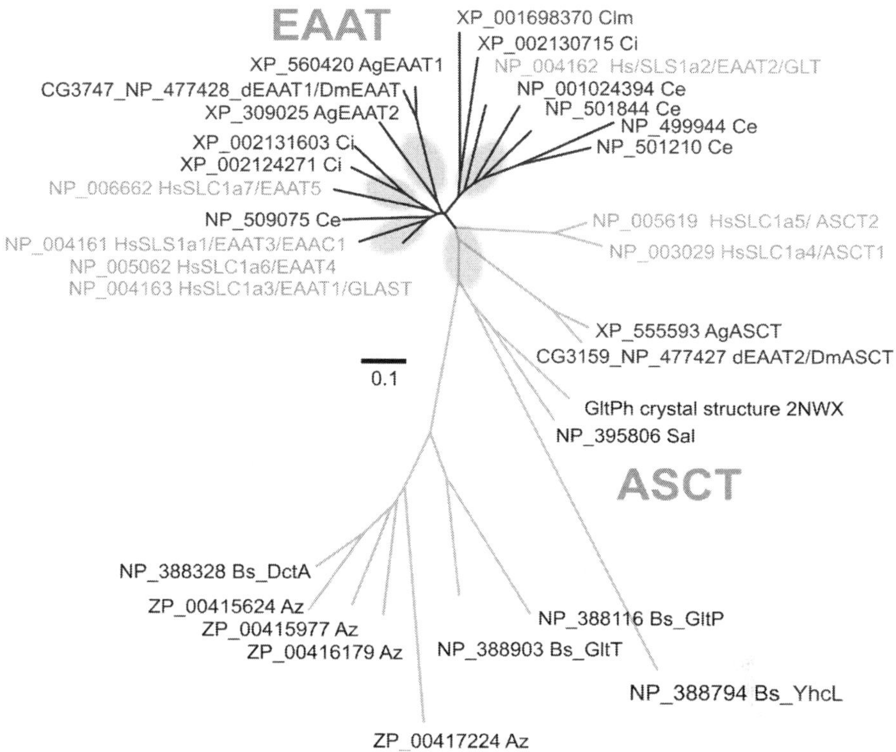

Fig. 4 Evolutionary relationships of 32 SLC1 members from nine representative genomes. The evolutionary history was inferred using the Neighbor-Joining method [448]. The optimal tree with the sum of branch length = 11.28 is shown. The tree is drawn to scale, with branch lengths in the same units as those of the evolutionary distances used to infer the phylogenetic tree. The evolutionary distances were computed using the Poisson correction method [449] and are in the units of the number of amino acid substitutions per site. All positions containing gaps and missing data were eliminated from the dataset (Complete deletion option). There were a total of 247 positions in the final dataset. Phylogenetic analyses were conducted using MEGA4 software [450]. Species abbreviations included in the sequence strings: Ag, *A. gambiae*; Az, *A. vinelandii*; Bs, *B. subtilis*; Ce, *C. elegance;* Ci, *C. intestinalis*; d, *D. melanogaster*; Clm, *Chlamydomonas reinhardtii*; Hs, *H. sapience*; Ph, *P. horikoshii*. HUGO SLC numbers and conventional nomenclature are shown for human transporters

mammals may reflect an increased sophistication in glutamatergic neuronal signaling, which is involved in the long-term potentiation and formation of memory [156–162].

Basal members of SLC1 include mammalian system ASC or ASCT transporters and serve as sodium ion-driven substrate providers of dispensable L-alanine, L-serine, L-cysteine, and L-threonine (SLC1a4, ASCT1) [163]; plus L-glutamine and L-asparagine (SLC1a5, ASCT2) [164] in tissues with a low capacity for the production of these amino acids [163, 165–167], as well as in rapidly proliferating and developing tissues, including tumors [164, 168]. Mammalian ASCTs also differ in their expression patterns [163, 164, 169]. They act as obligatory Na^+-dependent substrate exchangers [170], and can mediate the low affinity transport of acidic amino acids and some neutral amino acids [165].

Mammalian ASCTs are in close phylogenetic proximity to bacterial and archean SLC1 members compared to the more distant acidic amino acid transporters of the EAAT group (Fig. 4). However, the archean SLC1, GltPh, has been characterized as an acidic amino acid transporter mediating the sodium-coupled uptake of glutamate and aspartate, with a higher affinity for the latter [171]. The interaction of the GltPh with neutral amino acids remains unknown. In contrast, analysis of the closest bacterial homolog to archean GltPH and mammalian ASCTs, the *B. subtilis* symporter YhcL, suggests a role for this transporter in the uptake of L-cystine ($K_m = 0.6\,\mu$M) and several other substrates [154].

Intriguingly, many transporters of ASCT groups were identified as target-recognition receptors of retroviruses [172]. For example, the carboxyl-terminal region of extracellular loop 2 in the SLC1a5 (ASCT2) mammalian transporters comprises a cell-surface binding site and membrane gateway for retroviral infection [173]. The comparative phylogenetic tree (Fig. 4) shows that the vertebrate and dipteran relatives of ASCTs' phenotypes are closer to bacterial transporters. It is reasonable to propose that ASCTs represent an early evolutionary trend in the SLC1 family for primordial phenotypes and physiological roles in heterotrophic cells, acquiring nutrient carbon, nitrogen, and amino acids from prebiotic and earlier biotic pools of nutrients. The SLC1 family has undergone several characteristic expansions (Fig. 4; clusters with different background colors). These expansions may correspond to functional specializations, narrowing the substrate spectra and enhancing the acquisition of progressively decaying environmental nutrients (e.g. expansion in *A. vinelandii* (Az) and *B. subtilis* (Bs) groups: Fig. 4), as well as evolving new functions in eukaryotes and metazoan organisms (e.g. roles in the absorption of the neurotransmitter Glu in metazoan lineages).

Interestingly, the ASCT relatives are absent from the nematode (*C. elegance*) and urochordata (*Ciona intestinalis*) genomes, which instead possess paralogous expansions that overlap with the mammalian EAAT group (four Ce gene cluster and two Ci gene duplication) (Fig. 4). This may suggest that in these two organisms ASCT functions have been compensated for by other transport mechanisms. The functions of the Ce and Ci SLC1 genes remain to be analyzed.

EAATs comprises two established genes in *D. melanogaster* and three predicted genes in *A. gambiae*, which split into two paralogous clusters in the dipterans, and have no justifiable orthologs among the five mammalian transporters (Fig. 3). EAAT ($X^-{}_{AG}$) maintains local glutamate/aspartate concentrations, including those in the central excitatory insect synapses [174], glial tissue [175–178], visual and olfactory afferents [179–181], macrophages [182] and osteocytes [183]. Mammalian EAAT genes appear to have diverged through adaptations for particular physiological conditions [184, 185] and functions other than the tuning of local concentrations of glutamate, and beyond the uptake of excitatory amino acids at membranes, which surround synaptic clefts. For example, EAAT2 and EAAT3 may serve as the facultative providers of cysteine for glutathione synthesis [186], and EAAT5 may act as a leak channel and/or ligand-gated chloride channel [180, 187]. EAATs that share transport specificity are differentially expressed in the neuronal–astroglial interface of mammalian synaptic clefts [188]. The EAAT2 and EAAT1 are located on the plasma membrane of astrocytes that interact with excitatory synapses, while EAAT3 and EAAT4 are predominantly associated with postsynaptic membranes [189] (although, EAAT4 is more selectively expressed in GABA-ergic neurons, including Purkinje cells [190]), while EAAT5 is specifically expressed in the photoreceptors and bipolar cells of the retina [180].

Acidic amino acids have been implicated in many important physiological processes in mammals, including neuronal development, synaptic plasticity, and long-term potentiation (LTP). Glutamate also serves as a substrate in the glutamic acid decarboxylase (GAD)-mediated synthesis of the inhibitory neurotransmitter GABA in GABA-ergic neurons. EAATs play central roles in the action of glutamate by absorbing extracellular neurotransmitters and ensuring postsynaptic inactivation. In addition, glial and neuronal EAATs prevent neuronal damage caused by excessive glutamate-induced postsynaptic activation and neurotoxicity. Hence, EAATs perform the fine-tuning of intracellular and extracellular glutamate activities that are critical for neuronal functions and homeostasis. On the one hand, EAATs hyperactivity-mediated misbalance of synaptic transmissions has been inferred in schizophrenia and some other mental illnesses [191]. On the other hand, EAAT function deficiency in mediating glutamate-induced neurotoxicity has been implicated in a variety of neuropathological conditions and neurological disorders, including epilepsy, schizophrenia, Huntington's disease, Alzheimer's disease, amyotrophic lateral sclerosis (ASL), and Parkinson's disease dementia complex [189, 192].

Because SLC1 members play critical roles in neuronal and metabolic functions, genetic diseases associated with their dysfunctions are expected to be rare. The HsSLC1a1 (EAAC1) gene has been mapped to 9p24, inferred with mutations that may be responsible for familial dicarboxylicamino aciduria (ALS), a separate form of ASL due to a mutation in the SOD1 gene (147450) on chromosome 21 [193]. A N206S mutation of the potential N-linked glycosylation site in the EAAT2 gene was inferred as a reducing protein in glycosylation, surface expression, and transport function mutation [194, 195] in sporadic heterozygous ALS patients. Glutamate excitotoxicity induces oligodendrocyte and tissue damage in multiple sclerosis (MS), and it has been suggested that functionally compromising glutamate homeostasis polymorphisms in the EAAT2 gene's promoter may contribute to relapsing MS [196]. However, a heterozygote mutation in SLC1a3 has been associated with episodic ataxia, seizures, and hemiplegia, but was only observed in one patient [197].

Essential roles of glutamate in the insect CNS have been deduced from extensive immunohistochemical labeling of the ligand, its receptors, and transporters. L-Glutamate is also the most abundant amino acid in the insect CNS, acting as both an excitatory and an inhibitory neurotransmitter [198, 199] via high affinity binding/activation of postsynaptic receptors, which are classified as metabotropic and ionotropic subtypes [200, 201]. It also acts as the principal excitatory transmitter at neuromuscular junctions [202, 203], binding to specific receptors on somatic muscles [204, 205]. Obviously, neuronal glutamate in insects mimics processes known in the mammalian CNS, including afferent signaling and memory (e.g. it appears to mediate central mechanisms that are associated with long-term olfactory memory in the honeybee [206], and to mediate NMDA-type glutamate receptor-coupled regulatory pathways in the biosynthesis of juvenile hormones [207]). Several SLC1 members have been cloned and characterized from the nervous systems of insects, including two from fruit flies [179, 208], one from caterpillars [209], one from cockroaches [174], one from yellow fever mosquitoes [210], and one from honey bees [181]. They share up to 45% of their protein sequence identity and have conserved putative secondary structures, which are characterized by eight putative alpha-helical transmembrane domains and two to four beta-pleated domains, resembling the beta-barrel structures of pore-forming ion channels [208]. Demonstrated substrates for these insect EAATs include glutamate and aspartate, which may sometimes act as a co-transmitter with glutamate. In some cases aspartate is the preferable substrate (e.g. for dEAAT2 [211]). In addition to L/D-aspartate, dEAAT2 transports taurine [212], a nonproteinogenic amino acid that is essential in some organisms (e.g. cats) and physiological processes. Biophysical studies with vertebrate transporters have shown that the SLC1 transporters are electrogenic, the uptake of one amino acid$^-$ anion coupled to the co-transport of three Na^+ cations, one H^+ cation, and the counter-transport of one K^+ cation (thus, there is a net inward flux of two positive charges per cycle).

A novel property of vertebrate EAATs is the activity of substrate-gated anion channels, as suggested by the presence of channel-like structures in the C-terminal portion of the protein. For example, when *Drosophila* EAAT1 protein is expressed in frog oocytes it induces a chloride flux that is consistent with this postulated channel-like activity [208].

In situ hybridization detected the expression of both dEAAT1 and dEAAT2 in neuronal cells; however, with distinct patterns observed for each transcript [179]. Similar patterns of EAAT expression have been detected in the adult brain of the honeybee *Apis mellifera*, where it was associated with the mushroom bodies, within which glial cells are rarely observed [181]. In contrast, in situ hybridization, using *Trichoplusia ni* EAAT cDNA probes on whole mount caterpillar ganglia, identified a population of interfacial glial cells [213]. An immunohistochemical analysis on EAAT localization, using anti-peptide antibodies raised against the *T. ni* EAAT amino acid sequence, produced selective staining in the ganglionic neuropile, as well as in the extensions of glial sheath cells wrapped around the nerve processes that enervate skeletal muscles [214]. In contrast, an anti-peptide antibody raised against the *Aedes aegypti* EAAT stained only the thoracic ganglia, and not the muscles [210].

Comparisons between insect and vertebrate EAATs, based on structure, pharmacology, expression and phylogenetic proximity, support the paralogous divergence of vertebrate and insect transporter EAATs [181, 208, 209]. Some adaptations in these two groups are prominent and can be used as a basis for selective pharmacological alterations. For example, the fruit fly dEAAT2 possesses an unusually strong preference for aspartate as a substrate [211]. Sequence comparison among insect SLC1 (Fig. 3) reveals insect-specific clusters of putative transporters with similarity to both dEAAT1 and dEAAT2. Among the remaining glutamate transporters in the tree, there are no patterns related to the localization of expression, as there are examples of expression in both glial cells and neurons (dEAAT1 is neuronal whereas trnEAAT is glial). Several putative EAATs (also compared in Fig. 3) are derived from the recently completed mosquito (*A. gambiae*) and honeybee (*A. mellifera*) genomes. Both genomes contain three predicted EAATs. In contrast, the fruit fly genome contains only two dEAATs. On the basis of the phylogenetic differences in SLC1 complements between selected model genomes, it seems reasonable to propose that some significant differences in specific EAAT functions may exist across organisms.

The structure of the first SLC1 transporter from a single-member of the SLC1 family in the hyper-thermophilic archaebacterium *Pyrococcus horikoshii* has been solved (Glt_{ph} Accession # NP_143181; PDB ID, 2NWX) [171, 215]. The observation of significant protein sequence conservation between prokaryotic [216] and eukaryotic members of the SLC1 family, i.e. GLT and GLAST [217], supports notions of strong structural conservation and possibly may reveal the molecular architecture and mechanics of neuronal glutamate transporters, as well as other proteins in the family [215, 218]. It also provides a solid basis for the structural and pharmacological research of the eukaryotic members of SLC1 using in silico homology modeling, substrate docking, and molecular dynamic simulations. For example, a homology model of AgSLC1a1, which was built based on the 2NWL-A and 2NWW-C PDB coordinate files as a template, demonstrates key structural traits of the SLC1 family (Fig. 6). On the basis of the Glt_{Ph} crystal study, a monomer of the SLC1 protein possesses a conserved eight TMD structure forming a transmembrane channel, which is closed by two hairpin domains between TMDs 6 and 7 and TMDs 7 and 8 in a substrate-binding state (Fig. 4a, b, c). SLC1 proteins may form a homo-trimeric quaternary structure (Fig. 4d, e) similar to that observed in the GltPH crystal study [215]. This quaternary structure is essential to form a half-membrane-deep cup-shaped extracellular basin [215], which generates additional motive forces upon the interaction of the transporter with sodium ions (Fig. 4d, e). A freeze-fracture study of EAAT3 expressed in the plasma membrane of *Xenopus* oocytes suggested a homo-pentameric quaternary structure of this member of the SLC1 family [219]. Each Glt_{Ph} subunit formed an identical substrate-binding site that is comprised of two helical hairpins close to the bottom edge of the cup (Fig. 4d). It has been proposed that substrate translocation occurs upon a cation motive force-induced conformational movement of the transporter [171]. Substrates, identified in the crystallized structure, resembled an occluded state of the transporter and a key intermediate transporter conformation between closed and open states, and formed multiple interactions with HP1, HP2, TMD7, and TMD8 (Fig. 4f). The HP2 movement induced an internal gate opening conformation and allowed the diffusion of substrates to the cytosolic side of the plasma membrane. After that the transporter recycled itself to an open state, plausibly similar to that which was observed in the crystal study of the GltPH with TBOA (dl-threo-β-benzyloxyaspartate) inhibitor [171], which blocked transport by preventing the movement of HP2 to the closed state conformation (for the molecular logic of the transporter see for detailed discussion of the conformation events).

A detailed description and discussion of the challenges associated with the structural and conformational studies of the bacterial SLC1 and SLC6 family members is available in the recent review by the principal author [218]. More recently, Glt_{ph} was used in a combined molecular dynamic simulation and site-directed mutagenesis study, reporting the first time-resolved description of substrate-binding events, as well as refining additional details of the substrate–transporter interaction mechanism [220].

In summary, SLC1 is a key family in the active absorption of small neutral and acidic amino acids involved in metabolic and neuronal integration in organisms. This family may have emerged very early in life's history to absorb amino acids from abiotic and earlier biotic environments. It then diverged with the acquisition of new substrate selectivity and specific functions. The SLC1 family carries strongly conserved sequences, functions, and structures (i.e. pairwise identity of SLC1 selected for the phylogenetic tree in Fig. 4 is 29.5%, and for the alignment in Fig. 5 is 38.6%). Therefore, we are provided with a good opportunity for homology modeling and pharmaceutical drug discovery based on the resolved structures of prokaryotic transporters. Together with molecular and biophysical analysis of heterologously accessible members of the family, this provides a unique chance to explore the molecular mechanics of EAAT and ASCT actions, as well as the structural basis of SLC1s evolution and adaptation Fig. 6.

3.3 SLC6; Neurotransmitter and Nutrient Amino and Acid Transporters (NTTs and NATs)

The Sodium Neurotransmitter symporter Family (SNF, a.k.a. SLC6) is one of the largest and most functionally diverged families of secondary transporters. The members of SLC6 mediate a number of critical processes in metazoan organisms, including absorption of neurotransmitters: dopamine (DA), norepinephrine (NE), epinephrine, octopamine (OA, a counterpart of NE in invertebrate organisms), serotonin (5-HT), GABA; amino acid neuromodulators: glycine and proline; intracellular osmolites: taurine and betaine; and intracellular energy substrates: creatine and proline [221]. This family also includes a declining number of "orphan" members with yet to be identified phenotypes and physiological roles. For example, recent cloning of several orphans of mammalian SLC6 led to the discovery of molecular identities of classical B^0 and IMINO systems [222]. A functional expression of a few new lineage-specific orphan transporters, along with phylogenic analysis of their homologs in the completed genomes of vertebrate and invertebrate organisms, led to the discovery of a lineage-specific subfamily of Nutrient

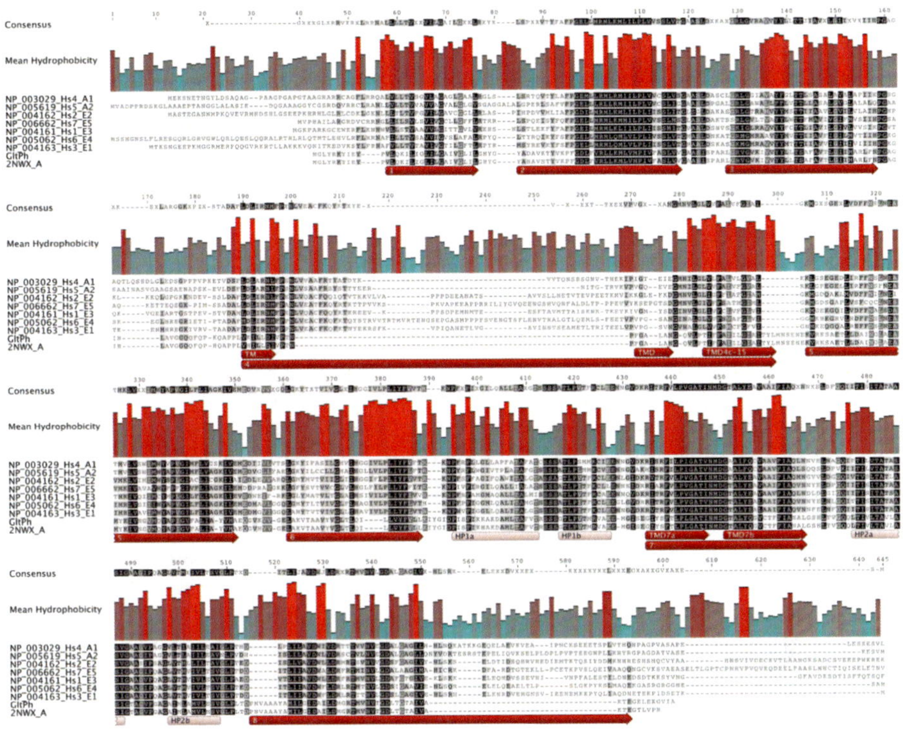

Fig. 5 Sequence alignment of selected members of the SLC1 family. A protein sequence alignment of all human SLC1 family members vs. Glt_{Ph} (glutamate transporter from *P. horikoshii*) and 2NWX_A and Glt_{Ph} structure sequences; (GB accession)#_Hs(SLCa)#_E(AAT)#) sequences; a sequence homology is indicated as a *dark background intensity*; a Clustal X [451] was used for the alignment following minor manual corrections). A mean hydrophobicity plot is generated using a three residual sliding window, *red signals* indicate high hydrophobicity. Transmembrane domain (TMDs) and helical hairpin motifs (HPs) interacting with substrates are indicated as *dark red arrows* and *light pink bands*, respectively

Amino acid Transporters (NAT), which mediate the monovalent cation (Na^+ or K^+)-dependent absorption of neutral and cationic amino acids [95, 221, 223]. Members of SLC6 exist in prokaryotes, archaea, and Metazoa. Basal prokaryotes usually have only one SLC6 per genome. However, advanced prokaryotes with complex life styles and developed symbiotic adaptations may have up to four SLC6 members per genome (Fig. 7). The available functional data suggest that bacterial members carry on sodium-dependent absorption of nutrient amino acids with a broad, for example LeuT from *Aquifex aeolicous* [224] or narrow, i.e. TnaT from [225] substrate selectivity profile. The active transport by SLC6 family members comprises a major mechanism for the accumulation of metabolically expensive and conditionally essential large neutral amino acids in prokaryotes. However, the SLC6 family is absent in the genomes of free-living and plant-associated symbiotic diazotrophs, as well as in the currently available genomes of intracellular parasites (Fig. 7). It is possible that nitrogen fixation and the evolution of comprehensive amino acid anabolism in diazotrophs relaxed the environmental acquisition of infrequent large neutral amino acids and, respectively, disabled the evolution of SLC6. On the other hand, the evolution of a comprehensive amino acid absorption cascade, such as the SLC6 family, together with a biologically induced escalation of environmental fluxes of amino acids, reduced demand for the intrinsic nitrogen fixation that may lead to the broad extinction of the energetically expensive diazotrophy. The requirement for nutrient absorption and redistribution of essential amino acids in metazoans led to additional expansions of SLC6. Together, the acquisition of stable ecological sources of amino acids and ability to regulate cellular concentration via the energy-efficient active transport may have disabled energetically costly synthesis cascades in eukaryotic heterotrophs, starting with the most metabolically expensive aromatic amino acids. Parasitic prokaryotes receiving nutrient amino acids from a concentrated intracellular or systemic pool of the associated hosts have little need for active sodium-coupled mechanisms, using instead simple and non energy-requiring facilitated substrate selective gating mechanisms (e.g. amino acid permeases). The low sodium environment introduces additional roadblocks for the evolution of sodium-dependent SLC6 mechanisms in intracellular parasites. However, the situation may be different in eukaryotic and metazoan pathogenic organisms, which still need to absorb amino acids during free-living stages and manage the distribution of essential amino acids between cells. They may need to keep precise downstream gradients of amino acids up to the intracellular metabolizing center. Those organisms require active transporters similar to SLC6 family members to handle the thermodynamically unfavorable delivery of amino acids.

The number of SLC6 members per genome varies across major groups of prokaryotes (Fig. 7). SLC6 is absent in diazotrophs and prokaryotes exploring the parasitic route. In contrast, in prokaryotes with broad environmental adaptations and sophisticated metabolism up to four SLC6 members may be present. It has been frequently stated that plants and fungi lack the SLC6 family [226, 227]. Indeed, current plant genomes including angiosperm, unicellular alga, and red algae show no traces of the SLC6 family. It is not surprising: these organisms have a virtually infinite supply of energy from photosynthesis, powering the anabolism of a complete set of amino acids, which together allow plants to be independent from the environmental acquisition of proteinogenic amino acids. Terrestrial plants and freshwater alga also use a very limited quantity of sodium ions in systemic circulations. However, the situation in heterotrophic fungi is not so explicit. The apparent drawback of the above statement became evident with the first screened genomes of yeasts, which appeared to lack SLC6 members [226]. However, baker's yeast, *Saccharomyces cerevisiae,* and beverage-fermenting fission yeast, *Schizosaccharomyces pombe*, exemplify organisms with adaptations to develop in the nutrient reach media. In addition, yeast cells utilize endocytosis for nutrient absorption that relies on vacuolar transport rather then SLC6-type plasma membrane transport mechanisms. Nonetheless, the BLAST query of a SLC6 consensus sequence vs. current fungi database (taxid: 4751) reveals a number of predicted proteins with an EXPECT value ranging from e-42 up to e-21. This unequivocally suggests the presence and important role of SLC6 in fungal metabolism and evolution. Moreover, the BLAST analysis suggests that the genomes of two agriculturally and economically devastating plant pathogens, *Magnaporthe grisea* and *Gibberella zeae* (a.k.a. rice blast and wheat head blight fungi) include at list three SLC6 members per genome (Fig. 6). Some SLC6 members found in the fungi genomes have been annotated, i.e. sodium-dependent noradrenaline transporter XP_001938178, and renal osmotic stress-induced Na-Cl organic solute co-transporter, XP_001934385. However, XP_001934385 and XP_001938178, together with the rest of the fungi-specific SL6 transporters, form a paralogous cluster in the comparative SLC6 tree (Fig. 8). The proximity of bacterial and metazoan transporters that specialized in sodium-nutrient amino acid absorption suggests similar functions in the fungal SLC6-NAT members; however, experimental proof supporting the functional annotation of these transporters has yet to be acquired.

Protozoans lack and synthetic mechanisms, therefore require environmental absorption of essential amino acids (Sect. 2.9). However, instead of using plasma membrane absorption of diluted amino acids which had already become infrequentin the protozoan biota, these organisms evolved a nutrition mode based on endocytosis and vacuolar digestion/transport. It has little significance for the SLC6-type mechanisms. The genome of free-living social amoeba, *Dictyostelium discoideum*, has no trace of the SLC6 family. Evolution of many protozoan organisms focused on "easy" access of nutrient amino acids via endo-parasitic adaptations (considering the majority of sequenced protozoan genomes, which is reasonably focused on the most important pathogenic organisms). That type of nutrition has little requirement for SLC6-NATs. However, concomitant parasitic protozoa diverged from a free-living ancestor and many of them have to pass an ecological gap between two hosts including a primary host and vectoring host, which actually may require some host-independent nutrients. For example, *Apicomplexa* (including cryptosporidium and the malaria parasite) exploring the endo-parasitic route over 500 Mia have a trace of SLC6, i.e. a single gene per genome with EXPECT > 1e-3 (Fig. 8). In contrast, the free-living ciliated protozoan, *Tetrahymena thermophila,* possesses three definite SLC6 transporters with EXPECT < 3e-40, which clearly form a paralogous cluster vs. apicomplexan members (Fig. 6). The function of the free-living protozoan NATs in essential amino acid absorption is most probable, but this remains highly tentative until at least one of the protozoan SLC6 members has been characterized. The situation with putative apicomplexan transporters is even less clear, since they descend in the extensively diverged collection of "orphan" SLC6 members with unknown transport function, for exampleDmBLOT *bloated tubules* (*blot*) gene encoding SLC6 that is involved in the organization of the apical cytocortex in *D. melanogaster* [228].

SLC6 is expanded dramatically in Metazoa and may include >20 members per genome in upper metazoan genera, including vertebrates and insects (Fig. 8). Metazoan SLC6 members form two major and clearly physiologically and phylogenetically distinct populations: the Neurotransmitter Transporters (NTTs) and

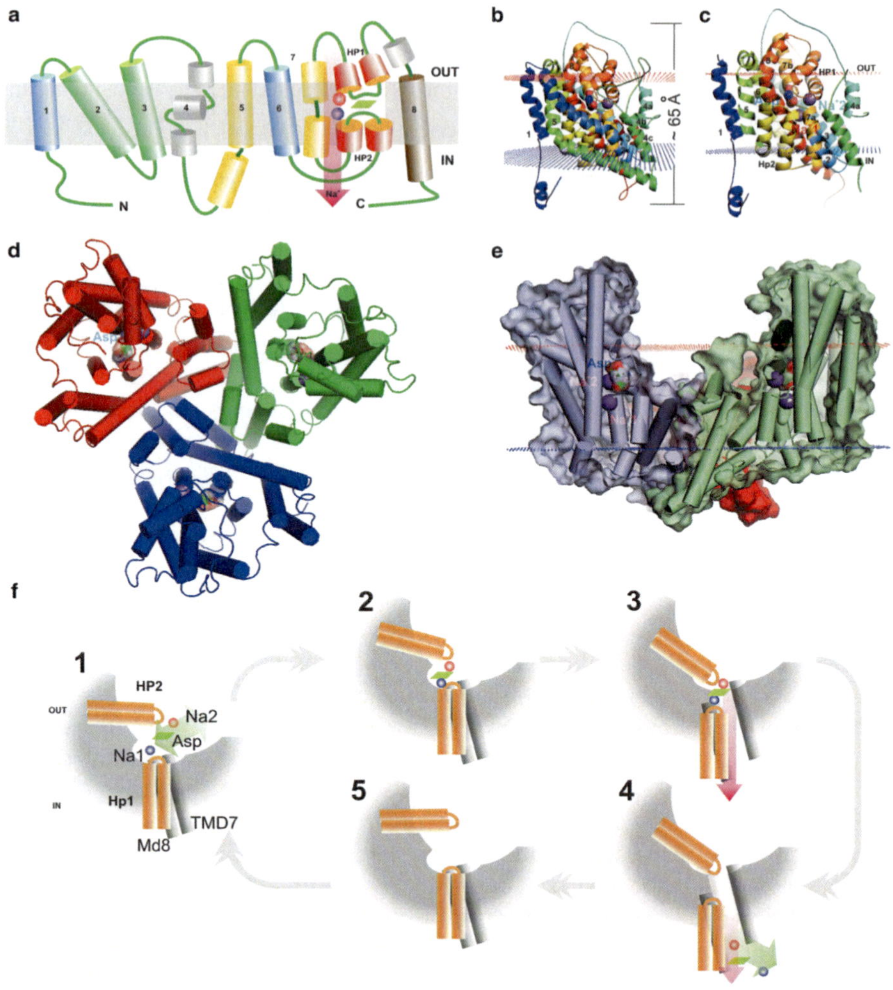

Fig. 6 Structure and presumable functional mechanics of SLC1 transporter. Representative 2D model of SLC1 transporters, cations, and organic substrates are shown as *spheres* and *parallelograms*, respectively. **b**, **c** A homology model of the AgEAAT1 monomer, built using the 2NWX template and Yasara Structure software (http://www.yasara.org/). **d**, **e** Top and side views of Glt_{ph} (based on 2NWX coordinate file [215]), *red* and *blue dots* represent outer and inner membrane surfaces, substrates are rendered as *molecular surfaces* (Asp^-) and *spheres* (Na^+). Sodium ions and aspartate are rendered as balls and *molecular surfaces*, respectively. Images were prepared from 2NWX PDB files using MacPymol (DeLano Scientific, http://www.delanoscientific.com/) and Corel Draw 4. **f** Putative model of SLC1 function assimilated from [171]: *1*. Initial binding of stochastically diffused substrates; *2*. Short-term complete coordination of the cargo unit ($2Na^+Asp^-$) in the substrate-binding pocket, this is the occlusion stage corresponding to the solved crystal structure of the transporter with substrates; *3*. Hypothetical initial conformational changes induced by sodium motive forces (*magenta arrow*); *4*. A final conformation leading to cytosolic substrate diffusion; *5*. A hypothetical steady-state open-to-outside conformation corresponding to the solved transporter with a TBOA inhibitor structure (adopted from [171])

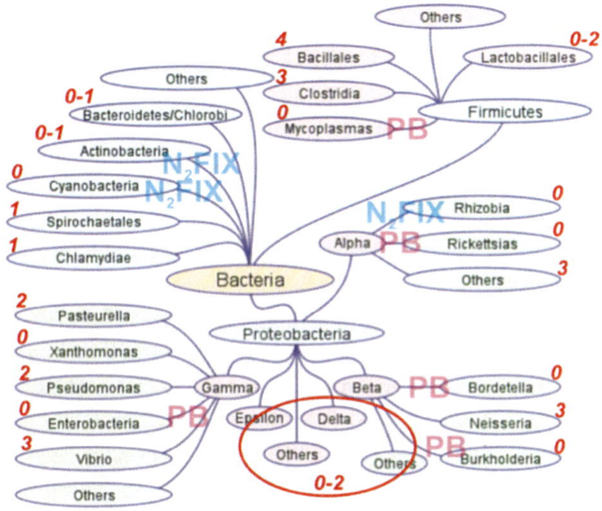

Fig. 7 Extent of SLC6 family members across bacterial diversity. Specific groups of bacteria are shown along with the typical number of SLC6 genes per lineage. N_2FIX and PB indicate diazotrophs and parasitic forms, respectively

Nutrient Amino acid Transporters (NATs) (Fig. 8). NTTs comprise strongly orthologous substrate-specific groups of SLC6. These members are associated with the absorption of specific amino acids and amino acid derivatives that serve signaling functions, primary but not exclusively, in the central nervous system (CNS). NTTs were the first cloned SLC6 members that led to the historical names of the family as Sodium Neurotransmitter Symporters (SNF) or Sodium-dependent Symporters of Neurotransmitters (SSN) [227, 229]. In spite of numerous annotations of new SLC6 members from basal genera as neurotransmitter symporters, the NTTs group is absent in the prokaryotes, protozoans, and fungi, probably because these organisms have none or very limited requirements for intercellular coordination of metabolism and obviously have no brain. NTTs represent a strongly orthologous and functionally conserved group of transporters mediating high affinity and selectivity transport mechanisms for GABA, monoamine, and Gly. However, some more recent paralogous divisions have been identified in the monoamine group of neurotransmitter transporters representing an interesting case of neurochemical divergence of protostome and deuterostome organisms (see Norepinephrine and Octopamine transporters). The NTTs branches (or branch) had to separate earlier in metazoan evolution to serve intercellular signaling functions, which are essential for the metabolic integration of metazoan organisms. A fraction of these transporters are expressed in specialized neuronal cells and were recruited to serve neurochemical signaling. Therefore, NATs with narrow substrate spectra and specific neuronal expression profiles may be viewed as templates representing the origin and evolution of neurotransmitter transporters. In contrast to orthologous NTTs, the NATs group comprises genera-specific and strongly paralogous clusters of transporters displaying a variety of substrate profiles, including broad and narrow substrate spectra (i.e. B^{0+}, B0, IMINO systems, and narrow specificity insect B0 system-like transporters; Fig. 6). The metazoan NATs cluster may include 4–15 genes per linage. Each of these genes may have unique temporal and spatial expression patterns, unique position in the amino acid traffic network, and a unique set of substrates; although, the functional consensus of presently characterized metazoan NATs strongly suggests that NATs members act in synergy as physiologically critical mechanisms for the monovalent cation active absorption of entire sets of essential and conditionally essential amino acids [95, 124, 146].

In summary, SLC6 includes two principal divisions with different physiological and evolutionary significance, NTTs and NATs, which can be specified as phylogenetic and functional subfamilies of SLC6. Yet, SLC 6 may be further divided into seven physiologically specific clusters of transporters, some of which (but not all) are present across all metazoan genera and also form phylogenetically separated clusters. This includes: (1) inhibitory neurotransmitter transporters for GABA and glycine; (2) monoamine neurotransmitter transporters, which can be divided into serotonin, dopamine, and norepinephrine plus paralogous octopamine specific clusters; (3) the transporters for osmolites betaine and taurine; (4) the transporters for metabolic intermediate creatine; (5) amino acid transporters for proline and glycine and the B^{0+} system; (6) the inebriated type, which represents a single gene per genome and that so far appear to be specific to insects, Echinoderms (i.e. sea urchin, *Strongylocentrotus purpuratus*), and Urochordata (i.e. sea squirts, *C. intestinalis*); and (7) recently identified Nutrient Amino acid Transporters (NATs) comprising the largest SLC6 subfamily with a mutual role in the absorption of essential and conditionally essential substrates in all Metazoa but also in a broader phylogenetic diversity of heterotrophic organisms. In addition, metazoan and protozoan but not prokaryotic organisms have at least one member falling into the category of "orphan" transporters, for which no substrates or functions remainunknown and many of whichhave to some extent disordered sequence-structure patterns compared to functional transporters of the SLC6 family (Fig. 8).

Fig. 8 Evolutionary relationships of 103 SLC6 members from ten complete genomes of representative prokaryotic and eukaryotic organisms. The evolutionary history was inferred using the Neighbor-Joining method [448]. The optimal tree with the sum of branch length = 25.19 is shown. The percentage of replicate trees in which the associated taxa clustered together in the bootstrap test (2,000 replicates) are shown next to the branches [452]. The tree is drawn to scale, with branch lengths in the same units as those of the evolutionary distances used to infer the phylogenetic tree. All positions containing alignment gaps and missing data were eliminated only in pairwise sequence comparisons (Pairwise deletion option). There were a total of 763 positions in the final dataset. Phylogenetic analyses were conducted in MEGA4 [450]. Leaves are SLC6 sequence accession, followed by species abbreviation and conventional nomenclature (for some sequences). Abbreviations: Aa, *A. aeolicus* VF5 *A. gambiae*; Ba, *Bacillus anthracis*; Ce, *C. elegance;* Dm, *D. melanogaster*; Gz, *G. zeae* PH-1; Mg, *M. grisea*; Ms, *Mus musculus*; Pberg, *Plasmodium berghei* str. ANKA; Ptr, *Pyrenophora tritici-repentis* Pt-1C-BFP; Pvix, *Plasmodium vivax* Sal-1; St, *S. thermophilum*; Tt, *T. thermophila*; Vib, *Vibrio cholerae* O1 biovar eltor str. N16961; HUGO SLC numbers and conventional nomenclature are shown for Ms transporters

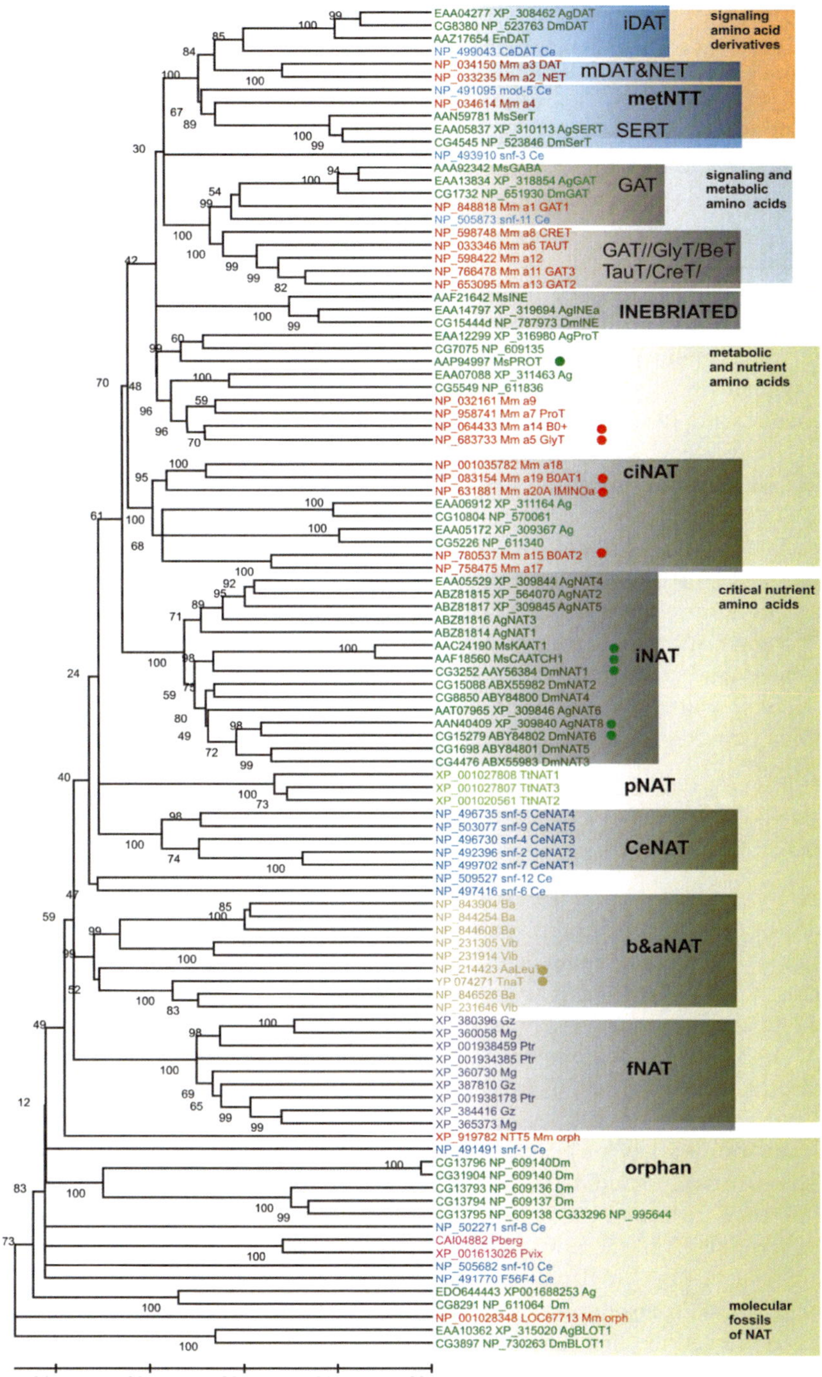

Fig. 8 (continued)

3.4 SLC6-NTTs Neurotransmitter Transporters

The SLC6-NTTs subfamily includes transporters for primary inhibitory neurotransmitters GABA and glycine and modulatory neurotransmitter monoamines. These transporters are expressed in the plasma membrane of neuronal and glial cells, and terminate the action of neurotransmitters performing neurotransmitter resorption that rapidly reduces their concentrations in the synaptic cliff below the affinity constant for postsynaptic receptor binding. The essential role of neurotransmitter uptake in the synaptic transmission and thus integrative neuronal functions [230] was established long before the cloning of the first transporter [231]. The existence of a family of plasma membrane transporters responsible for Na^+ and Cl^--dependent transport in the nervous system [232] first became evident when similarity was observed between two mammalian transporter cDNAs isolated based on protein purification and sequencing [231] and on expression cloning [233]. The similarity between these two transporters, one for GABA [231, 234] and the other for norepinephrine [233], was subsequently used to clone a variety of other transporters in different metazoan organisms, which transport different neurotransmitters as well as diverse substrates in the nervous system and other tissues [232]. All NTTs possess common features of structure and function, but exhibit remarkable differences in their interaction with transmembrane electrochemical gradients for Na^+, K^+, and Cl^- as well as substrate selectivity, affinity, and carrier capacity. The protein sequence of hydrophobicity and conformation profiles predict quite similar 12-TMD structures for these transporters, with cytosolic N- and C-terminals and an extended extracellular loop between TMDs 3 and 4 having a variable number of N-glycosylation sites.

3.4.1 GABA Transporters

γ-Aminobutyric acid (GABA) transporters (GATs) evolved to maintain low subcellular levels of GABA in neuronal and neuronal tissues, and to absorb synaptic and extrasynaptic GABA in the brain to ensure inactivation of GABA-ergic neurochemical signals [235–237]. Similar as in the vertebrate CNS, GABA serves primary inhibitory functions in the invertebrate nervous system [238]. Immunocytochemical studies of GABA-ergic transmission sites using antibodies against GABA, glutamic acid decarboxylase (the essential enzyme for GABA synthesis), and various GABA receptor subunits have revealed an extensive pattern of GABA activity in the nervous systems of a variety of insect species. This pattern implies roles for GABA in diverse processes, including visual and olfactory processing, as well as learning and memory (reviewed in [213, 239]). The main mechanism by which GABA activity is terminated on the postsynaptic membrane is through reuptake into neurons and glial cells by high-affinity transporters [238]. Partial protein sequencing of the purified GABA transporter allowed the cloning of the first GABA transporter that also represented the first member of the new family of Na+-dependent neurotransmitter transporters [231]. High-affinity insect GABA transporters (GATs) have been cloned and characterized from the hawkmoth, *Manduca sexta* [240] and the cabbage looper, *T. ni* [241]. In each species, Northern blot analysis demonstrated expression of a single mRNA transcript within the nervous system. The hawkmoth masGAT protein has also been localized via immunocytochemistry [239]. These results indicated that transporter immunoreactivity closely coincides with previously determined patterns of GABA-ergic activity, including staining of the brain areas involved in processing of visual and olfactory information. The two characterized lepidopteran GATs show ionic stoichiometry and kinetics similar to those of the vertebrate GAT1 subtype [240, 241]. However, the pharmacological profiles of the lepidopteran transporters are distinct from any known vertebrate subtype, particularly in their insensitivity to several nipecotic acid derivatives, which are potent inhibitors of GAT1. Only one GAT gene per genome with strong orthology to mammalian GAT1 (SLC6a1) is present in representative invertebrates (Fig. 8). In contrast, vertebrates exhibit greater diversity of GAT. There are three vertebrate GATs with unique cellular expression patterns, affinities for GABA, and particular functional specializations, which are enumerated as GAT1 (SLC6a1), 2 (SLC6a13), and 3 (SLC6a11). In addition, a set of transporters phylogenetically related to GATs, were identified in vertebrate genomes. These transporters demonstrate affinities for excessively metabolized amino acids that are not involved in the protein synthesis, including taurine, creatine, and betaine. The presence of multiple GATs has been reported in *Drosophila* [242], which is in contrast to the apparently unified phenotype studied in two lepidopteran models [240, 241]. Nevertheless, fruit fly and another dipteran representative, mosquitoes, carry a single GAT gene of the SLC6 family per genome. Consequently, any diversity in GAT transcripts or proteins must be a result from alternative splicing of a single gene. To date, no genes or transcripts that appear to encode taurine or creatine transporters have been identified in insects.

3.4.2 Monoamine Transporters

Key monoamine neurotransmitters serotonin (5-HT), dopamine (DA), norepinephrine (NE), and octopamine (AO) are derived from the most metabolically expensive and ecologically rare aromatic amino acids, tryptophan and phenylalanine/tyrosine. Two major signaling pathways, serotonergic and dopaminergic, share a common evolutionary heritage dating back to an earlier metazoan organism, perhaps the ancestor of the Cnidaria and the Bilateria [243]. Two additional monoaminergic pathways, (1) based on the phenolamines tyramine (TA) and octopamine (OA) [244, 245], and (2) based on the catecholamine norepinephrine (NE) and epinephrine (E) (Vincent et al., 1998), diverged in the bilaterian nervous system with specific restriction to protostomes and deuterostomes, respectively (summarized in [246]). Thus, neuroactive monoamines in many groups of invertebrates, including arthropods and nematodes, differ from those vertebrates in which octopamine appears to substitute functionally for norepinephrine. Another interesting—and important for antidepressant drug discovery—characteristic of invertebrate transporters comes into view upon cloning of cocaine-insensitive serotonin and dopamine transporters from tobacco silkworm, *Bombyx mori*, and coca plant pest, *Eloria noyesi*, BmSERT [247] and EnDAT [248], respectively.

Serotonin transporter (SERT; 5HTT). Indolamine 5-HT is a key modulator of cognitive and emotional mechanisms in mammals [249] that include appetite [250], memory and learning [251], aggression and antisocial behavior [252], anxiety and depression [253, 254]. Serotonergic systems of invertebrate and vertebrate brains act in similar ways; for example it reciprocally interacts with specific neuronal circuitry supporting general behavioral arousal [255]. Serotonin uptake in the insect nervous system was first demonstrated in the nerve fibers of the locust brain [256], followed by the discovery of high-affinity serotonin transport activities in many other insect species. In insects serotonin is also involved in neuromuscular transmission [257] and epithelial transport processes, such as fluid balance [258–260], as well as in modulation of acid/base balance and alkalinization in the alimentary canal [261–264]. It also serves in the integration of complex behavior patterns in insects, including: aggression and defense [265, 266], feeding [267] and courtship arousal [268], circadian rhythms [269, 270], and memory [271]. The first insect serotonin transporter cDNA (dSERT) was cloned and characterized from *D. melanogaster* [272, 273]. The fruit fly transporter has pharmacological properties that are distinct from those of mammalian SERTs, often expressing a minor sensitivity to antidepressant antagonists. A SERT that was recently cloned and characterized from the hawkmoth, *M. sexta*, was less sensitive to cocaine binding than other SERTs [274]. Structural modification of the cocaine-sensitive human SERT, by using chimera which mimic fragments of the insect molecule, revealed important elements involved in the interaction with cocaine [274]. This finding is relevant for human health and important for understanding serotonin transporter biology. It also illustrates the value of arthropod models in uncovering structure-function relationships of complex transport proteins through comparative studies. The serotonin transporter comprises a single gene in insects [246] and many other analyzed metazoan organisms (Fig. 8) which is likely to be a universal pattern across the animal Kingdom.

Dopamine Transporter (DAT). Dopamine is the most abundant monoamine in the metazoan nervous system, with dopamine-containing neurons being widely distributed throughout the nervous systems of a variety of species [275]. In contrast, only trace quantities of norepinephrine that appears to be the second vertebrate catecholamine neurotransmitter, can be detected in the invertebrate nervous systems. Dopaminergic neurons modulate both visceral and skeletal muscle activity, with dopamine affecting flight motor patterns in *M. sexta* and the escape circuit in cockroaches [275]. Dopamine release from peripheral nerve endings also stimulates salivary gland secretions in many species [276, 277]. The first insect dopamine transporter (DAT) was cloned and characterized from the fruit fly [278]. This transporter exhibited substrate selectivity similar to vertebrate DATs (dopamine and tyramine are preferred substrates), but showed a distinct pharmacological profile that is reminiscent of mammalian norepinephrine transporters (with high affinities for tri-cyclic antidepressants). A second DAT, isolated from the cabbage looper moth has similar properties [279]. These transporters cluster closely together, along with another putative DAT sequence present in the mosquito genome (Fig. 6). It has been proposed that these invertebrate DATs represent conserved copies of a primordial catecholamine transporter that existed before the divergence of mammalian catecholamine carriers into subtypes responsible for dopamine and norepinephrine uptake [246].

Norepinephrine transporter (NET). Norepinephrine (NE) is the principal messenger of central noradrenergic and peripheral sympathetic synapses of vertebrate organisms, as well as a key stress hormone involved in many regulatory actions including acute "fight or flight" hyperarousal response. It is present in the neuronal and peripheral tissues of gastropod mollusks and may induce complex behavioral responses [280–282]. However, its role as a neurotransmitter in the molluscan synapse is unclear. It has been shown that chordate NET is a result of the proximal duplication of an ancestral monoamine gene following the loss of an invertebrate-type DAT gene in a basal protochordate (i.e. *C. intestinalis.*) NE is synthesized from DA, which is balanced by the availability of essential phenylalanine-tryptophan precursors, as well as NE and DA recycling. The defects of NE regulation have been implicated in attention-deficit/hyperactivity disorder [283], depression, and hypo/hypertension [284–286]. The norepinephrine transporter (NET) is mainly expressed in the presynaptic membrane of noradrenergic synapses. Thus, by reabsorbing neurotransmitters, NET not only terminates postsynaptic activity of NE, but also plays a very important role in presynaptic recycling of NE and neuronal homeostasis [287–290]. The complex regulation of NET upon numerous physiological and metabolic factors has been recently reviewed [286].

Octopamine transporter (OAT). Octopamine is the monohydroxylic analogue of norepinephrine in insects. It is present in high concentrations in a range of insect tissues, including the nervous system, where it may act as a neurotransmitter, neurohormone, and neuromodulator [291]. It modulates afferent and efferent signaling in locusts [292, 293]. Na^+-dependent (and independent) octopamine uptake has been reported in various cockroach tissues [294] and the firefly lantern [295, 296]. In the honeybee it mediates consolidation of olfactory memory [297] and hygienic behavior [298], as well as complex social behavior, including labor establishment [299]. Despite evident neurotransmitter roles for octopamine, examination of completed dipteran genomes reveals no alternative genes, which may encode a Na^+/Cl^--dependent transporter for octopamine uptake. In these insects, it has been proposed that octopamine may be dispersed by an alternate mechanism, or perhaps taken up by a member of an unrelated transporter family [278]. A high-affinity octopamine transporter has been cloned from the central nervous system of cabbage looper *T. ni* [300], leading to the identification of two functionally distinct dopamine and octopamine transporters in the CNS of the moth [279]. The TnOAT demonstrated a relaxed ionic dependency relative to other insect monoamine transporters and was generally resistant to pharmacological inhibitors of monoamine transporters. Comparison of the OAT sequence with other insect Na^+/Cl^--dependent transporters shows that it clusters with mammalian norepinephrine transporters, almost equidistant from the serotonin and dopamine transporters but not other insect transporters, including those for monoamines [146]. A recent study of the subject suggests that invertebrate octopamine and vertebrate norepinephrine transporters are paralogs resulting from the duplication of an ancestral monoamine transporter.

3.4.3 Glycine and Proline Transporters

In vertebrates, a distinct subfamily of Na^+/Cl^--dependent NTTs transports the amino acids glycine (GLYT) and proline (PROT) [301]. Multiple subtypes of transporters for the inhibitory amino acid glycine are expressed in both the CNS and in peripheral tissues. Although arthropods apparently lack receptors for glycine [302], sequence comparison with vertebrate GLYTs reveals potential Gly-type transporters in both the *D. melanogaster* and *A. gambiae* genomes (CG5549 or NP_611836 and EAA07088, respectively). Vertebrates only have a single high affinity proline transporter type (it is brain specific), which is thought to be associated with modulating activities at glutamatergic synapses [301]. In insects, proline is an alternative energy source and major osmolite, but its role in the nervous system remains unproven. Nevertheless, a Na^+/Cl^--dependent proline transporter was recently cloned from the caterpillar, *M. sexta*, where it is predominantly expressed in the nervous system [303]. This protein exclusively transported proline when heterologously expressed in *Xenopus* oocytes. The sequence comparison shows that the caterpillar PROT is related to the known vertebrate glycine and proline transporters [303]. Both the *D. melanogaster* and *A. gambiae* genomes contain a related sequence (CG7075 and its counterpart, Fig. 7) representing putative orthologs.

3.4.4 SLC6-NAT, Nutrient (Essential) Amino Acid Transporters

The NATs represent the largest subfamily of SLC6 with a remarkable paralogous divergence of its members but with an evident functional consensus of all presently cloned members. Within a large comparative frame for example in Fig. 6, NATs comprise a set of paralogous (lineage-specific) clusters. NATs exist in each organism that has an SLC6 family and they are always basal in a generic phylogenetic tree of the family. Furthermore, in organisms with up to four SLC6 members, NAT represents an exclusive phenotype; thus in bacteria, protozoa, and fungi the neurotransmitter sodium symporter family (NSS, SNF, SLC6) is the nutrient amino acid monovalent cation transporter (NAT) family per se.

Several NAT members were cloned and functionally analyzed [171] including NATs from archaea (one), bacteria (one), insects (eight from four species: *M. sexta*, *A. gambiae*, *A. aegypti*, and *D. melanogaster*), and mammals (four from three species) (indicated by arrow in Fig. 6). The phylogenetically detectable counterparts of the characterized NATs exist in all concomitant heterotrophs including Fungi, Protozoa, as well as basal invertebrate and vertebrate lineages (based on the screening of available complete genomes). The characterized NATs represent greatly diverged transport mechanisms with distinct substrate spectra and ion-cooling schemes. But all of these transporters aid in the absorption of essential or conditionally essential amino acids and some derived metabolites, the environmental fluxes and availability of which experienced changes upon the evolution of biotic synthesis within the Biosphere. The sets of NAT substrates predominantly include large aliphatic amino acids and aromatic amino acids resembling the set that is specific for a broad substrate spectra B^0 system of amino acids (green font color in Fig. 2). Evidently, all characterized NATs discriminate against the transport of small size neutral and acidic amino acids (red font color in Fig. 2), but may include charged derivatives of amino acids, i.e. L-DOPA or 5-HTP and large cationic amino acids, which may be chirally specific or nonspecific. Thus, a size selection of substrates apparently is the most universally selected trait in the evolution of these transport mechanisms. The role of such a selectivity mode would be especially important to balance intracellular amino acid pools vs. the imbalanced pool of nutrient amino acids in which large neutral and aromatic amino acids represent a minority. The characteristics of several representative NATs are briefly summarized in retrospective order.

MsKAAT1 (K^+ Amino Acid Transporter), a potassium (and sodium) ion coupled neutral amino acid transporter, was the first NAT to be cloned [304]. It has been identified upon iterative screening of a caterpillar cDNA library in *Xenopus* oocytes. Oocytes injected with KAAT1 transcript induced neutral amino acid uptake but not charged ones or GABA, resembling in this profile the mammalian system B0 [130, 223] as well as the earlier reported amino acid uptake in brush-border vesicles from the midgut epithelium of *M. sexta* [305, 306]. As reported from the vesicle studies, msKAAT1 utilizes not only Na^+ but the K^+ gradient as well. Unexpectedly, at 1 mM Leu, MsKAAT1 displayed $K_m{}^{K+} = 32 \pm 2.8$ mM, five times higher than $K_m{}^{Na+}$. The high expression of msKAAT1 in absorptive columnar cells of the posterior midgut, which was confirmed by Northern blots and in situ hybridization [304], as well as an unusual mechanism which is adapted for K^+ rather than Na^+ motive forces for amino acid translocation, fits with the physiology of caterpillar larvae nutrition that is based on a low Na^+ and high K^+ plant diet. In addition, studies of pH dependency showed that MsKAAT1 transport efficiency is maximal in the alkaline media which is physiologically normal for caterpillar midgut [307]. Reciprocal ligand inhibition experiments show that MsKAAT1 substrates to be transported require the presence of the carboxylic group and the alpha-amino group, as well as the uncharged side chain [308]. The following studies of MsKAAT1 led to several important findings, including the ability of heterologous MsKAAT1 to induce carrier coupled and uncoupled ion fluxes with specific ion selectivity profiles [309, 310], the key role of the Y147 site in substrate selectivity and affinity [311], the critical role of the E93 site in the functional folding of MsKAAT1 [312], and the irreversible pharmacological inhibition of MsKAAT1 by modification of the R76 residue with an arginine-modifying reagent, phenylglyoxal [313], the oligomerization of MsNATs [314], and some structure-function aspects of substrate specificity in the MsKAAT1 [315].

MsCAATCH1 (Cation Amino Acid Transporter Channel), a nutrient amino acid transporter that also acts as an ion channel, was cloned using degenerate primers binding conserved sites of the SLC6 family but avoiding MsKAAT1 specific sites [316]. In contrast to MsKAAT1, MsCAATCH1 was Cl^--independent but combines a carrier-associated current with a carrier-independent leakage current that was readily blocked upon interaction of the transporter with L-isomers of methionine, alanine, leucine, histidine, glycine, and threonine. In addition, msCAATCH1 exhibits different substrate selectivity depending upon substrate cations, for example preferring threonine in the presence of K^+ but preferring proline in the presence of Na^+ with apparent $K_m^{[Na+]Pro} = 330$ vs. $K_m^{[Na+]Tre} = 35$ and $K_m^{[K+]Pro} = 1900$ vs. $K_m^{[K+]Tre} = 235$ in μM, respectively. Simultaneous assay of isotope-labeled amino acid uptake and voltage-clamped currents revealed that transport and ligand-gated channel mechanisms are thermodynamically uncoupled [317]. MsCAATCH1 and MsKAAT1 share 90.6% of peptide sequence identity. MsNATs share the identical role of Y147 in their ligand affinity [318] along with the analogous role of Y140 in GAT1 [319]. Despite this striking similarity in structures and some site functions, the two MsNATs are remarkably different in physiological and electrochemical properties. Hence, the high sequence identity is unlikely regulated from genetic polymorphisms in the utilized template or alternative splicing. Apparently, it is a very interesting case of recent gene duplication that also suggests an extremely rapid functional delineation of the NAT phenotypes. The screening of MsNAT homologs in the initial draft of the *Bombix mori* genome suggests that indeed two virtually identical NAT genes exist out of a total of four putative BmNATs. They are located proximally in a single annotated contiguous sequence, and are even predicted as a single protein tandem. Recently, a natural deletion of one BmNAT gene was identified that induces resistance in the larvae to the lethal *Bombyx* parvo-like virus. This study clearly shows specific NAT selection for an offensive strategy against viral infection [320].

AeAAT1, represents the first NAT cloned from the dipteran organism yellow fever vector mosquito, *A. aegypti* [95, 321]. The heterologous expression and electrochemical analysis revealed that AeAAT1 acts as a broad spectra transporter with highest efficiency in the absorption of phenylalanine [95, 321]. Despite remarkable sequence similarity with MsKAAT1 and MsCAATCH1, the substrate and electrochemical profiles of AeAAT1 differ from the two lepidopteran transporters. For example, the Cl^- dependency and ability to utilize inward K^+ motive forces, that are characteristic to MsKAAT1, were low or absent in AeAAT1 (Fig. 12). Heterologous expression/characterization in *Xenopus* oocytes showed that AeAAT1 is a mild-affinity/high-throughput transporter (substrate affinity profile for specific amino acids F >> C > H > A ≈ S > M > I > Y > T ≈ G > N≈ P > L-DOPA > GABA > L; $K_m^{Phe} = 0.45$ mM; K_m^{Na+}=38 mM; $\eta Na^+/\eta Phe \approx 2/1$). Whole mount in situ hybridization of mosquito larval midgut confirmed an increased accumulation of AeAAT1 transcripts in the posterior midgut, salivary glands, cardia, and specific basal cells of gastric caeca. The posterior midgut of dipteran larvae was associated with apical amino acid absorption [322]; however, the elevated AeNAT1 function in the anterior components of the alimentary canal, which include cardia and salivary gland, was unexpected, because these structures are unlikely to contribute to the apical absorption of the substrate and may alternatively conduct some apical secretory functions, for example secretion of salivary enzymes and components of the peritrophic membrane. The AeAAT1 shares some properties with the mammalian B^0, the molecular identities of which were not available at that time, and to a lesser extent with the B^{0+} system, the sole existing molecular representative cloned from human [323]. The apparent differences in substrate affinities and electrochemical profiles, which are the anticipated results of adaptive divergence, make uncertain the possibility of defining orthologs between insect and mammalian NATs. Two other mammalian transporters that also mediate neutral amino acid uptake (selective for aromatic amino acids) are members of the Na^+-independent TAT transport systems [324] and the heterodimeric LAT$^+$4F2hc group [325, 326], which are characterized as members of the mammal SLC16 and SLC7$^+$SLC3 phylogenetic families, respectively. An increasing number of studies indicate that TAT and heterodimeric LAT transporters participate in the basolateral epithelial uptake/exchange of various organic solutes in growing cells [327]. By contrast, AeAAT1 and other NATs appear to play a prominent role in the primary uptake of essential nutrient amino acids from the lumen into the epithelial cells through the apical brush-border membranes in the posterior midgut. AeAAT1 identified in the cardia and gastric caeca was proposed as an amino acid absorption mechanism acting at the basal membrane. The basal AeAAT1 action may be critical in supporting extensive secretions in these tissues. AeAAT1 has a high affinity to phenylalanine, which suggests a role as substrate provider in phenylalanine-consuming pathways. The cloning and characterization of the first dipteran transporters, together with the existing information about the functions of two lepidopteran MsNATs [311, 317] and the phylogenetic identification of lineage-specific NAT expansions in the SLC6 family of key genomic model organisms (e.g. archaeon, bacterium, worm, fly, mosquito, rat, and human), led to the discovery of the NAT subfamily – with a hypothetical synopsis that the NAT cluster is acting and evolving as the key mechanism that specializes in the absorption and redistribution of essential amino acids [95].

Mammalian NATs of B^0 and IMINO systems transporters. The study of insect NATs and the identification of NAT expansions in the basal part of the SLC6 family coincide with the cloning and characterizing of two orthologous proteins from a putative mammalian NAT cluster. Analysis of these mammalian transporters led to the discovery of the molecular identity of the B^0 system of the mouse (MmSLC6a19) [328] and human (HsSLC6a19) [329] transporters (a.k.a. $B^0$1 or B^0AT1). Later, four additional representatives of rodent NATs were cloned and characterized as an additional B^0 transporter (a.k.a. $B^0$2 or B^0AT2) in the mouse [137], and as IMINO system transporters in the mouse [330], rat [331], and opossum [332]. This set of new transporters shares 1:1 sodium-coupling stoichiometries. However, IMINO is chloride-dependent, a trait that is weak or absent in the B^0 group. The key differences among mammalian NATs are in their substrate spectra. B^0AT 1 and B^0AT2 have distinct affinity profiles. Specifically, B^0AT1 transports all neutral amino acids, preferring Leu, Met, Ile, and Val, whereas B^0AT2 transports a similar set as well as Pro, for which it apparently demonstrates a higher affinity. In contrast, mouse IMINO transporters expressed in *Xenopus laevis* oocytes were strictly selective to imino group substrates, including a canonical amino acid Pro [330]. Intriguingly, through reliable transcript silencing in the cultured opossum kidney cells and heterologously expressed *Opossum*SLC6a20 in *X. laevis* oocytes, it has been demonstrated that this mechanism transports a notable fraction of B^0 substrates [332]. B^0AT1 (SLC6a19) has been implicated in the Hartnup aminoacidurias, an autosomal recessive metabolic disorder affecting the absorption of neutral amino acids [329, 333]. The dietary tryptophan does not increase plasma levels in Hartnup cases [334], suggesting that SLC6a19 is the only transporter for the active absorption of essential tryptophan in the human alimentary canal [222]. In addition to intestinal and kidney locations, mammalian NATs were present in a variety of tissues, including brain tissue. In the absorptive regions of the alimentary canal NATs may have similar luminal but slightly different axial distribution [138]. Several detailed reviews elucidating different aspects of mammalian NATs were published describing phylogenetic and functional specificity of this transporter protein cluster [149, 222, 335]. Presently, the identified mammalian NATs comprise five recognizable orthologous groups representing SLC6a15, a16, a17, a18, a 19, and a20 transporters. Each of those groups includes rodents and human transporters. In a lager phylogenetic frame the mammalian NAT-SLC6 branch also includes unidentified insects but not nematode transporters. The identified insect transporters form two groups that are clearly orthologous between insects but appear to be paralogous relative to mammalian NATs (Fig. 8). The members of this NAT cluster exist in *Urochordata* and fish (not shown). Therefore, the biologically correct definition for this cluster would be the chordate-insect NAT cluster (ciNAT2, Fig. 8)

AgAAT8 and AgNAT6, representing two unusual aromatic substrate-specific transporters, were cloned from the malaria mosquito *A. gambiae* [124, 126, 336]. They comprise a pair of the closest phylogenetic relatives of the anopheles iNAT-SLC6 subfamily (52.9% pairwise sequence similarity with 51.7% of identical sites, total 343 residuals). Both transporters belong to the insect-specific NAT-SLC6 subfamily that counts seven members in the *A. gambiae* genome and represents an earlier evolutionary branch of the SLC6 family, which is extinct in the chordates and appears to be paralogous to nematode NATs (Fig. 8). Interestingly, these two genes represent the first upstream and last downstream genes of all seven NATs gene clusters (Fig. 9) that are located on the positive strand of the 3L chromosome between the 12.09–12.16 Mb positions, based on the genome annotation [337]. Similar patterns of tight genomic clustering are characteristic of NAT populations throughout insect species (Fig. 9). The paralogous diversification, together with the strong genomic clustering of NATs, provides strong support to the notion of extensive gene cluster duplication in the evolution of the insect NAT populations. Such duplications would relax trait selection pressure and may allow for generations of new phenotypes consequently increasing ecological plasticity and physiological fitness of the integrated nutrient amino acid absorption mechanism. It was anticipated that the phylogenetic proximity of AgNAT6 and AgNAT8 would resemble recently diverged phenotypes of NATs transporters, providing some ideas on the degree of phenotype conservation and functional plasticity in the subfamily.

AgNAT6 and AgNAT8 were cloned and heterologously expressed in the *Xenopus* oocytes, and relative distribution and systemic expression was analyzed using transporter-specific immunolabeling and in-situ hybridization. Like many other characterized NATs, the aromatic AgNATs comprise a rectifying, normally

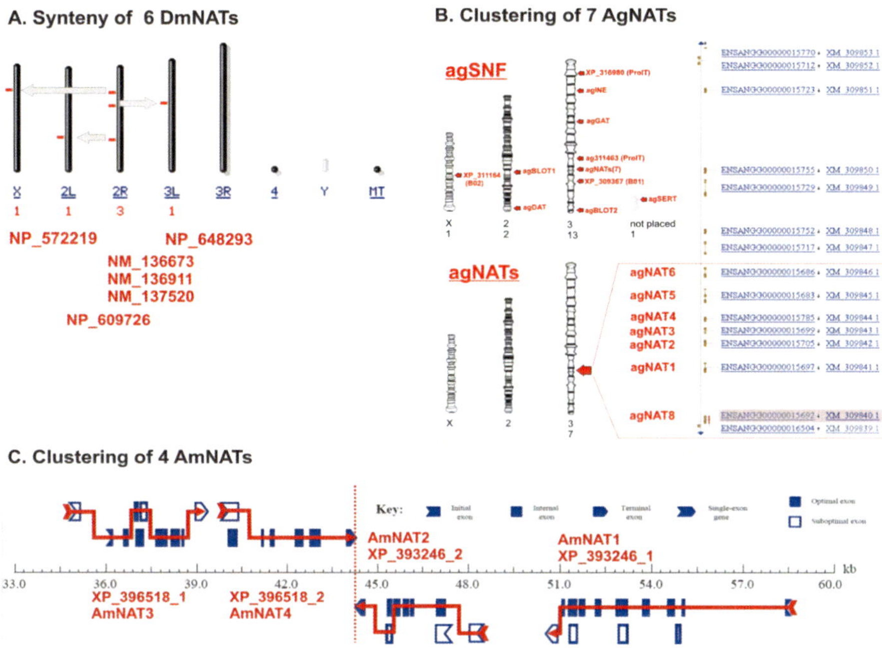

Fig. 9 Genome distribution of NAT populations in different insects. **a** Distribution and possible rearrangement of DmNAT duplicates in the landscape of Dm genome. **b** Chromosomal distribution of AgSLC6 member (*top*) and 7 AgNATs (*bottom*) are shown along with a high-resolution map of the NAT-specific locus (EMBL gene accession and NCBI transcript map accession numbers are shown). **c** Unfold duplication of AmNATs as it was detected in the contiguous sequence of genomic fragment. Abbreviations: Ag, *A. gambiae*; Dm, *D. melanogaster*; Am; *A. mellifera*

irreversible mechanism (Fig. 10a–i) with generalized 2/1 cation/amino acid substrate coupling stoichiometries. They use Na^+ motive forces; however, they could also utilize K^+ motive forces (Fig. 10d, e) with appropriate inward directions, resembling in this respect a transient phenotype between sodium-specific and potassium-preferring NATs, such as the mammalian B^0ATs [328, 338] and caterpillar KAAT1 [304]. The ion preferences of mammalian and caterpillar NATs reflect adaptations of these transporters to high-sodium (in mammals) and trace-sodium/high-potassium (in caterpillars) environments. Respectively, the capacity of AgNATs to use Na^+ and K^+ was explained as an adaptation to the environment with variable availability of both cations. Freshwater larvae deal with low concentrations of both ions in the environment. A cation pool being depleted by NATs activity in the absorptive region of the posterior midgut may be internally recycled in mosquito larvae via active apical efflux of these cations or through the previously proposed anterior midgut alkalinization pathways [264]. Nevertheless, a long-term internal pool of available Na^+ and K^+ ions may depend on environmental availability of these ions. Both ions could be randomly disproportional and very limited in the freshwater larval inhabitant. The availability of Na^+ and K^+ potassium is also quite disproportional over the life cycle of the adult mosquito. The Na^+/K^+ balance could be very low upon adult feeding on plant fluids, but dramatically increased after a blood meal. Both aromatic NATs were sensitive to extracellular Cl^-, when present outside it increases net current via the transporters (Fig. 10d, e). However, it appears to be a relaxed attribute of the NAT function for example some other transporters in this group have none or low requirements for extracellular Cl^-. This result may suggest that external Cl^- acts as a NAT modulator because [Cl^--Na^+–AA^0] symport mechanisms will reduce but not increase the apparent current, and putative antiport mechanisms must demonstrate an opposite trend upon the reduction of extracellular Cl^- concentrations. Another possibility, which fits the data, is that AgNAT6 function is facilitated by electroneutral (bidirectional) Cl^- transport. The structural basis of Cl^- interaction with SLC6 members has been analyzed, suggesting a possible substitution of Cl^- with negatively charged residuals in the chloride-independent members of SLC6 [339, 340], however exact roles of this anion in SLC6 function remains enigmatic. Surprisingly enough, in contrast to B^0-like substrate profile in the AeAAT1 and both MsNATs, AgAAT8 and AgNAT6, AgNAT8 mediates transport with strictly narrow and unique substrate selectivity. AgNAT8 has high apparent affinity for phenylalanine and products of phenylalanine catabolism, including tyrosine and L-DOPA ($K_m{}^Y > 0.3$ mM) but not DA or OA; and tryptophan and tryptophan derivatives 5-HTP (Fig. 10 g). Despite an obvious similarity with AgNAT8, the electrochemical properties of AgNAT6 have some unique traits. The most notable difference is in organic substrate specialization with strong and narrow selection of indole-branched substrates (Fig. 10 g). All collected data suggests that AgNAT6 is specialized to transport tryptophan with minimal interference from phenylalanine and other neutral amino acids, which together could be in much higher concentrations in larval and adult nutrient digests.

In situ hybridization with transcript-specific probes and real-time quantitative RCR reveals high expression and respective functions of insect NATs in the absorptive and secretory regions of the alimentary canal [126, 336] (e.g. see a representative panel of an AgNAT8 hybridization pattern in Fig. 11). In addition, cell-specific hybridization identified both aromatic NATs ((Fig. 11d, f–j); and [126]) in the insect brain and sensory afferents. The comparative study revealed

Fig. 10 Electrochemical properties of representative SLC6-NAT members. **a–f**, Representative collections of IV plots obtained from heterologously expressed insect NATs in *X. laevis* oocytes; adopted from published works [95, 124, 304, 316, 336, 342, 453]. Each plot includes several IVs that were acquired after applications of specified substrates. Background currents without substrate application were subtracted; except for **b**. IVs indicated by distinct line types represent specific ion currents (**a**) or basal solution conditions (**c–f**; see *inserts*). **g** Apparent normalized affinity values of AgNAT 6 interaction with representative substrates (*bars* are means ± STDV, n > 3) and Hill constants (*circles* are means ± STDV) for selected organic substrates. **h** Apparent affinity constant of AgNAT8 interaction with substrates. **i** Normalized L- and D- substrate-induced responses in DmNAT1 injected oocytes. **j** Apparent affinity of DmNAT1 interaction with representative substrates

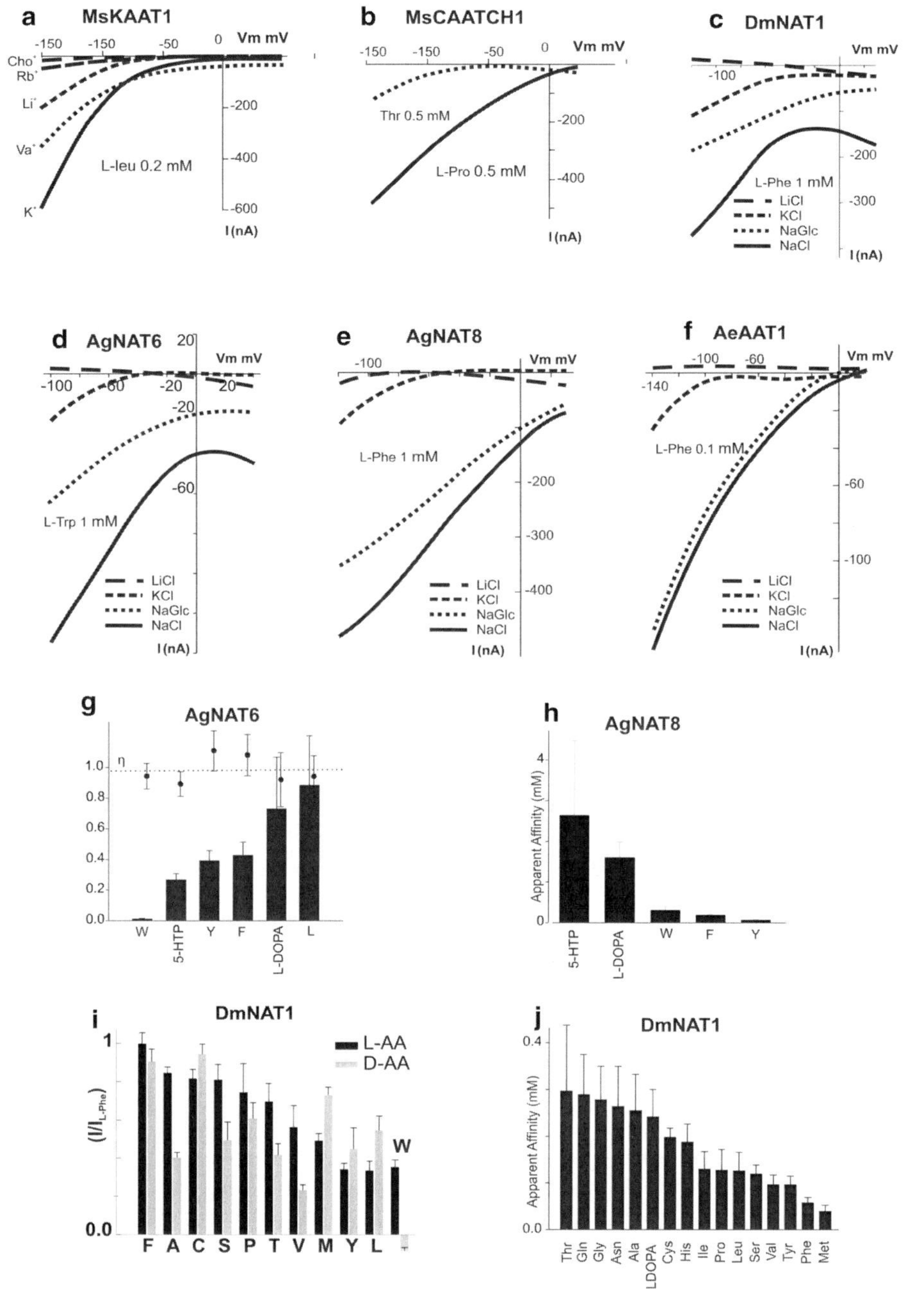

Fig. 10 (continued)

homodimer formation by both transporters in vivo and in a heterologous system and also clarified some important differences in the polar and spatial distribution of this transporter [125]. Specifically, AgNAT6 occupies a broad area of the larval gut with some notable differences in the polar distribution in the gastric caeca and anterior midgut area [125] (e.g. panels in Figs. 12 and 13respectively summarizing a relative spatial and polar distribution of AgNAT 6 and AgNAT8). This varied distribution may be essential for the specific functions of different NATs. We may anticipate that cell- and tissue-specific expression of sodium-dependent transporters in the alimentary canal can reduce competition of transporters for the energy of transmembrane electrochemical gradients, while in the CNS it may support cell-specific metabolic processes for example synthesis of indole- and catechol-amine neurotransmitters in the case of AgNAT6 and AgNAT8. The presently characterized aromatic amino acid specific NATs include one bacterial transporter TnaT [225]. The *tnaT* gene encoding a unique sodium-dependent mechanism with strong tryptophan selectivity has been cloned from *Symbiobacterium thermophilum* [225], a microorganism whose growth depends on commensalisms and whose genome includes a number of transporters for peptides and amino acids [341] but only a sole SLC6 member, *tnaT*. Apparently, *tnaT* acts as a pipeline for delivery of symbiotic substrates in synergy with other members of the *tna* operon of *S. thermophilum* surviving on tryptophan-specific media [341]. Together with the data about insect NAT it may suggest that the mechanism for narrow substrate selection evolved in the NAT population many times. The usage of aromatic amino acids varies dramatically in different organisms, developmental stages, tissues, and individual cells. Their availability also varies upon ecological and nutrient adaptations. For example, freshwater mosquito larvae filter-feed on microorganisms and organic debris, and mammals digesting protein-rich food may have a different profile and concentrations of essential amino acids in the alimentary canal and systemic circulations. Mosquito larvae require high quantities of aromatic amino acids for cuticle formation and tanning upon successive ecdysis as well as rapid protein accumulation between ecdysis events. A high-throughput

Fig. 11 Spatial expression of AgNAT8 in the alimentary canal, central and peripheral nervous system of *A. gambiae* larvae. **a** In situ hybridization of AgNAT8 in the whole-mount midgut from 4th instar larvae (*dark brown signal*). Magnified images of the salivary gland (**b**), cardia (**c**), isolated fragment of ventral neuronal cord that includes three thoracic ganglia, (**d**) rectal gland. **f, g** Dorsal and lateral aspects of AgNAT8 hybridization pattern in the larval stemma (eye); this labeling is associated with primary photosensitive neurons (*green arrows*). **h** An example of hybridization of sensory neurons associated with vibration-sensitive hair sensilla on the larval head (*black arrow*). **i** Labeling in the basal part of the peg sensillum at the head capsule (*white arrow*); this structure has an unknown sensory role in larvae, but it is morphologically similar to temperature-sensitive structures in adults. **j** Hybridization of specific populations of chemosensory neurons in maxillary pulp (MP) and labium afferents (LB) (*open arrows*). AMG, anterior midgut; CA, cardia, a sub-esophageal invagination; GC, gastric caeca; MT, Malpighian tubules; PMG, posterior midgut. *Bars* represent scale in micrometers. Adopted with modifications from [336]

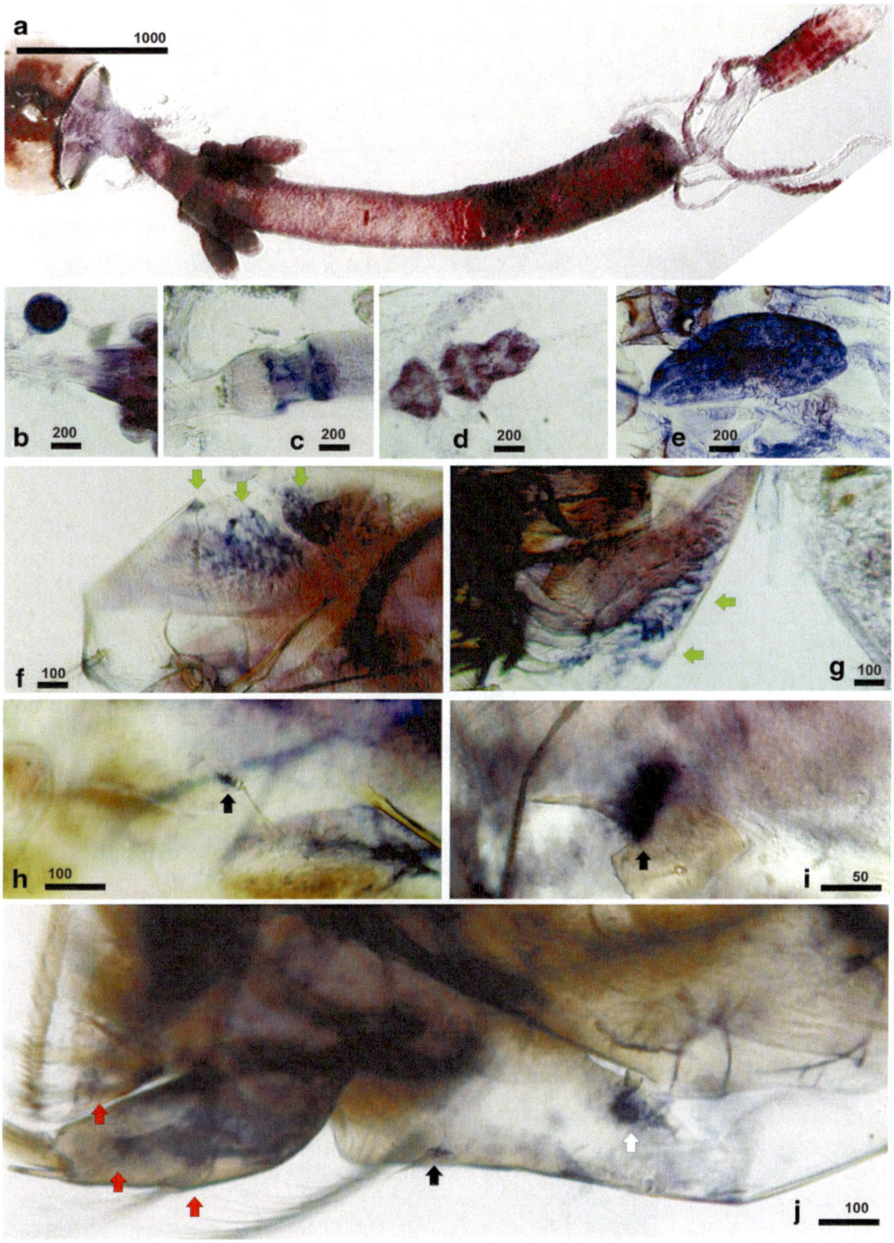

Fig. 11 (continued)

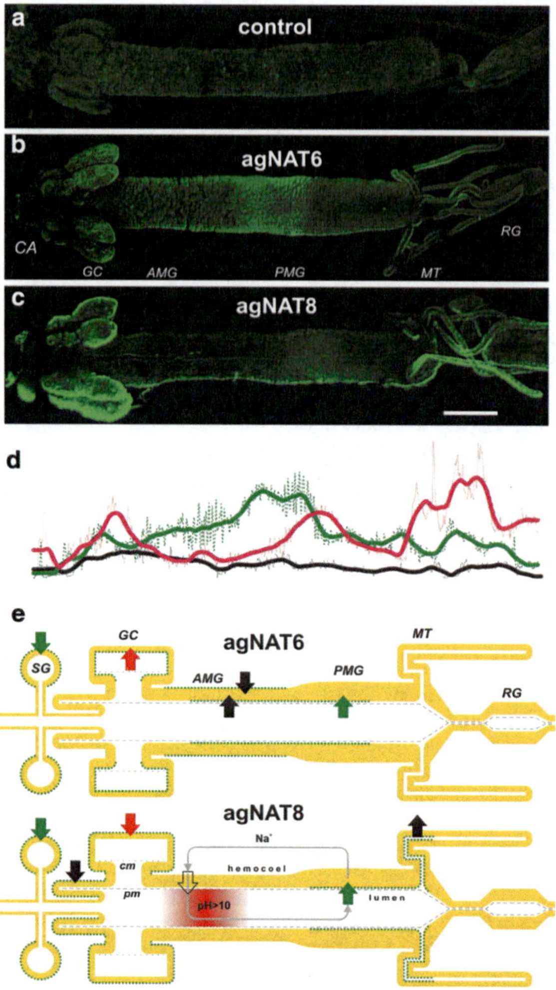

Fig. 12 Relative spatial distribution of AgNAT6 and AgNAT8 in the larval alimentary canal. Isolated alimentary canal of fourth instar *A. gambiae* larvae were labeled with (**a**) preimmune serum, (**b**) AgNAT6- and (**c**) AgNAT8-specific antibodies. Abbreviations: CA, cardia; GC, gastric caeca; AMG, anterior midgut; PMG, posterior midgut; MT, Malpighian tubes; RG, rectal gland. (**d**) Relative fluorescence intensities are shown along the larval alimentary canal (scans from **a, b, c** are represented by *black, green and magenta lines*, respectively). The *scale bar* is 1 mm. A summary diagram of AgNAT6 and AgNAT8 distribution patterns in the larval alimentary canal. *Green dotted lines* indicate membrane docking of the transporters. *Color arrows* display direction of transport at specified locations. The colors show ether identical (*green*), specific (*black*), or opposed directions of substrate transport by AgNAT6 (*top part*) vs. AgNAT8 (*bottom part*). *Gray arrows* show a putative circuit of cation recycling via alkalinization (*purple gradient region*) and NAT-coupled pathway. The *empty arrow* indicates location of a putative H^+ V-ATPase and cation exchanger coupled mechanism for cation translocation in the AMG. Abbreviation: SG, salivary gland; CA, cardia; GC, gastric caeca; AMG, anterior midgut, PMG, posterior midgut; MT, Malpighian tubes; RG, rectal gland; cm, caecal membrane; pm, peritrophic membrane. Adopted with modification from [125]

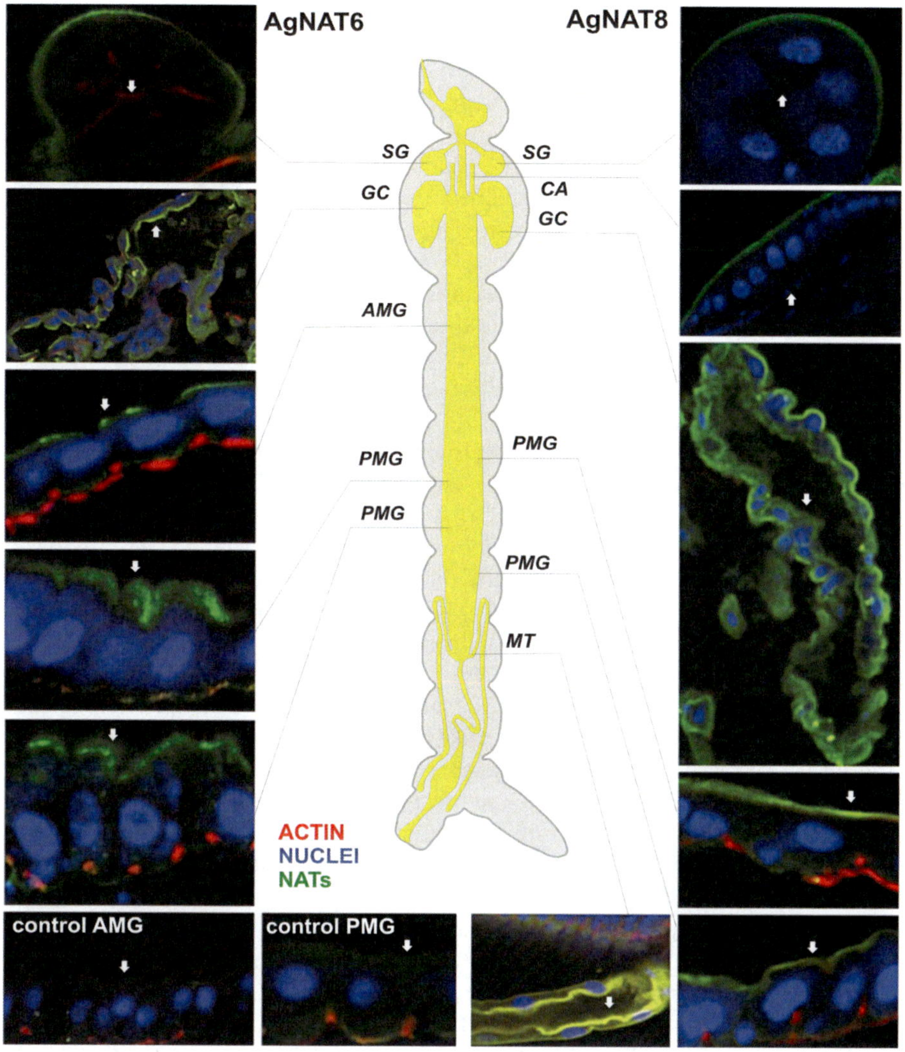

Fig. 13 Relative polar distribution of AgNATs in the alimentary epithelia. Immunolabeling of frozen sections of larval alimentary canal with AgNAT6 and AgNAT8 antibodies (*left and right side* of the panel, respectively). *Green*, *red*, and *blue* channels represent NAT, actin, and nuclei-specific labeling. The actin and nuclei channels were turned off on a few sections to improve overall visualization. Relative positions of the apical membrane are indicated by *white arrows*. Approximate positions of individual sections are shown on the *central diagram*. SG, salivary gland; CA, cardia; GC, gastric caeca; AMG, anterior midgut; PMG, posterior midgut; MT, Malpighian tubes. Control sections of the AMG (control AMG) and PMG (control PMG) incubated with pre-bleed serum are shown in the *bottom left corner*. *Scale bars* 50 μm. Adopted with modification from [125]

delivery of aromatic amino acids from the digestive system to the ovaries is critical in the egg development and chorion formation. The presently identified aromatic NAT is perfectly suited to acquire most underrepresented essential amino acids from the larval environment as well as to normalize the supply of such amino acids for specific metabolic processes such as oogenesis or neurotransmitter synthesis.

DmNAT1 is another interesting member of SLC6 contributing to the diversity of iNAT phenotypes. It has been cloned from the classic genetic and genomic model organism fruit fly, *D. melanogaster* [342]. The CG3252 gene product, DmNAT1, represents the first Nutrient Amino acid Transporter of the six gene DmNATs (Fig. 8). In contrast to all previously characterized insect and mammalian NATs it absorbs a broader set of neutral amino acids. Specifically, in addition to B^0-specific L-substrates, DmNAT1 equally or more effectively transports D-amino acids with sub millimolar affinities and 1:1 sodium:amino acid transport stoichiometries (Fig. 10i–g). DmNAT1 is transcribed in the absorptive and secretory regions of the larval alimentary canal and larval brain, strongly supporting its roles in the primary absorption and redistribution of large neutral L-amino acids as well as corresponding D-isomers. The absorption of D-amino acids supposedly promotes the acquisition of fermented and symbiotic products, and may support the previously identified and unique capacity of fruit fly larvae to survive upon the substitution of essential amino acids by D-isomers [343]. The strong transcription of DmNAT1 was described in the larval fruit fly brain, and suggests a role for it in neuronal nutrition and broad clearance of L/D enantiomers of neutral amino acids from the fly brain, which is specifically important for neurochemical signaling occurring between neuronal cells in the neuropile area of the insect brain. Neuronal and glial DmNAT1 may also absorb synaptic D-serine and modulate NMDA receptor-coupled signal transduction. The characterization of the first invertebrate B^0-like transporter extends the biological roles of the SLC6 family, revealing adaptations for the absorption of D-isomers of the essential amino acids. These findings suggest that some members of the NAT-SLC6 subfamily evolved specific properties which contributed to nutrient symbiotic relationships and neuronal functions [342].

Similar to members of the SLC1 family, the SLC6 transporters share notable conservation in their sequences and structures (Fig. 14a). Most of SLC6s have 12 transmembrane domains and intracellular N, C termini (Fig. 14b). The crystal structure of the first NAT from the archaebacterium *Aquifex aeolicus* VF5 (2A65) has been solved [224]. This prokaryotic NAT shares approx. 20% sequence identity with metazoan SLC6 members. The conservation degree increases in TMDs, being especially high in the TMDs responsible for the formation of the substrate-binding pocket, thus regionally satisfying reciprocal molecular modeling of these transporters with a focus on substrate-binding envelopes. Identification and characterization of new NAT phenotypes makes it possible to improve spatial restraint models and the overall quality of in silico modeling. Characterization of new NATs in combination with in silico structural analysis will aid in the identification of mutations and structural aspects that drive substrate selectivity and electrochemical adaptations in NAT phenotypes. A number of models of SLC6 members being built are based on reciprocal structural homologies and homologies with a crystallized

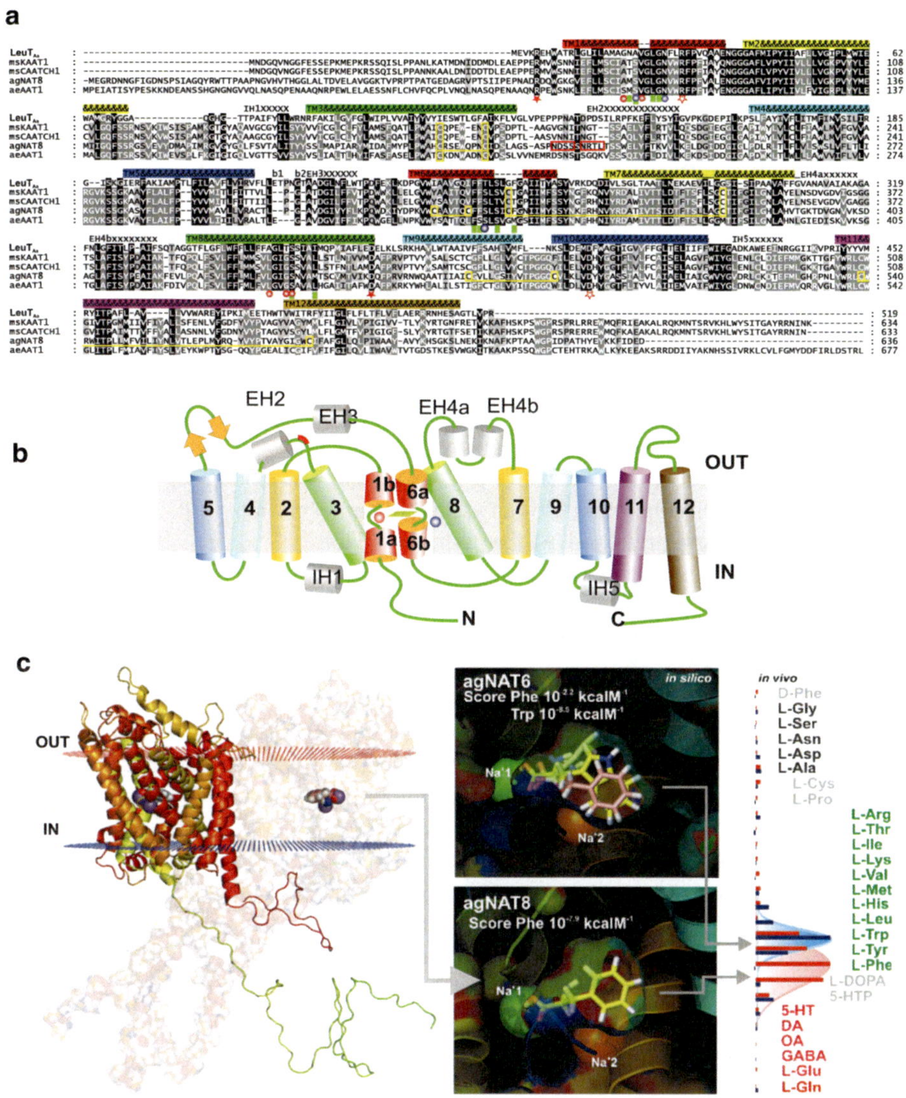

Fig. 14 An example of in silico and in vivo analyses of AgNATs. *Right*. A complete 3D model of agNAT8 homodimer (inner and outer surfaces of the plasma membrane are indicated by *blue and red dot* matrixes). *Middle*. In silico modeling and analysis of agNAT6/8 substrate binding pocket. *Left* Substrate transport efficiency for agNAT6 and agNAT8 in a heterologous expression system. *Text color* indicates substrate categories

bacterial transporter (e.g. Fig. 14d). Using 3D models of homologous transporters it will be possible to identify key structural correlates of functional specificity and perform de novo predictions of substrate profiles as well as in silico screening of ligand libraries with an aim to identify potent and specific modulators of NAT functions.

3.4.5 "Inebriated" (Ine) Cluster

The identification of Na^+/Cl^--dependent transporter family members based on structural similarity has resulted in the discovery of many putative transporters for which the substrates are not known. Many of these so-called "orphan" transporters are larger than other Na^+/Cl^--dependent transporters, with extended fourth and sixth extracellular loops and larger C-termini. The expression patterns of these orphan transporters are often diverse, with extensive nervous system and peripheral distributions. Insect examples include the *Drosophila* blot (bloated tubules) and *ine* (inebriated) proteins, and the *M. sexta ine* protein (ms*ine*). The "*ine*" locus, relating to a characteristic "inebriated" behavior phenotype [344], was initially cloned from *D. melanogaster* and its expression was mapped by in situ hybridization in the posterior hindgut, Malpighian tubules, anal plate, garland cells, and a subset of cells in the CNS of developing embryo [345]. It has been proposed by analogy with other NTTs that the neuronal hyper-excitability characteristic of the *ine* phenotype is a result of failure in some neurotransmitter uptake due to *ine* mutations [345]; however, the identity of any such substrates remains elusive. Additional transgenic characterization led to the hypothesis that a neuronal dm*ine* isoform mediates down-regulation of sodium channels via a neurotransmitter uptake [346], whereas an epithelial dm*ine* isoform compensates osmotic load in the hindgut and Malpighian tubules via an osmolite carrier [347]. Substantial progress in characterization of *ine* has been made via the heterologous expression of an ine-orthologous transporter from *M. sexta* in *Xenopus* oocytes [348]. That study revealed that *ine* responds to hyper-osmotic stimuli via the inositol trisphosphate-coupled release of intracellular Ca^{2+}, which subsequently activates Ca^{2+}-dependent K^+ channels. Our phylogenetic analysis indicates a single *ine* gene in *A. gambiae* and two splicoforms of a single *ine* gene in *D. melanogaster* (CG15444, located on the L2 chromosome). A single INE gene exists in each completed genome of insects as well as the representative genomes of Urochordates and Echinoderms, but is absent in the vertebrate and nematode genomes. The insect *ine* group, with undefined ligand and transient functions, represents an evolutionary root of the NTTs. It may provide primordial genetic sources in the evolution of GLYT, GAT, and monoamine neurotransmitter transporters, but also may reflect a case of the ultimate adaptation of a transporter to the signaling function. Therefore, the reasons why substrates for these putative transporters have remained elusive may not only be technical, but biological as well. The primary function of some of these proteins may be to regulate transmembrane ion fluxes in different tissues, including the CNS [348].

3.4.6 Bloated Tubules (Blot) Group

Few other orphan genes, named by similarity with a mammalian orphan group, have been analyzed in *Drosophila* [228]. This gene, named bloated tubules (*blot*), exhibits complex and dynamic expression patterns during embryogenesis and supports processes that are critical for cell division, differentiation, and epithelial morphogenesis. *blot* transcripts are expressed in ectodermally derived epithelia including

Malpighian tubules and in the nervous system. *blot* mutants are larval lethal showing swollen Malpighian tubule cells with an enlarged and disorganized apical surface. Embryos lacking the maternal blot expression die during early organogenesis, showing an inability to form actin filaments in the apical cortex, which results in impaired syncytial nuclear divisions, severe cytoskeleton defects, and a failure to cellularize stages of development. Drosophila *blot* form a putative orthologous cluster together with a single SLC6 member from other dipteran genomes, i.e. putative AgBLOT1. It is grouped together with several other genes in the basal portion of the SLC6 family tree; however, the lack of notable clustering and the long phylogenetic branches suggest a long or extensive divergent history of these orphan transporters.

In summary: The broad diversity of phenotypes, together with their apparent rapid specialization in substrate and ion specificity, suggests remarkable adaptive plasticity within NAT-SLC6 mechanisms via alterations of a few identifiable sites in the substrate-binding pocket. A likely reason for the additional expansion of NATs in invertebrates with narrow substrate transporters is the amplification of the transport network for the acquisition and redistribution of environmentally faint substrates. It is reasonable to propose that metazoan organisms competing for an identical set of essential amino acids evolved distinct strategies and mechanisms for their absorption and redistribution.

3.5 SLC7 + SLC3, Cationic and Heterodimeric Transporters (CAT and HAT)

SLC7 integrates two functionally and structurally different groups of transporters: the cationic amino acid transporter (CAT) subfamily and the heterodimeric amino acid transporter (HAT) subfamily, based on phylogenetic, structural, and physiological properties [327] (Fig. 15a, b). SLC7-related transporter heterogeneity has been partially explored in insects and one representative HAT-homologous sequence (AeaLAT) has recently been cloned from *A. aegypti* [349]. Phylogenetic analysis of insect transporters revealed three specific clusters. Two large and distal clusters apparently comprise insect versions of CAT and HAT subfamilies, respectively. The third, small, cluster is parental to the CAT/HAT cluster (Fig. 14b), which by its phylogenetic loci represents transporters similar to mammalian orphan SLC7A4, with an as yet indefinite function [350]. In mammals SLC7 genes encode a broad variety of transport systems including sodium-independent b^+, y^+, y^+L (for cationic AA) as well as $b^{0,+}$ and a sodium-dependent system $B^{0,+}$; y^+L specific for neutral amino acids [327, 351].

3.5.1 Cationic Amino Acid Transporters (CATs)

The cationic amino acid transporter (CAT, SLC7A1–4) subfamily comprises 14 TMD transporters which are typically located on the plasma membrane [350]. Most of these transporters regulate local concentrations of cationic amino acids through

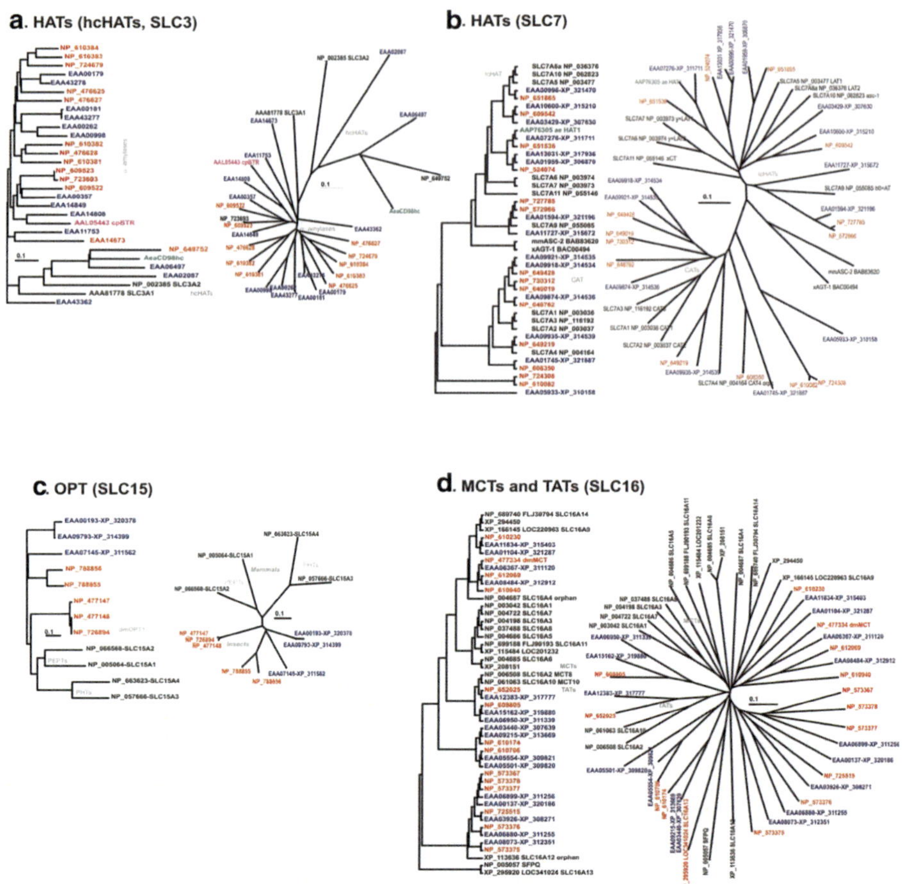

Fig. 15 **a** Phylogenetic proximity of mammalian HATs (hcHATs, SLC3) with dipteran orthologs of heavy chain subunits with amylases. Unrooted phylogram (*left*) and dendrogram (*right*) of putative heavy chain subunit of HATs and amylases from dipteran genomes, along with known orthologs from *H. sapiens* genome (*black font*). Other *font colors*: *green*, heavy chain subunit of heterodimeric amino acid transporter from *A. aegypti*; *blue*, *A. gambiae* sequences; *red*, *D. melanogaster* sequences; pink, a binary toxin-binding alpha-glucosidase from *C. pipiens*. **b** Phylogenetic proximity of dipteran and mammalian cationic amino acid transporter (CATs) and light chain subunits of heterodimeric amino acid transporter lcHATs (SLC7). Cladogram rooted with an EAA05933-XP_310158 *A. gambiae* sequence (*left*) and unrooted dendrogram (*right*) of putative dipteran CATs and lcHATs along with identified and partly characterized orthologs from *H. sapiens* genome. **c** A phylogenetic distribution of dipteran and mammalian oligopeptide transporters (OPTs, SLC15). Unrooted phylogram (*left*) and dendrogram (*right*). *Font colors*: *blue*, *A. gambiae* sequences; *red*, *D. melanogaster* sequences; black, *H. sapiens* sequences. Other abbreviations: dmOPTs, *D. melanogaster* oligopeptide transporters; PEPTs, *H. sapiens* peptide transporters; PHTs, *H. sapiens* peptide/histidine transporters. **d** Phylogenetic distribution of dipteran and mammalian monocarboxilate and T-system amino acid transporters (MCTs and TATs; SLC16). Rooted cladogram (*left*) and unrooted dendrogram (*right*). Human orphan member of SLC16 (SLC16A13) and PTB associated splicing factor (NP_005057) are selected as outgroup. *Font colors*: *blue*, *A. gambiae* sequences; *red*, *D. melanogaster* sequences; *black*, *H. sapiens* sequences. *Scale bars* are evolutionary distances in mutations per site

(1) facilitated diffusion and/or (2) differential amino acid exchange between relatively saturated physiological domains in organisms, for example growing and/or actively metabolizing cells, as well as nutrient and secretory epithelia [327]. The first CAT was identified as the receptor for murine ecotropic leukemiaviruses in mouse, MuLVR or mmCAT1, which mimics a y^+-transport system (sodium ion-independent L-arginine, L-lysine, and L-ornithine uptake) phenotype in *Xenopus* oocytes [352, 353]. Now mammalian CAT subtypes 1–3 (SLC7A1-3 respectively) represent a large group of Na^+-independent transporters. CATs in nonepithelial tissues predominantly function as y^+-system nutrient cationic amino acid providers. In epithelial tissues CATs mediate basolateral transport which apparently is an essential component of transepithelial transport of amino acids. Thermodynamically CAT Types 2 and 3 mediate downhill diffusion (uniport) of selected cationic amino acids with extremely weak coupling or no coupling to other substrates. Since its substrates carry charge at physiological pH, CAT transport can be electrophoretic as well. In contrast CAT Type 1 is strongly *trans*-stimulated by intracellular substrates and may act as a thermodynamically coupled exchanger of specific intracellular and extracellular amino acids. Neutral amino acids can be recognized by most CATs with rather low apparent affinity [351]. Mammalian CATs exhibit abundant and sub-type-specific patterns of spatial expression and immunolabeling. They appear to harbor genetic defects that are factors of several metabolic and immune diseases, including cystinuria and lysinuric protein intolerance [354]. In addition to their relatively universal function in absorption and emission of cationic amino acids including arginine, CATs serve as rate-limiting substrate providers for enzymatic production of nitric oxide (NO) in endothelial cells and macrophages [355–357]. CATs undergo complex regulation at transcriptional [358] and post-transcriptional [359] levels, which include modulation by substrates and nontransported ligands, as well as by cytoskeleton proteins [360].

3.5.2 Heterodimeric Amino Acid Transporters (HATs)

The heterodimeric amino acid transporter (HAT, mammalian SLC7A5–11) subfamily comprises 12-TMD transport proteins (also referred to as light chain catalytic subunit, approx. 40 kDa) whose functional expression requires association with its ancillary counterpart peptide, a heavy chain (approx. 80 kDa) subunit [327]. The heavy chain subunits (mammalian SLC3) are currently represented by two proteins with apparent α-glucosidase homology: 4F2hc, the heavy chain of the lymphocyte activation antigen 4F2 [361] and rBAT, a Type II membrane glycoprotein related to the $b^{0,+}$ amino acid transport phenotype [362–364]. Heavy chain proteins have a single membrane spanning domain and an extracellular globular C-terminal domain in their secondary structure. Their interaction with light chain subunits is stabilized via 4F2hc-specific disulfide covalent binding [365] or rBAT-specific noncovalent interaction [366]. The suggested role of the heavy chain SLC3 component is in coordinated delivery of the heterodimeric complex to the plasma membrane whereas the light chain SLC7 of HATs serves as a carrier mechanism [327, 367].

In contrast to facilitators and nonobligatory exchangers like CATs, the obligatory exchangers, HATs, are associated with a great variety of transport systems which are diverse in terms of distribution and substrate selectivity, including: L system for large neutral amino acids; asc system for small neutral amino acids, for example Ala, Ser, Cys; x_c^- system for anionic amino acids and $y^+L/b^{0,+}$ system which combines carriers for cationic and neutral amino acids [327]. Virtually all cloned HATs are obligatory exchangers of amino acids, except for the sodium ion driven y^+L transport of neutral amino acids by SLC7A6 or SLC7A7 and SLC3A3 HATs [368, 369]. The varied physiological roles of HATs have been inferred from their activity in heterologous expression systems, distribution, and earlier brush-border vesicle analyses. For example, in nutritive epithelial tissues they serve as basolateral complements of apical B^0 and $B^{0,+}$ transporters, which are both essential for transepithelial absorption of amino acids [370]. They also may serve as tertiary active transport mechanisms for epithelial reabsorption of cystine and dibasic amino acids [371]. In nonpolarized cells, heterodimeric transporters serve in several unique mechanisms in which nutrient neutral amino acid uptake is thermodynamically coupled to emission of metabolized or redundant intracellular neutral amino acids [372]. Notable examples of HAT function are neuronal and glial activity of aSLC7A11/xCT transporter which is heterologously associated with a 4F2hc subunit [373] and mediates 1:1 exchange of extracellular cystine for intracellular glutamate. This transporter, therefore, represents a rate-limiting component in GSH (glutathione) synthesis pathways. GSH is a tri-peptide of glycine, glutamate, and cysteine which is crucial for cells and tissues that tolerate endogenous free-radical stress, for example: activated macrophages, hypothalamic neurons, a variety of glial cells and pancreas, as well as several cell culture lines [374–378].

Transporters structurally similar to SLC7 have been reported throughout all life kingdoms, however with quite specific patterns of taxonomic proximity for particular species (Table 2). The insect CAT cluster, defined by phylogenetic reciprocity to mammalian CATs, includes six genes in both dipteran models (Fig. 14b). However, insect CAT genes do not form exact orthological matches with mammalian CATs, except for the SLC7A4 (NP_004164) gene, which clusters with insect NP_649219 and EAA09935 sequences (Fig. 15). In contrast, insect CATs exhibit better interspecific homology by forming four orthologous clusters when they are analyzed within frames of dipteran lineages (Fig. 3). It is most likely that proximal insect CATs are functional equivalents of mammalian CATs 1–3, which evolved via lineage-specific gene duplications; however, exact functions remain elusive, pending cloning and functional characterization of such gene transcripts. In addition to the four CATs characterized in mammalian models, both insect genomes include two more, apparently parental, genes that are basal phylogenetic level relatives to CATs and HATs.

Insect HATs comprise nine putative genes in *A. gambiae* but only five genes in *D. melanogaster* which together occupy five apparently orthologous clusters (Figs. 3 and 14b). Only one such group forms a cluster with mammalian SLC7A9 (NP_055085) transporters (Fig. 14). Several *A. gambiae* genes are not mapped, so the question about the exact number of HAT genes in this model remains to be clarified. One HAT gene cloned and characterized from *A. aegypti* [349], AAP76305, appears to be orthologous to EAA07276 and NP_651536 genes in *A. gambiae* and *D. melanogaster*, respectively.

3.5.3 Auxiliary Subunits of SLC7 Comprising SLC3 Family

SLC3 proteins (heavy chain subunit of HAT; Fig. 14a) serve as obligatory cofactors of SLC7 transport proteins, but do not appear to be phylogenetically related to some membrane solute carrier families. In contrast, this subunit is homologous to the α-amylase or glycosyl hydrolase Family 13, all of which bear a unique catalytic $(\beta/\alpha)_8$-barrel domain with the active site being at the C-terminal end of the barrel beta-strands. All of them share a functional signature in that they catalyze a variety of glycoside bond cleavage/formation reactions [379]. The extracellular domain of SLC3 members has up to 30–40% peptide identity with insect maltases and bacterial α-glucosidases [367] which in turn indicates a possible route of evolution common for α-amylase family proteins [380]. Nevertheless, rBAT (see below) lacks amylase-specific activity when heterologously expressed in *Xenopus* oocytes [364]; it appears to lack amylase-specific catalytic residues [365] when compared to *Bacillus cereus* oligo1,6-glucosidase (O1,6G), which is available in the Protein Data Bank (PDB) database [381].

Presently, two members of mammalian SLC3 which form functional HATs are characterized [367]. First, rBAT (SLC3A1), initially was identified by expression cloning in *Xenopus* oocytes [362–364]. Second, 4F2hc (SLC3A2), was initially identified as a lymphocyte activation antigen [361]. Both of these proteins appear to be widespread throughout the Metazoa. They also share remarkable homology with a number of putative genes from bacteria and fungi. However, the affiliation of these bacterial and fungal proteins with HAT forming proteins requires a detailed comparison of sequences and secondary structures as well as functional assay, because of their apparent orthology with amylases. Intriguingly, SLC7 homologous proteins are present in plants and archaea which lack SLC3 counterparts. This fact makes dialectic the obligatory requirement of SLC3 and CLC7 association at least in some lineages.

Two putative genes resembling the mammalian SLC3A2 heavy chain subunit are currently found in the *A. gambiae* genome (EAA06497, located on the X chromosome; and EAA02087, not yet mapped) whereas only one gene is present in *D. melanogaster* (NP_649753, located on the 3R chromosome). All other dipteran sequences with apparent SLC3 homology reside in theα-amylase-specific gene cluster. Intriguingly, some products of these genes are recognized as BT receptors, for example AAL05443 which encodes the receptor of *Bacillus sphaericus* binary toxin (BT) in *Culex pipiens* midgut [382]. Another important finding from analysis of dipteran sequences is that heavy chain proteins apparently evolved not from α-amylases but from some common primordial ancestor (Fig. 14a).

Two complementary subunits (AeaLAT and AeaCD98hc, light chain and heavy chain subunits, respectively) of an insect HAT have been cloned recently from the mosquito *A. aegypti*. Like mammalian HATs, insect transporters require expression of complementary subunits to produce a functional L-amino acid transporter phenotype in *Xenopus* oocytes. Heterologously expressed mosquito AeaLAT+AeaCD98hc subunits mediate uptake of large neutral and basic amino acids, partly resembling $y^{+}L$ system transporters. Small acidic and neutral amino acids appear to be poor substrates for this transporter [349]. Like mammalian HAT, these insect transporters can utilize a Na^{+} gradient (up-regulated approx. 44%) for the uptake of neutral amino acids (L-leucine; $K_m{}^{leu} = 67\,\mu M$), but not basic amino acids (L-lysine), which is virtually Na^{+}-independent at pH of 7.4 [349]. It has been suggested, that the electrochemical gradient for amino acids at high extracellular concentrations appears to be the major factor that enables amino acid transport by AeaLAT+AeaCD98hc. AeaLAT showed high level expression in the gastric caeca, Malpighian tubules, and hindgut of larvae. In caeca and hindgut, expression was in the apical cell membrane. However, in Malpighian tubules and midgut (which showed low level expression), the transporter was detected in the basolateral membrane. This expression profile supports the conclusion that this AeaLAT+AeaCD98hc pair mediates a nutrient amino acid emission from posterior midgut epithelial cells to hemolymph [349]. A combination of similar basal $y^{+}L$ HATs with apical $B^{0,+}$ transporters, such as MsKAAT1, MsCAATCH1 and (recently cloned from mosquito larvae midguts) AeAAT1 and agAAT8 may comprise a complete transepithelial transport metabolon for nutrient amino acid uptake. It has been stated that the apical location of AeaLAT in the gastric caeca implies its involvement in the absorption of amino acids from the lumen [349]. Similarly, its basal location in the midgut suggests that it may participate in the emission of amino acids or their derivatives to the hemolymph for circulation to the gastric caeca and cardia lumen where they are required for assembly of the peritrophic matrix.

3.6 SLC15, Oligopeptide Transporters (OPTs)

Oligopeptide transporters (OPTs, a.k.a. Peptide Transporters (PTRs) [383]) evolved as an efficient, energy-saving uptake mechanism complementing or replacing amino acid carriers in special cases for example (1) they enable high-throughput intracellular loading of nutrient amino acids in peptide form and (2) they remove specific and nonspecific oligopeptides, which may interfere with cellular signaling and cellular interactions, from systemic and sub-cellular circulations. In mammals OPTs mediate H^{+}-coupled uptake of small peptides or similar substrates into intestinal and renal epithelial cells, bile duct epithelium, glial and epithelia cells of the choroid plexus, lung and mammary gland [384]. OPTs mainly utilize transmembrane proton motive forces and appear to be reversible when heterologously expressed in HeLa cells or *Xenopus* oocytes [385, 386] and in vivo [387].

The first proton-coupled peptide transporter PEPT1 (SLC15A1) was identified by expression-cloning from rabbit small intestine cDNA in the Hediger laboratory [385]. It mediates the uptake of small peptides, being independent of extracellular

Na^+, K^+, and Cl^- gradients and, surprisingly, independent of the membrane potential as well. PEPT1 is transcribed in intestine, kidney, and liver with small amounts in brain. Simultaneously, PEPT1 was cloned by Murer and colleagues in attempts to define the molecular basis for proton-dependent peptide transport in BBMV from rabbit small intestinal epithelia [388]. This group also cloned an orthologous transporter PEPT2 (SLC15A2) from kidney cortex epithelia which, in contrast to low-affinity high-throughput PepT1, mediates high-affinity low-throughput transport of a virtually identical subset of substrates [386]. A large number of PepT1- and PEPT2-related OPTs have been identified by homology cloning in subsequent years [389–393], including one from insects, *D. melanogaster* OPT-1 [394]. More recently two other OPTs were identified in mammals: a brain-specific peptide histidine transporter 1 (rat PHT1 a.k.a. human PTR4, SLC15A4) [395] and peptide histidine transporter 2 (rPHT2 a.k.a. hPTR3 (SLC15A3) isolated from the rat lymphatic tissue [396]. In contrast to the broad substrate affinity of PEPT1 and 2, PHT1 and 2 have more specific sets of substrates including histidine and some model di- and tri-peptides. These transporters apparently play different, more specific physiological roles, since they are strongly expressed in specialized tissues such as brain for PTH1 [395] and lymphatic system, lung, spleen and thymus for PHT 2 [396]. OPTs range from 450 to 700 amino acid residues, most of the eukaryotic members having approx. 650 amino acid residues and 12 predicted TMDs with cytoplasmic N- and C-termini [384].

SLC15 transporters are present in most life kingdoms, except for the Archaea (Table 2), being especially abundant in plants where they are in phylogenetic proximity to nitrate transporters. The yeast *S. cerevisiae* has one member of the SLC15 family that is specialized in the absorption of di-/tripeptides (the PTR system). OPTs comprise three putative genes in the *A. gambiae* genome and three genes in the *D. melanogaster* genome with three different spiciformis known for the single characterized example of a dmOPT1 gene (Fig. 15c). Within the framework of mammalian and dipteran lineages, OPTs form separate paralogous branches, indicating relatively recent gene duplication. Respectively, the function of different dipteran OPTs is difficult to deduce solely from the phylogenetic analysis. The only insect OPT characterized to date is from *D. melanogaster* [394]. The *opt* gene encodes a protein that mediates proton-dependent di- and tri-peptide transport in intestinal and Malpighian tubular ("renal") epithelia. Like mammalian OPTs, fruit fly OPT1 has 12 TMDs and a long (200 amino acid residue) fifth external loop. It is most similar to human renal PEPT2 and intestinal PEPT1 proteins, containing the highly conserved histidine residue, thought to be involved in peptide binding and transport, at position 88.

Up to 30% of total amino acids in the hemolymph of adult fruit flies are present in the form of di- or tri-peptides which serve in osmoregulation [397, 398]. Protein digestion in *Drosophila* occurs mainly in the midgut where resulting peptides and amino acids are rapidly taken up by the epithelium [399] and metabolized or transported to the hemolymph. Hemolymph amino acids are secreted into the Malpighian tubular lumen (equivalent to vertebrate glomerular filtration) but are reabsorbed in the hindgut. Compared to mammalian peptide transporters discussed above, little was known about the absorption of peptides in insects prior to the cloning of dmOPT1. The *optl* gene encodes a high-affinity di- and tri-peptide transporter that is proton-dependent. It is transcribed in germinal and somatic tissues of both males and females where it is most highly expressed in nurse cells of the ovary. It is also expressed in epithelial cells of the midgut and rectum and is mildly expressed in neurons. Uptake was assayed in HeLa cells transfected with pCMVOPT1 or pCMV2 using lipofectamine. Tritiated L-alanylalanine uptake was measured in buffer with high NaCl and low KCl at pH 6.0. Evidence that the peptide transport is proton-coupled was a 10-fold decrease in uptake when extracellular pH was increased from 6 to 7 and a 20-fold decrease when it was increased from 6 to 8. Uptake was virtually abolished by FCCP (25 μM), which collapses the proton gradient. The K_m for OPT1-dependent alanylalanine transport at external pH = 6 was 48.8 μM [394].

3.7 SLC16, T System Amino Acid Transporter (TAT)

SLC16 represent a large family of H^+-coupled transporters and apparent uniporters [400], which currently includes 14 members [401] (Fig. 14d). Only five mammalian SLC16 members have been characterized (Fig. 15d). Four characterized members encode MonoCarboxilate Transporters (MCTs) and one encodes T system Amino acid Transporter (TAT) mediating transport aromatic amino acids [401]. Specifically TAT1 (SLC16A10) mediates H^+ and Na^+ uncoupled, sub-millimolar affinity, active transport or facilitated diffusion of aromatic amino acids and some related metabolites (L-phenylalanine, L-tryptophan, L-tyrosine, and L-DOPA). It is expressed in basolateral membranes of intestinal and renal epithelial cells as well as plasma membranes of various nonpolarized cells for example in placenta, liver, and muscle [324, 402, 403]. (SLC16A3 TATs should not be confused with sulfate transporter Tat1 (SLC23A11) [404, 405] or bacteria/yeast specific TAT1 permeases [406]). The tissue distributions of a few other members of SLC16 have been analyzed [407] but no electrochemical and ligand data are yet available. A recent comprehensive review aids our SLC16 description by mentioning that heterologous expression of a TAT1-related SLC16A2 transporter in *Xenopus* oocytes revealed a thyroid hormone (T4 and T3) transporter phenotype, which has micromolar apparent substrate affinity and appears to be independent of sodium- and proton-motive forces [401].

All characterized SLC16 members have a 12 TMD structure with a characteristic cytosolic loop between TMDs 6 and 7. At least some members require an auxiliary protein for trafficking and functional coordination in plasma membranes, for example MCT2 interacts with an unknown protein factor [408] and MCT1 and MCT4 interact with CD147 (also known as OX-47, extracellular matrix metalloproteinase inducer (EMMPRIN), HT7, or basigin) [401, 408, 409] which was confirmed using a Fluorescence Resonance Energy Transfer (FRET) technique [410] and by reduction of transport by antisense silencing of a CD147 ortholog in *Xenopus* oocytes [411].

Phylogenetic analysis discloses 15 putative genes in *A. gambiae* and 14 genes, corresponding with the number of identified transcripts, in *D. melanogaster* (Figs. 8 and 14). One member of *D. melanogaster* MCTs was identified before annotation of the genome (CAB42050); however, it does not cluster with mammalian MCT1 (SLC16A1) but instead fits into an apparently orthologous phylogenetic cluster with four other putative members of dipteran MCTs (Fig. 15). In fact, most MCTs form mammalian- or dipteran-lineage specific, paralogous, clusters except for orphans SLC16A9 and SLC16A14 and for characterized aromatic amino acid transporter, TAT (SLC16A10), and a putative thyroid hormone transporter (SLC16A2), which shows proximity with dipteran transporters and would reasonably be expected to have similar physiological roles (Fig. 14d).

3.8 Vesicular Glutamate and Phosphate Transporters, vGLUTs and PiTs (SLC17)

In mammals the SLC17 family comprises two physiologically distinct groups of characterized transporters: Type I inorganic phosphate transporters (P_iTs) and vesicular glutamate transporters (vGLUTs) (Fig. 16a). The PiT family, which was initially restricted to transporters of inorganic phosphates across plasma membranes, has now been expanded to include transporters of various organic and inorganic anions across both plasma membranes and endomembranes. The vGluT family comprises extensively analyzed vesicular glutamate transporters, which are specialized to serve excitatory neurotransmission that enables endomembrane traffic and vesicular accumulation of excitatory amino acids. Since no SLC17-related insect transporter has yet been characterized, we include only a brief description of known SLC17 transporters from a recent comprehensive review [412].

Before their molecular identities were established, P_iTs and vGluTs had been identified by comparative studies of vesicular glutamate uptake in BBMVs from neuronal tissues of several vertebrates and invertebrates (goldfish, frogs, turtles, pigeons, rats, fruit fly, and crayfish). Their properties, including millimolar Glu affinity, as well as V-ATPase energization and chloride stimulation, appear to be extremely conserved over 350–400 million years [413]. Phylogenetic analysis indicates a single gene copy in *D. melanogaster* and *A. gambiae* genomes with close proximity to three, apparently paralogous, vGluT gene clusters in mammals (Fig. 16a). Two additional *A. gambiae* genes and one *D. melanogaster* gene are more distantly related to the vGluTs. No dipteran genes with proximity to P_iT transporters have been found to date. The earlier defined dmP_iT transporter gene (NP_572188) now appears to be phylogenetically separated from P_iT gene clusters; its function cannot be deduced from the phylogenetic pattern alone. Intriguingly, dipteran genomes include several new SLC17 gene clusters, which are absent in mammals and have no matches with characterized genes inother lineages. Two of these clusters indicate remarkable gene duplications in *D. melanogaster* (Fig. 16a).

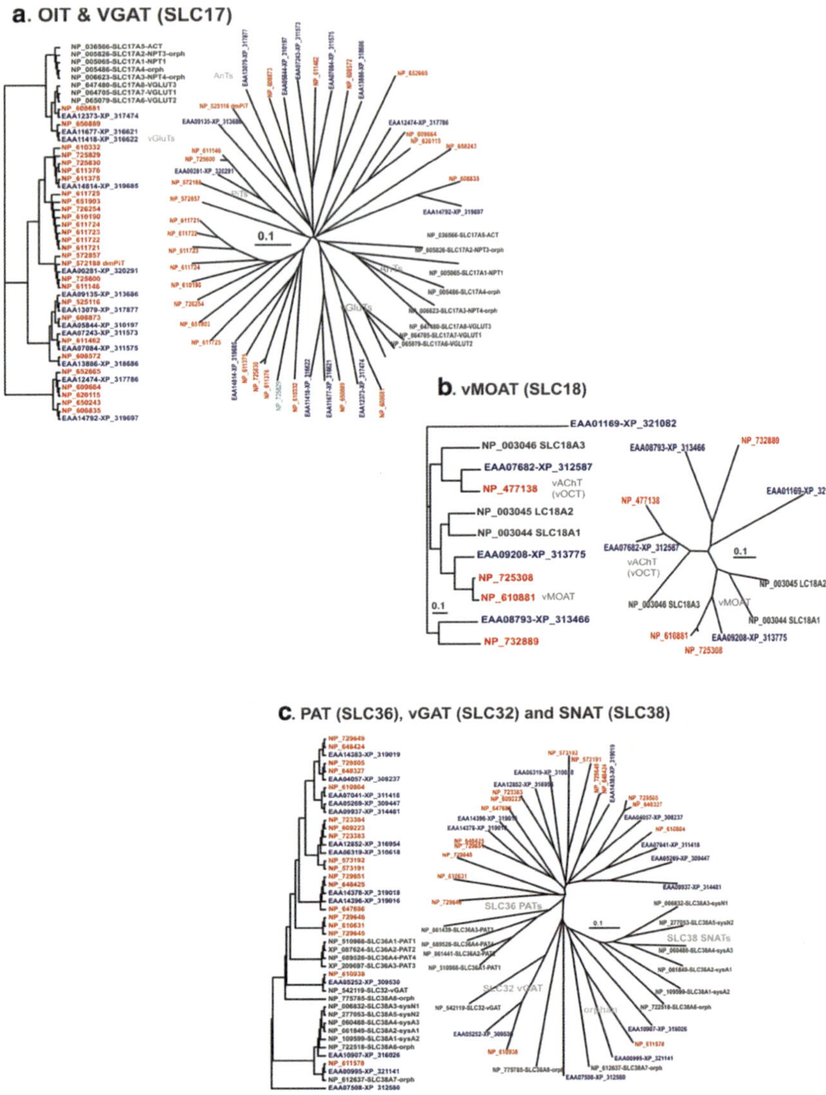

Fig. 16 **a** Phylogenetic proximity of organic and inorganic anion (PiTs) and vesicular glutamate transporters (vGluTs) from dipteran insect and selected mammalian models (SLC17). Unrooted cladogram (*left*) and dendrogram (*right*). Other abbreviations: AST, anion/sugar transporter; NPT, sodium inorganic phosphate transporter. **b** Phylogenetic proximity of dipteran and mammalian vesicular monoamine transporters (vMOATs; SLC18). Unrooted phylogram (*left*) and dendrogram (*right*). *Font colors*: *blue*, *A. gambiae* sequences; *red*, *D. melanogaster* sequences; *black*, *H. sapiens* sequences. **c** Phylogenetic transitions between dipteran and mammalian sodium neutral amino acid transporters (SNATs, SLC38), proton amino acid transporters (PATs, SLC36), and vesicular GABA transporter (vGAT, SLC32). Unrooted phylogram (*left*) and dendrogram (*right*). *Font colors*: *blue*, *A. gambiae* sequences; *red*, *D. melanogaster* sequences; *black*, *H. sapiens* sequences. Other abbreviations: sysA and sysN, systems A and system N transporters, respectively. *Scale bars* are evolutionary distances in mutations per site

In summary, relatively conserved VGluTs seem to have diverged through lineage-specific gene duplications that reflect increases in functional and morphological complexity of the CNS. Hence, a role for the genes that encode transporters related to the SLC17 family can be expected in insect vesicular glutamate homeostasis. In contrast, Pi absorption and homeostasis, which are essential for all organisms, appears to occur in insects through phylogenetically unrelated roots of transporters that are distinct from P_iTs of SLC20, SLC34, and even SLC17.

3.8.1 Inorganic Phosphate Transporter, Type I PiTs (SLC17A1)

The first member of mammalian SLC17A1 (NaP_i-1) was identified by expression cloning from a rabbit kidney cortex cDNA library, based on its ability to induce Na^+-dependent phosphate transport in *Xenopus* oocytes, with 750-fold higher apparent uptake than controls injected with total poly(A) RNA [414]. However, sub-millimolar substrate affinities of rabbit NaP_i-1 and the next-cloned ortholog from *H. sapiens* [415], as well as specific electrochemical properties of their heterologous expression in *Xenopus* oocytes, for example conductance for mineral and organic anions [416–418] and low pH sensitivity [417], do not match characteristics of Na^+-dependent P_i uptake in brush-border membrane vesicles [419] and do not support a primary role for these transporters in high-affinity phosphate transport [412, 420]. Indeed, two different P_iT types have been identified [421, 422], which are unrelated to SLC17 Na^+/P_i^- co-transporters. One new type is a SLC34 family, Type II P_iT, whose members mediate apical P_i uptake in various polarized epithelia [423]. Transporters of the other new type were originally described as retroviral receptors [424] but are now recognized as Type III P_iTs (SLC20), which appear to be basolateral components of Pi homeostasis [425]. A recent revision of P_iTs tends to support the earlier proposal that some members of the SLC17 family are facultative P_iTs, with major roles in the rapid acquisition of electrochemical and pH balance via anion secretion [419]. In fact they may utilize a variety of electrochemical mechanisms ranging from facilitated diffusion to electrophoretic Na^+-coupled symport [412].

Intriguingly, despite a crucial role of P_i homeostasis in all organisms, and the presence of homologous proteins in other metazoans, as well as in bacteria, fungi and plant lineages, there is no trace of SLC20 and SLC34 families in dipteran genomes and they appear to be absent in arthropods in general. Conversely, mammalian Type I P_iTs do not include any dipteran sequences (Fig. 16a), which leaves uncertain the presence of SLC17-related P_iTs in dipteran genomes.

3.8.2 Sialin (SLC17A5)

Extensive genetic analysis and BBMV studies revealed a physiological role for another SLC17 member, Sialin (SLC17A5), which apparently mediates lysosomal H^+-coupled (1:1) emission of amino carbohydrates (sialic acid) from lysosomal catabolism of glycosylated membrane proteins, as well as serving endomembrane traffic of some monocarboxylic acids [412].

3.8.3 vGlutT (SLC17A6, 7, 8)

Three members of mammalian SLC17 (SLC17A6, A7, and A8, a.k.a. vGluT1, 2, and 3) are recruited in excitatory neurotransmission, enabling vesicular storage of L-glutamate. Initially defined as plasma membrane, Na^+ coupled PiTs, vGluT1 [426] and vGluT2 [427] transport phenotypes have since been shown to be H^+-coupled carriers that mediate accumulation of L-glutamate into synaptic vesicles of neurons [428] and glutamatergic endocrine cells [429, 430]. All characterized vGluTs have been morphologically associated with glutamatergic neurons of the postnatal brain and especially abundant at membranes of synaptic vesicles, although spatial differences in subtype distribution in the brain are prominent [431–433]. The neuronal specificity of vGluTs makes them useful markers of excitatory neuronal components, for example recent demonstration of the existence of peripheral glutamatergic components in the gastroenteropancreatic system and testis [434]. The expression of vGluTs occurs very early in embryogenesis [435]. High-resolution immuno-electron microscopy also demonstrates a specific subcellular distribution of vGluT types, for example vGluT1 and vGluT2 are associated mainly with nerve terminals; in contrast, vGluT3 is present in vesicular structures of astrocytes and neuronal dendrites [432].

Apparently all vGluTs share highly conserved, pmf-driven, electrochemical mechanisms with nearly-millimolar apparent $K_m{}^{Glu}$. Like PiTs, vGluTs enable Cl^- and some other anionic conductances in BBMV and heterologous models [428, 436, 437]. Vesicular glutamate accumulation has been shown to be inhibited by the broad specificity blocker of chloride conductance DIDS (4,4'-diisothiocyanatostilbene-2,2'-disulfonic acid). Like all other endomembrane transporters, electrophoretic vGluTs are energized by electrogenic V-ATPase.

3.9 SLC36, Proton Amino Acid Transporters (PAT)

Proton Amino acid Transporters (PATs) comprise a group of electrophoretic H^+/amino acid (1:1) symporters that mediate cytosolic acquisition of small neutral amino acids from intracellular (lysosomal) and extracellular proteolysis. (The term "proton" is used loosely in Biology as shorthand for hydronium ion.) In mammals the PAT group is represented by four genes that encode two well-characterized transporters, PAT1/LYAAT-1 and PAT2, and two orphan transporters [438]. The first orphan has been characterized as a rat brain-specific lysosomal amino acid transporter (rLYAAT-1) [439] and also as a mouse intestinal epithelium-specific plasma membrane proton/amino acid transporter (mPAT1) which induces H^+ driven, low affinity ($K_m{}^{G,\,A,\,P} = 7,\ 7.5,\ 2.8\,mM$) transport of Gly, Ala, Pro and several structurally related derivatives in *Xenopus* oocytes [440]. In contrast PAT2 has a narrower substrate selectivity and higher affinity ($K_m{}^{G,A,P} = 0.59,\ 0.26,\ 0.12\,mM$) [440]. PATs are differentially and broadly expressed in mammalian tissues except for the orphan PAT3 which so far has been found only in adult mice testis [441].

Immunocytochemical studies show that rLYAAT-1 is associated with other cellular membranes for example Golgi apparatus, ER, and occasionally with the plasma membrane [442].

Dipteran genomes comprise nine putative genes in *A. gambiae* and 11 genes, encoding 16 identified transcripts, in *D. melanogaster*. These transporters split into five orthologous groups within the dipteran framework; one additional paralogous group is present in *D. melanogaster* (Fig. 16c). However, the entire population of these dipteran transporters is paralogous to mammalian PATs.

The phylogeny of SNATs and PATs apparently reflects transitions of electrochemical properties observed among characterized transporters in these families. Such transitions were suggested earlier as an evolutionary trend from a H^+-driven ancestral carrier, for example in PATs and vGAT, toward acquisition of a Na^+-driven mechanism for example in SNATs [438, 443]. An inverse evolutionary trend, from a Na^+-dependent ancestral mechanism toward a H^+-driven mechanism, is apparent from the phylogenetic analysis. Although H^+-driven mechanisms are indeed older; the SNATs and PATs population is self-rooted and apparently shares a common ancestor with the Na^+-driven SNF (SLC6) transporters (Fig. 6). This fact means that Na^+-driven mechanisms were already established before the SNAT and PAT families diverged. In addition, a Na^+-driven mechanism is encoded in the older A system of SLC38 transporters which reside on the plasma membrane (Figs. 6 and 9). The ability to modify properties under low intracellular Na^+ demand (N system SLC38) and directly utilize proton motive forces (SLC36) presumably was acquired later during the establishment of phagocytosis and lysosomal digestion in earlier eukaryotes. SLC36 members residing in various endomembranes correspond with lysosomal traffic for which the apical intestinal membrane is just a single possible position. Paralogous clustering in insect and mammalian SLC36 is not surprising, since this group is under the pressure of diverse nutrient requirements in different tissues and cell types. In contrast, vGAT (SLC 32) serving in inhibitory neurotransmission, evolved under a stabilizing pressure, which is reflected by its *trans*-specific orthology. Despite structural similarity, SLC32 and SLC36 exhibit striking differences in their kinetic mechanisms: SLC32 vGAT is an exchanger (antiporter) utilizing H^+ motive forces from synergetic activity modulated by Cl^- channels and V-ATPase for ligand accumulation [444], whereas SLC36 is a symporter utilizing both H^+ and ligand electrochemical motive forces for ligand secretion through lysosomal membranes and/or ligand uptake through plasma membranes [441].

3.10 SLC38, Sodium Neutral Amino acid Transporters (SNATs)

This cluster represents the parental group in the SNAT+PAT superfamily, whose members fall into several consecutive phylogenetic levels in dipteran models (Fig. 16c). In mammals characterized SNATs comprise System A and N transporters, each of which mediates a low affinity transport ($0.3 < K_m{}^{\text{substrate}} < 2$ mM) of varied subsets of small, aliphatic amino acids [445]. System A transporters appear to be broadly expressed in all mammalian tissues and to mediate Na^+-driven, electrophoretic, and pH-sensitive neutral amino acid uptake through plasma membranes. In contrast, system N transporters have narrower substrate profiles and utilize proton motive forces, which may facilitate a reverse amino acid transport and consequently an active emission of catabolic amino acids [446, 447]. Both systems are tightly regulated via hormonal and intracellular signaling pathways and pH-dependent allosteric modifications. SNATs are implicated in many physiological processes associated with reciprocal *trans*-cellular amino acid exchange, cellular nutrition, and metabolite secretion/detoxification pathways [445].

4 Perspectives

The postgenomic era facilitates the identification of virtually all transport proteins in the comprehensive selection of genomic models. The emerging era of crystallization of membrane proteins provides essential structural information. Together with systematic molecular cloning and the characterization of new transporters, these methods advance the accumulation and systematization of molecular and physiological data, which ultimately makes it possible to deduce the function of particular transporters by their phylogenetic proximity and homology modeling. Amino acids are precursors in a variety of catabolic and anabolic pathways in which amino acid transporters serve as rate-limiting components. Metazoan amino acid transporters include a set of transport mechanisms, which supply essential amino acids in amino acid utility pathways, including protein, neurotransmitter, and hormone syntheses.

Acknowledgments This work was supported by NIH Research grant R01 AI-30464.

References

1. Saier MH. Vectorial metabolism and the evolution of transport systems. J Bacteriol 2000;182:5029–35.
2. Kvenvolden KA, Lawless JG, Ponnamperuma C. Nonprotein Amino Acids in the Murchison Meteorite. Proc Natl Acad Sci USA 1971;68:486–90.
3. Bell EA. ‘Uncommon’ amino acids in plants. FEBS Lett 1976;64:29–35.
4. Sulser H, Sager F. Identification of uncommon amino acids in the lentil seed (Lens culinaris Med.). In: Experientia; 1976:422–3.
5. Cone JE, Del Rio RM, Davis JN, Stadtman TC. Chemical characterization of the selenoprotein component of clostridial glycine reductase: identification of selenocysteine as the organoselenium moiety. Proc Natl Acad Sci USA 1976;73:2659–63.
6. Atkins JF, Gesteland R. Biochemistry. The 22nd amino acid. Science 2002;296:1409–10.
7. Fujii N. D-amino acids in living higher organisms. Orig Life Evol Biosph 2002;32:103–27.
8. Torres AM, Tsampazi C, Geraghty DP, Bansal PS, Alewood PF, Kuchel PW. D-amino acid residue in a defensin-like peptide from platypus venom: effect on structure and chromatographic properties. Biochem J 2005;391:215–20.
9. Mitchell JB, Smith J. D-amino acid residues in peptides and proteins. Proteins 2003;50:563–71.
10. Patrick ML, Bradley TJ. The physiology of salinity tolerance in larvae of two species of Culex mosquitoes: the role of compatible solutes. J Exp Biol 2000;203:821–30.

11. Balboni E. A proline shuttle in insect flight muscle. Biochem Biophys Res Commun 1978;85:1090–6.
12. Gade G, Auerswald L. Beetles' choice—proline for energy output: control by AKHs. Comp Biochem Physiol B Biochem Mol Biol 2002;132:117–29.
13. Oparin AI. The Origin of Life. Moscow: Moscow Worker publisher; 1924.
14. Miller SL. A production of amino acids under possible primitive earth conditions. Science 1953;117:528–9.
15. Miyakawa S, Yamanashi H, Kobayashi K, Cleaves HJ, Miller SL. Prebiotic synthesis from CO atmospheres: implications for the origins of life. Proc Natl Acad Sci USA 2002;99:14628–31.
16. Nuevo M, Auger G, Blanot D, d'Hendecourt L. A Detailed Study of the Amino Acids Produced from the Vacuum UV Irradiation of Interstellar Ice Analogs. Orig Life Evol Biosph J Intl Soc Study Origin Life 2008;38:37–56.
17. Rode BM. Peptides and the origin of life. Peptides 1999;20:773–86.
18. Thaxton CB, Bradley WL, Olsen RL. The Mystery of Life's Origin: Reassessing Current Theories. New York: Philosophical Library, Inc.; 1984.
19. Schwarttz AW. From Big Bang to Primordial Planet. Setting the Stage for the Origin of Life. In: Schopf JW, ed. Life's origin : the beginnings of biological evolution. Berkeley: University of California Press; 2002:208 p.
20. Oro J, Gibert J, Lichtenstein H, Wikstrom S, Flory DA. Amino-acids, aliphatic and aromatic hydrocarbons in the Murchison Meteorite. Nature 1971;230:105–6.
21. Oro J, Mills T. Chemical evolution of primitive solar system bodies. Adv Space Res 1989;9:105–20.
22. Lawless JG, Kvenvolden KA, Peterson E, Ponnamperuma C, Moore C. Amino acids indigenous to the Murray meteorite. Science 1971;173:626–7.
23. Encrenaz T. Search for organic molecules in the outer solar system. Adv Space Res 1986;6:237–46.
24. Bernstein MP, Dworkin JP, Sandford SA, Cooper GW, Allamandola LJ. Racemic amino acids from the ultraviolet photolysis of interstellar ice analogues. Nature 2002;416:401–3.
25. Munoz Caro GM, Meierhenrich UJ, Schutte WA, et al. Amino acids from ultraviolet irradiation of interstellar ice analogues. Nature 2002;416:403–6.
26. Kuan Y, Charnley S, Huang H, Tseng W, Kisiel Z. Interstellar glycine. Astrophys J 2003;593:848–67.
27. Oro J, Berry JM. Comets and life. Adv Space Res 1987;7:23–32.
28. Kobayashi K, Kasamatsu T, Kaneko T, et al. Formation of amino acid precursors in cometary ice environments by cosmic radiation. Adv Space Res 1995;16:21–6.
29. Seckbach J. Life in the universe : from the Miller experiment to the search for life on other worlds. Dordrecht ; Boston: Kluwer Academic Publishers; 2004.
30. Kasting JF. Impacts and the origin of life. Earth Miner Sci 1990;59:37–42.
31. Kasting JF. Earth's early atmosphere. Science 1993;259:920–6.
32. Chyba C, Sagan C. Endogenous production, exogenous delivery and impact-shock synthesis of organic molecules: an inventory for the origins of life. Nature 1992;355: 125–32.
33. Mukhin LM. Volcanic processes and synthesis of simple organic compounds on primitive earth. Orig Life 1976;7:355–68.
34. Markhinin EK, Podkletnov NE. The phenomenon of formation of prebiological compounds in volcanic processes. Orig Life 1977;8:225–35.
35. Podkletnov NE, Markhinin EK. New data on abiogenic synthesis of prebiological compounds in volcanic processes. Orig Life 1981;11:303–15.
36. Schwartz AW. Did minerals perform prebiotic combinatorial chemistry? Chem Biol 1996;3:515–8.
37. Holm NG, Cairns-Smith AG, Daniel RM, et al. Marine hydrothermal systems and the origin of life: future research. Orig Life Evol Biosph 1992;22:181–242.

38. Ferris JP. Chemical markers of prebiotic chemistry in hydrothermal systems. Orig Life Evol Biosph 1992;22:109–34, 91–242.
39. Amend JP, Shock EL. Energetics of amino acid synthesis in hydrothermal ecosystems. Science 1998;281:1659–62.
40. Wachtershauser G. Origin of life. Life as we don't know it. Science 2000;289:1307–8.
41. Huber C, Wachtershauser G. alpha-Hydroxy and alpha-amino acids under possible Hadean, volcanic origin-of-life conditions. Science 2006;314:630–2.
42. Miller SL, Laszcano A. Formation of the building Blocks of Life. In: Schopf JW, ed. Life's origin : the beginnings of biological evolution. Berkeley: University of California Press; 2002:78–108.
43. Wong JT. Coevolution theory of the genetic code at age thirty. Bioessays 2005;27:416–25.
44. Oparin AI. Origin and evolution of metabolism. Comp Biochem Physiol 1962;4:371–7.
45. Gilbert W. The RNA World. Nature 1986;319:618.
46. Woese CR. The genetic code; the molecular basis for genetic expression. New York: Harper & Row; 1967.
47. Crick FH. The origin of the genetic code. J Mol Biol 1968;38:367–79.
48. Orgel LE. Evolution of the genetic apparatus. J Mol Biol 1968;38:381–93.
49. Ikehara K, Niihara Y. Origin and evolutionary process of the genetic code. Curr Med Chem 2007;14:3221–31.
50. Oba T, Fukushima J, Maruyama M, Iwamoto R, Ikehara K. Catalytic activities of [GADV]-peptides. Formation and establishment of [GADV]-protein world for the emergence of life. Orig Life Evol Biosph 2005;35:447–60.
51. Ikehara K. Possible steps to the emergence of life: the [GADV]-protein world hypothesis. Chem Rec 2005;5:107–18.
52. Ikehara K. Origins of gene, genetic code, protein and life: comprehensive view of life systems from a GNC-SNS primitive genetic code hypothesis. J Biosci 2002;27:165–86.
53. Brooks DJ, Fresco JR, Lesk AM, Singh M. Evolution of amino acid frequencies in proteins over deep time: inferred order of introduction of amino acids into the genetic code. In: Mol Biol Evol; 2002:1645–55.
54. Brooks DJ, Fresco JR. Increased frequency of cysteine, tyrosine, and phenylalanine residues since the last universal ancestor. In: Mol Cell Proteomics; 2002:125–31.
55. Jordan IK, Kondrashov FA, Adzhubei IA, et al. A universal trend of amino acid gain and loss in protein evolution. Nature 2005;433:633–8.
56. Cunchillos C, Lecointre G. Evolution of amino acid metabolism inferred through cladistic analysis. J Biol Chem 2003;278:47960–70.
57. Cunchillos C, Lecointre G. Early steps of metabolism evolution inferred by cladistic analysis of amino acid catabolic pathways. C R Biol 2002;325:119–29.
58. Voet D, Voet JG. Solutions manual to accompany biochemistry. 3rd ed. Hoboken, N.J.: John Wiley & Sons; 2004.
59. Michal G. Biochemical pathways : an atlas of biochemistry and molecular biology. New York, Heidelberg: John Wiley ; Spektrum Akademischer Verlag; 1999.
60. Young VR, Ajami AM. Glutamine: the emperor or his clothes? J Nutr 2001;131:2449S–59S; discussion 86S–7S.
61. Horowitz NH. On the Evolution of Biochemical Syntheses. Proc Natl Acad Sci USA 1945;31:153–7.
62. Cordón F. Tratado Evolucionista de Biología. Madrid, Spain: Aguilar; 1990.
63. Keefe AD, Lazcano A, Miller SL. Evolution of the biosynthesis of the branched-chain amino acids. Origins of life and evolution of the biosphere : the journal of the International Society for the Study of the Origin of Life 1995;25:99–110.
64. Jensen RA. Enzyme recruitment in evolution of new function. Annu Rev Microbiol 1976;30:409–25.
65. Caetano-Anolles G, Yafremava LS, Gee H, Caetano-Anolles D, Kim HS, Mittenthal JE. The origin and evolution of modern metabolism. Int J Biochem Cell Biol 2008.

66. Trifonov EN. Consensus temporal order of amino acids and evolution of the triplet code. Gene 2000;261:139–51.
67. Wiltschi B, Budisa N. Natural history and experimental evolution of the genetic code. Appl Microbiol Biotechnol 2007;74:739–53.
68. Berman-Frank I, Lundgren P, Chen YB, et al. Segregation of nitrogen fixation and oxygenic photosynthesis in the marine cyanobacterium Trichodesmium. Science 2001;294:1534–7.
69. Kasting JF, Siefert JL. Biogeochemistry. The nitrogen fix. Nature 2001;412:26–7.
70. Raymond J, Siefert JL, Staples CR, Blankenship RE. The natural history of nitrogen fixation. Mol Biol Evol 2004;21:541–54.
71. Di Giulio M. The universal ancestor and the ancestors of Archaea and Bacteria were anaerobes whereas the ancestor of the Eukarya domain was an aerobe. J Evol Biol 2007;20:543–8.
72. Blank CE. Evolutionary timing of the origins of mesophilic sulphate reduction and oxygenic photosynthesis: a phylogenomic dating approach. Geobiology 2004;2:1–20.
73. Falkowski Pg. Evolution of the nitrogen cycle and its influence on the biological sequestration of CO 2 in the ocean. Science 1997;387:272–5.
74. Gonzalez-Lopez J, Rodelas B, Pozo C, Salmeron-Lopez V, Martinez-Toledo MV, Salmeron V. Liberation of amino acids by heterotrophic nitrogen fixing bacteria. Amino Acids 2005;28:363–7.
75. Thorneley RN, Ashby GA. Oxidation of nitrogenase iron protein by dioxygen without inactivation could contribute to high respiration rates of Azotobacter species and facilitate nitrogen fixation in other aerobic environments. Biochem J 1989;261:181–7.
76. van Rhijn P, Vanderleyden J. The Rhizobium-plant symbiosis. Microbiol Rev 1995;59:124–42.
77. Smil V. Enriching the earth : Fritz Haber, Carl Bosch, and the transformation of world food production. Cambridge, Mass.: MIT Press; 2001.
78. Berman-Frank I, Lundgren P, Falkowski P. Nitrogen fixation and photosynthetic oxygen evolution in cyanobacteria. Res Microbiol 2003;154:157–64.
79. Jékely G. Did the last common ancestor have a biological membrane? . Biology Direct 2006;1:1–4.
80. Wachtershauser G. Evolution of the first metabolic cycles. Proc Natl Acad Sci USA 1990;87:200–4.
81. Graf J, Ruby EG. Host-derived amino acids support the proliferation of symbiotic bacteria. Proc Natl Acad Sci USA 1998;95:1818–22.
82. Moya A, Pereto J, Gil R, Latorre A. Learning how to live together: genomic insights into prokaryote-animal symbioses. Nat Rev Genet 2008;9:218–29.
83. Dunn MS, Shankman S, Camien MN, Block H. The amino acid requirements of twenty-three lactic acid bacteria. J Biol Chem 1947.
84. Watrin L, Martin-Jezequel V, Prieur D. Minimal amino acid requirements of the hyperthermophilic archaeon pyrococcus abyssi, isolated from deep-sea hydrothermal vents. Appl Environ Microbiol 1995;61:2069.
85. Payne SH, Loomis WF. Retention and loss of amino acid biosynthetic pathways based on analysis of whole-genome sequences. Eukaryotic Cell 2006;5:272–6.
86. Singh J, Aneja KR. From ethnomycology to fungal biotechnology : exploiting fungi from natural resources for novel products. New York: Kluwer Academic/Plenum Publishers; 1999.
87. White JF. Clavicipitalean fungi : evolutionary biology, chemistry, biocontrol, and cultural impacts. New York: M. Dekker; 2003.
88. Kavanagh K. Fungi : biology and applications. Chichester, West Sussex, England ; ; Hoboken, NJ: John Wiley & Sons; 2005.
89. Heizer EM, Jr., Raiford DW, Raymer ML, Doom TE, Miller RV, Krane DE. Amino acid cost and codon-usage biases in 6 prokaryotic genomes: a whole-genome analysis. Mol Biol Evol 2006;23:1670–80.

90. Akashi H, Gojobori T. Metabolic efficiency and amino acid composition in the proteomes of Escherichia coli and Bacillus subtilis. Proc Natl Acad Sci USA 2002;99:3695–700.
91. Craig CL, Weber RS. Selection costs of amino acid substitutions in ColE1 and ColIa gene clusters harbored by Escherichia coli. Mol Biol Evol 1998;15:774–6.
92. Swire J. Selection on synthesis cost affects interprotein amino acid usage in all three domains of life. J Mol Evol 2007;64:558–71.
93. Langenbuch M, Portner HO. Changes in metabolic rate and N excretion in the marine invertebrate Sipunculus nudus under conditions of environmental hypercapnia: identifying effective acid-base variables. J Exp Biol 2002;205:1153–60.
94. Coulson RA, Hernandez T. Increase in metabolic rate of the alligator fed proteins or amino acids. J Nutr 1979;109:538–50.
95. Boudko DY, Kohn AB, Meleshkevitch EA, et al. Ancestry and progeny of nutrient amino acid transporters. Proc Natl Acad Sci USA 2005;102:1360–5.
96. Bush DR. Amino Acid Transport. In: Singh B, ed. Plant amino acids : biochemistry and biotechnology. New York: Marcel Dekker; 1999:305–18.
97. Ortiz-Lopez A, Chang H, Bush DR. Amino acid transporters in plants. Biochim Biophys Acta 2000;1465:275–80.
98. Chen L, Bush DR. LHT1, a lysine- and histidine-specific amino acid transporter in arabidopsis. Plant Physiol 1997;115:1127–34.
99. Birschwilks M, Haupt S, Hofius D, Neumann S. Transfer of phloem-mobile substances from the host plants to the holoparasite Cuscuta sp. J Exp Bot 2006;57:911–21.
100. Schulze W, Frommer WB, Ward JM. Transporters for ammonium, amino acids and peptides are expressed in pitchers of the carnivorous plant Nepenthes. Plant J 1999;17:637–46.
101. Ellison AM, Gotelli NJ. Nitrogen availability alters the expression of carnivory in the northern pitcher plant, Sarracenia purpurea. Proc Natl Acad Sci USA 2002;99:4409–12.
102. Liu X, Bush DR. Expression and transcriptional regulation of amino acid transporters in plants. Amino Acids 2006;30:113–20.
103. Wipf D, Ludewig U, Tegeder M, Rentsch D, Koch W, Frommer WB. Conservation of amino acid transporters in fungi, plants and animals. Trends Biochem Sci 2002;27:139–47.
104. Rentsch D, Boorer KJ, Frommer WB. Structure and function of plasma membrane amino acid, oligopeptide and sucrose transporters from higher plants. J Membr Biol 1998;162:177–90.
105. AGI. Analysis of the genome sequence of the flowering plant Arabidopsis thaliana. Nature 2000;408:796–815.
106. Fischer WN, Andre B, Rentsch D, et al. Amino acid transport in plants. Trends Plant Sci 1998;3:188–95.
107. Su YH, Frommer WB, Ludewig U. Molecular and functional characterization of a family of amino acid transporters from Arabidopsis. Plant Physiol 2004;136:3104–13.
108. McNeal JR, Kuehl JV, Boore JL, de Pamphilis CW. Complete plastid genome sequences suggest strong selection for retention of photosynthetic genes in the parasitic plant genus Cuscuta. BMC Plant Biol 2007;7:57.
109. Revill MJ, Stanley S, Hibberd JM. Plastid genome structure and loss of photosynthetic ability in the parasitic genus Cuscuta. J Exp Bot 2005;56:2477–86.
110. Runyon JB, Mescher MC, De Moraes CM. Volatile chemical cues guide host location and host selection by parasitic plants. Science 2006;313:1964–7.
111. Merchant SS, Prochnik SE, Vallon O, et al. The Chlamydomonas genome reveals the evolution of key animal and plant functions. Science 2007;318:245–50.
112. Fitzgerald LM, Szmant AM. Biosynthesis of 'essential' amino acids by scleractinian corals. Biochem J 1997;322 (Pt 1):213–21.
113. Dewel RA. Colonial origin for Emetazoa: major morphological transitions and the origin of bilaterian complexity. J Morphol 2000;243:35–74.
114. Glansdorff N, Xu Y, Labedan B. The Last Universal Common Ancestor: emergence, constitution and genetic legacy of an elusive forerunner. Biol Direct 2008;3:29.

115. Rumpho ME, Summer EJ, Manhart JR. Solar-powered sea slugs. Mollusc/algal chloroplast symbiosis. Plant Physiol 2000;123:29–38.
116. Zhang J, Leontovich A, Sarras MP, Jr. Molecular and functional evidence for early divergence of an endothelin-like system during metazoan evolution: analysis of the Cnidarian, hydra. Development 2001;128:1607–15.
117. De Marais DJ. Evolution. When did photosynthesis emerge on Earth? Science 2000;289:1703–5.
118. Nelson N, Ben-Shem A. The structure of photosystem I and evolution of photosynthesis. Bioessays 2005;27:914–22.
119. Wong JT. A co-evolution theory of the genetic code. In: Proc Natl Acad Sci USA; 1975:1909–12.
120. Daniel H. Molecular and integrative physiology of intestinal peptide transport. Annu Rev Physiol 2004;66:361–84.
121. Hosie AHF, Poole PS. Bacterial ABC transporters of amino acids. Res Microbiol 2001;152:259–70.
122. Saier MH, Jr. Families of transmembrane transporters selective for amino acids and their derivatives. Microbiology 2000;146 (Pt 8):1775–95.
123. Burkovski A, Kramer R. Bacterial amino acid transport proteins: occurrence, functions, and significance for biotechnological applications. Appl Microbiol Biotechnol 2002;58:265–74.
124. Boudko DY, Meleshkevitch EA, Harvey WR. Novel transport phenotypes in the sodium neurotransmitter symporter family. Faseb J 2005;19:A748–A.
125. Okech BA, Meleshkevitch EA, Miller MM, Popova LB, Harvey WR, Boudko DY. Synergy and specificity of two Na+-aromatic amino acid symporters in the model alimentary canal of mosquito larvae. J Exp Biol 2008;211:1594–602.
126. Meleshkevitch EA, Robinson M, Popova LB, M.M. M, W.R. H, D.Y. B. Cloning and functional expression of the eukaryotic sodium-tryptophan symporter. JEB 2008;in press.
127. Christensen HN. Transport System Serving for Mono- + Diamino Acids. Proc Natl Acad Sci USA 1964;51:337.
128. Gerencser GA, Stevens BR. Thermodynamics of symport and antiport catalyzed by cloned or native transporters. J Exp Biol 1994;196:59–75.
129. Christensen HN. Methods for distinguishing amino acid transport systems of a given cell or tissue. Fed Proc 1966;25:850–3.
130. Bannai S, Christensen HN, Vadgama JV, et al. Amino acid transport systems. Nature 1984;311:308.
131. Kilberg MS, Häussinger D. Mammalian amino acid transport : mechanisms and control. New York: Plenum Press; 1992.
132. Christensen HN, Albritton LM, Kakuda DK, MacLeod CL. Gene-product designations for amino acid transporters. J Exp Biol 1994;196:51–7.
133. Broer S. Adaptation of plasma membrane amino acid transport mechanisms to physiological demands. Pflugers Arch 2002;444:457–66.
134. Broer S. Amino acid transport across mammalian intestinal and renal epithelia. Physiol Rev 2008;88:249–86.
135. Hyde R, Taylor PM, Hundal HS. Amino acid transporters: roles in amino acid sensing and signalling in animal cells. Biochem J 2003;373:1–18.
136. Bohmer C, Broer A, Munzinger M, et al. Characterization of mouse amino acid transporter B(0)AT1 (slc6a19). Biochem J 2005;389:745–51.
137. Broer A, Tietze N, Kowalczuk S, et al. The orphan transporter v7-3 (slc6a15) is a Na+-dependent neutral amino acid transporter (B0AT2). Biochem J 2006;393:421–30.
138. Romeo E, Dave MH, Bacic D, et al. Luminal kidney and intestine SLC6 amino acid transporters of B0AT-cluster and their tissue distribution in Mus musculus. Am J Physiol Renal Physiol 2006;290:F376–83.
139. Saier MH. Genome archeology leading to the characterization and classification of transport proteins. Curr Opin Microbiol 1999;2:555–61.

140. Saier MH, Jr. A functional-phylogenetic classification system for transmembrane solute transporters. Microbiol Mol Biol Rev 2000;64:354–411.
141. Busch W, Saier MH, Jr. The IUBMB-Endorsed Transporter Classification System. In: Yan Q, ed. Membrane transporters : methods and protocols Methods in molecular biology. Totowa, N.J.: Humana Press; 2003:xii, 369 p.
142. Saier MH, Jr. Families of Transporters and Their Classification. In: Quick MW, ed. Transmembrane transporters. New York: Wiley-Liss; 2002:1–18.
143. Busch W, Saier MH, Jr. The IUBMB-endorsed transporter classification system. Mol Biotechnol 2004;27:253–62.
144. Saier MH, Jr., Tran CV, Barabote RD. TCDB: the Transporter Classification Database for membrane transport protein analyses and information. Nucleic Acids Res 2006;34: D181–6.
145. Jack DL, Yang NM, Saier MH, Jr. The drug/metabolite transporter superfamily. Eur J Biochem 2001;268:3620–39.
146. Boudko DY, Stevens BR, Donly BC, Harvey WR. Nutrient Amino acid and Neurotransmitter transporters. In: Lawrence I. Gilbert KIaSSG, ed. Comprehensive Molecular Insect Science. First edition ed. Amsterdam: Elsevier; 2005:255–309.
147. Hediger MA, Romero MF, Peng JB, Rolfs A, Takanaga H, Bruford EA. The ABCs of solute carriers: physiological, pathological and therapeutic implications of human membrane transport proteins. Pflugers Arch 2004;447:465–8.
148. Palacin M, Estevez R, Bertran J, Zorzano A. Molecular biology of mammalian plasma membrane amino acid transporters. Physiol Rev 1998;78:969–1054.
149. Bergeron MJ, Simonin A, Burzle M, Hediger MA. Inherited epithelial transporter disorders-an overview. J Inherit Metab Dis 2008.
150. Kanai Y, Endou H. Functional properties of multispecific amino acid transporters and their implications to transporter-mediated toxicity. J Toxicol Sci 2003;28:1–17.
151. Asai K, Baik SH, Kasahara Y, Moriya S, Ogasawara N. Regulation of the transport system for C4-dicarboxylic acids in Bacillus subtilis. Microbiology 2000;146 (Pt 2):263–71.
152. Krom BP, Warner JB, Konings WN, Lolkema JS. Transporters involved in uptake of di- and tricarboxylates in Bacillus subtilis. Antonie Van Leeuwenhoek 2003;84:69–80.
153. Tolner B, Ubbink-Kok T, Poolman B, Konings WN. Characterization of the proton/glutamate symport protein of Bacillus subtilis and its functional expression in Escherichia coli. J Bacteriol 1995;177:2863–9.
154. Burguiere P, Auger S, Hullo MF, Danchin A, Martin-Verstraete I. Three different systems participate in L-cystine uptake in Bacillus subtilis. J Bacteriol 2004;186:4875–84.
155. Kanai Y, Hediger MA. The glutamate and neutral amino acid transporter family: physiological and pharmacological implications. Eur J Pharmacol 2003;479:237–47.
156. Pita-Almenar JD, Collado MS, Colbert CM, Eskin A. Different mechanisms exist for the plasticity of glutamate reuptake during early long-term potentiation (LTP) and late LTP. J Neurosci 2006;26:10461–71.
157. Wang ZY, Zhang YQ, Zhao ZQ. Inhibition of tetanically sciatic stimulation-induced LTP of spinal neurons and Fos expression by disrupting glutamate transporter GLT-1. Neuropharmacology 2006;51:764–72.
158. Shen Y, Linden DJ. Long-term potentiation of neuronal glutamate transporters. Neuron 2005;46:715–22.
159. Hughes EG, Maguire JL, McMinn MT, Scholz RE, Sutherland ML. Loss of glial fibrillary acidic protein results in decreased glutamate transport and inhibition of PKA-induced EAAT2 cell surface trafficking. Brain Res Mol Brain Res 2004;124:114–23.
160. Levenson J, Weeber E, Selcher JC, Kategaya LS, Sweatt JD, Eskin A. Long-term potentiation and contextual fear conditioning increase neuronal glutamate uptake. Nat Neurosci 2002;5:155–61.
161. Luscher C, Malenka RC, Nicoll RA. Monitoring glutamate release during LTP with glial transporter currents. Neuron 1998;21:435–41.

162. Diamond JS, Bergles DE, Jahr CE. Glutamate release monitored with astrocyte transporter currents during LTP. Neuron 1998;21:425–33.
163. Arriza JL, Kavanaugh MP, Fairman WA, et al. Cloning and expression of a human neutral amino acid transporter with structural similarity to the glutamate transporter gene family. J Biol Chem 1993;268:15329–32.
164. Kekuda R, Prasad PD, Fei YJ, et al. Cloning of the sodium-dependent, broad-scope, neutral amino acid transporter Bo from a human placental choriocarcinoma cell line. J Biol Chem 1996;271:18657–61.
165. Utsunomiya-Tate N, Endou H, Kanai Y. Cloning and functional characterization of a system ASC-like Na+-dependent neutral amino acid transporter. J Biol Chem 1996;271:14883–90.
166. Wolfgang CL, Lin C, Meng Q, Karinch AM, Vary TC, Pan M. Epidermal growth factor activation of intestinal glutamine transport is mediated by mitogen-activated protein kinases. J Gastrointest Surg 2003;7:149–56.
167. Furuya S, Watanabe M. Novel neuroglial and glioglial relationships mediated by L-serine metabolism. Arch Histol Cytol 2003;66:109–21.
168. Li R, Younes M, Frolov A, et al. Expression of neutral amino acid transporter ASCT2 in human prostate. Anticancer Res 2003;23:3413–8.
169. Kanai Y, Hediger MA. The glutamate/neutral amino acid transporter family SLC1: molecular, physiological and pharmacological aspects. Pflugers Arch 2004;447:469–79.
170. Broer A, Wagner C, Lang F, Broer S. Neutral amino acid transporter ASCT2 displays substrate-induced Na+ exchange and a substrate-gated anion conductance. Biochem J 2000;346:705–10.
171. Boudker O, Ryan RM, Yernool D, Shimamoto K, Gouaux E. Coupling substrate and ion binding to extracellular gate of a sodium-dependent aspartate transporter. Nature 2007;445:387–93.
172. Rasko JEJ, Battini J-L, Gottschalk RJ, Mazo I, Miller AD. The RD114/simian type D retrovirus receptor is a neutral amino acid transporter. PNAS 1999;96:2129–34.
173. Marin M, Lavillette D, Kelly SM, Kabat D. N-linked glycosylation and sequence changes in a critical negative control region of the ASCT1 and ASCT2 neutral amino acid transporters determine their retroviral receptor functions. J Virol 2003;77:2936–45.
174. Donly C, Jevnikar J, McLean H, Caveney S. Substrate-stereoselectivity of a high-affinity glutamate transporter cloned from the CNS of the cockroach Diploptera punctata. Insect Biochem Mol Biol 2000;30:369–76.
175. Kirschner MA, Copeland NG, Gilbert DJ, Jenkins NA, Amara SG. Mouse excitatory amino acid transporter EAAT2: isolation, characterization, and proximity to neuroexcitability loci on mouse chromosome 2. Genomics 1994;24:218–24.
176. Kanai Y, Smith CP, Hediger MA. The elusive transporters with a high affinity for glutamate. Trends Neurosci 1993;16:365–70.
177. Kanai Y, Hediger MA. Primary structure and functional characterization of a high-affinity glutamate transporter. Nature 1992;360:467–71.
178. Suchak SK, Baloyianni NV, Perkinton MS, Williams RJ, Meldrum BS, Rattray M. The 'glial' glutamate transporter, EAAT2 (Glt-1) accounts for high affinity glutamate uptake into adult rodent nerve endings. J Neurochem 2003;84:522–32.
179. Besson MT, Soustelle L, Birman S. Identification and structural characterization of two genes encoding glutamate transporter homologues differently expressed in the nervous system of *Drosophila melanogaster*. FEBS Lett 1999;443:97–104.
180. Arriza JL, Eliasof S, Kavanaugh MP, Amara SG. Excitatory amino acid transporter 5, a retinal glutamate transporter coupled to a chloride conductance. Proc Natl Acad Sci USA 1997;94:4155–60.
181. Kucharski R, Ball EE, Hayward DC, Maleszka R. Molecular cloning and expression analysis of a cDNA encoding a glutamate transporter in the honeybee brain. Gene 2000;242:399–405.
182. Rimaniol AC, Haik S, Martin M, et al. Na+-dependent high-affinity glutamate transport in macrophages. J Immunol 2000;164:5430–8.

183. Huggett JF, Mustafa A, O'Neal L, Mason DJ. The glutamate transporter GLAST-1 (EAAT-1) is expressed in the plasma membrane of osteocytes and is responsive to extracellular glutamate concentration. Biochem Soc Trans 2002;30:890–3.
184. Amara SG, Fontana AC. Excitatory amino acid transporters: keeping up with glutamate. Neurochem Int 2002;41:313–8.
185. Arriza JL, Fairman WA, Wadiche JI, Murdoch GH, Kavanaugh MP, Amara SG. Functional comparisons of three glutamate transporter subtypes cloned from human motor cortex. J Neurosci 1994;14:5559–69.
186. Chen Y, Swanson RA. The glutamate transporters EAAT2 and EAAT3 mediate cysteine uptake in cortical neuron cultures. J Neurochem 2003a;84:1332–9.
187. Fairman WA, Vandenberg RJ, Arriza JL, Kavanaugh MP, Amara SG. An excitatory amino-acid transporter with properties of a ligand-gated chloride channel. Nature 1995;375: 599–603.
188. Torres GE, Amara SG. Glutamate and monoamine transporters: new visions of form and function. Curr Opin Neurobiol 2007;17:304–12.
189. Danbolt NC. Glutamate uptake. Prog Neurobiol 2001;65:1–105.
190. Furuta A, Martin LJ, Lin CL, Dykes-Hoberg M, Rothstein JD. Cellular and synaptic localization of the neuronal glutamate transporters excitatory amino acid transporter 3 and 4. Neuroscience 1997;81:1031–42.
191. Ganel R, Rothstein JD. Glutamate Transporter Dysfunction and Neuronal Death. In: Adelmann G, Jonas P, Monyer H, eds. Ionotropic glutamate receptors in the CNS Handbook of experimental pharmacology v 141. Berlin ; New York: Springer; 2000:xxii, 535 p.
192. Yi JH, Hazell AS. Excitotoxic mechanisms and the role of astrocytic glutamate transporters in traumatic brain injury. Neurochem Int 2006;48:394–403.
193. Smith CP, Weremowicz S, Kanai Y, Stelzner M, Morton CC, Hediger MA. Assignment of the gene coding for the human high-affinity glutamate transporter EAAC1 to 9p24: potential role in dicarboxylic aminoaciduria and neurodegenerative disorders. Genomics 1994;20: 335–6.
194. Aoki M, Lin CL, Rothstein JD, et al. Mutations in the glutamate transporter EAAT2 gene do not cause abnormal EAAT2 transcripts in amyotrophic lateral sclerosis. Ann Neurol 1998;43:645–53.
195. Trotti D, Aoki M, Pasinelli P, et al. Amyotrophic lateral sclerosis-linked glutamate transporter mutant has impaired glutamate clearance capacity. J Biol Chem 2001;276:576–82.
196. Pampliega O, Domercq M, Villoslada P, Sepulcre J, Rodriguez-Antiguedad A, Matute C. Association of an EAAT2 polymorphism with higher glutamate concentration in relapsing multiple sclerosis. J Neuroimmunol 2008;195:194–8.
197. Jen JC, Wan J, Palos TP, Howard BD, Baloh RW. Mutation in the glutamate transporter EAAT1 causes episodic ataxia, hemiplegia, and seizures. Neurology 2005;65:529–34.
198. Usherwood PNR. Insect glutamate receptors. *Adv Insect Physiol* 1994;24:309–41.
199. Bicker G, Schafer S, Ottersen OP, Storm-Mathisen J. Glutamate-like immunoreactivity in identified neuronal populations of insect nervous systems. J Neurosci 1988;8:2108–22.
200. Monaghan DT, Bridges RJ, Cotman CW. The excitatory amino acid receptors: their classes, pharmacology, and distinct properties in the function of the central nervous system. Annu Rev Pharmacol Toxicol 1989;29:365–402.
201. Gasic GP, Hollmann M. Molecular neurobiology of glutamate receptors. Annu Rev Physiol 1992;54:507–36.
202. Jan LY, Jan YN. Properties of the larval neuromuscular junction in Drosophila melanogaster. J Physiol 1976b;262:189–214.
203. Jan LY, Jan YN. L-glutamate as an excitatory transmitter at the Drosophila larval neuromuscular junction. J Physiol 1976a;262:215–36.
204. Petersen SA, Fetter RD, Noordermeer JN, Goodman CS, DiAntonio A. Genetic analysis of glutamate receptors in drosophila reveals a retrograde signal regulating presynaptic transmitter release. Neuron 1997;19:1237–48.

205. Schuster CM, Ultsch A, Schloss P, Cox JA, Schmitt B, Betz H. Molecular cloning of an invertebrate glutamate receptor subunit expression in Drosophila muscle. Science 1991;254:112–4.
206. Maleszka R, Helliwell P, Kucharski R. Pharmacological interference with glutamate re-uptake impairs long-term memory in the honeybee, apis mellifera. Behav Brain Res 2000;115:49–53.
207. Chiang AS, Lin WY, Liu HP, et al. Insect NMDA receptors mediate juvenile hormone biosynthesis. Proc Natl Acad Sci USA 2002;99:37–42.
208. Seal RP, Daniels GM, Wolfgang WJ, Forte M, Amara S. Identification and characterization of a cDNA encoding a neuronal glutamate transporter from *Drosophila melanogaster*. Recept Channel 1998;6:51–64.
209. Donly BC, Richman A, Hawkins E, McLean H, Caveney S. Molecular cloning and functional expression of an insect high-affinity Na+-dependent glutamate transporter. *Eur J Biochem* 1997;248:535–42.
210. Umesh A, Cohen BN, Ross LS, Gill SS. Functional characterization of a glutamate/aspartate transporter from the mosquito Aedes aegypti. J Exp Biol 2003;206:2241–55.
211. Besson MT, Soustelle L, Birman S. Selective high-affinity transport of aspartate by a *Drosophila* homologue of the excitatory amino-acid transporters. Curr Biol 2000;10: 207–10.
212. Besson MT, Re DB, Moulin M, Birman S. High affinity transport of taurine by the Drosophila aspartate transporter dEAAT2. J Biol Chem 2005;280:6621–6.
213. Caveney S, Donly BC. Neurotransmitter transporters in the insect nervous system. Adv Insect Physiol 2002;29:55–149.
214. Gardiner RB, Ullensvang K, Danbolt NC, Caveney S, Donly BC. Cellular distribution of a high-affinity glutamate transporter in the nervous system of the cabbage looper Trichoplusia ni. J Exp Biol 2002;205:2605–14.
215. Yernool D, Boudker O, Jin Y, Gouaux E. Structure of a glutamate transporter homologue from Pyrococcus horikoshii. Nature 2004;431:811–8.
216. Yernool D, Boudker O, Folta-Stogniew E, Gouaux E. Trimeric subunit stoichiometry of the glutamate transporters from Bacillus caldotenax and Bacillus stearothermophilus. Biochemistry 2003;42:12981–8.
217. Haugeto O, Ullensvang K, Levy LM, et al. Brain glutamate transporter proteins form homomultimers. J Biol Chem 1996;271:27715–22.
218. Gouaux E. Review. The molecular logic of sodium-coupled neurotransmitter transporters. Philos Trans R Soc Lond B Biol Sci 2008.
219. Eskandari S, Kreman M, Kavanaugh MP, Wright EM, Zampighi GA. Pentameric assembly of a neuronal glutamate transporter. Proc Natl Acad Sci USA 2000;97:8641–6.
220. Shrivastava IH, Jiang J, Amara SG, Bahar I. Time-resolved mechanism of extracellular gate opening and substrate binding in a glutamate transporter. J Biol Chem 2008;283: 28680–90.
221. Chen YB, Durnford DG, Koblizek M, Falkowski PG. Plastid regulation of Lhcb1 transcription in the chlorophyte alga Dunaliella tertiolecta. Plant Physiol 2004;136:3737–50.
222. Broer S. The SLC6 orphans are forming a family of amino acid transporters. Neurochemistry International Glutamate in CNS Metabolism and Neurotransmission: Interactions at the Inter and Intracellular Level 2006;48:559–67.
223. Castagna M, Shayakul C, Trotti D, Sacchi VF, Harvey WR, Hediger MA. Molecular characteristics of mammalian and insect amino acid transporters: implications for amino acid homeostasis. J Exp Biol 1997;200 (Pt 2):269–86.
224. Yamashita A, Singh SK, Kawate T, Jin Y, Gouaux E. Crystal structure of a bacterial homologue of Na+/Cl−-dependent neurotransmitter transporters. Nature 2005;437:215–23.
225. Androutsellis-Theotokis A, Goldberg NR, Ueda K, et al. Characterization of a functional bacterial homologue of sodium-dependent neurotransmitter transporters. J Biol Chem 2003;278:12703–9.

226. Hoglund PJ, Adzic D, Scicluna SJ, Lindblom J, Fredriksson R. The repertoire of solute carriers of family 6: identification of new human and rodent genes. Biochem Biophys Res Commun 2005;336:175–89.
227. Lill H, Nelson N. Homologies and family relationships among Na+/Cl− neurotransmitter transporters. Meth Enzymol 1998;296:425–36.
228. Johnson K, Knust E, Skaer H. bloated tubules (blot) encodes a Drosophila member of the neurotransmitter transporter family required for organisation of the apical cytocortex. Dev Biol 1999;212:440–54.
229. Chen NH, Reith ME, Quick MW. Synaptic uptake and beyond: the sodium- and chloride-dependent neurotransmitter transporter family SLC6. Pflugers Arch 2003;29:29.
230. Levi G, Raiteri M. Exchange of Neurotransmitter Amino-Acid at Nerve-Endings Can Simulate High Affinity Uptake. Nature 1974;250:735–7.
231. Guastella J, Nelson N, Nelson H, et al. Cloning and expression of a rat brain GABA transporter. Science 1990;249:1303–6.
232. Liu QR, Mandiyan S, Nelson H, Nelson N. A family of genes encoding neurotransmitter transporters. Proc Natl Acad Sci USA 1992;89:6639–43.
233. Pacholczyk T, Blakely RD, Amara SD. Expression cloning of a cocaine- and antidepressant-sensitive human noradrenaline transporter. Nature 1991;350:350–4.
234. Nelson H, Mandiyan S, Nelson N. Cloning of the human brain GABA transporter. FEBS Lett 1990;269:181–4.
235. Borden LA, Smith KE, Gustafson EL, Branchek TA, Weinshank RL. Cloning and expression of a betaine/GABA transporter from human brain. J Neurochem 1995;64:977–84.
236. Lill H, Nelson N. Homologies and family relationships among Na+/Cl− neurotransmitter transporters. In: Amara SG, ed. Methods Enzymol Neurotransmitter Transporter. San Diego: Academic Press; 1998: 425–36.
237. Kanner BI. Structure and function of sodium-coupled GABA and glutamate transporters. J Membr Biol 2006;213:89–100.
238. Callec JJ. Synaptic transmission in the central nervous system. In: Kerkut GA, Gilbert LI, eds. Comprehensive Insect Physiology, Biochemistry and Pharmacology. Oxford: Pergamon Press; 1985:139–79.
239. Umesh A, Gill SS. Immunocytochemical localization of a Manduca sexta gamma-aminobutyric acid transporter. J Compar Neurol 2002;448:388–98.
240. Mbungu D, Ross LS, Gill SS. Cloning, functional expression, and pharmacology of a GABA transporter from Manduca sexta. Arch Biochem Biophys 1995;318:489–97.
241. Gao X, McLean H, Caveney S, Donly C. Molecular cloning and functional characterization of a GABA transporter from the CNS of the cabbage looper, Trichoplusia ni. Insect Biochem Mol Biol 1999;29:609–23.
242. Neckameyer WS. Dopamine and mushroom bodies in Drosophila: experience-dependent and -independent aspects of sexual behavior. Learn Mem 1998;5:157–65.
243. Raible F, Arendt D. Metazoan evolution: some animals are more equal than others. Curr Biol 2004;14:R106–8.
244. Pflüger H-J, Stevenson PA. Evolutionary aspects of octopaminergic systems with emphasis on the arthropods. Arthropod Structure and Development. Arthropod Struct Dev 2005;34:379–96.
245. Evans PD, Maqueira B. Insect octopamine receptors: a new classification scheme based on studies of cloned Drosophila G-protein coupled receptors. Invert Neurosci 2005;5:111–8.
246. Caveney S, Cladman W, Verellen L, Donly C. Ancestry of neuronal monoamine transporters in the Metazoa. J Exp Biol 2006;209:4858–68.
247. Gu HH, Wu X, Han DD. Conserved serine residues in serotonin transporter contribute to high-affinity cocaine binding. Biochem Biophys Res Commun 2006;343:1179–85.
248. Chen R, Wu X, Wei H, Han DD, Gu HH. Molecular cloning and functional characterization of the dopamine transporter from Eloria noyesi, a caterpillar pest of cocaine-rich coca plants. Gene 2006;366:152–60.

249. Haddley K, Vasiliou AS, Ali FR, Paredes UM, Bubb VJ, Quinn JP. Molecular genetics of monoamine transporters: relevance to brain disorders. Neurochem Res 2008;33:652–67.
250. Zis AP, Goodwin FK. Novel antidepressants and the biogenic amine hypothesis of depression. The case for iprindole and mianserin. Arch Gen Psychiatry 1979;36:1097–107.
251. Shirahata T, Tsunoda M, Santa T, Kirino Y, Watanabe S. Depletion of serotonin selectively impairs short-term memory without affecting long-term memory in odor learning in the terrestrial slug Limax valentianus. Learn Mem 2006;13:267–70.
252. King MV, Marsden CA, Fone KC. A role for the 5-HT(1A), 5-HT(4) and 5-HT(6) receptors in learning and memory. Trends Pharmacol Sci 2008;29:482–92.
253. Eison MS. Serotonin: a common neurobiologic substrate in anxiety and depression. J Clin Psychopharmacol 1990;10:26S–30S.
254. Weizman A, Weizman R. Serotonin transporter polymorphism and response to SSRIs in major depression and relevance to anxiety disorders and substance abuse. Pharmacogenomics 2000;1:335–41.
255. Gillette R. Evolution and Function in Serotonergic Systems. Integr Comp Biol 2006;46:838–46.
256. Klemm N, Schneider L. Selective uptake of indolamine into nervous fibers in the brain of the desert locust, *Schistocerca gregaria* Forskal (Insecta). A fluorescence and electron microscopic investigation. Comp Biochem Physiol 1975;50C:177–82.
257. McDonald TJ. Neuromuscular pharmacology of insects. Annu Rev Entomol 1975;20:151–66.
258. Maddrell SH. A Diuretic Hormone in Rhodnius Prolixus Stal. Nature 1962;194:605.
259. O'Donnell MJ, Maddrell SH. Secretion by the Malpighian tubules of Rhodnius prolixus stal: electrical events. J Exp Biol 1984;110:275–90.
260. Brown CS, Nestler C. Catecholamines and indolalkylamines. In: Kerkut GA, Gilbert LI, eds. *Comprehensive Insect Physiology,* Biochemistry and Pharmacology. Oxford: Pergamon Press; 1985:436–97.
261. Clark TM, Hutchinson MJ, Huegel KL, Moffett SB, Moffett DF. Additional morphological and physiological heterogeneity within the midgut of larval Aedes aegypti (Diptera : Culicidae) revealed by histology, electrophysiology, and effects of Bacillus thuringiensis endotoxin. Tissue Cell 2005;37:457–68.
262. Schewe B, Schmalzlin E, Walz B. Intracellular pH homeostasis and serotonin-induced pH changes in Calliphora salivary glands: the contribution of V-ATPase and carbonic anhydrase. J Exp Biol 2008;211:805–15.
263. Boudko DY, Moroz LL, Harvey WR, Linser PJ. Alkalinization by chloride/bicarbonate pathway in larval mosquito midgut. Proc Natl Acad Sci USA 2001;98:15354–9.
264. Okech BA, Boudko DY, Linser PJ, Harvey WR. Cationic pathway of pH regulation in larvae of Anopheles gambiae. J Exp Biol 2008;211:957–68.
265. Baier A, Wittek B, Brembs B. Drosophila as a new model organism for the neurobiology of aggression? J Exp Biol 2002;205:1233–40.
266. Kostowski W, Tarchalska-Krynska B, Markowska L. Aggressive behavior and brain serotonin and catecholamines in ants (Formica rufa). Pharmacol Biochem Behav 1975;3:717–9.
267. Novak MG, Rowley WA. Serotonin depletion affects blood-feeding but not host-seeking ability in Aedes triseriatus (Diptera: Culicidae). J Med Entomol 1994;31:600–6.
268. Yellman C, Tao H, He B, Hirsh J. Conserved and sexually dimorphic behavioral responses to biogenic amines in decapitated Drosophila. Proc Natl Acad Sci USA 1997;94:4131–6.
269. Muszynska-Pytel M, Cymboroski B. The role of serotonin in regulation the circadian rhythm of locomotor activity in the cricket (Acheta domesticus L.) II. Distribution of serotonin and variations in different brain structure. Comp Biochem Physiol C 1978;59:17–20.
270. Muzynska-Pytel M, Cymborowski B. The role of serotonin in regulation of the circadian rhythms of locomotor activity in the cricket (Acheta domesticus L.) I. Circadian variations in serotonin concentration in the brain and hemolymph. Comp Biochem Physiol C 1978;59: 13–5.

271. Carew TJ. Molecular enhancement of memory formation. Neuron 1996;16:5–8.
272. Demchyshyn LL, Pristupa ZB, Sugamori KS, et al. Cloning, expression, and localization of a chloride-facilitated, cocaine-sensitive serotonin transporter from *Drosophila melanogaster*. Proc Natl Acad Sci USA 1994;91:5158–62.
273. Corey JL, Quick MW, Davidson N, Lester HA, Guastella J. A cocaine-sensitive *Drosophila* serotonin transporter: Cloning, expression, and electrophysiological characterization. Proc Natl Acad Sci USA 1994;91:1188–92.
274. Sandhu SK, Ross LS, Gill SS. A cocaine insensitive chimeric insect serotonin transporter reveals domains critical for cocaine interaction. Eur J Biochem 2002b;269:3934–44.
275. Osborne RH. Insect neurotransmission: Neurotransmitters and their receptors. Pharmacol Ther 1996;69:117–42.
276. Kanner BI. Sodium-coupled neurotransmitter transport: structure, function and regulation. J Exp Biol 1994;196:237–49.
277. Giros B, el Mestikawy S, Godinot N, et al. Cloning, pharmacological characterization, and chromosome assignment of the human dopamine transporter. Mol Pharmacol 1992;42:383–90.
278. Pörzgen P, Park SK, Hirsch J, Sonders MS, Amara SG. The antidepressant-sensitive dopamine transporter in *Drosophila melanogaster:* A primordial carrier for catecholamines. Molec Pharmacol 2001;59:83–95.
279. Gallant P, Malutan T, McLean H, Verellen L, Caveney S, Donly C. Functionally distinct dopamine and octopamine transporters in the CNS of the cabbage looper moth. Eur J Biochem 2003;270:664–74.
280. Croll RP, Boudko DY, Pires A, Hadfield MG. Transmitter contents of cells and fibers in the cephalic sensory organs of the gastropod mollusc Phestilla sibogae. Cell Tissue Res 2003;314:437–48.
281. Croll RP, Boudko DY, Hadfield MG. Histochemical survey of transmitters in the central ganglia of the gastropod mollusc Phestilla sibogae. Cell Tissue Res 2001;305:417–32.
282. Braubach OR, Dickinson AJ, Evans CC, Croll RP. Neural control of the velum in larvae of the gastropod, Ilyanassa obsoleta. J Exp Biol 2006;209:4676–89.
283. Kim CH, Waldman ID, Blakely RD, Kim KS. Functional gene variation in the human norepinephrine transporter: association with attention deficit hyperactivity disorder. Ann NY Acad Sci 2008;1129:256–60.
284. Robertson D, Flattem N, Tellioglu T, et al. Familial orthostatic tachycardia due to norepinephrine transporter deficiency. Ann NY Acad Sci 2001;940:527–43.
285. Zhou J. Norepinephrine transporter inhibitors and their therapeutic potential. Drugs Future 2004;29:1235–44.
286. Bonisch H, Bruss M. The norepinephrine transporter in physiology and disease. Handb Exp Pharmacol 2006:485–524.
287. Blakely RD, De Felice LJ, Hartzell HC. Molecular physiology of norepinephrine and serotonin transporters. J Exp Biol 1994;196:263–81.
288. Axelrod J, Kopin IJ. The uptake, storage, release and metabolism of noradrenaline in sympathetic nerves. Prog Brain Res 1969;31:21–32.
289. Iversen LL, Jarrott B, Simmonds MA. Differences in the uptake, storage and metabolism of (+)- and (−)-noradrenaline. Br J Pharmacol 1971;43:845–55.
290. Zahniser NR, Doolen S. Chronic and acute regulation of Na+/Cl− -dependent neurotransmitter transporters: drugs, substrates, presynaptic receptors, and signaling systems. Pharmacol Ther 2001;92:21–55.
291. Roeder T. Octopamine in invertebrates. Progr Neurobiol 1999;59:533–61.
292. Homberg U. Neurotransmitters and neuropeptides in the brain of the locust. Microsc Res Tech 2002;56:189–209.
293. Leitch B, Judge S, Pitman RM. Octopaminergic modulation of synaptic transmission between an identified sensory afferent and flight motoneuron in the locust. J Comp Neurol 2003;462:55–70.

294. Sinakevitch IG, Geffard M, Pelhate M, Lapied B. Octopamine-Like Immunoreactivity in the Dorsal Unpaired Median (Dum) Neurons Innervating the Accessory-Gland of the Male Cockroach Periplaneta-Americana. Cell and Tissue Research 1994;276:15–21.
295. Copeland J, Robertson HA. Octopamine as the Transmitter at the Firefly Lantern – Presence of an Octopamine-Sensitive and a Dopamine-Sensitive Adenylate-Cyclase. Comp Biochem Physiol C-Pharmacol Toxicol Endocrinol 1982;72:125–7.
296. Robertson HA, Carlson AD. Octopamine – Presence in Firefly Lantern Suggests a Transmitter Role. J Exp Zool 1976;195:159–64.
297. Farooqui T, Robinson K, Vaessin H, Smith BH. Modulation of early olfactory processing by an octopaminergic reinforcement pathway in the honeybee. J Neurosci 2003;23:5370–80.
298. Spivak M, Masterman R, Ross R, Mesce KA. Hygienic behavior in the honey bee (Apis mellifera L.) and the modulatory role of octopamine. J Neurobiol 2003;55:341–54.
299. Schulz DJ, Barron AB, Robinson GE. A role for octopamine in honey bee division of labor. Brain Behav Evol 2002;60:350–9.
300. Malutan T, McLean H, Caveney S, Donly C. A high-affinity octopamine transporter cloned from the central nervous system of cabbage looper. Trichoplusia ni. Insect Biochem Mol Biol 2002;32:343–57.
301. Deken SL, Fremeau RTJ, Quick MW. Family of Sodium-Coupled Transporters for GABA, Glycine, Proline, Betaine, Taurine, and Creatine. In: Reith MEA, ed. Neurotransmitter Transporters: Structure, Function and Regulation. Totowa, New Jersey: Humana Press Inc; 2002:193–234.
302. Witte I, Kreienkamp H-J, Gewecke M, Roeder T. Putative histamine-gated chloride channel subunits of the insect visual system and thoracic ganglion. J Neurochem 2002;83:504–14.
303. Sandhu SK, Ross LS, Gill SS. Molecular cloning and functional expression of a proline transporter from *Manduca sexta*. Insect Biochem Mol Biol 2002a;32:1391–400.
304. Castagna M, Shayakul C, Trotti D, Sacchi VF, Harvey WR, Hediger MA. Cloning and characterization of a potassium-coupled amino acid transporter. Proc Natl Acad Sci USA 1998;95:5395–400.
305. Hennigan BB, Wolfersberger MG, Harvey WR. Neutral amino acid symport in larval Manduca sexta midgut brush-border membrane vesicles deduced from cation-dependent uptake of leucine, alanine, and phenylalanine. Biochim Biophys Acta 1993a;1148:216–22.
306. Hennigan BB, Wolfersberger MG, Parthasarathy R, Harvey WR. Cation-dependent leucine, alanine, and phenylalanine uptake at pH 10 in brush-border membrane vesicles from larval Manduca sexta midgut. Biochim Biophys Acta 1993b;1148:209–15.
307. Peres A, Bossi E. Effects of pH on the uncoupled, coupled and pre-steady-state currents at the amino acid transporter KAAT1 expressed in Xenopus oocytes. J Physiol Lond 2000;525:83–9.
308. Vincenti S, Castagna M, Peres A, Sacchi VF. Substrate selectivity and pH dependence of KAAT1 expressed in Xenopus laevis oocytes. J Membr Biol 2000;174:213–24.
309. Bossi E, Centinaio E, Castagna M, et al. Ion binding and permeation through the lepidopteran amino acid transporter KAAT1 expressed in Xenopus oocytes. J Physiol-Lond 1999;515:729–42.
310. Bossi E, Sacchi VF, Peres A. Ionic selectivity of the coupled and uncoupled currents carried by the amino acid transporter KAAT1. Pflugers Archiv Eur J Physiol 1999;438:788–96.
311. Liu Z, Stevens BR, Feldman DH, Hediger MA, Harvey WR. K+ amino acid transporter KAAT1 mutant Y147F has increased transport activity and altered substrate selectivity. J Exp Biol 2003;206:245–54.
312. Sacchi VF, Castagna M, Mari SA, Perego C, Bossi E, Peres A. Glutamate 59 is critical for transport function of the amino acid cotransporter KAAT1. Am J Physiol Cell Physiol 2003;285:C623–32.
313. Castagna M, Vincenti S, Marciani P, Sacchi VF. Inhibition of the lepidopteran amino acid co-transporter KAAT1 by phenylglyoxal: role of arginine 76. Insect Mol Biol 2002;11: 283–9.

314. Bossi E, Soragna A, Miszner A, Giovannardi S, Frangione V, Peres A. Oligomeric structure of the neutral amino acid transporters Kaat1 And Caatch1. 10.1152/ajpcell.00473.2006. Am J Physiol Cell Physiol 2006:00473.2006.
315. Miszner A, Peres A, Castagna M, et al. Structural and functional basis of amino acid specificity in the invertebrate cotransporter KAAT1. J Physiol 2007;581:899–913.
316. Feldman DH, Harvey WR, Stevens BR. A novel electrogenic amino acid transporter is activated by K+ or Na+, is alkaline pH-dependent, and is Cl−-independent. J Biol Chem 2000;275:24518–26.
317. Quick M, Stevens BR. Amino acid transporter CAATCH1 is also an amino acid-gated cation channel. J Biol Chem 2001;276:33413–8.
318. Stevens BR, Feldman DH, Liu ZL, Harvey WR. Conserved tyrosine-147 plays a critical role in the ligand-gated current of the epithelial cation/amino acid transporter/channel CAATCH1. J Exp Biol 2002;205:2545–53.
319. Bismuth Y, Kavanaugh MP, Kanner BI. Tyrosine 140 of the gamma-aminobutyric acid transporter GAT-1 plays a critical role in neurotransmitter recognition. J Biol Chem 1997;272:16096-102.
320. Ito K, Kidokoro K, Sezutsu H, et al. Deletion of a gene encoding an amino acid transporter in the midgut membrane causes resistance to a Bombyx parvo-like virus. Proc Natl Acad Sci USA 2008;105:7523–7.
321. Harvey WR, Pung L, Meleshkevitch EA, Kohn A, Boudko DY. Molecular and electrochemical integration of nutrient amino acid uptake in mosquito larvae. In: Sixth International Congress of Comparative Physiology and Biochemistry; 2003; Mt. Buller, Australia; 2003.
322. Clements AN. The Biology of Mosquitoes. London: Chapman and Hall Press; 1992.
323. Sloan JL, Mager S. Cloning and Functional Expression of a Human Na+ and Cl− -dependent Neutral and Cationic Amino Acid Transporter B0+. J Biol Chem 1999;274: 23740–5.
324. Kim DK, Kanai Y, Chairoungdua A, Matsuo H, Cha SH, Endou H. Expression cloning of a Na+-independent aromatic amino acid transporter with structural similarity to H+/monocarboxylate transporters. J Biolo Chem 2001;276:17221–8.
325. Mastroberardino L, Spindler B, Pfeiffer R, et al. Amino-acid transport by heterodimers of 4F2hc/CD98 and members of a permease family. Nature 1998;395:288–91.
326. Rossier G, Meier C, Bauch C, et al. LAT2, a New Basolateral 4F2hc/CD98-associated Amino Acid Transporter of Kidney and Intestine. J Biol Chem 1999;274:34948–54.
327. Verrey F, Closs EI, Wagner CA, Palacin M, Endou H, Kanai Y. CATs and HATs: the SLC7 family of amino acid transporters. Pflugers Arch 2003;11:11.
328. Broer A, Klingel K, Kowalczuk S, Rasko JEJ, Cavanaugh J, Broer S. Molecular cloning of mouse amino acid transport system B-0, a neutral amino acid transporter related to Hartnup disorder. J Biol Chem 2004;279:24467–76.
329. Seow HF, Broer S, Broer A, et al. Hartnup disorder is caused by mutations in the gene encoding the neutral amino acid transporter SLC6A19. Nat Genet 2004;36:1003–7.
330. Kowalczuk S, Broer A, Munzinger M, Tietze N, Klingel K, Broer S. Molecular cloning of the mouse IMINO system: an Na+- and Cl− -dependent proline transporter. Biochem J 2005;386:417–22.
331. Takanaga H, Mackenzie B, Suzuki Y, Hediger MA. Identification of mammalian proline transporter SIT1 (SLC6A20) with characteristics of classical system imino. J Biol Chem 2005;280:8974–84.
332. Ristic Z, Camargo SM, Romeo E, et al. Neutral amino acid transport mediated by ortholog of imino acid transporter SIT1/SLC6A20 in opossum kidney cells. Am J Physiol Renal Physiol 2006;290:F880–7.
333. Kleta R, Romeo E, Ristic Z, et al. Mutations in SLC6A19, encoding B0AT1, cause Hartnup disorder. Nat Genet 2004;36:999–1002.
334. Jonas AJ, Butler IJ. Circumvention of defective neutral amino acid transport in Hartnup disease using tryptophan ethyl ester. J Clin Invest 1989;84:200–4.

335. Verrey F, Ristic Z, Romeo E, et al. Novel renal amino acid transporters. Annu Rev Physiol 2005;67:557–72.
336. Meleshkevitch EA, Assis-Nascimento P, Popova LB, et al. Molecular characterization of the first aromatic nutrient transporter from the sodium neurotransmitter symporter family. J Exp Biol 2006;209:3183–98.
337. Holt RA, Subramanian GM, Halpern A, al. e. The genome sequence of the malaria mosquito *Anopheles gambiae*. Science 2002;298:129–49.
338. Broer A, Cavanaugh JA, Rasko JEJ, Broer S. The molecular basis of neutral aminoacidurias. Pflugers Archiv Eur J Physiol 2006;451:511–7.
339. Zomot E, Bendahan A, Quick M, Zhao Y, Javitch JA, Kanner BI. Mechanism of chloride interaction with neurotransmitter:sodium symporters. Nature 2007;449:726–30.
340. Forrest LR, Tavoulari S, Zhang YW, Rudnick G, Honig B. Identification of a chloride ion binding site in Na+/Cl -dependent transporters. Proc Natl Acad Sci USA 2007;104:12761–6.
341. Ueda K, Yamashita A, Ishikawa J, et al. Genome sequence of Symbiobacterium thermophilum, an uncultivable bacterium that depends on microbial commensalism. Nucleic Acids Res 2004;32:4937–44.
342. Miller MM, Popova LB, Meleshkevitch EA, Tran PV, Boudko DY. The invertebrate B(0) system transporter, D. melanogaster NAT1, has unique d-amino acid affinity and mediates gut and brain functions. Insect Biochem Mol Biol 2008;38:923–31.
343. Geer BW. Utilization of D-amino acids for growth by Drosophila melanogaster larvae. J Nutr 1966;90:31–9.
344. Stern M, Ganetzky B. Identification and characterization of inebriated, a gene affecting neuronal excitability in Drosophila. J Neurogenet 1992;8:157–72.
345. Soehnge H, Huang X, Becker M, Whitley P, Conover D, Stern M. A neurotransmitter transporter encoded by the Drosophila inebriated gene. Proc Natl Acad Sci USA 1996;93:13262–7.
346. Huang Y, Stern M. In vivo properties of the Drosophila inebriated-encoded neurotransmitter transporter. J Neurosci 2002;22:1698–708.
347. Huang X, Huang Y, Chinnappan R, Bocchini C, Gustin MC, Stern M. The Drosophila inebriated-encoded neurotransmitter/osmolyte transporter: dual roles in the control of neuronal excitability and the osmotic stress response. Genetics 2002;160:561–9.
348. Chiu C-S, Ross LS, Cohen BN, Lester HA, Gill SS. The transporter-like protein inebriated mediates hyperosmotic stimuli through intracellular signalling. J Exp Biol 2000;203:3531–46.
349. Jin X, Aimanova K, Ross LS, Gill SS. Identification, functional characterization and expression of a LAT type amino acid transporter from the mosquito Aedes aegypti. Insect Biochem Mol Biol 2003;33:815–27.
350. Wolf S, Janzen A, Vekony N, Martine U, Strand D, Closs EI. Expression of solute carrier 7A4 (SLC7A4) in the plasma membrane is not sufficient to mediate amino acid transport activity. Biochem J 2002;364:767–75.
351. Closs EI. Expression, regulation and function of carrier proteins for cationic amino acids. Curr Opin Nephrol Hypertens 2002;11:99–107.
352. Albritton LM, Tseng L, Scadden D, Cunningham JM. A Putative Murine Ecotropic Retrovirus Receptor Gene Encodes a Multiple Membrane-Spanning Protein and Confers Susceptibility to Virus-Infection. Cell 1989;57:659–66.
353. Kim JW, Closs EI, Albritton LM, Cunningham JM. Transport of cationic amino acids by the mouse ecotropic retrovirus receptor. Nature 1991;352:725–8.
354. Simell O. Lysinuric protein intolerance and other cationic amino acidurias. In: Scriver CR, Beaudet AL, Sly SW, Valle D, eds. Metabolic and molecular bases of inherited diseases,. 8th ed. New York: McGraw-Hill; 2001:4933–56.
355. Zharikov SI, Block ER. Characterization of L-arginine uptake by plasma membrane vesicles isolated from cultured pulmonary artery endothelial cells. Biochim Biophys Acta 1998;1369:173–83.

356. Rothenberg ME, Doepker MP, Lewkowich IP, et al. Cationic amino acid transporter 2 regulates inflammatory homeostasis in the lung. Proc Natl Acad Sci USA 2006;103:14895–900.
357. Boudko DY, Cooper BY, Harvey WR, Moroz LL. High-resolution microanalysis of nitrite and nitrate in neuronal tissues by capillary electrophoresis with conductivity detection. J Chromatogr B Analyt Technol Biomed Life Sci 2002;774:97–104.
358. Fernandez J, Bode B, Koromilas A, et al. Translation Mediated by the Internal Ribosome Entry Site of the cat-1 mRNA Is Regulated by Glucose Availability in a PERK Kinase-dependent Manner. J Biol Chem 2002;277:11780–7.
359. Graf P, Forstermann U, Closs EI. The transport activity of the human cationic amino acid transporter hCAT-1 is downregulated by activation of protein kinase C. Br J Pharmacol 2001;132:1193–200.
360. Zharikov SI, Sigova AA, Chen S, Bubb MR, Block ER. Cytoskeletal regulation of the L-arginine/NO pathway in pulmonary artery endothelial cells. Am J Physiol Lung Cell Mol Physiol 2001;280:L465–73.
361. Hemler ME, Strominger JL. Characterization of antigen recognized by the monoclonal antibody (4F2): different molecular forms on human T and B lymphoblastoid cell lines. J Immunol 1982;129:623–8.
362. Bertran J, Werner A, Moore ML, et al. Expression cloning of a cDNA from rabbit kidney cortex that induces a single transport system for cystine and dibasic and neutral amino acids. Proc Natl Acad Sci USA 1992;89:5601–5.
363. Tate SS, Yan N, Udenfriend S. Expression cloning of a Na(+)-independent neutral amino acid transporter from rat kidney. Proc Natl Acad Sci USA 1992;89:1–5.
364. Wells RG, Hediger MA. Cloning of a rat kidney cDNA that stimulates dibasic and neutral amino acid transport and has sequence similarity to glucosidases. Proc Natl Acad Sci USA 1992;89:5596–600.
365. Chillaron J, Roca R, Valencia A, Zorzano A, Palacin M. Heteromeric amino acid transporters: biochemistry, genetics, and physiology. Am J Physiol Renal Physiol 2001;281:F995–1018.
366. Pfeiffer R, Spindler B, Loffing J, Skelly PJ, Shoemaker CB, Verrey F. Functional heterodimeric amino acid transporters lacking cysteine residues involved in disulfide bond. FEBS Lett 1998;439:157–62.
367. Palacin M, Kanai Y. The ancillary proteins of HATs: SLC3 family of amino acid transporters. Pflugers Arch 2003;6:6.
368. Broer A, Wagner CA, Lang F, Broer S. The heterodimeric amino acid transporter 4F2hc/y+LAT2 mediates arginine efflux in exchange with glutamine. Biochem J 2000;349:787–95.
369. Pfeiffer R, Rossier G, Spindler B, Meier C, Kuhn L, Verrey F. Amino acid transport of y+L-type by heterodimers of 4F2hc/CD98 and members of the glycoprotein-associated amino acid transporter family. Embo J 1999;18:49–57.
370. Fernandez E, Torrents D, Chillaron J, Martin Del Rio R, Zorzano A, Palacin M. Basolateral LAT-2 has a major role in the transepithelial flux of L-cystine in the renal proximal tubule cell line OK. J Am Soc Nephrol 2003;14:837–47.
371. Chillaron J, Estevez R, Mora C, et al. Obligatory amino acid exchange via systems bo,+-like and y+L-like. A tertiary active transport mechanism for renal reabsorption of cystine and dibasic amino acids. J Biol Chem 1996;271:17761–70.
372. Meier C, Ristic Z, Klauser S, Verrey F. Activation of system L heterodimeric amino acid exchangers by intracellular substrates. Embo J 2002;21:580–9.
373. Sato H, Tamba M, Ishii T, Bannai S. Cloning and expression of a plasma membrane cystine/glutamate exchange transporter composed of two distinct proteins. J Biol Chem 1999;274:11455–8.
374. Bridges CC, Kekuda R, Wang HP, et al. Structure, function, and regulation of human cystine/glutamate transporter in retinal pigment epithelial cells. Investig Ophthalmol Visual Sci 2001;42:47–54.

375. Tomi M, Funaki T, Abukawa H, et al. Expression and regulation of L-cystine transporter, system x(c)(–), in the newly developed rat retinal Muller cell line (TR-MUL). Glia 2003;43:208–17.
376. Bassi MT, Gasol E, Manzoni M, et al. Identification and characterisation of human xCT that co-expresses, with 4F2 heavy chain, the amino acid transport activity system x(c)(–). Pflugers Archiv Eur J Physiol 2001;442:286–96.
377. Sato H, Tamba M, Okuno S, et al. Distribution of cystine/glutamate exchange transporter, system x(c)(–), in the mouse brain. J Neurosci 2002;22:8028–33.
378. Wang HY, Tamba M, Kimata M, Sakamoto K, Bannai S, Sato H. Expression of the activity of cystine/glutamate exchange transporter, system x(c)(–), by xCT and rBAT. Biochem Biophys Res Commun 2003;305:611–8.
379. MacGregor EA, Janecek S, Svensson B. Relationship of sequence and structure to specificity in the alpha-amylase family of enzymes. Biochim Biophys Acta 2001;1546:1–20.
380. Janecek S. Sequence Similarities and Evolutionary Relationships of Microbial, Plant and Animal Alpha-Amylases. Eur J Biochem 1994;224:519–24.
381. Watanabe K, Hata Y, Kizaki H, Katsube Y, Suzuki Y. The refined crystal structure of Bacillus cereus oligo-1,6-glucosidase at 2.0 angstrom resolution: Structural characterization of proline-substitution sites for protein thermostabilization. J Mol Biol 1997;269:142–53.
382. Darboux I, Nielsen-LeRoux C, Charles JF, Pauron D. The receptor of Bacillus sphaericus binary toxin in Culex pipiens (Diptera: Culicidae) midgut: molecular cloning and expression. Insect Biochem Mol Biol 2001;31:981–90.
383. Steiner HY, Naider F, Becker JM. The PTR family: a new group of peptide transporters. Mol Microbiol 1995;16:825–34.
384. Daniel H, Kottra G. The proton oligopeptide cotransporter family SLC15 in physiology and pharmacology. Pflugers Arch 2003;7:7.
385. Fei YJ, Kanai Y, Nussberger S, et al. Expression cloning of a mammalian proton-coupled oligopeptide transporter. Nature 1994;368:563–6.
386. Boll M, Herget M, Wagener M, et al. Expression cloning and functional characterization of the kidney cortex high-affinity proton-coupled peptide transporter. Proc Nat Acad Sci USA 1996;93:284–9.
387. Tomita Y, Takano M, Yasuhara M, Hori R, Inui KI. Transport of Oral Cephalosporins by the H+/Dipeptide Cotransporter and Distribution of the Transport Activity in Isolated Rabbit Intestinal Epithelial-Cells. J Pharmacol Exp Therapeut 1995;272:63–9.
388. Boll M, Markovich D, Weber WM, Korte H, Daniel H, Murer H. Expression Cloning of a Cdna from Rabbit Small-Intestine Related to Proton-Coupled Transport of Peptides, Beta-Lactam Antibiotics and Ace-Inhibitors. Pflugers Archiv Eur J Physiol 1994;429: 146–9.
389. Liang R, Fei YJ, Prasad PD, et al. Human intestinal H+/peptide cotransporter. Cloning, functional expression, and chromosomal localization. J Biol Chem 1995;270:6456–63.
390. Saito H, Terada T, Okuda M, Sasaki S, Inui K. Molecular cloning and tissue distribution of rat peptide transporter PEPT2. Biochimica Et Biophysica Acta-Biomembranes 1996;1280:173–7.
391. Rubio-Aliaga I, Boll M, Daniel H. Cloning and characterization of the gene encoding the mouse peptide transporter PEPT2. Biochem Biophys Res Commun 2000;276:734–41.
392. Chen H, Pan YX, Wong EA, Bloomquist JR, Webb KE. Molecular cloning and functional expression of a chicken intestinal peptide transporter (cPepT1) in Xenopus oocytes and Chinese hamster ovary cells. J Nutr 2002;132:387–93.
393. Verri T, Kottra G, Romano A, et al. Molecular and functional characterisation of the zebrafish (Danio rerio) PEPT1-type peptide transporter. Febs Lett 2003;549:115–22.
394. Roman G, Meller V, Wu KH, Davis RL. The opt1 gene of Drosophila melanogaster encodes a proton-dependent dipeptide transporter. Am J Physiol Cell Physiol 1998;44:C857–C69.
395. Yamashita T, Shimada S, Guo W, et al. Cloning and functional expression of a brain peptide/histidine transporter. J Biol Chem 1997;272:10205–11.

396. Sakata K, Yamashita T, Maeda M, Moriyama Y, Shimada S, Tohyama M. Cloning of a lymphatic peptide/histidine transporter. Biochem J 2001;356:53–60.
397. Chen PS. Amino acid and protein metabolism in insect development. Adv Insect Physiol 1966;3:53–132.
398. Collett JI. Small peptides, a life-long store of amino acid in adult *Drosophila* and *Calliphora*. J Insect Physiol 1976;22:433–1440.
399. Law JH, Dunn PE, Kramer KJ. Insect proteases and peptidases. Adv Enzymol Relat Areas Mol Biol 1977;45:389–425.
400. Bonen A. Lactate transporters (MCT proteins) in heart and skeletal muscles. Med Sci Sports Exercise 2000;32:778–89.
401. Halestrap AP, Meredith D. The SLC16 gene family-from monocarboxylate transporters (MCTs) to aromatic amino acid transporters and beyond. Pflugers Arch 2003;9:9.
402. Kim do K, Kanai Y, Matsuo H, et al. The human T-type amino acid transporter-1: characterization, gene organization, and chromosomal location. Genomics 2002;79: 95–103.
403. Kim DK, Kanai Y, Chairoungdua A, et al. Identification and characterization of a novel epithelial aromatic amino acid transporter TAT1. Jpn J Pharmacol 2002;88:223P.
404. Toure A, Morin L, Pineau C, Becq F, Dorseuil O, Gacon G. Tat1, a novel sulfate transporter specifically expressed in human male germ cells and potentially linked to RhoGTPase signaling. J Biol Chem 2001;276:20309–15.
405. Vincourt JB, Jullien D, Amalric F, Girard JP. Molecular and functional characterization of SLC26A11, a sodium-independent sulfate transporter from high endothelial venules. Faseb J 2003;17.
406. Bajmoczi M, Sneve M, Eide DJ, Drewes LR. TATI encodes a low-affinity histidine transporter in Saccharomyces cerevisiae. Biochem Biophys Res Commun 1998;243:205–9.
407. Price NT, Jackson VN, Halestrap AP. Cloning and sequencing of four new mammalian monocarboxylate transporter (MCT) homologues confirms the existence of a transporter family with an ancient past. Biochem J 1998;329:321–8.
408. Kirk P, Wilson MC, Heddle C, Brown MH, Barclay AN, Halestrap AP. CD147 is tightly associated with lactate transporters MCT1 and MCT4 and facilitates their cell surface expression. Embo J 2000;19:3896–904.
409. Zhao C, Wilson MC, Schuit F, Halestrap AP, Rutter GA. Expression and distribution of lactate/monocarboxylate transporter isoforms in pancreatic islets and the exocrine pancreas. Diabetes 2001;50:361–6.
410. Wilson MC, Meredith D, Halestrap AP. Fluorescence resonance energy transfer studies on the interaction between the lactate transporter MCT1 and CD147 provide information on the topology and stoichiometry of the complex in situ. J Biol Chem 2002;277: 3666–72.
411. Meredith D, Halestrap AP. OX47 (basigin) may act as a chaperone for expression of the monocarboxylate transporter MCT1 at the plasma membrane of Xenopus laevis oocytes. J Physiol Lond 2000;526:23P–4P.
412. Reimer RJ, Edwards RH. Organic anion transport is the primary function of the SLC17/type I phosphate transporter family. Pflugers Arch 2003;17:17.
413. Tabb JS, Ueda T. Phylogenetic studies on the synaptic vesicle glutamate transport system. J Neurosci 1991;11:1822–8.
414. Werner A, Moore ML, Mantei N, Biber J, Semenza G, Murer H. Cloning and expression of cDNA for a Na/Pi cotransport system of kidney cortex. Proc Natl Acad Sci USA 1991;88:9608–12.
415. Miyamoto K, Tatsumi S, Sonoda T, et al. Cloning and functional expression of a Na(+)-dependent phosphate co-transporter from human kidney: cDNA cloning and functional expression. Biochem J 1995;305:81–5.
416. Busch AE, Biber J, Murer H, Lang F. Electrophysiological insights of type I and II Na/Pi transporters. Kidney Int 1996;49:986–7.

417. Busch AE, Schuster A, Waldegger S, et al. Expression of a renal type I sodium/phosphate transporter (NaPi-1) induces a conductance in Xenopus oocytes permeable for organic and inorganic anions. Proc Natl Acad Sci USA 1996;93:5347–51.
418. Broer S, Schuster A, Wagner CA, et al. Chloride conductance and Pi transport are separate functions induced by the expression of NaPi-1 in Xenopus oocytes. J Membr Biol 1998;164:71–7.
419. Amstutz M, Mohrmann M, Gmaj P, Murer H. Effect of pH on phosphate transport in rat renal brush border membrane vesicles. Am J Physiol 1985;248:F705–10.
420. Murer H, Hernando N, Forster I, Biber J. Proximal tubular phosphate reabsorption: molecular mechanisms. Physiol Rev 2000;80:1373–409.
421. Magagnin S, Werner A, Markovich D, et al. Expression Cloning of Human and Rat Renal Cortex Na/Pi Cotransport. PNAS 1993;90:5979–83.
422. Kavanaugh MP, Miller DG, Zhang W, et al. Cell-surface receptors for gibbon ape leukemia virus and amphotropic murine retrovirus are inducible sodium-dependent phosphate symporters. Proc Natl Acad Sci USA 1994;91:7071–5.
423. Murer H, Forster I, Biber J. The sodium phosphate cotransporter family SLC34. Pflugers Arch 2003;16:16.
424. Olah Z, Lehel C, Anderson WB, Eiden MV, Wilson CA. The cellular receptor for gibbon ape leukemia virus is a novel high affinity sodium-dependent phosphate transporter. J Biol Chem 1994;269:25426–31.
425. Collins JF, Bai L, Ghishan FK. The SLC20 family of proteins: dual functions as sodium-phosphate cotransporters and viral receptors. Pflugers Arch 2003;21:21.
426. Ni B, Rosteck PR, Jr., Nadi NS, Paul SM. Cloning and expression of a cDNA encoding a brain-specific Na(+)-dependent inorganic phosphate cotransporter. Proc Natl Acad Sci USA 1994;91:5607–11.
427. Aihara Y, Mashima H, Onda H, et al. Molecular cloning of a novel brain-type Na(+)-dependent inorganic phosphate cotransporter. J Neurochem 2000;74:2622–5.
428. Bellocchio EE, Reimer RJ, Fremeau RT, Jr., Edwards RH. Uptake of glutamate into synaptic vesicles by an inorganic phosphate transporter. Science 2000;289:957–60.
429. Hayashi M, Otsuka M, Morimoto R, et al. Differentiation-associated Na+-dependent inorganic phosphate cotransporter (DNPI) is a vesicular glutamate transporter in endocrine glutamatergic systems. J Biol Chem 2001;276:43400–6.
430. Takamori S, Rhee JS, Rosenmund C, Jahn R. Identification of differentiation-associated brain-specific phosphate transporter as a second vesicular glutamate transporter (VGLUT2). J Neurosci 2001;21:RC182.
431. Bellocchio EE, Hu H, Pohorille A, Chan J, Pickel VM, Edwards RH. The localization of the brain-specific inorganic phosphate transporter suggests a specific presynaptic role in glutamatergic transmission. J Neurosci 1998;18:8648–59.
432. Fremeau RT, Jr., Burman J, Qureshi T, et al. The identification of vesicular glutamate transporter 3 suggests novel modes of signaling by glutamate. Proc Natl Acad Sci USA 2002;99:14488–93.
433. Varoqui H, Schafer MK, Zhu H, Weihe E, Erickson JD. Identification of the differentiation-associated Na+/PI transporter as a novel vesicular glutamate transporter expressed in a distinct set of glutamatergic synapses. J Neurosci 2002;22:142–55.
434. Hayashi M, Morimoto R, Yamamoto A, Moriyama Y. Expression and localization of vesicular glutamate transporters in pancreatic islets, upper gastrointestinal tract, and testis. J Histochem Cytochem 2003;51:1375–90.
435. Gleason KK, Dondeti VR, Hsia HL, Cochran ER, Gumulak-Smith J, Saha MS. The vesicular glutamate transporter 1 (xVGlut1) is expressed in discrete regions of the developing Xenopus laevis nervous system. Gene Expr Patterns 2003;3:503–7.
436. Fremeau RT, Troyer MD, Pahner I, et al. The expression of vesicular glutamate transporters defines two classes of excitatory synapse. Neuron 2001;31:247–60.

437. Varoqui H, Schafer MKH, Zhu HM, Weihe E, Erickson JD. Identification of the differentiation-associated Na+/P-I transporter as a novel vesicular glutamate transporter expressed in a distinct set of glutamatergic synapses. J Neurosci 2002;22:142–55.
438. Boll M, Foltz M, Anderson CM, et al. Substrate recognition by the mammalian proton-dependent amino acid transporter PAT1. Mol Membr Biol 2003;20:261–9.
439. Sagne C, Agulhon C, Ravassard P, et al. Identification and characterization of a lysosomal transporter for small neutral amino acids. Proc Natl Acad Sci USA 2001;98:7206–11.
440. Boll M, Foltz M, Rubio-Aliaga I, Kottra G, Daniel H. Functional characterization of two novel mammalian electrogenic proton-dependent amino acid cotransporters. J Biol Chem 2002;277:22966–73.
441. Boll M, Foltz M, Rubio-Aliaga I, Daniel H. A cluster of proton/amino acid transporter genes in the human and mouse genomes. Genomics 2003b;82:47–56.
442. Agulhon C, Rostaing P, Ravassard P, Sagne C, Triller A, Giros B. Lysosomal amino acid transporter LYAAT-1 in the rat central nervous system: an in situ hybridization and immunohistochemical study. J Comp Neurol 2003;462:71–89.
443. Chen Z, Fei YJ, Anderson CM, et al. Structure, function and immunolocalization of a proton-coupled amino acid transporter (hPAT1) in the human intestinal cell line Caco-2. J Physiol 2003b;546:349–61.
444. Gasnier B. The SLC32 transporter, a key protein for the synaptic release of inhibitory amino acids. Pflugers Arch 2003;16:16.
445. Mackenzie B, Erickson JD. Sodium-coupled neutral amino acid (System N/A) transporters of the SLC38 gene family. Pflugers Arch 2003;4:4.
446. Chaudhry FA, Krizaj D, Larsson P, et al. Coupled and uncoupled proton movement by amino acid transport system N. Embo J 2001;20:7041–51.
447. Broer A, Albers A, Setiawan I, et al. Regulation of the glutamine transporter SN1 by extracellular pH and intracellular sodium ions. J Physiol Lond 2002;539:3–14.
448. Saitou N, Nei M. The neighbor-joining method: a new method for reconstructing phylogenetic trees. Mol Biol Evol 1987;4:406–25.
449. Zuckerkandl E, Pauling L. Evolutionary divergence and convergence in proteins. In: Vogel VBaHJ, ed. Evolving Genes and Proteins. New York.: Academic Press, 1965:97–166.
450. Tamura K, Dudley J, Nei M, Kumar S. MEGA4: Molecular Evolutionary Genetics Analysis (MEGA) software version 4.0. Mol Biol Evol 2007;24:1596–9.
451. Thompson JD, Gibson TJ, Plewniak F, Jeanmougin F, Higgins DG. The CLUSTAL_X windows interface: flexible strategies for multiple sequence alignment aided by quality analysis tools. Nucleic Acids Res 1997;25:4876–82.
452. Felsenstein J. Confidence limits on phylogenies: An approach using the bootstrap. Evolution 1985;39:783–91.
453. Meleshkevitch EA, Robinson M, Popova LB, M.M. M, W.R. H, D.Y. B. Cloning and functional expression of the eukaryotic sodium-tryptophan symporter. JEB 2009;in press.
454. Neidhardt FC, Ingraham JL, Schaechter M. Physiology of the bacterial cell : a molecular approach. Sunderland, Mass., U.S.A.: Sinauer Associates; 1990.
455. Stryer L. Molecular design of life. New York: W.H. Freeman and Co.; 1989.
456. Ugawa S, Sunouchi Y, Ueda T, Takahashi E, Saishin Y, Shimada S. Characterization of a mouse colonic system B0+ amino acid transporter related to amino acid absorption in colon. Am J Physiol Gastrointest Liver Physiol 2001;281:G365–G70.
457. Eiden LE, Schafer MK, Weihe E, Schutz B. The vesicular amine transporter family (SLC18): amine/proton antiporters required for vesicular accumulation and regulated exocytotic secretion of monoamines and acetylcholine. Pflugers Arch 2003;24:24.
458. Boll M, Daniel H, Gasnier B. The SLC36 family: proton-coupled transporters for the absorption of selected amino acids from extracellular and intracellular proteolysis. Pflugers Arch 2004;447:776–9.
459. Bodoy S, Martin L, Zorzano A, Palacin M, Estevez R, Bertran J. Identification of LAT4, a novel amino acid transporter with system L activity. J Biol Chem 2005;280:12002–11.

Index

G.A. Gerencser (ed.), *Epithelial Transport Physiology*,
DOI 10.1007/978-1-60327-229-2,

D

G

H

R

S